Quantum Coherence, Correlation and Decoherence in Semiconductor Nanostructures

Quantum Coherence, Correlation and Decoherence in Semiconductor Nanostructures

Edited by
T. Takagahara
*Dept. of Electronics and Information Science,
Kyoto Institute of Technology,
Kyoto, Japan*

ACADEMIC PRESS
An imprint of Elsevier Science
Amsterdam • Boston • London • New York • Oxford • Paris
San Diego • San Francisco • Singapore • Sydney • Tokyo

This book is printed on acid-free paper.

Copyright © 2003, Elsevier Science (USA)

All rights reserved. No part of this publication may be reproduced or transmitted in any form or by any means, electronic or mechanical, including photocopying, recording, or any information storage and retrieval system, without permission in writing from the publisher.

Academic Press
An Imprint of Elsevier Science
84 Theobald's Road, London WC1X 8RR, UK
http://www.academicpress.com

Academic Press
An Imprint of Elsevier Science
525 B Street, Suite 1900, San Diego, California 92101-4495, USA
http://www.academicpress.com

ISBN 0-12-682225-5

A catalogue record for this book is available from the Library of Congress

A catalogue record for this book is available from the British Library

Typeset by Replika Press Pvt. Ltd 100% EOU, India.
Printed and bound in Great Britain by MPG Books Bodmin, Cornwall

03 04 05 06 07 MP 9 8 7 6 5 4 3 2 1

Preface

Semiconductor nanostructures occupy an intermediate position between bulk materials and molecules and provide ample opportunities to discover unexpected phenomena and to reveal new physical aspects. The discrete energy spectra of electrons, holes and excitons due to the quantum confinement lead to the suppression of relaxation processes that dominate in bulk materials, resulting in reduced homogeneous linewidth of relevant transitions. The quantum confinement leads also to the relatively large oscillator strength of the excitonic transitions. The wavefunction is extended coherently over a nanostructure and all the atomic transition dipole moments within the wavefunction are added up coherently. It is well-known that the Sommerfeld factor which represents the relative importance of the continuum (unbound) exciton states is monotonically decreasing with reducing the dimensionality of the system. Thus the discrete exciton states play the essential role in the optical properties of semiconductor nanostructures. Thus we have sharp optical transitions with large oscillator strength. This is the physical basis of the recent remarkable progress in the linear and nonlinear optical spectroscopies of semiconductor quantum dots.

Semiconductor nanostructures are attracting much interest as the most promising device to implement the quantum information processing and the quantum computation. The quantum coherence in nanostructures can be most elegantly manipulated by optical means. Thus the excitons and multi-excitons in semiconductor nanostructures are the most elementary objects in the quantum state control. Great progress has been made in the last decade in the research of optical properties, relaxation and decoherence processes in nanostructures. At the same time, the electron spin or the nuclear spin in semiconductors is very promising to manipulate the quantum coherence due to their long coherence times. The electron spin is reflected in the exciton polarization properties. The hyperfine interaction between the electron spin and the nuclear spin would be enhanced due to the quantum confinement of the electron wavefunction. Many interesting phenomena related to this interaction have been discovered recently.

In addition to the enhanced exciton effect, the Coulomb interaction in semiconductor nanostructures leads to strong correlation among electrons and holes. Recently, it has been revealed that many-body interactions among carriers have specific spectral and temporal signatures in nonlinear optical responses. As a consequence, new ultrafast spectroscopic techniques combined with microscopic many-body theory have brought dramatic progress in the study of many-body Coulomb interactions both in bulk and in quantum confined structures. In particular, the role of Coulomb correlation among more than two particles has been beautifully revealed in the last several years.

Another recent highlight in the nanostructure optics is the physics of microcavity polariton. The striking aspect is the tunability of the exciton-photon coupling strength from the weak perturbative regime to the strong normal mode coupling regime. Recent experiments have revealed many interesting aspects of the nonlinear optical processes associated with microcavity polaritons. Especially, the composite nanocrystal-microcavity system is promising to realize the entangled states among several nanocrystals through the whispering gallery modes in the strong coupling regime. Such an entangled state is an indispensable step toward the quantum information processing. On the other hand, in the weak or intermediate coupling regime of the light-matter interaction, there appears an interesting feature in the resonant secondary emission, e.g. the coexistence of the coherent part (Rayleigh scattering) and the incoherent part (photoluminescence) and the interplay of disorder and polaritonic effects.

Last but not least, there is a fundamental interest in the ultrafast coherent phenomena both in bulk semiconductors and in semiconductor nanostructures. The coherent nonlinear pulse propagation and the carrier-wave Rabi flopping have been observed successfully in the case of extremely strong pulse intensities. These studies have revealed a new paradigm of the coherent light-matter interaction. Fundamentally new high-field effects in semiconductor nanostructures have been predicted and are now being tested experimentally.

The understanding of fundamental physics aspects of the optical coherence and the spin coherence is now being revolutionized. This revolutional progress would bring about breakthroughs in the field of the quantum information processing. In view of these rapidly growing fields of research, we believe that it is timely to publish a book which surveys the present status of our understanding of the quantum coherence, correlation and decoherence in semiconductor nanostructures, putting emphasis on the basic physics aspects.

Kyoto, Japan T. Takagahara
October 2002

List of Contributors

Numbers in parentheses indicate the pages on which the authors' contributions begin.

S.R. Bolton (166), *Physics Department, Williams College, 33 Lab Campus Drive, Williamstown, MA 01267, USA*

A.S. Bracker (207), *Naval Research Laboratory, Washington, DC 20375, USA*

Gang Chen (281), *Harrison M. Randall Laboratory of Physics, University of Michigan, Ann Arbor, MI 48109-1120, USA*

Pochung Chen (281), *Department of Physics, University of California San Diego, La Jolla, CA 92093, USA*

Al.L. Efros (207), *Naval Research Laboratory, Washington, DC 20375, USA*

J. Förstner (1), *Institut für Theoretische Physik, Technische Universität Berlin, D-10623 Berlin, Germany*

D. Gammon (207, 281), *Naval Research Laboratory, Washington, DC 20375, USA*

H. Giessen (1), *Department of Physics and Material Sciences Center, Philipps-Universität, D-35032 Marburg, Germany*

J.R. Guest (281), *Harrison M. Randall Laboratory of Physics, University of Michigan, Ann Arbor, MI 48109-1120, USA*

S. Hughes (40), *Department of Physics, University of Surrey, Guildford, Surrey GU2 7XH, UK*

A. Knorr (1), *Institut für Theoretische Physik, Technische Universität Berlin, D-10623 Berlin, Germany*

S.W. Koch (1), *Department of Physics and Material Sciences Center, Philipps-Universität, D-35032 Marburg, Germany*

V.L. Korenev (207), *A.F. Ioffe Institute, St Petersburg, Russia*

J. Kuhl (1), *Max-Planck-Institut für Festkörperforschung, D-70569 Stuttgart, Germany*

S. Linden (1), *Max-Planck-Institut für Festkörperforschung, D-70569 Stuttgart, Germany*

I.A. Merkulov (207), *A.F. Ioffe Institute, St Petersburg, Russia*

O.D. Mücke (23), *Institut für Angewandte Physik, Universität Karlsruhe (TH), Wolfgang-Gaede-Straße 1, 76131 Karlsruhe, Germany*

N.C. Nielsen (1), *Max-Planck-Institut für Festkörperforschung, D-70569 Stuttgart, Germany*

C. Piermarocchi (281), *Department of Physics, University of California San Diego, La Jolla, CA 92093, USA*

E. Runge (89), *Humboldt University Berlin, Hausvogteiplatz 5-7, D-10117 Berlin, Germany; Max Planck Institute for the Physics of Complex Systems, Nöthnitzer Str. 38, D-01187 Dresden, Germany*

V. Savona (89), *Humboldt University Berlin, Hausvogteiplatz 5-7, D-10117 Berlin, Germany*

L.J. Sham (281), *Department of Physics, University of California San Diego, La Jolla, CA 92093, USA*

D.G. Steel (281), *Harrison M. Randall Laboratory of Physics, University of Michigan, Ann Arbor, MI 48109-1120, USA*

T.H. Stievater (281), *Harrison M. Randall Laboratory of Physics, University of Michigan, Ann Arbor, MI 48109-1120, USA*

T. Takagahara (395), *Department of Electronics and Information Science, Kyoto Institute of Technology, Matsugasaki, Sakyo-ku, Kyoto 606-8585, Japan*

J.G. Tischler (207), *Naval Research Laboratory, Washington, DC 20375, USA*

T. Tritschler (23), *Institut für Angewandte Physik, Universität Karlsruhe (TH), Wolfgang-Gaede-Straße 1, 76131 Karlsruhe, Germany*

M. Wegener (23), *Institut für Angewandte Physik, Universität Karlsruhe (TH), Wolfgang-Gaede-Straße 1, 76131 Karlsruhe, Germany*

Hailin Wang (366), *Department of Physics and Oregon Center for Optics, University of Oregon, Eugene, OR 97403, USA*

R. Zimmermann (89), *Humboldt University Berlin, Hausvogteiplatz 5-7, D-10117 Berlin, Germany*

Contents

Preface .... *v*
List of Contributors .... *vii*

Chapter 1
Coherent nonlinear pulse propagation on a free-exciton resonance in a semiconductor

N.C. Nielsen, S. Linden, J. Kuhl, J. Förstner, A. Knorr, S.W. Koch, and H. Giessen

1.1 Introduction 1
1.2 Theoretical background 2
1.3 Samples and experimental techniques 6
1.4 Results and discussion 8
      Excitation-induced suppression of temporal polariton beating 8
      Self-induced transmission and multiple pulse breakup 10
      Phonon-induced dephasing of the excitonic polarization 17
1.5 Conclusions 18
      Acknowledgments 19
      References 19

Chapter 2
Carrier-wave Rabi flopping in semiconductors

O.D. Mücke, T. Tritschler, and M. Wegener

2.1 Introduction 23
2.2 Carrier-wave Rabi flopping 25
      Experiments 27
      Theory 32
2.3 Conclusions 37

Acknowledgments 37
References 37

Chapter 3
High-field effects in semiconductor nanostructures

S. Hughes

3.1 Introduction 40
3.2 General theory 43
3.3 High-field electo-optics in quantum wells and wires 45
   Real space theoretical approach to electon–hole wave packets 47
   Electro-magneto-optical simulations in quantum wells 49
   Static Franz–Keldysh effect in quantum wires 53
   Dynamic Franz–Keldysh effect in quantum wires 59
3.4 Excitonic trapping, ultrafast population transfer, and Rabi flopping 63
   Theory of high optical field effects in quantum wells 63
   Excitonic trapping and ultrafast population transfer 65
3.5 Carrier wave Rabi flopping 71
   Theory and computation of sub-optical-carrier pulse propagation 72
   Breakdown of the area theorem in a two-level atom 73
   Carrier-wave Rabi flopping in semiconductors 79
3.6 Conclusions 83
   Acknowledgments 84
   References 84

Chapter 4
Theory of resonant secondary emission:
Rayleigh scattering versus luminescence

R. Zimmermann, E. Runge, and V. Savona

4.1 Introduction 89
4.2 Disorder eigenstates of excitons 94
4.3 Exciton Hamiltonian and density-matrix approach 101
4.4 Exciton kinetics with acoustic phonon scattering 108
4.5 Coherent and incoherent emission in the time domain 113
4.6 Speckle measurement and interferometry 117

4.7	Frequency-resolved secondary emission	120
4.8	Signatures of level repulsion	123
4.9	Enhanced resonant backscattering	132
4.10	Spin- and polarization-dependent emission	139
4.11	Polariton effects in the secondary emission	147
	Appendix A: Potential variance	155
	Appendix B: Weak-memory and Markov approximation	156
	Appendix C: Radiative rates	157
	References	159

Chapter 5
Higher-order Coulomb correlation effects in semiconductors
S.R. Bolton

5.1	Introduction	166
5.2	Ultrafast spectroscopy of semiconductor nanostructures as probes of Coulomb correlations	167
	Overview of the semiconductor equations of motion with optical excitation	169
	Non-interacting and Hartree–Fock approximations	170
	Beyond the coherent SBE: screening and scattering	172
	Ultrafast optical measurement techniques	173
5.3	Beyond the screened HF approximation – theoretical approaches to many-body correlations	178
	Biexcitons and few-level theories	178
	The dynamics-controlled truncation scheme	180
	The coherent limit	182
	Interpreting and solving the equations of the DCT	182
	The effective polarization model	185
	Phonons	186
5.4	Experimental studies of high-order Coulomb correlations	187
	The fully coherent regime	187
	Contributions from incoherent densities	197
	Contributions beyond the four-particle level	199
	Contributions beyond the $\chi^{(3)}$ truncation	201
5.5	Future directions	201
	References	203

Chapter 6
Electronic and nuclear spin in the optical spectra of semiconductor quantum dots

D. Gammon, Al.L. Efros, J.G. Tischler, A.S. Bracker, V.L. Korenev, and I.A. Merkulov

6.1 Introduction to spin in the optical spectrum 207
6.2 Photoluminescence spectroscopy of quantum dots 211
    Natural (interface fluctuation) QDs 211
    Photoluminescence spectroscopy of single QDs 215
    PL excitation spectroscopy of single QDs 217
6.3 Exciton fine-structure (spin and sublevels) 219
    Exchange interaction 219
    Long-range exchange interaction 222
    Zeeman interaction 227
    Pseudo-spin model 232
    Relaxation 237
    Polarization including finite relaxation 238
    Hanle effect 239
6.4 Trions (singly charged excitons) 240
    Trions in natural QDs 240
    Fine structure in single trion spectroscopy 243
    Optical orientation of negatively charged excitons 248
6.5 Hyperfine interaction 252
    Hyperfine interaction: static and dynamic 256
    Dynamical polarization of nuclei: Overhauser effect 258
    Nuclear dipole–dipole interactions 259
    Optical nuclear magnetic resonance 261
6.6 Spin relaxation 263
    Spin relaxation: spin–orbit interactions 263
    Spin relaxation: hyperfine interaction 265
    Hanle effect for localized electrons 268
6.7 Conclusions 271
    Acknowledgments 273
    Appendix Relaxation of the nuclear spin due to the fluctuating electronic spin 273
    References 275

Chapter 7
Coherent optical spectroscopy and manipulation of single quantum dots

Gang Chen, T.H. Stievater, J.R. Guest, D.G. Steel, D. Gammon, Pochung Chen, C. Piermarocchi, and L.J. Sham

7.1 Introduction	282
Semiconductor QDs	282
Excitons and biexcitons	283
Modeling single QDs	285
Quantum coherence and quantum computing based on optically driven QDs	287
Single QD optical spectroscopy	288
7.2 Single exciton optical spectroscopy	293
PL and PLE	293
Linear absorption from single QD excitons	295
CW and transient nonlinear optical response from single QD excitons	298
Magneto-excitons	304
7.3 Coherent optical control of single exciton states	306
7.4 Rabi oscillations of single quantum dots	310
Rabi oscillation theory for two-level systems	311
Strong-field differential transmission: Rabi oscillations of single QD excitons	313
Understanding the decay: coupling to delocalized excitons	316
7.5 Biexcitons in single QDs	318
Excitation of single QD biexcitons using CW fields	319
Dephasing of biexcitons	325
Direct measurement of biexciton lifetime	326
Biexcitonic transition dipole moment	329
Optical selection rules	330
7.6 Optically induced two exciton-state entanglement	330
7.7 Single quantum dot as a prototype quantum computer	338
Basic operations for quantum computation	339
The Deutsch–Jozsa problem	340
Fast quantum computing by pulse shaping	343
Examples of pulse design	347
Fast control applied to the Deutsch–Jozsa algorithm	350

7.8 Summary 354
      References 355

Chapter 8
Cavity QED of quantum dots with dielectric microspheres

Hailin Wang

8.1 Introduction 366
8.2 Whispering gallery modes in a dielectric microsphere 368
8.3 Composite system of dielectric microsphere and
     MBE-grown nanostructure 372
8.4 Composite system of dielectric microsphere and semiconductor
     nanocrystals 375
      Coupling nanocrystals to a dielectric microsphere: low-Q regime 376
      Coupling nanocrystals to a dielectric microsphere: high-Q regime 381
      Dephasing in semiconductor nanocrystals 385
8.5 Summary 390
      Acknowledgments 391
      References 392

Chapter 9
Theory of exciton coherence and decoherence in semiconductor quantum dots

T. Takagahara

9.1 Intoduction 395
9.2 Exciton Rabi splitting in a single quantum dot 396
9.3 Dressed exciton state 401
9.4 Exciton Rabi oscillation in a single quantum dot 402
9.5 Bloch vector model 406
9.6 Numerical results and discussion 410
9.7 Wave packet interferometry 420
9.8 Effect of two-photon coherence 424
9.9 Exciton dephasing in semiconductor quantum dots 431
9.10 Green function formalism of exciton dephasing rate 436

9.11	Exciton–phonon interactions	442
9.12	Excitons in anisotropic quantum disk	444
9.13	Temperature-dependence of the exciton dephasing rate	445
9.14	Elementary processes of exciton pure dephasing	450
9.15	Mechanisms of population decay of excitons	452
	Phonon-assisted population relaxation	452
	Phonon-assisted exciton migration	453
9.16	Recent progress in studies on exciton decoherence	459
9.17	Theory of dephasing of nonradiative coherence	459
9.18	Summary	465
	Acknowledgments	466
	References	466
Index		471

Quantum Coherence, Correlation and Decoherence in
Semiconductor Nanostructures
T. Takagahara (Ed.)
Copyright © 2001 American Physical Society. All rights reserved.

Chapter 1
Coherent nonlinear pulse propagation on a free-exciton resonance in a semiconductor*

N.C. Nielsen, S. Linden, and J. Kuhl

Max-Planck-Institut für Festkörperforschung, D-70569 Stuttgart, Germany

J. Förstner and A. Knorr

Institut für Theoretische Physik, Technische Universität Berlin, D-10623 Berlin, Germany

S.W. Koch and H. Giessen

*Department of Physics and Material Sciences Center, Philipps-Universität,
D-35032 Marburg, Germany*

Abstract

The coherent exciton-light coupling in pulse propagation experiments on the A-exciton resonance in bulk CdSe is investigated over a broad intensity range. At low light intensities, polariton propagation beats due to interference between excited states on both polariton branches are observed. In an intermediate intensity regime, the temporal polariton beating is suppressed in consequence of exciton–exciton interaction. At the highest light intensities, self-induced transmission and multiple pulse breakup are identified as a signature for carrier density Rabi flopping. Exciton–phonon scattering is shown to gradually eliminate coherent nonlinear propagation effects due to enhanced dephasing of the excitonic polarization. Calculations using the semiconductor Maxwell–Bloch equations are in qualitative agreement with the experimental data.

1.1   Introduction

The investigation of pulse propagation through opaque materials is of great importance for the understanding of coherent nonlinear light–matter interaction. While the associated

---

*This article was originally published in *Physical Review B* **64**, 245202 (2001). © American Physical Society 2001. Reprinted with permission.

optical effects are well established in atomic and molecular vapors – which can be modeled by noninteracting two-level systems [1,2] – the situation is substantially modified in semiconductors due to the Coulomb interaction between optically generated electron–hole excitations [3–5]. However, many of the familiar coherent transient two-level phenomena such as free induction decay, photon echo, and Rabi flopping have been rediscovered in semiconductors [3,4,6]. For bound excitons in CdS, which are well approximated by noninteracting two-level systems, even the appearance of self-induced transparency (SIT) has been detected several years ago [7]. On the other hand, spatial dispersion [8] and excitation-induced nonlinearities of the free-exciton resonance [6] give rise to striking differences of the optical response in semiconductors when compared with idealized two-level systems. In particular, numerical studies of pulse propagation in semiconductors came to the conclusion that these many-body effects may prevent the establishment of complete SIT on free-exciton resonances in condensed matter [9]. Only within simplified model systems can the phenomenon of excitonic SIT in resonantly excited semiconductors be investigated analytically [10]. Nevertheless, Rabi flopping of the carrier density, coherent long-distance propagation, and a high degree of transmission have been predicted [11,12]. This so-called self-induced transmission regime was recently discovered on the free-exciton resonance in CdSe [13].

In this chapter, we present a comprehensive analysis of subpicosecond pulse propagation on the A-exciton resonance of CdSe. Our work clearly identifies coherent exciton–light coupling over a broad intensity range and permits comparison with numerical calculations based on the semiconductor Maxwell–Bloch equations. The increase of the signal-to-noise ratio by approximately one order of magnitude as compared to the data of Ref. 13 reveals interesting novel features of coherent light–matter interaction. In particular, we were able to determine the pulse delay and the effective propagation velocity in dependence on the pulse intensity and to measure the increasing suppression of coherent nonlinear pulse propagation in the presence of phase-destroying exciton–phonon scattering.

1.2 Theoretical background

In this section, we summarize known facts on pulse propagation in semiconductor bulk material and outline their theoretical description. The transition from linear to nonlinear optical phenomena around the band edge of a semiconductor occurs if the density of optically generated excitons is large enough to allow interaction processes between them.

At low light intensities, the interaction between the optically generated excitons can be neglected. In this case, the exciton–light system forms new quasiparticles, so-called exciton-polaritons [14]. The exciton–radiation coupling causes an anticrossing between the dispersion relations of the exciton and light, thus splitting the polariton dispersion into an upper and a lower branch. If the frequency spectrum of a resonant short pulse coherently excites a broad range of modes on both branches, the interference of the excited polaritons at the end of the sample results in the formation of a pulse tail whose shape exhibits a pronounced nonperiodic temporal beating. Several periods of this polariton beating have been observed experimentally and proved excellent agreement with linear dispersion theory for the excitonic resonance [15]. However, at increased pulse intensities, the polariton beating is found to be suppressed due to incoherent exciton–exciton interaction which yields dephasing of the excitonic polarization. Experimentally, this suppression can be realized via a faster decay of the propagated pulse tail as well as a reduction of the beat modulation depth [16]. Upon further increase of the intensity, a new type of propagation-induced pulse shape oscillations occurs. These oscillations are due to Rabi flopping of the carrier density. The corresponding temporally interchanging absorption and gain during a full flopping period cause modulations on the initial pulse shape [13].

To describe the pulse shape modulations over the whole intensity regime, the transition between the linear and the nonlinear optical response must be treated theoretically. This can be done by using material equations that describe the temporal and spatial evolution of the material polarization and adding a wave equation for the calculation of the optical field. While the dynamics of the optical field can be calculated within the full wave [17] or the reduced wave equation [9,18], the corresponding polarization, which acts as a source term in the wave equation, is calculated by the semiconductor Bloch equations [3,4]. The semiconductor Bloch equations have been developed on different levels of complexity for the interaction of optically excited electron–hole pairs. They contain mean-field effects and correlations in the second-order Born approximation (SOBA) (Refs 19–27) or in the coherent dynamics-controlled truncation scheme (DCTS) [28–30]. A description of the related quantum kinetic phenomena can be found in Ref. 4.

The DCTS is typically used in the weakly nonlinear intensity range to describe coherent phenomena, especially bound states, whereas the SOBA can be applied to a broad intensity range including the description of optical gain and Rabi flopping, at the expense of higher-order correlations such as biexcitons. Here, we use both techniques where they are appropriate in the description of pulse propagation phenomena. One should note that the overlap between both sets of equations occurs in the low-density regime, if higher-order correlation functions are neglected in the DCTS and the low-density limit is applied to the SOBA.

In the following we discuss the similarities and the differences of two-level systems and semiconductors more formally in terms of the corresponding equations of motion (compare Refs 1–3). The reduced wave equation $(\partial/\partial z) \Omega(t, z) = -i \beta P(t, z)$ [9] is used for the description of the field envelope E in the form of the Rabi frequency $\Omega(t, z) = (d/\hbar) E(t, z)$ (d is the dipole moment). The source of the field is the polarization envelope P and β is a constant determined by material properties such as dipole moments and the refractive index. P is given by the off-diagonal density matrix element σ_{21} in the case of a two-level system $P = n_0 d\, \sigma_{21}$, where n_0 is the number density, and by all wave number transitions P_k in the case of a semiconductor $P = (d/V) \sum_k P_k$, where V is the sample volume. In the case of linear optics, σ_{21} is determined by a simple oscillator equation: $\dot{\sigma}_{21} = i\Delta\sigma_{21} + \frac{1}{2} i\Omega$ (Δ is the detuning). For semiconductors in the same limit, the equations for P_k can be diagonalized within the exciton basis to a similar form, where all exciton states λ contribute: $\dot{P}_\lambda = i\Delta_\lambda(z)P_\lambda + i\Omega_\lambda$. Restricting to the basic excitonic state $\lambda = 1s$, the material equations are formally identical, despite the fact that the dispersion $\Delta_{1s}(z)$ of the $1s$ exciton has to be taken into account. Thus, in the case of linear optics, the material equations for two-level systems and excitons are formally similar and their coupling to the wave equation yields the discussed dispersion anticrossing of two oscillators. Despite modifications due to spatial dispersion in a semiconductor, both the two-level and semiconductor dynamics contain the temporal beating in the pulse tail described above [31].

In the case of nonlinear optics of two-level systems, the only relevant nonlinearity is the Pauli-blocking nonlinearity $\dot{\sigma}_{21}|_{nl} = -i\Omega\sigma_{22}$, where σ_{22} is the transient occupation of the upper level. Without additional dephasing terms, the wave equation and the density matrix σ describe the vanishing polariton beating and the formation of pulse breakup due to Rabi flopping in two-level systems. Here, Rabi flopping is the transient oscillation of σ_{22} between occupation zero and one: $\sigma_{22}(t) = \sin^2[\Theta(t)/2]$. The pulse area $\Theta(t, z)$ is defined by the temporal integral over the Rabi frequency $\Omega(t, z)$: $\Theta(t, z) = \int_{-\infty}^{t} \Omega(t', z)dt' = (d/\hbar) \int_{-\infty}^{t} E(t', z)dt'$. The process of Rabi oscillation of the density leads to a temporal interplay of absorption and amplification for pulses having an area larger than or equal to 2π (high-intensity regime). In view of this, the coherent propagation of a resonant light pulse in a noninteracting two-level system is determined by the area theorem of McCall and Hahn [32,33]. According to the area theorem, a light pulse with area $\Theta = 2\pi$ and hyperbolic secant envelope exhibits lossless soliton propagation on the resonance (SIT) due to an exact cancellation of the absorption and amplification process (one complete Rabi flop). For input areas larger than 3π, repeated Rabi oscillations cause pulse breakup into separate pulses with area of 2π. SIT and multiple pulse breakup were pioneered in atomic vapor about 30 years ago and have been thoroughly investigated in the 1970's [34–36].

Concerning the nonlinear optics of free excitons in a semiconductor, many-body effects have to be taken into account, which can be divided into mean-field and correlation effects. Mean-field effects consist not only of the Pauli-blocking contributions, but yield for the semiconductor coherent Coulomb renormalizations of the field as well as of the single-particle energies. For instance, the field renormalization is given by the Coulomb-mediated emission of all wave number transitions $\hbar\Omega_k = \frac{1}{2}dE + \Sigma_q V_{|k-q|}P_q$, where Ω_k is the generalized Rabi frequency and V_k is the Coulomb interaction. For subpicosecond to picosecond pulses in semiconductors such as CdSe or GaAs, this field correction almost doubles the Rabi frequency compared to two-level systems [11,12]. As opposed to a noninteracting two-level system, one would therefore expect almost twice as many Rabi oscillations of the carrier density for a given pulse area. However, it must be noted that the number of Rabi flops for a 2π pulse in a semiconductor (yielding always one complete Rabi flop in a two-level system) varies as a function of the ratio of exciton binding energy and peak Rabi frequency [37] and care has to be taken in the interpretation of experiments. Numerical investigations of the corresponding semiconductor Bloch equations in mean-field approximation for the polarization coupled to Maxwell's wave equation for the propagating light field demonstrated Rabi flopping, coherent long-distance propagation, and a high degree of transmission already for pulse areas larger than π [9,11,12]. Lossless propagation – as in the case of SIT in two-level systems – was, however, excluded, even under idealized conditions where only mean-field effects are taken into account.

The next step of sophistication in treating the semiconductor material equations arises at the level where correlation effects are included. In principle, these effects cannot be neglected in nonlinear optics because they compensate some of the mean-field effects [19–27,38], especially energy shifts in stationary spectra. Using the SOBA, it can be recognized that within the course of time the correlations drive the system into a quasiequilibrium by carrier–carrier scattering and excitation-induced dephasing of the macroscopic polarization. For instance, in the weakly nonlinear regime and for a single circularly polarized pulse, the coupling of the one-exciton states to the free two-exciton continuum yields optical dephasing. At high densities, correlation effects reduce the coherent interaction effects such as exciton–polariton formation and Rabi oscillations. However, several experiments and theories have shown that coherent mean-field effects dominate the incoherent effects due to carrier-carrier scattering on sufficiently short time scales (compare the recent Ref. 39). As complete Rabi oscillations and SIT seem not to be possible in semiconductors, the associated pulse propagation phenomena are referred to as self-induced transmission [13].

1.3 Samples and experimental techniques

The experiments were performed on two CdSe bulk crystals grown by hot-wall epitaxy on transparent BaF$_2$ substrates [40]. We used an optically thin sample with Beer absorption length $\alpha L = 1.7$ to study propagation effects in the linear and the weakly nonlinear regime and an optically thick sample with $\alpha L = 6.5$ to investigate the characteristics of high-intensity coherent nonlinear transmission. The c axis of CdSe was oriented perpendicularly to the substrate. Thus, both the intrinsic A- and B-exciton resonances could be excited with linearly polarized light normally incident to the samples ($\mathbf{E} \perp \mathbf{c}$). Figures 1.1(a) and 1.1(b) show the corresponding linear absorption spectra at $T = 8$ K. The thin CdSe sample exhibits a well-defined A-exciton resonance at $\lambda_A = 684.6$ nm with a full width at half maximum of approximately 3 nm. The large A-exciton binding energy $E_x = 22.5$ meV (offset between A exciton and continuum) implies a large transition dipole moment and only weak interaction with the continuum states. The pronounced

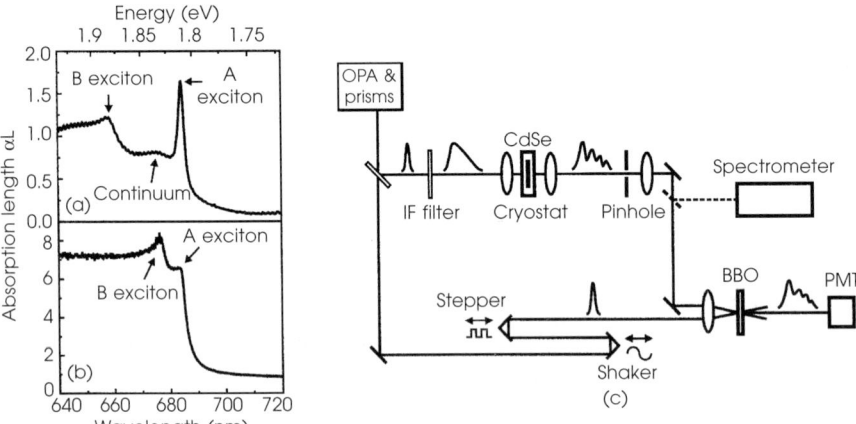

Fig. 1.1(a), (b) Linear absorption spectra of the two CdSe samples at $T = 8$ K. (a) Thin sample: $\alpha L = 1.7$ at $\lambda_A = 684.6$ nm. (b) Thick sample: $\alpha L = 6.5$ at $\lambda_A = 683.8$ nm. (c) Experimental setup using 60–70 fs pulses tunable around 684 nm from an optical parametric amplifier (OPA). The propagated pulses are time-resolved by cross correlation in a 2 mm thick β-barium-borate (BBO) crystal and detected with a photomultiplier tube (PMT). [From N.C. Nielsen *et al.*, Phys. Rev. B **64**, 245202 (2001).]

splitting of 72 meV between the *A* and *B* excitons can be attributed to strain in the CdSe layer due to lattice mismatch with the BaF_2 substrate and thermal expansion during the growth process [41]. In the thick sample, however, strain relaxation prevented the occurrence of increased *A-B* exciton splitting. Additionally, the *A*-exciton resonance, situated at λ_A = 683.8 nm, shows substantial inhomogeneous broadening.

Figure 1.1(c) illustrates the experimental setup. We used 60–70 fs pulses tunable around 684 nm with pulse energies of about 90 nJ from an optical parametric amplifier (OPA) (Ref. 42) pumped by a regenerative Ti:sapphire amplifier (COHERENT REGA) at a repetition rate of 200 kHz. Careful alignment of a prism compressor [43] made sure that the pulses were almost chirp-free, which was essential for the subsequent propagation experiment. The experimental configuration involves splitting of the linearly polarized OPA output into two portions: One part (67%) was additionally attenuated and focused with a *f* = 25 mm microscope objective (Ealing, *NA* = 0.15) onto the CdSe samples in a cold finger cryostat (*T* = 8 K), while the second part (33%) passed through a variable delay line. We determined the spot size on the sample via knife-edge test and assumed an uncertainty of 20%. A first part of the transmitted pulses was spectrally recorded and a second part time-resolved by cross correlation with output pulses from the OPA in a 2 mm thick β-barium-borate (BBO) crystal cut for type I phase matching. The intensity cross-correlation signal may be written as

$$I_{sig}(\tau) \propto \int_{-\infty}^{\infty} I_{trans}(t) I_{del}(t-\tau) dt, \qquad (1.1)$$

where $I_{trans}(t)$ and $I_{del}(t)$ are the temporal profiles of the transmitted and the delayed pulses, and τ is the pulse delay. A fast-scan sampling technique was adopted to measure the cross-correlation signals: By means of the discrete translation of a stepper, we could calibrate the oscillation of a shaker, which in turn periodically modulated the pulse delay at a frequency of 70 Hz [44]. Thus, we achieved a high signal-to-noise ratio due to short measurement cycles and averaging over many (up to 4000) scans. Careful spectral and spatial filtering of the signal was found to be indispensable for the propagation experiment. Insertion of narrow bandpass filters (IF filters) before the cryostat adjusted the broad spectrum of the ultrashort OPA pulses to the *A*-exciton resonance of the CdSe samples. Thereby, the incident pulses were stretched up to 220 fs for 3 nm bandwidth and 800 fs for 1 nm bandwidth with roughly Lorentzian spectral shape. The theoretical assumption of spatially homogeneous wave fronts was approximated by imaging the transmitted beam onto a pinhole, cutting out a region of constant intensity.

8 N.C. Nielsen et al.

1.4 Results and discussion

In this section, the transition from linear to nonlinear pulse propagation on the free-exciton resonance is discussed. Within this transition, two basic nonlinear effects are observed: the exciton–exciton interaction induced damping of the temporal polariton beating and the Rabi flopping related occurrence of temporal pulse breakup.

Excitation-induced suppression of temporal polariton beating

Figure 1.2(a) shows the experimental results of the propagation of 220 fs pulses resonantly tuned to the A-exciton resonance of the thin CdSe sample with $\alpha L = 1.7$ at $T = 8$ K. The left column illustrates the temporal cross-correlation traces of the transmitted pulses, whereas the right column shows the corresponding transmitted spectra behind the sample. The lowest traces characterize the input pulse as measured after propagation through the substrate alone. For producing the input pulse, we made use of a spectral filter with 3 nm bandwidth to ensure that both branches of the polariton dispersion were coherently

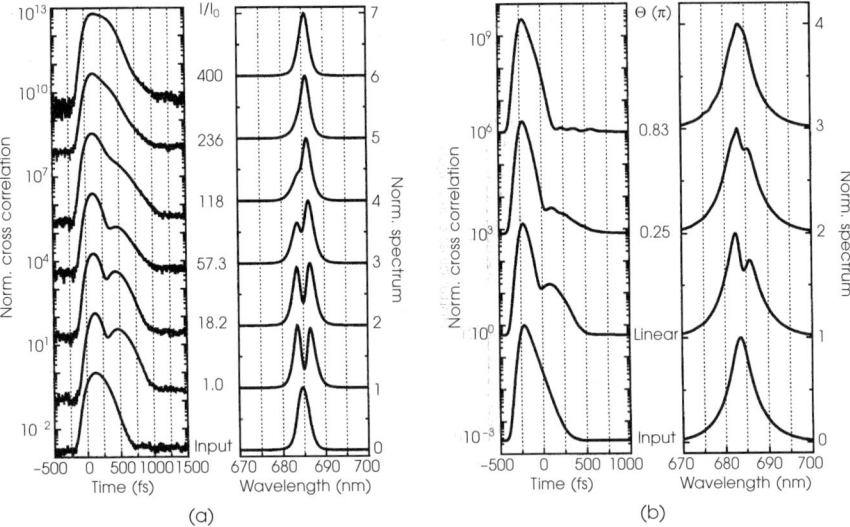

Fig. 1.2(a) Propagation of 220 fs pulses through the thin CdSe sample with $\alpha L = 1.7$ for increasing intensities I. $\lambda = 684.6$ nm, $I_0 = 0.06$ MW/cm^2, and $T = 8$ K. The normalized cross-correlation traces are shown on a logarithmic scale at the left and the normalized transmitted spectra are plotted on a linear scale at the right. (b) Numerical simulation using the semiconductor Maxwell–Bloch equations in the dynamics-controlled truncation scheme. [From N.C. Nielsen *et al.*, Phys. Rev. B **64**, 245202 (2001).]

excited. In this manner, we obtained the above mentioned 220 fs pulses with roughly single-sided exponential envelope. Going from bottom to top in the figure, the pulse intensity is increasing from $I_0 = 0.06$ MW/cm^2 to $400 \times I_0 = 24$ MW/cm^2. The transmitted pulses indicate the expected temporal polariton beating in the linear propagation regime: Two distinct pulses, i.e., one full beat period, can be observed which are separated by approximately 340 fs. Further polariton beats are suppressed because of inhomogeneous broadening of the resonance and the short sample length [17]. With increasing input intensity, the temporal beating decreases and almost vanishes at $118 \times I_0 = 7$ MW/cm^2. Simultaneously, the spectral dip (right column), which originates from the excitonic absorption, fades away. At an intensity of $236 \times I_0 = 14$ MW/cm^2, a shoulder with a delay of 270 fs with respect to the pulse maximum develops on the trailing edge of the transmitted pulse. Upon further increase of the intensity to $400 \times I_0 = 24$ MW/cm^2, the delay of the shoulder shortens to approximately 180 fs. At the same time, the transmitted spectra resemble the input spectrum and do not broaden, ruling out any perturbing influence of self-phase modulation (SPM) (Ref. 45) from off-resonant states or the substrate.

A theoretical model capable of explaining our experimental observations, especially the suppression of the polariton beating, was developed recently [17] on the basis of the semiconductor Maxwell–Bloch equations in the DCTS (see section 1.2). This model can only be applied for the weakly nonlinear regime discussed here. The material equations were evaluated for plane-wave propagation using a tight-binding approximation for the band structure and one-dimensional Coulomb interaction [38]. Inhomogeneous broadening of the resonance due to sample strain was additionally included in the model by averaging the polarization calculated for Gaussian-distributed band gap energies. Parameters for the CdSe material are given in Ref. 46. For comparison with the experiment, the computed time-resolved signals were convolved with a 50 fs Gaussian pulse according to Eq. (1.1). The initial pulse had a duration of 180 fs with a single-sided exponential envelope. Results of the numerical simulation are depicted in Fig. 1.2(b). For linear excitation, a single polariton beat can be observed about 310 fs after the pulse maximum. At higher excitation densities, the theoretical cross-correlation traces exhibit the anticipated suppression of the polariton beating both temporally and spectrally in full agreement with our experiment.

The theoretical analysis shows that only one polariton beat period is found at low intensities because beating at longer times is suppressed due to the inhomogeneous broadening of the resonance. At increasing intensities, the excitonic polarization is dephased caused by the coupling of excitons to unbound two-exciton states, i.e., to an interaction-generated continuum. The observed shoulder at highest intensities

(Fig. 1.2(a)) indicates a regime with higher-order nonlinearities. The forming of the shoulder is a manifestation of the beginning of a first carrier density Rabi flop, with the latter occurring at a higher frequency when the light field is intensified. Indeed, we find a shorter delay of the shoulder with respect to the pulse maximum for the higher pulse intensity. However, for increasing intensities, coherent pulse breakup cannot reveal itself clearly in such a thin sample [1,47] and the low-intensity DCTS breaks down at sufficiently high carrier densities. The cross correlation of a 200 fs pulse with a 50 fs pulse is another limiting factor since it smears out the fine structure when Rabi flopping induced pulse breakup exhibits more than one minimum. Furthermore, the effects of Rabi flopping induced pulse breakup become much more pronounced for longer propagation distances [13]. For these reasons, we choose a thicker sample as well as a longer pulse duration and higher pulse intensities to investigate pulse breakup due to Rabi flopping. Correspondingly, the theoretical description is carried out within the SOBA.

Self-induced transmission and multiple pulse breakup

Using the thick sample with $\alpha L = 6.5$ in the following studies, we turn our attention to the high-excitation regime exclusively. Low-intensity light will no longer propagate through the sample since the linear transmission is only 0.15%. We utilized a 1 nm bandwidth filter in order to excite a narrow distribution of transitions within the A-exciton resonance, thus reducing the influence of inhomogeneous broadening and of the cross-correlation measurement. Figure 1.3 demonstrates that coherent nonlinear propagation is indeed observable on the initial pulse shape at high pulse intensities. Again, the plot shows temporal cross-correlation traces for increasing intensities from bottom to top. The lowest trace illustrates the roughly single-sided exponential input pulse with 800 fs duration. At an intensity $I_0 = 105$ MW/cm^2, the pulse transmitted through the sample is already steepened, shortened, and shows a higher symmetry compared with the original pulse shape. On closer inspection, the trace exhibits a slight shoulder structure about 550 fs after the maximum. In addition, a large delay of approximately 700 fs is measured with respect to the input pulse (evaluated at 50% of the normalized signal height between the rising pulse edges). Increasing the intensity to $1.9 \times I_0 = 200$ MW/cm^2, the pulse reshaping and the shoulder structure become even more pronounced. Both the temporal distance between the two pulse components and the pulse delay reduce to approximately 400 and 470 fs, respectively. Upon further increase of the intensity to $2.9 \times I_0 = 305$ MW/cm^2 and $4.2 \times I_0 = 440$ MW/cm^2, distinct pulse breakups into several individual pulses are observed. At $2.9 \times I_0$, the intensified second pulse is emitted about 320 fs after the main pulse component, and approximately 1000 fs after the maximum, a small third shoulder develops. At $4.2 \times I_0$, three succeeding peaks are clearly visible, which appear about 350, 690, and 1040 fs after the first peak. As shown later, the pulse breakup is related to Rabi oscillations. In a simple model based on the two-level solution for σ_{22} and using $E \propto \sqrt{I}$, one expects the dynamics of the pulse breakup to be roughly twice as fast at four times the intensity. The number of peaks is correct, and the reemission frequency grows by a factor of 1.6 within the covered intensity range. Moreover, the experimental results show that the delay between the transmitted and the original pulses further reduces with increasing intensity. This subject will be quantitatively discussed below. At the input intensity of $4.2 \times I_0$, we measured a total nonlinear transmission through the sample (corrected for surface reflectivity) greater than 25%. Behind the pinhole, which was inserted to restrict the signal detection to directly propagated and spatially homogeneous wavefronts, the transmission degree reached 6.25% compared to a linear transmission of merely 0.15% for $\alpha L = 6.5$. In combination with these nonlinear transmission values, the data of Fig. 1.3 prove the presence of self-induced transmission and multiple pulse breakup due to coherent carrier density Rabi flopping.

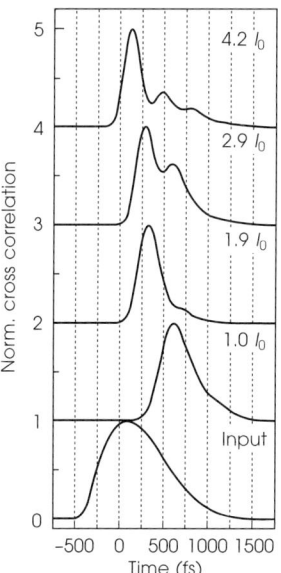

Fig. 1.3 Propagation of 800 fs pulses through the thick CdSe sample with $\alpha L = 6.5$ for increasing intensities I. $\lambda = 683$ nm, $I_0 = 105$ MW/cm^2, and $T = 8$ K. The normalized cross-correlation traces are shown on a linear scale. [From N.C. Nielsen et al., Phys. Rev. B **64**, 245202 (2001).]

In order to analyze the high-intensity dynamics in more detail, we plotted traces transmitted through the thick sample on a logarithmic scale (Fig. 1.4(a)). Note that this set of traces differs from that in Fig. 1.3, but demonstrates the excellent reproducibility of our experimental data. Figure 1.4(b) shows the results of numerical calculations based on the semiconductor Maxwell–Bloch equations. Here, we applied the slowly varying envelope approximation (SVEA) of the field equation for numerical simplicity [9], whereas the material equations include mean-field and correlation effects (diagonal and nondiagonal dephasing as well as nonlinear polarization scattering) in the SOBA for polarization and carrier distribution [19–27]. The resulting material equations are a standard tool in semiconductor optics. For the CdSe material, we used parameters given in Ref. 46, except for the inhomogeneous broadening which is numerically not tractable because of the long propagation distance and the more involved material equations. However, we believe that due to the longer pulse width, inhomogeneous broadening is of minor importance. The theoretical analysis shows that the pulse breakup can be

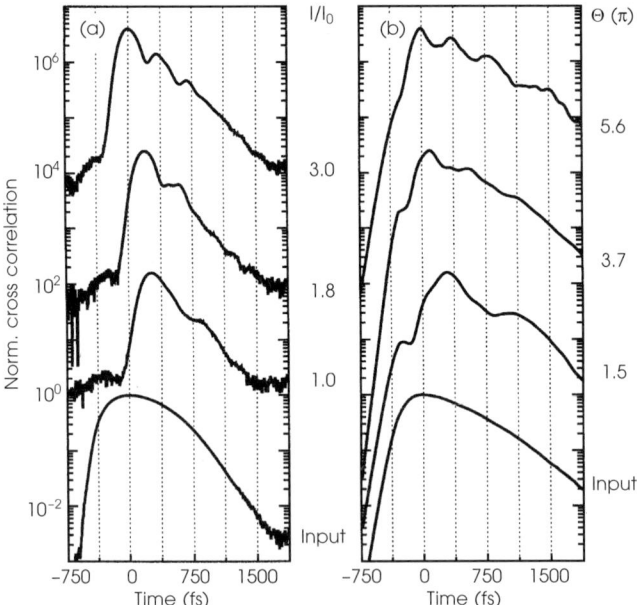

Fig. 1.4 (a) Propagation of 800 fs pulses through the thick CdSe sample with $\alpha L = 6.5$ for increasing intensities I. $\lambda = 683$ nm, $I_0 = 150$ MW/cm^2, and $T = 8$ K. The normalized cross-correlation traces are shown on a logarithmic scale. (b) Numerical simulation using the semiconductor Maxwell–Bloch equations in the second-order Born approximation. [From N.C. Nielsen *et al.*, Phys. Rev. B **64**, 245202 (2001).]

traced back to Rabi flopping of the carrier density in a highly excited semiconductor. Increasing the intensity corresponds to a higher Rabi oscillation frequency and therefore to a faster interplay of absorption and emission in the course of time, which modulates the pulse shape. Our theoretical investigations yield that carrier–carrier scattering on the time scale of the pulse duration is of minor importance in comparison to mean-field effects which cover the main physics, similar to the recent results presented in Ref. 39.

In Fig. 1.4(a), we observe intensity-dependent multiple pulse breakup comparable to the behavior in Fig. 1.3. The lowest input intensity is $I_0 = 150$ MW/cm^2. Interestingly, the second Rabi flop for $1.8 \times I_0 = 270$ MW/cm^2 shows a double substructure which is reproduced by the numerical simulation involving many-body interactions for a pulse area $\Theta = 3.7\pi$. Note that only the curves for $1.8 \times I_0 = 270$ MW/cm^2 and $\Theta = 3.7\pi$ are directly comparable in this figure. According to $\Theta \propto E \propto \sqrt{I}$, $\Theta = 1.5\pi$ corresponds to $0.3 \times I_0$ and 5.6π corresponds to $4.1 \times I_0$. Thus, the calculated traces represent the limits wherein the experimental curves are set. The breakup into two pulses is reproduced for the lowest input intensity $I_0 = 150$ MW/cm^2 and an area of 1.5π, respectively. Owing to the fact that the input pulse shape of the numerical model differs slightly from the experimental input pulse, the second peak is more pronounced in the theory. Also, the temporal distance with respect to the maximum is smaller in the experiment because of the higher excitation density, which results in a larger Rabi frequency and therefore faster reemission from the sample. For a pulse area $\Theta = 5.6\pi$, the calculated transmitted pulse shows four separate peaks, whereas the experimental trace indicates breakup into only three individual pulses. Most likely, the fourth pulse is already eliminated due to increased incoherence in the sample, contrary to the measurement series depicted in Fig. 1.3. The temporal spacing between the maximum and the succeeding peaks is roughly the same as can be seen when comparing the upper curves in Figs 1.4(a) and 1.4(b), showing that the many-body calculation is able to produce the correct pulse velocity over a long propagation distance in a broad range of intensities. Precursors, which are purely propagation induced, can be seen both experimentally and theoretically for each excitation density.

Clearly observable in experiment and theory is the reduced delay between transmitted and input pulses for increased light intensities, indicating that the dynamical response of the matter emerges earlier due to the larger pulse area and higher Rabi frequency. In turn, this implies the definition of a higher effective propagation velocity v_{eff} through the sample: $v_{\text{eff}} = L/(\tau_{1/2} + L/c_0)$. Here, $\tau_{1/2}$ is the pulse delay measured at 50% of the maximum signal height between the rising edges of the transmitted and the input pulses and L is the sample thickness derived from the absorption length αL. Assuming an absorption coefficient $\alpha \approx 1$ μm^{-1} [48], L is approximately 6.5 μm for the thick sample

($\alpha L = 6.5$). The pulse delay and the effective velocity are plotted versus input intensity in Fig. 1.5. The delay decreases from 840 fs for $I = 78$ MW/cm^2 to below 400 fs for $I = 400$ MW/cm^2. The corresponding effective velocity v_{eff} is ranging from 0.025 c_0 to 0.052 c_0. In the literature for coherent nonlinear propagation in isolated two-level systems [1,33], we find an estimate for the propagation velocity of 2π solitons $v_{\text{eff}} = c_0/(n + \frac{1}{2}\alpha\tau_p c_0)$. For a pulse duration $\tau_p = 800$ fs, the absorption coefficient from above, and the refractive index $n = \sqrt{\varepsilon_b}$ with the background dielectric constant $\varepsilon_b \approx 9$ for our material [48], a velocity of $v_{\text{eff}} = 0.0081\, c_0$ is calculated in this model. Considering the mean-field correction for the excitonic many-body system (see section 1.2), we expect a comparable velocity for a π pulse in our semiconductor system (one complete Rabi flop). Extrapolating the intensity-dependent velocity curve to around 25 MW/cm^2, corresponding to an area of about π (as shown below), one would obtain a velocity in the range of 0.015 c_0 to 0.02 c_0. This velocity is quite low and means that the light is absorbed into the excited state and reemitted subsequently with an intensity dependent delay given by the Rabi frequency. More elaborate, but different experiments in atomic three-level systems have recently perfected the slowdown to a complete halt of the light [49,50].

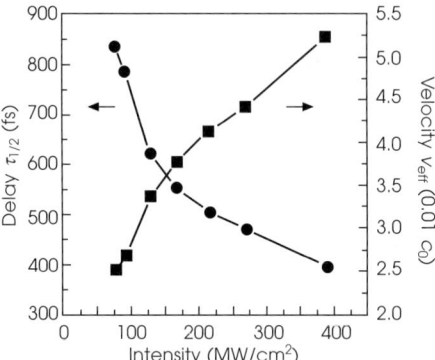

Fig. 1.5 Delay $\tau_{1/2}$ and effective propagation velocity v_{eff} for the propagation of 800 fs pulses through the thick CdSe sample with $\alpha L = 6.5$ at various input intensities I. $\lambda = 683$ nm and $T = 8$ K. [From N.C. Nielsen et al., Phys. Rev. B **64**, 245202 (2001).]

The simplified model which describes the dynamics of the pulse breakup allows one to directly relate measured pulse intensities I to calculated pulse areas Θ. Due to the mean-field correction, breakup into two (four) individual pulse components is supposed to occur for incident 2π (4π) pulses. With respect to the measurement series depicted in Fig. 1.3, using $\Theta \propto \sqrt{I}$, we obtain an intensity of approximately 27 MW/cm^2 for an area of π. Comparing experimental and simulated pulse traces in Fig. 1.4, where the features of the $1.8 \times I_0 = 270$ MW/cm^2 curve are very well reproduced by the theoretical curve calculated for a pulse area of 3.7π, we deduce $I \approx 20$ MW/cm^2 for $\Theta = \pi$. Thus, an intensity of about 25 MW/cm^2 corresponds to an area of π for an 800 fs pulse in a CdSe sample with $\alpha L = 6.5$. Since the measured external intensity $I = \frac{1}{2}\sqrt{\varepsilon_0 \mu_0}\,(E_{\text{out}})^2$ and the internal pulse area $\Theta \approx (d/\hbar)\,E_{\text{in}}\,\tau_p$, where the relation between the electric field envelope inside and outside the sample is given by the Fresnel formula $E_{\text{in}}/E_{\text{out}} = 2/(n+1)$ with the refractive index $n \approx 3$, the dipole moment amounts to $d = 3.8$ eÅ. Taking the literature value for the longitudinal-transverse splitting energy of the A exciton in CdSe $\Delta_{\text{LT}} = 0.9$ meV (compare Ref. 48) and using the relation $\Delta_{\text{LT}} = (2d^2)/(\varepsilon_b \varepsilon_0 a_B^3)$ (derived from Eq. (11.10) in Ref. 3) with the exciton Bohr radius $a_B = 53$ Å for an A-exciton binding energy $E_x = 15$ meV and the background dielectric constant $\varepsilon_b = 9$, the dipole moment is calculated to be $d = 1.8$ eÅ. Considering the uncertainty in quantifying the applied intensity and in relating this intensity to pulse area, the agreement is fairly good. It will be interesting to investigate the exact influence of the mean-field correction on the pulse area required for self-induced transmission [37].

Next, we discuss the influence of self-induced transmission on the spectral shape of the propagated pulses. Figures 1.6(a) and 1.6(b) depict the transmitted spectra for low and high input intensities (150 and 450 MW/cm^2 with $\Theta = 1.5\pi$ and 5.6π, respectively). At low intensities, the spectra represent symmetrical peaks in the experiment as well as in theory. While the transmitted spectrum roughly coincides with the input spectrum in theory, the experimental curve is slightly shifted and broadened towards the low-energy side. The shift is certainly due to detuning of the excitation against the resonance in the experiment as the maximum of the transmitted spectrum is found to be located exactly on the edge of the inhomogeneously broadened A-exciton resonance of the thick sample ($\lambda_A = 683.8$ nm). The spectral broadening emerges from spectral components at the low-energy side of the resonance that do not interact with the system. At high intensities, the transmitted spectra are asymmetrically broadened: The high- and low-energy sides show a slowly decaying tail and a steep edge, respectively. Modulations occur in the wings of the spectra, probably caused by the temporal modulation which leads to the pulse breakup. In contrast to the good agreement of the spectral shape between experiment and theory, the peak shifts of the spectra have opposite sign. We believe that this behavior originates from the detuning towards higher energies in the experiment. At low intensities, reemission of light following the temporal evolution of the excitonic polarization occurs on the edge of the A-exciton resonance, featuring the highest transition dipole moment. However, for increasing intensity, the exciting light field governs the spectral range within the inhomogeneously broadened exciton resonance where coherent Rabi

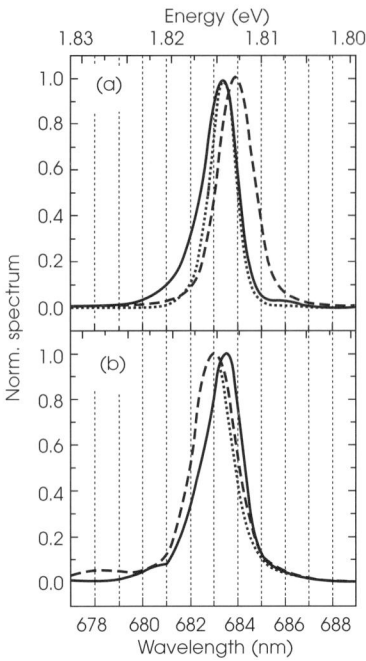

Fig. 1.6 Transmitted spectra corresponding to the resonant propagation of 800 fs pulses through CdSe at low and high intensities in the regime of self-induced transmission. The normalized spectra are plotted on a linear scale with respect to the input spectrum (dotted line). (a) Experiment: Thick sample with $\alpha L = 6.5$ at $T = 8$ K, $\lambda = 683.5$ nm, $I = 150$ MW/cm^2 (dashed line) and 450 MW/cm^2 (solid line). (b) Theory: Numerical calculations using the semiconductor Maxwell–Bloch equations in the second-order Born approximation, $\Theta = 1.5\pi$ (dashed line) and 5.6π (solid line). [From N.C. Nielsen et al., Phys. Rev. B **64**, 245202 (2001).]

flopping proceeds. Thus, the theoretically predicted redshift could not be observed in the experiment.

The fact that we do not observe spectral broadening at higher intensities and in the case of pulse breakup is a strong indication against SPM due to other transitions than the considered free excitons. This can be quantified by using the simplified formula $\Delta\Phi = (2\pi/\lambda)\, n_2 IL$ [45] with the nonlinear refractive index $n_2 \approx 10^{-12}$ cm^2/W for the CdSe bulk material [51], $I \approx 25$ MW/cm^2 for a pulse area $\Theta = \pi$, $L \approx 6.5$ μm for the thick sample, and $\lambda = 683.8$ nm. We obtain a nonlinear phase shift of $\Delta\Phi \approx 1.5$ mrad, which is two orders of magnitude less than the phase shift of 0.2 rad required for SPM [52].

The data presented in this chapter provide clear evidence for long-distance coherent pulse propagation and high nonlinear transmission due to Rabi oscillations on the *A*-exciton resonance in CdSe for pulse areas up to a multiple of π. In agreement with theoretical predictions, this finding demonstrates that even for excitation intensities in the range of 100 MW/cm^2, substantial coherence between the exciting laser field and the excitonic polarization is maintained over several hundred femtoseconds despite many-body interaction effects. Consequently, excitation-induced dephasing is less important in progress of the applied pulses on the ultrashort time scale considered here.

Phonon-induced dephasing of the excitonic polarization

In order to investigate the influence of enhanced phase relaxation on the features of self-induced transmission, we have performed pulse propagation experiments at varying sample temperature. If the temperature is increased, phase-destroying electron/hole-phonon scattering occurs that, for our purposes, primarily reduces the amplitude of the excitonic polarization and the coherent interaction during the duration of the pulse. The contribution of phonon scattering to phase relaxation grows linearly with T in the range where acoustic phonon scattering dominates and even superlinearly for temperatures which permit optical phonon scattering [53,54].

Figure 1.7 presents cross-correlation traces for 800 fs pulses transmitted through

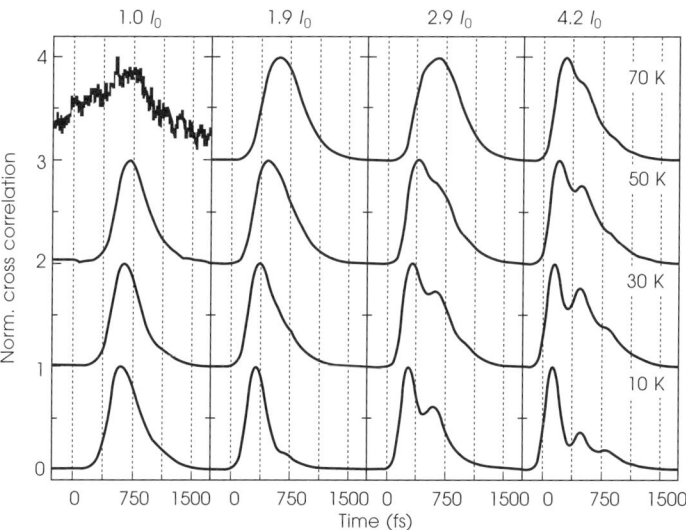

Fig. 1.7 Propagation of 800 fs pulses through the thick CdSe sample with $\alpha L = 6.5$ for increasing intensities I (horizontal direction) and increasing temperature T (vertical direction). $\lambda = 683$ nm and $I_0 = 105$ MW/cm^2. The normalized cross-correlation traces are shown on a linear scale. [From N.C. Nielsen *et al.*, Phys. Rev. B **64**, 245202 (2001).]

the thick CdSe sample ($\alpha L = 6.5$) for varying intensities and temperatures. In the horizontal direction, the intensity is varied from $I_0 = 105$ MW/cm^2 to $4.2 \times I_0 = 440$ MW/cm^2, corresponding to a doubling of the external electric field and the Rabi frequency, respectively. In the vertical direction, the temperature is raised from bottom to top from 10 to 70 K. The set of data in the bottom line depicts the development of coherent multiple pulse breakup as already discussed above. The right column (highest intensity) shows a remarkable reduction of the modulation depth with rising temperature. Notice that the decay of the modulation depth with time caused by dephasing processes is strongly enhanced for higher temperatures because of the additional contribution of exciton–phonon scattering. For the highest intensity trace at $T = 70$ K (upper right), the pulse is still transmitted, but the pulse breakup has almost completely vanished. Only a weak second peak is barely visible. The coherently propagated amount of light evidently has dropped compared to the value at $T = 10$ K owing to the interaction with phonons which dephases the coherent polarization and supports the buildup of an incoherent exciton population [55]. Thus, the dominant part of the transmission originates from incoherent bleaching of the exciton transition, which explains the increased unmodulated background level. Similarly, the dominance of the exciton-phonon coupling over the exciton–light coupling arises from the growing delay between the input and the transmitted pulses for elevated temperatures. The effect is reproduced in every single column, however, phase relaxation due to phonon scattering manifests itself more strongly at lower input intensities where the importance of the exciton–phonon interaction is enhanced in comparison with the exciton-light coupling.

The observations discussed in this section are certainly not caused by the redshift of the A-exciton resonance of approximately 9 meV within the viewed temperature range. Due to the inhomogeneously broadened resonance of the thick sample, detuning experiments at $T = 10$ K with the excitation shifted towards higher energies by more than 7 meV showed only little influence on the features of coherent multiple pulse breakup.

1.5 Conclusions

We have presented a comprehensive study of subpicosecond pulse propagation on the A-exciton resonance in bulk CdSe. At low pulse intensities, polariton formation and the corresponding interference effects result in temporal polariton beating. The linear polariton concept begins to fail at higher excitation densities where the polariton beating is suppressed in consequence of incoherent exciton–exciton interaction which causes dephasing of the excitonic polarization. Self-induced transmission and coherent multiple pulse breakup due to Rabi flopping of the carrier density are prominent at pulse areas beyond π. At intensities on the order of 100 MW/cm^2, tight coherent control of the temporal evolution of the excitonic polarization by the applied ultrashort pulse results in a large amount of coherent nonlinear transmission and a high contrast ratio of the pulse breakup. The experiments can be described theoretically using the semiconductor Maxwell–Bloch equations, which accomplish the transition from linear to nonlinear optics by taking into account many-body interactions consisting of mean-field and correlation effects. Further findings such as the intensity to pulse area relation as well as pulse delays and effective propagation velocities in dependence on the pulse intensity yield quantitative agreement between the experiment and the semiconductor Maxwell–Bloch theory. Increasing the phase relaxation rate by introducing exciton–phonon scattering at elevated sample temperatures greatly diminishes the amplitude of the coherent excitonic polarization, thus gradually destroying the contrast ratio of the Rabi flopping induced pulse breakup.

Acknowledgments

We would like to thank W.W. Rühle and B. Hillebrands for continuous support. We are grateful to F. Widulle for a critical reading of the manuscript. The high-quality samples were grown by M. Grün and M. Hetterich in the group of C. Klingshirn. Financial support came from the Deutsche Forschungsgemeinschaft through the Graduiertenkolleg "Optoelektronik mesoskopischer Halbleiter" and through the Quantenkohärenzschwerpunkt. H. Giessen's present address is the Institute of Applied Physics, University of Bonn, D-53115 Bonn, Germany.

References

1. Allen, L., and Eberly, J.H. (1987). *Optical Resonance and Two-Level Atoms* (Dover Publications, New York).
2. Scully, M.O., and Zubairy, M.S. (1997). *Quantum Optics* (Cambridge University Press, New York).
3. Haug, H., and Koch, S.W. (1993). *Quantum Theory of the Optical and Electronic Properties of Semiconductors* (World Scientific Publishing, Singapore).
4. Haug, H., and Jauho, A.P. (1996). *Quantum Kinetics in Transport and Optics of Semiconductors* (Springer-Verlag, Berlin).
5. Klingshirn, C.F. (1997). *Semiconductor Optics* (Springer-Verlag, Berlin).
6. *Phys. Status Solidi B* **159**, (1990); **173**, (1992); **188**, (1995); **206**, (1998); **221**, (2000).

7. Jütte, M., Stolz, H., and von der Osten, W. (1996). Linear and Nonlinear Pulse Propagation at Bound Excitons in CdS, *J. Opt. Soc. Am.* B **13**, 1205–1209.
8. Agranovich, V.M., and Ginzburg, V.L. (1984). Crystal Optics with Spatial Dispersion and Excitons, Vol. 42 of Springer Series in Solid-State Sciences (Springer-Verlag, Berlin).
9. Knorr, A., Binder, R., Lindberg, M., and Koch, S.W. (1992). Theoretical Study of Resonant Ultrashort-Pulse Propagation in Semiconductors, *Phys. Rev. A* **46**, 7179–7186.
10. Talanina, I., Burak, D., Binder, R., Giessen, H., and Peyghambarian, N. (1998). Theoretical Study of Solitonlike Propagation of Picosecond Light Pulses Interacting with Wannier Excitons, *Phys. Rev. E* **58**, 1074–1080.
11. Binder, R., Koch, S.W., Lindberg, M., Peyghambarian, N., and Schäfer, W. (1990) Ultrafast Adiabatic Following in Semiconductors, *Phys. Rev. Lett.* **65**, 899–902.
12. Koch, S.W., Knorr, A., Binder, R., and Lindberg, M. (1992). Microscopic Theory of Rabi Flopping, Photon Echo, and Resonant Pulse Propagation in Semiconductors, *Phys. Status Solidi B* **173**, 177–187.
13. Giessen, H., Knorr, A., Haas, S., Koch, S.W., Linden, S., Kuhl, J., Hetterich, M., Grün, M., and Klingshirn, C. (1998). Self-Induced Transmission on a Free Exciton Resonance in a Semiconductor, *Phys. Rev. Lett.* **81**, 4260–4263.
14. Hopfield, J.J. (1958). Theory of the Contribution of Excitons to the Complex Dielectric Constant of Crystals, *Phys. Rev.* **112**, 1555–1567.
15. Fröhlich, D., Kulik, A., Uebbing, B., Mysyrowicz, A., Langer, V., Stolz, H., and von der Osten, W. (1991). Coherent Propagation and Quantum Beats of Quadrupole Polaritons in Cu_2O, *Phys. Rev. Lett.* **67**, 2343–2346.
16. Fröhlich, D., Kulik, A., Uebbing, B., Langer, V., Stolz, H., and von der Osten, W. (1992). Propagation Beats of Quadrupole Polaritons in Cu_2O, *Phys. Status Solidi B* **173**, 31–40.
17. Förstner, J., Knorr, A., Kuckenburg, S., Meier, T., Koch, S.W., Giessen, H., Linden, S., and Kuhl, J. (2000). Nonlinear Polariton Pulse Propagation in Bulk Semiconductors, *Phys. Status Solidi B* **221**, 453–457.
18. Kim, D.S., Shah, J., Miller, D.A.B., Damen, T.C., Schäfer, W., and Pfeiffer, L. (1993). Femtosecond-Pulse Distortion in Quantum Wells, *Phys. Rev. B* **48**, 17 902–17 905.
19. Lindberg, M., and Koch, S.W. (1988). Effective Bloch Equations for Semiconductors, *Phys. Rev. B* **38**, 3342–3350.
20. Rappen, T., Peter, U.G., Wegener, M., and Schäfer, W. (1994). Polarization Dependence of Dephasing Processes: A Probe for Many-Body Effects, *Phys. Rev. B* **49**, 10 774–10 777.
21. Schäfer, W., Brener, I., and Knox, W. (1994). *Many-Body Effects at the Fermi Edge of Modulation Doped Semiconductors: A Numerical Study*, in *Coherent Optical Interactions in Semiconductors*, edited by R.T. Phillips (Plenum Press, New York), pp. 343–347.
22. Rossi, F., Haas, S., and Kuhn, T. (1994). Ultrafast Relaxation of Photoexcited Carriers: The Role of Coherence in the Generation Process, *Phys. Rev. Lett.* **72**, 152–155.
23. Heiner, E. (1989). Screening of the Coulomb Potential in the Athermal Stage in Highly Excited Semiconductors, *Phys. Status Solidi B* **153**, 295–305.
24. Pötz, W. (1996). Microscopic Theory of Coherent Carrier Dynamics and Phase Breaking in Semiconductors, *Phys. Rev. B* **54**, 5647–5664.
25. Jahnke, F., Kira, M., Koch, S.W., Khitrova, G., Lindmark, E.K., Nelson, T.R., Wick, Jr., D.V.,

Berger, J.D., Lyngnes, O., Gibbs, H.M., and Tai, K. (1996). Excitonic Nonlinearities of Semiconductor Microcavities in the Nonperturbative Regime, *Phys. Rev. Lett.* **77**, 5257–5260.
26. Knorr, A., Hughes, S., Stroucken, T., and Koch, S.W. (1996). Theory of Ultrafast Spatio-Temporal Dynamics in Semiconductor Heterostructures, *Chem. Phys.* **210**, 27–47.
27. Jahnke, F., Kira, M., and Koch, S.W. (1997). Linear and Nonlinear Optical Properties of Excitons in Semiconductor Quantum Wells and Microcavities, *Z. Phys. B* **104**, 559–572.
28. Axt, V.M., and Stahl, A. (1994). A Dynamics-Controlled Truncation Scheme for the Hierarchy of Density Matrices in Semiconductor Optics, *Z. Phys. B* **93**, 195–204.
29. Lindberg, M., Hu, Y.Z., Binder, R., and Koch, S.W. (1994). $\chi^{(3)}$ Formalism in Optically Excited Semiconductors and its Applications in Four-Wave-Mixing Spectroscopy, *Phys. Rev. B* **50**, 18 060–18 072.
30. Schäfer, W., Kim, D.S., Shah, J., Damen, T.C., Cunningham, J.E., Goossen, K.W., Pfeiffer, L.N., and Köhler, K. (1996). Femtosecond Coherent Fields Induced by Many-Particle Correlations in Transient Four-Wave Mixing, *Phys. Rev. B* **53**, 16 429–16 443.
31. Burnham, D.C., and Chiao, R.Y. (1969). Coherent Resonance Fluorescence Excited by Short Light Pulses, *Phys. Rev.* **188**, 667–675.
32. McCall, S.L., and Hahn, E.L. (1967). Self-Induced Transparency by Pulsed Coherent Light, *Phys. Rev. Lett.* **18**, 908–911.
33. McCall, S.L., and Hahn, E.L. (1969). Self-Induced Transparency, *Phys. Rev.* **183**, 457–485.
34. Gibbs, H.M., and Slusher, R.E. (1970). Peak Amplification and Breakup of a Coherent Optical Pulse in a Simple Atomic Absorber, *Phys. Rev. Lett.* **24**, 638–641.
35. Slusher, R.E., and Gibbs, H.M. (1972). Self-Induced Transparency in Atomic Rubidium, *Phys. Rev. A* **5**, 1634–1659.
36. Slusher, R.E., and Gibbs, H.M. (1972). Self-Induced Transparency in Atomic Rubidium, *Phys. Rev. A* **6**, 1255(E)–1257(E).
37. Östreich, Th., and Knorr, A. (1993). Various Appearances of Rabi Oscillations for 2π-Pulse Excitation in a Semiconductor, *Phys. Rev. B* **48**, 17 811–17 817.
38. Sieh, C., Meier, T., Knorr, A., Jahnke, F., Thomas, P., and Koch, S.W. (1999). Influence of Carrier Correlations on the Excitonic Optical Response Including Disorder and Microcavity Effects, *Eur. Phys. J. B* **11**, 407–421.
39. Saba, M., Quochi, F., Ciuti, C., Martin, D., Staehli, J.L., Deveaud, B., Mura, A., and Bongiovanni, G. (2000). Direct Observation of the Excitonic AC Stark Splitting in a Quantum Well, *Phys. Rev. B* **62**, R16 322–R16 325.
40. Grün, M., Hetterich, M., Becker, U., Giessen, H., and Klingshirn, C. (1994). Wurtzite-Type CdS and CdSe Epitaxial Layers – I. Growth and Characterization, *J. Cryst. Growth* **141**, 68–74.
41. Becker, U., Giessen, H., Zhou, F., Gilsdorf, Th., Loidolt, J., Müller, M., Grün, M., and Klingshirn, C. (1992). Shift of the Excitonic Resonances by Thermal Strain and Lattice Mismatch in CdS Thin Epitaxial Layers, *J. Cryst. Growth* **125**, 384–387.
42. Reed, M.K., Steiner-Shepard, M.K., Armas, M.S., and Negus, D.K. (1995). Microjoule-Energy Ultrafast Optical Parametric Amplifiers, *J. Opt. Soc. Am. B* **12**, 2229–2236.
43. Fork, R.L., Martinez, O.E., and Gordon, J.P. (1984). Negative Dispersion Using Pairs of Prisms, *Opt. Lett.* **9**, 150–152.

44. Edelstein, D.C., Romney, R.B., and Scheuermann, M. (1991). Rapid Programmable 300 ps Optical Delay Scanner and Signal-Averaging System for Ultrafast Measurements, *Rev. Sci. Instrum.* **62**, 579–583.
45. Diels, J.C., and Rudolph, W. (1996). *Ultrashort Laser Pulse Phenomena* (Academic Press, San Diego), pp. 140–141.
46. The CdSe material was modelled with a gap energy E_g = 1.85 eV, a tight binding band width energy Δ_p = 35.2 meV, an exciton binding energy E_x = 15 meV, a longitudinal-transverse splitting energy Δ_{LT} = 1 meV, an effective electron and hole mass m_e = 0.125 m_0 and m_h = 0.431 m_0, a background dielectric constant ε_b = 9, and an absorption coefficient α = 1 µm^{-1}. A value of 6 meV was assumed for the inhomogeneous broadening of the resonance.
47. Lamb, Jr., G.L. (1971). Analytical Descriptions of Ultrashort Optical Pulse Propagation in a Resonant Medium, *Rev. Mod. Phys.* **43**, 99–124.
48. Landolt-Börnstein, *Numerical Data and Functional Relationships in Science and Technology*, New Series, Group III, Vol. 17b, edited by I. Broser, R. Broser, and A. Hoffmann (Springer-Verlag, Berlin, 1982), pp. 202–224, 442–457.
49. Phillips, D.F., Fleischhauer, A., Mair, A., Walsworth, R.L., and Lukin, M.D. (2001). Storage of Light in Atomic Vapor, *Phys. Rev. Lett.* **86**, 783–786.
50. Liu, C., Dutton, Z., Behroozi, C.H., and Hau, L.V. (2001). Observation of Coherent Optical Information Storage in an Atomic Medium Using Halted Light Pulses, *Nature* (London) **409**, 490–493.
51. Wherrett, B.S., Walker, A.C., and Tooley, F.A.P. (1988). *Nonlinear Refraction for CW Optical Bistability*, in *Optical Nonlinearities and Instabilities in Semiconductors*, edited by H. Haug (Academic Press, New York), p. 244.
52. Eggleton, B.J., Lenz, G., Slusher, R.E., and Litchinitser, N.M. (1998). Compression of Optical Pulses Spectrally Broadened by Self-Phase Modulation with a Fiber Bragg Grating in Transmission, *Appl. Opt.* **37**, 7055–7061.
53. Schultheis, L., Honold, A., Kuhl, J., Köhler, K., and Tu, C.W. (1986). Optical Dephasing of Homogeneously Broadened Two-Dimensional Exciton Transitions in GaAs Quantum Wells, *Phys. Rev. B* **34**, 9027–9030.
54. Lee, J., Koteles, E.S., and Vassell, M.O. (1986). Luminescence Linewidths of Excitons in GaAs Quantum Wells below 150 K, *Phys. Rev. B* **33**, 5512–5516.
55. Thränhardt, A., Kuckenburg, S., Knorr, A., Meier, T., and Koch, S.W. (2000). Quantum Theory of Phonon-Assisted Exciton Formation and Luminescence in Semiconductor Quantum Wells, *Phys. Rev. B* **62**, 2706–2720.

Quantum Coherence, Correlation and Decoherence in
Semiconductor Nanostructures
T. Takagahara (Ed.)
Copyright © 2003 Elsevier Science (USA). All rights reserved.

Chapter 2
Carrier-wave Rabi flopping in semiconductors

O.D. Mücke, T. Tritschler, and M. Wegener

*Institut für Angewandte Physik, Universität Karlsruhe (TH),
Wolfgang-Gaede-Straße 1, 76131 Karlsruhe, Germany*

Abstract

Carrier-wave Rabi flopping occurs when the Rabi frequency becomes comparable with the light frequency, while maintaining electronic coherence. Exciting the model semiconductor GaAs, which has a band gap period of 2.9 fs, with optical pulses which are both, extremely short (5 fs) and extremely intense (estimated Rabi periods <3 fs), we can meet this highly unusual condition. After reviewing corresponding recently published experimental spectra around the third harmonic of the band gap, we present additional data on the transmitted fundamental wave and compare all with theory. The relevance of these results for exploiting coherent effects in semiconductor saturable absorbers for femtosecond mode-locked lasers is discussed.

2.1 Introduction

If a two-level system is excited by a resonant light field, a periodic oscillation of the inversion can result [1,2]. This periodic oscillation between absorption and inversion is known as Rabi oscillation and requires coherence of the two-level system. The frequency of this oscillation, the Rabi frequency $\Omega_R = \hbar^{-1} d\tilde{E}$, is proportional to the envelope of the light field $\tilde{E}$ and to the dipole matrix element d of the optical transition. What happens if the light intensity is so large that the period of one Rabi oscillation becomes as short as a cycle of light (2.9 fs = h/E_g for the band edge of GaAs)? What happens if one uses light pulses containing only one or two cycles of light?

These questions [3,4] are both of scientific as well as of some technical interest. Scientifically, they bring us into a highly unusual regime of light–matter interaction in solids as well as into a completely unexplored regime of nonlinear optics. The situation, Rabi frequency equal to light frequency, for GaAs parameters actually means that the semiconductor turns into a metal after half an optical cycle (1.45 fs = 2.9/2 fs), returning to a semiconductor after another 1.45 fs. Moreover, for semiconductors, this carrier-wave regime connects two effects which are usually thought of as being unrelated, namely (envelope) Rabi flopping on the one hand and Zener tunneling [5] on the other hand. For envelope Rabi flopping, one assumes that the light frequency ω is much larger than the Rabi frequency Ω_R, i.e. $\Omega_R \ll \omega$, which allows us to treat the problem within a frame which rotates with the frequency of light (rotating wave approximation) [1,2]. For a semiconductor, this can lead to a periodic oscillation of the occupation of electron states within the conduction (valence) band. Zener tunneling, on the other hand, occurs for static electric fields, i.e. for $\omega = 0$ or, equivalently, for $\Omega_R \gg \omega$. In the presence of a large electric field, an electron can tunnel from the valence band to the conduction band. In the regime of carrier-wave Rabi flopping, i.e. for $\Omega_R \approx \omega$, these two known pictures merge.

For parameters of the model semiconductor GaAs, the condition $\Omega_R = \omega$ corresponds to electric field envelopes on the order of 2×10^9 V/m, equivalent to about one Volt per lattice constant! The potential drop over one unit cell of the lattice is comparable to the band gap (E_g = 1.42 eV for GaAs at room temperature) as well as to the width of the conduction and valence band, respectively. Obviously, at this point, the light field can no longer be considered as a perturbation.

What is the potential technical interest? Semiconductors are already widely used as saturable absorbers in femtosecond lasers. This started with InGaAs multiple quantum well samples [6], which were introduced into a solid state color center laser [7]. Usually, semiconductor saturable absorbers work in the incoherent regime, i.e. one takes advantage of the bleaching of absorption associated with the filling of available phase space in the conduction band of the semiconductor. On a timescale of several tens of femtoseconds and under these high carrier density conditions, coherent effects usually play only a very minor role, if any at all. However, if one is interested in the generation of optical pulses on a sub-10 fs scale, coherent effects will come into play – no matter whether desired or not. Coherence effects in saturable absorption are obviously nothing else but Rabi oscillations. The Rabi oscillation might lead to an attractive side effect. In order to avoid Q-switching [8], one does not want bleaching to monotonously increase with light fluence, but rather exhibit a maximum and decrease thereafter. For Rabi flopping, such maximum automatically occurs at an envelope pulse area of 2π. If one wants to generate optical pulses of only a few optical cycles in duration along these lines, one has to understand the regime of carrier-wave Rabi flopping.

Let us quickly go through the history of conventional Rabi flopping in semiconductors before we address carrier-wave Rabi flopping. Ref. [9] discussed Rabi flopping in the framework of the semiconductor Bloch equations. Two important changes [2] with respect to the optical Bloch equations of the well-known two-level system [1] arise. First, one has to deal with bands of states rather than discrete states. This aspect is intrinsic to semiconductors and is somewhat similar to what is called inhomogeneous broadening in atomic or molecular systems. Second, the Coulomb interaction among charged carriers couples the various optical transitions. Within the mean-field (Hartree–Fock) approximation, this interaction can be thought of in terms of an internal field. The total electric field seen by the carriers is the sum of the external laser field and an internal field which is given by a sum over all optical polarizations. We have called this phenomenon local field effect [10]. This internal field is not necessarily small, it can indeed become as large as the laser field itself [9] or even exceed it. In addition to this, the Coulomb interaction also leads to scattering, i.e. to energy relaxation as well as to phase relaxation [11,12]. For times of only few tens of femtoseconds, these processes can no longer be described by simple time constants, electron-phonon [13–15] as well as carrier–carrier [16,17] scattering become non-Markovian. In other words, the system dynamics approaches again a Hamiltonian one. Using pulsed excitation, Rabi flopping has been observed experimentally on excitons in semiconductors and semiconductor quantum wells [18–20] as well as in microcavities [21]. All these Rabi floppings exhibited periods in the range from 50 fs to 1 ps. Further theoretical [22] and experimental work [23] focused on Rabi flopping of continuum states rather than excitons. Only recently, Rabi flopping in single quantum dots was also discussed [24].

2.2 Carrier-wave Rabi flopping

The notion of carrier-wave Rabi flopping was first used by S. Hughes, who discussed an ensemble of identical uncoupled two-level systems [3]. As one is interested in the system's dynamics on a timescale of one period of light or less, both, the rotating wave approximation and the slowly varying envelope approximation [1] must obviously *not* be used. His theoretical work as well as that of others [25] is based on the theoretical framework of Ref. [26].

What are the anticipated signatures of carrier-wave Rabi flopping? The condition Rabi period equal to the light period corresponds to a huge intensity (for a solid). While it might be possible to reach this condition with pulses of several tens of femtoseconds in duration, it is not very likely that the electronic system will remain coherent meanwhile. Coherence, however, is a prerequisite for any type of Rabi flopping. Thus it seems favorable to study excitation with very short pulses, ideally with only one or two cycles of light in duration. Remember that, for GaAs parameters, the period of light corresponding to the room temperature band gap energy is 2.9 fs. To highlight the general aspects of carrier-wave Rabi flopping, let us first review the behavior for an ensemble of uncoupled and identical two-level systems, which is the level of sophistication of Refs [3,25, and 26]. For reference, Fig. 2.1(a) schematically depicts conventional Rabi flopping plotted on the Bloch sphere, i.e. the Rabi period is much larger than the light period. For clarity, we neglect any damping at this point. The components u and v of the Bloch vector (u, v, w) correspond to twice the real and imaginary part of the optical transition amplitude, respectively, w is the inversion of the two-level system [1]. In this representation, the optical oscillation corresponds to an orbiting of the Bloch vector parallel to the equatorial plane (uv plane) with the optical transition frequency Ω (here $\Omega = \omega = 2\pi/2.9$ fs), the oscillation of the inversion to a motion in the vw plane. For a square-shaped pulse with envelope pulse area $\Theta = 2\pi$ starting from the south pole, i.e. all electrons are in the ground state (valence band), the Bloch vector spirals up to the north pole, i.e. all electrons are in the excited state (conduction band) and back to the south pole. This leads to a modulation of the real part of the optical transition amplitude u (Fig. 2.1(a)), which is roughly similar to a quantum beating. Thus, the corresponding spectrum of the polarization would exhibit two peaks centered around the transition frequency. Figure

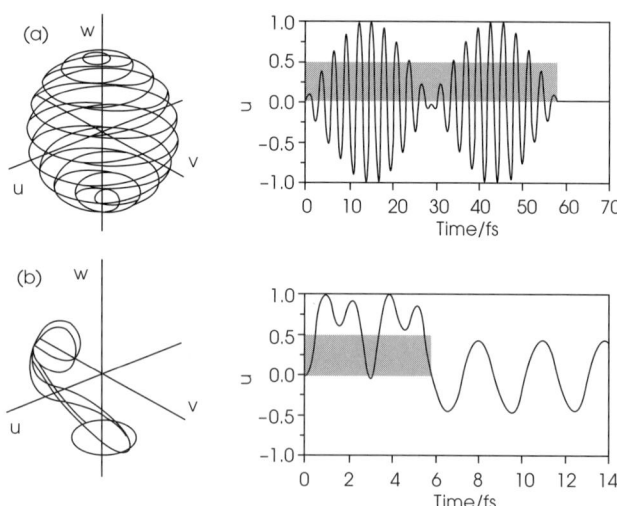

Fig. 2.1 (a) Scheme of the trace of the Bloch vector for conventional Rabi flopping. Pulse duration is 20 optical cycles, envelope pulse area is $\Theta = 2\pi$. (b) Same for carrier-wave Rabi flopping. Pulse duration is 2 optical cycles, $\Theta = 4\pi$. The optical pulse envelopes are indicated by the grey areas.

2.1(b) shows results for $\Theta = 4\pi$ and for a much shorter pulse, *such that the Rabi period equals the light period*. Two related aspects are obvious. First, though $\Theta = 4\pi$, the Bloch vector does not come back to the south pole. In this sense, the usual definition of the envelope pulse area Θ fails. Hence, also the area theorem of nonlinear optics, which is based on this definition, fails. Despite this failure, we quote Θ for reference in this article. Second, it is obvious that the optical polarization becomes strongly distorted during the two cycles of the optical pulse (see u versus time in Fig. 2.1(b)). Thus, harmonics are being generated, the most prominent of which, for an inversion symmetric medium, is the third harmonic. For low intensities, this is nothing but the resonantly enhanced third-harmonic generation. *For very high intensities, i.e. for carrier-wave Rabi flopping, one expects a double-peak structure around the third harmonic of the transition frequency.* Note that absolutely no harmonics are generated *after* the two cycles of the optical pulse (see Fig. 2.1(b)). Here one merely has a free oscillation of the optical polarization with the optical transition frequency of the two-level system Ω.

Experiments

From the above it has become clear that carrier-wave Rabi flopping needs short, i.e. one or two optical cycles long, and very intense optical pulses. From the anticipated signal levels and the anticipated damage thresholds of thin films of semiconductors, high repetition rate laser systems are strongly favored. Thus, we perform our experiments with 5 fs linearly polarized (p-polarization) optical pulses at 81 MHz (=1/12 ns) repetition rate, which have recently become available [27]. Our home-built copy of this laser system very nearly reproduces the pulse properties described in Ref. [27]. The typical average output power of the laser is 120 mW. Plate 1(a) shows a typical laser spectrum, which has been obtained via Fourier-transform of an interferogram taken with a pyroelectric detector, which is spectrally extremely flat. The Michelson interferometer used at this point and for all results throughout this article is carefully balanced and employs home-made beam splitters fabricated by evaporating a thin film of silver on a 100 µm thin glass substrate. The Michelson interferometer is actively stabilized by means of the Pancharatnam screw [28], which allows for continuous scanning of the time delay while maintaining active stabilization. The remaining fluctuations in the time delay between the two arms of the interferometer are around ±0.05 fs. The spectral wings which can be seen in Plate 1(a) result from the spectral characteristics of the output coupler. The measured interferometric autocorrelation depicted in Fig. 2.2(b) is very nearly identical to the one computed from the spectrum (Plate 1(a)) under the assumption of a constant spectral phase. This shows that the pulses are nearly transform-limited. The intensity profile computed under the same assumption is shown as an inset in Plate 1(b) and reveals a duration of about 5 fs. As a result of the strongly structured spectrum (a square-function to zeroth order), the intensity versus time shows satellites (a sinc^2-function to zeroth order). Using a high numerical aperture reflective microscope objective [29], we can tightly focus these pulses to a profile which is very roughly Gaussian with 1 µm radius. This value has carefully been measured by a knife-edge technique at the sample position (see Plate 4(a)). This sample position is equivalent to that of the second harmonic (SHG) crystal used for the autocorrelation in terms of group delay dispersion. In front of the sample, each arm of the interferometer typically has an average power of about 8 mW. The resulting peak intensity of one arm can be estimated as

$$I_0 = \frac{8\,\mathrm{mW}}{\pi(10^{-4}\,\mathrm{cm})^2}\frac{12\,\mathrm{ns}}{5\,\mathrm{fs}} = 0.6 \times 10^{12}\,\mathrm{W\,cm^{-2}}. \qquad (2.1)$$

This corresponds to a field envelope (in vacuum)

$$\tilde{E}_0 = \sqrt{2\sqrt{\frac{\mu_0}{\varepsilon_0}}I_0} = 2.1 \times 10^7\,\mathrm{V\,cm^{-1}}, \qquad (2.2)$$

or *about one Volt per lattice constant a* as mentioned in the introduction ($a \approx 0.5$ nm). To estimate the envelope pulse area Θ, one furthermore needs the dipole matrix element d of the optical dipole transition. From the literature for GaAs we find $d = 0.3e$ nm [30] and $d = 0.6e$ nm [20]. Choosing $d = 0.5e$ nm in this article, this translates into an envelope pulse area

$$\Theta = \hbar^{-1} d \tilde{E}_0 \times 5\,\mathrm{fs} = 8.1 > 2\pi \qquad (2.3)$$

for one arm ($I = 0.601 \times I_0$ corresponds to 2π pulse area), and $> 4\pi$ (two Rabi periods) for two constructively interfering arms of the interferometer. For a resonant 5 fs pulse and a 2.9 fs band gap period, this corresponds to a Rabi frequency which even slightly exceeds the light frequency. It is also interesting to give a very rough estimate for the excited carrier density under these conditions. The GaAs band-to-band absorption coefficient is $\alpha = 10^4$ cm^{-1}. If all the light was absorbed according to this number – certainly an upper limit – one arrives at a carrier density of

$$n_{eh} = \alpha I_0 \times 5\,\mathrm{fs}/1.42\,\mathrm{eV} = 1.3 \times 10^{20}\,\mathrm{cm^{-3}}. \qquad (2.4)$$

For constructive interference of the two arms of the interferometer, this number needs to be multiplied by a factor of four. Thus, we can safely conclude that the highest carrier densities approach 10^{20} cm^{-3}. In the experiment, we use a 0.6 µm thin film of GaAs clad between Al$_{0.3}$Ga$_{0.7}$As barriers, grown by metal-organic vapor phase epitaxy on a GaAs substrate. The sample is glued onto a 1 mm thick sapphire disk and the GaAs substrate is removed. Finally, a $\lambda/4$-antireflection coating is evaporated. The light emitted by this sample, held under ambient conditions, is collected by a second reflective microscope objective [29], is spectrally pre-filtered by a sequence of four fused-silica prisms, and is sent into a 0.25 m focal length grating spectrometer connected to a liquid-nitrogen cooled, back-illuminated, UV-enhanced charge-coupled-device (CCD) camera. For a second set of experiments the transmitted light is dispersed in a miniature spectrometer which allows to simultaneously cover the wavelength range from 500 nm to 1100 nm.

Let us first discuss results for single pulses only, i.e. we block one arm of the interferometer. Figure 2.2 shows spectra at the third harmonic for different pulse intensities I in multiples of I_0, as defined above. For the attenuation we have used metallic beam splitters on 100 μm thin fused silica substrates, the dispersion of which has carefully been compensated for by the extra-cavity sequence of four CaF_2 prisms [27]. At low intensity, i.e. for $I = 0.017 \times I_0$, we observe a single maximum around 300 nm wavelength which is interpreted as the usual third-harmonic generation which is resonantly enhanced by the GaAs band edge here. With increasing intensity, we find a second maximum emerging at the long wavelength side, which gains more and more weight. At the highest intensity, i.e. $I = 0.779 \times I_0$, the 10 × magnification reveals an additional smaller maximum around 340 nm wavelength. In Ref. [4] we have interpreted this overall behavior as a signature of carrier-wave Rabi flopping. Note that the intensities revealing a double-peak structure in the third-harmonic spectrum correspond very well to our

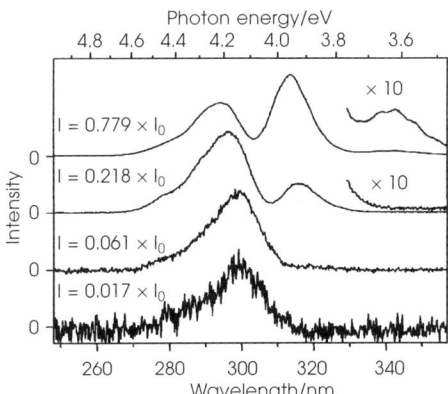

Fig. 2.2 Experiment: Spectra of light emitted into the forward direction around the third harmonic of the GaAs band gap frequency. The spectra are shown on a linear scale, vertically displaced and individually normalized (from top to bottom: maxima correspond to 5664, 439, 34, and 4 counts/s). Excitation with 5 fs pulses. The intensity I of the pulses is indicated.

above simple estimates, i.e. we estimated a full Rabi flop for an intensity of $I = 0.601 \times I_0$.

In the second set of experiments we study the third-harmonic spectra for excitation with phase-locked pulse pairs with time delay τ, i.e. we open both arms of the interferometer. It is interesting to note that Θ is the same for τ = 0 and for e.g. τ equal to two optical cycles – because the two optical fields simply add. Yet, the corresponding Rabi frequency is larger for τ = 0. For low intensities (Fig. 2.4(a)), i.e. for small Rabi frequency as compared to the light frequency, the third-harmonic spectrum is simply modulated as a function of τ due to interference of the laser pulses within the sample leading to a period of about 2.9 fs. In contrast to this, for higher intensities (Plate 2(b)–(d)) where the Rabi frequency becomes comparable to the light frequency, the shape of the spectra changes dramatically with time delay τ. For e.g. τ = 0 in Plate 2(b), the two pulses simply interfere constructively and we find the same spectral double maximum structure as in the single pulse experiments (Fig. 2.2). For larger τ, i.e. after one or two optical cycles, this double maximum disappears and is replaced by one prominent and much larger maximum. For the highest intensity, i.e. for Plate 2(d) – which corresponds to an envelope pulse area Θ of more than 4π – the behavior is quite involved with additional fine structure for |τ| < 1 fs. Note that the spectra for τ = 0 nicely reproduce the behavior seen in Fig. 2.2.

Beside the interference of the laser pulses in the sample, at larger time delays |τ| one additionally observes interference of the third-harmonic signals corresponding to the two phase-locked pulses on the detector leading to periods around one femtosecond in Plate 3. It can also be seen from Plate 3 that the splitting in the spectra gradually approaches zero for large time delays.

We have also deliberately introduced positive or negative group velocity dispersion by moving one of the extra-cavity CaF_2 prisms in or out of the beam with respect to the optimum position (not shown here, but depicted in Ref. [4]). Obviously, this leaves the amplitude spectrum of the laser pulses unaffected. We find [4] that one quickly gets out of the regime of carrier-wave Rabi flopping, i.e. both the splitting at τ = 0 as well as the dependence of the shape on the time delay τ, quickly disappear with increasing pulse chirp. This demonstrates that it is not just the large bandwidth of the pulses but the fact that they are short – two optical cycles – which is important for the observation of carrier-wave Rabi flopping.

In the theory section we will show that many aspects of our experiment are well explained by the theory of carrier-wave Rabi flopping in a semiconductor. However, one might argue that such splitting in the third-harmonic spectra (Fig. 2.2 and Plates 2 and 3) could possibly also arise due to a completely different effect, namely self-phase modulation of the laser pulse (which is known to result in spectral side maxima) within the 0.6 µm thin but finite GaAs sample and subsequent conventional (off-resonant) third-harmonic generation. This interpretation can easily be ruled out by measuring the transmitted laser spectrum – which constitutes the third set of experiments. Plate 4 schematically shows the geometry. To vary the excitation intensity without having to introduce filters (which would definitely require to change the dispersion compensation), we simply move the sample in the z-direction through the fixed focus ($z = 0$) of the microscope objective and collect the transmitted light with the fixed second microscope objective. Not even at the highest intensities achievable (Plate 4(b)), we find any indication for such effects (the laser spectrum is of course somewhat modified due to absorption for photon energies above the band gap, which lies around 870 nm wavelength). However, having done this, we have noticed some interesting details in the spectra. To enhance their visibility we define a differential transmission, $\Delta T/T$, as

$$\frac{\Delta T}{T} = \frac{I_t(z) - I_t(z = -\infty)}{I_t(z = -\infty)}, \qquad (2.5)$$

where $I_t(z)$ is the transmitted light intensity at sample position z. The condition $z = -\infty$ actually corresponds to $z = -20$ µm in the experiment, where the profile is so large that we can safely assume that linear optics applies. Plate 5 shows corresponding results for three different incident light intensities I in units of I_0 as defined above. First, all results are closely symmetric around $z = 0$, which indicates that changes in absorption dominate. Changes in the refractive index might lead to focusing or defocusing of the beam which would result in an asymmetric dependence on z (similar to the known so-called z-scan technique e.g. described in Ref. [31]). Second, one can see a large increase in transmission for wavelengths shorter than the GaAs band edge (approximately 870 nm) around $z = 0$ (Plate 5(a)). $z = 0$ corresponds to the highest intensity in each plot. The maximum around 670 nm wavelength results from bleaching of the band gap of the $Al_{0.3}Ga_{0.7}As$ barriers of the GaAs double heterostructure which accidentally coincides with the pronounced maximum in the laser spectrum (Plate 1(a)) also around 680 nm. For larger intensity, Plate 5(b), the transmission maximum around $z = 0$ flattens and we observe pronounced induced absorption for wavelengths longer than the GaAs band edge. For the highest intensity (Plate 5(c)), this induced absorption becomes the dominating feature throughout most of the spectral range. Note that little if any induced transparency is observed for wavelengths between 780 nm (170 meV above the unrenormalized band gap $E_g = 1.42$ eV) and 700 nm (350 meV above the unrenormalized band gap) while the laser spectrum (Plate 1(a)) still has significant amplitude there. This indicates that these states high up in the band-to-band continuum of GaAs must experience a much stronger damping (phase relaxation) and/or energy relaxation than those states near the band gap. This point will become very important in the theory section.

Theory

To describe our experiments, one has to solve Maxwell equations coupled to the material equations and investigate the light transmitted or emitted into the forward direction. Let us consider wave propagation into the z-direction with $\vec{E}$ and $\vec{D}$ fields being polarized along the x-direction with corresponding components $E(z, t)$ and $D(z, t)$, respectively. The $\vec{H}$ and the $\vec{B}$ fields are polarized along the y-direction with corresponding components $H(z, t)$ and $B(z, t)$. Under these conditions, Maxwell equations (in S.I. units) immediately give

$$\frac{\partial E(z,t)}{\partial z} = -\frac{\partial B(z,t)}{\partial t} \tag{2.6}$$

$$\frac{\partial H(z,t)}{\partial z} = -\frac{\partial D(z,t)}{\partial t}. \tag{2.7}$$

In the semiconductor we further have

$$B(z, t) = \mu_0 H(z, t) \tag{2.8}$$

and $D(z, t) = \varepsilon_0 E(z, t) + P(z, t),$ (2.9)

with the (real) medium polarization $P(z, t)$. The electric field impinging onto the sample from the vacuum on the left, i.e. from $z = -\infty$, is a plane wave and can be written as

$$E(z, t) = \tilde{E}(t - z/c_0) \cos(\omega_0 (t - z/c_0) + \phi) \tag{2.10}$$

with the vacuum velocity of light $c_0 = 1/\sqrt{\mu_0 \varepsilon_0}$, the (real) electric field envelope $\tilde{E}$, the laser center frequency ω_0 and the carrier-envelope offset (CEO) phase ϕ. Note that ϕ would drop out when using the rotating wave approximation (RWA) and/or the slowly varying envelope approximation (SVEA) [1]. In contrast to this, it is generally important in the carrier-wave regime. Except for Plate 7 where we discuss this aspect explicitly, we choose $\phi = 0$. The material enters via the polarization P which has to be computed microscopically from the underlying Hamiltonian H. Neglecting the Coulomb interaction of carriers, any type of intraband optical processes, phonons and their coupling to the carriers, suppressing spin indices and using the dipole approximation for the optical transitions from the valence (v) to the conduction (c) band at wave vector $\vec{k}$ we have [2]

$$\begin{aligned} H = &\sum_{\vec{k}} E_c(\vec{k}) c^{\dagger}_{c\vec{k}} c_{c\vec{k}} + \sum_{\vec{k}} E_v(\vec{k}) c^{\dagger}_{v\vec{k}} c_{v\vec{k}} \\ &- \sum_{\vec{k}} d_{cv}(\vec{k}) E(z, t) (c^{\dagger}_{c\vec{k}} c_{v\vec{k}} + c^{\dagger}_{v\vec{k}} c_{c\vec{k}}). \end{aligned} \tag{2.11}$$

Here $E_{c,v}(\vec{k})$ are the single particle energies of electrons in the conduction and valence band respectively (the band structure), and $d_{cv}(\vec{k})$ is the (real) dipole matrix element for an optical transition at electron wave vector $\vec{k}$. Note that in our above discussion we have used $d = d_{cv}(\vec{k})$. The creation $c^\dagger$ and annihilation c operators create and annihilate crystal electrons in the indicated band (c, v) at the indicated momentum $(\vec{k})$. The optical polarization is given by

$$P(z,t) = \frac{1}{V} \sum_{\vec{k}} d_{cv}(\vec{k})(p_{vc}(\vec{k}) + \text{c.c.}) + P_b(z,t) \quad (2.12)$$

where the optical transition amplitudes

$$p_{vc}(\vec{k}) = \langle c^\dagger_{v\vec{k}} c_{c\vec{k}} \rangle \quad (2.13)$$

depend on time t as well as parametrically on the propagation coordinate z. As usual, the sum in eq. (2.12) can be expressed via the combined density of states $D_{cv}(E)$ as $\sum_{\vec{k}} \ldots \to \int D_{cv}(E) \ldots dE$, which neglects all anisotropies. The background polarization $P_b(z,t) = \varepsilon_0 \chi_b(z) E(z,t) = \varepsilon_0 (\varepsilon_b(z) - 1) E(z,t)$ accounts for all high energy optical transitions not explicitly accounted for in eq. (2.11) and can be expressed in terms of the background dielectric constant $\varepsilon_b(z)$. The dynamics of $p_{vc}(\vec{k})$, as well as those of the occupation numbers in the conduction band

$$f_c(\vec{k}) = \langle c^\dagger_{c\vec{k}} c_{c\vec{k}} \rangle \quad (2.14)$$

and in the valence band

$$f_v(\vec{k}) = \langle c^\dagger_{v\vec{k}} c_{v\vec{k}} \rangle \quad (2.15)$$

are easily calculated from the Heisenberg equation of motion for any operator $\mathcal{O}$ according to

$$-i\hbar \frac{\partial}{\partial t} \mathcal{O} = [H, \mathcal{O}]. \quad (2.16)$$

Employing the usual anticommutation rules, i.e.

$$[c_{c\vec{k}}, c^\dagger_{c\vec{k}'}]_+ = \delta_{\vec{k}\vec{k}'}, \quad [c_{v\vec{k}}, c^\dagger_{v\vec{k}'}]_+ = \delta_{\vec{k}\vec{k}'}, \quad (2.17)$$

and that all other anticommutators are zero, this leads us to the known Bloch equations for the transition amplitude

$$\left(\frac{\partial}{\partial t} + i\Omega(\vec{k})\right) p_{vc}(\vec{k}) + \left(\frac{\partial}{\partial t} p_{vc}(\vec{k})\right)_{\text{rel}} = i\hbar^{-1} d_{cv}(\vec{k}) E(z,t)(f_v(\vec{k}) - f_c(\vec{k})), \quad (2.18)$$

with the optical transition energy $\hbar\Omega(\vec{k}) = E_c(\vec{k}) - E_v(\vec{k})$, and for the occupation in the conduction band

$$\frac{\partial}{\partial t}f_c(\vec{k}) + \left(\frac{\partial}{\partial t}f_c(\vec{k})\right)_{\text{rel}} = 2\hbar^{-1}d_{cv}(\vec{k})E(z,t)\,\text{Im}(p_{vc}(\vec{k})). \tag{2.19}$$

Here we have assumed a real dipole matrix element. $(1 - f_v(\vec{k}))$ can be interpreted as the occupation of holes and obeys an equation similar to $f_c(\vec{k})$. The terms with subscript "rel" have been added phenomenologically and describe dephasing and relaxation, respectively. They will be discussed later. Note that the transition amplitude $p_{vc}(\vec{k})$ and the occupation factors $f_c(\vec{k})$ and $f_v(\vec{k})$ are easily connected to the components of the Bloch vector (u, v, w) mentioned in the introduction via

$$\begin{pmatrix} u \\ v \\ w \end{pmatrix} = \begin{pmatrix} 2\,\text{Re}(p_{vc}(\vec{k})) \\ 2\,\text{Im}(p_{vc}(\vec{k})) \\ f_c(\vec{k}) - f_v(\vec{k}) \end{pmatrix} \tag{2.20}$$

In the modeling excitation is with $t_{\text{pulse}} = 5$ fs pulses, the envelope of which, $\tilde{E}(t)$, is sech (t/t_0)-shaped with $t_0 = 1/(2\,\text{arcosh}(\sqrt{2}))\,t_{\text{pulse}}$. Their center frequency is given by $\hbar\omega_0 = E_g$. The phase ϕ between envelope and carrier-wave of the pulses is chosen to be zero, i.e. the actual optical field is given by $E(z = 0, t) = \tilde{E}(t)\cos(\omega_0 t)$. Deviations from these choices (Plate 7) are indicated. The relaxation terms in eqs (2.18) and (2.19) are chosen as $\left(\frac{\partial}{\partial t}p_{vc}(\vec{k})\right)_{\text{rel}} = -\frac{1}{T_2}p_{vc}(\vec{k})$, with $T_2 = 50$ fs, and $\left(\frac{\partial}{\partial t}f_c(\vec{k})\right)_{\text{rel}} = 0$. For the Maxwell part, a 0.6 μm thin slice of this 'material' with additional background dielectric constant $\varepsilon_b = 10.9$ is sandwiched between a $\lambda/4$-antireflection coating (with $\lambda = c_0\,2.9\,\text{fs}/\sqrt[4]{\varepsilon_b}$) on the front side and a semi-infinite substrate with dielectric constant $\varepsilon = \varepsilon_b$ on the back. This corresponds to the sample geometry used in the experiment.

If an antireflection-coated sample is used, the light intensity $\propto nE^2$ in the medium is the same as the light intensity in vacuum. Thus, the electric field and the envelope pulse area inside the sample are attenuated by a factor of $\sqrt{n}$. The medium refractive index n is very nearly similar to the background refractive index n_b (for single two-level systems). Thus, for convenience of the reader, we quote the incident (vacuum) envelope pulse areas in multiples of $\sqrt{n_b}$. E.g. $\Theta = 2\pi\sqrt{n_b}$ in vacuum simply corresponds to 2π envelope pulse area in the semiconductor.

Figure 2.3 shows results for identical two-level systems with a density of 10^{18} cm^{-3} and with transition energy E_g. The single maximum around the third harmonic of the GaAs band gap splits into two maxima which shift symmetrically with respect to the center frequency. In the theory, the energetic separation of the two maxima is roughly given by the envelope pulse area, while it appears to be smaller in the experiment (Fig. 2.2) by about a factor of two. Beside these similarities there are obvious deviations as well: In the experiment, we also observe one maximum at small pulse areas. There, however, another maximum gradually grows on the long wavelength side as the pulse area increases (Fig. 2.2).

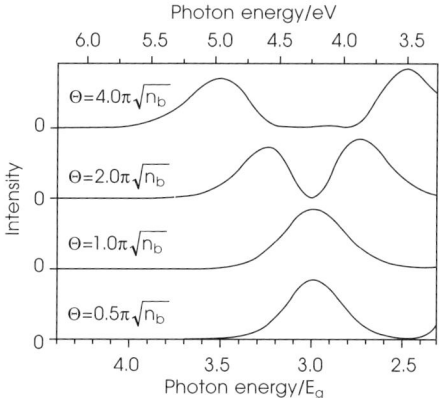

Fig. 2.3 Theory: Spectra of light emitted into the forward direction around the third harmonic of the transition frequency. The spectra are shown on a linear scale, vertically displaced and individually normalized. Resonant excitation with 5 fs pulses. The envelope pulse area Θ in front of the sample is indicated. E.g. $\Theta = 2\pi\sqrt{n_b}$, with $n_b = \sqrt{\varepsilon_b}$, corresponds to an envelope pulse area of 2π inside the sample for vanishing reflection losses.

In the following we demonstrate that the continuum of states of the semiconductor significantly changes the above picture. For clarity, we show solutions of the Bloch equations (no propagation effects) at this point. Corresponding results are depicted in Plate 6. Here, ω denotes the (spectrometer) photon frequency, ω_0 the laser center frequency and $\Omega = \Omega(\vec{k}) = \hbar^{-1}(E_c(\vec{k}) - E_v(\vec{k}))$ the transition frequency of one transition within the band. Without band gap renormalization, it is clear that there are no states below the band gap energy (dashed horizontal line); nevertheless, we depict these data. Again, the laser center frequency is centered at the band gap energy, i.e. we have $\hbar\omega_0 = E_g$. The laser spectrum is shown on the right hand side lower corner as the grey-shaded area. The spectrum for $\hbar\Omega = E_g$ is also depicted by the white line. Obviously, it resembles the results shown in Fig. 2.3 very closely, even though we do not account for propagation effects here. For small envelope pulse area, $\Theta = 0.5\pi$, we find a single rather narrow maximum around $\omega/\omega_0 = 3$ and $\Omega/\omega_0 = 1$. Its width correlates with the width of the laser spectrum. This single maximum is nothing but the usual, yet resonantly enhanced, third harmonic generation. It experiences a constriction for $\Theta = 1.0\pi$, which evolves into a shape that resembles an anticrossing for $\Theta = 2.0\pi$. Here, two separate peaks are only observed in a rather narrow region around $\hbar\Omega = \hbar\omega_0 = E_g$, while for larger $\hbar\Omega$ only a single maximum occurs. Also, we find that the contribution of larger frequency transitions is by no means small. For e.g. $\hbar\Omega = 2$ eV transition energy, the signal is actually larger than for the band gap, i.e. for $\hbar\Omega = 1.42$ eV. This trend continues for yet larger pulse areas (see $\Theta = 4.0\pi$ in Plate 6). While there is considerable resonant enhancement (as can be seen from Plate 6(a)), this enhancement becomes less important at large pulse areas because the resonant transitions are completely saturated.

The actual spectra (compare eq. (2.12)) are the integral over the individual contributions, multiplied with the combined density of states, over the relevant range of transition energies. The bands themselves clearly have contributions even at $\hbar\Omega = 5$ eV. If one would sum up all these contributions at e.g. $\Theta = 4.0\pi$ (Plate 6(d)), one no longer gets two maxima but rather a single maximum around $\omega/\omega_0 = 3$, which would no longer be in agreement with the experiments. Thus, there must be a reason why the high energy transitions do not contribute significantly. It was first pointed out to us by H. Haug [32] that the reason might be that the high energy transitions are likely to have much shorter dephasing times which significantly suppresses their contribution. Also, band gap renormalization becomes quite significant at these very large carrier densities. If one e.g. integrates the spectra from 1.2 to 1.6 eV transition energy $\hbar\Omega$ with a constant density of states (not shown), the experimental behavior is reproduced much better than in Fig. 2.3. In particular, one gets a gradual growth of a second spectral maximum rather than the sudden splitting observed for a single two-level system (Fig. 2.3). This interpretation of short dephasing times of high energy transitions as a result of the large excitation is consistent with our observations depicted in Plate 5, where we do not observe any bleaching for these states either. Also, these short dephasing times of high energy transitions might lead to induced absorption and, thus, negative values of $\Delta T/T$ at wavelengths below the band gap in Plate 5 as well.

Finally, we depict in Plate 7 results obtained for $sinc^2$-shaped pulses, which we have smoothed by a Gaussian, i.e. $E(t) \propto sinc\,(t/t_0)\,\exp[-t^2/(2\tau_{Gauss}^2)]\,\cos(\omega_0 t + \phi)$ with $t_0 = t_{pulse}/2.7831$ and $\tau_{Gauss} = 20$ fs. Here, we do find a dependence of the third-harmonic spectra on the CEO phase ϕ – in contrast to the $sech^2$-pulses of Fig. 2.3 and Plate 6. This influence becomes relevant when the contributions of the fundamental wave and the third harmonic, i.e. when the contributions of the optical polarization originating from $\omega/\omega_0 = 1$ and $\omega/\omega_0 = 3$, respectively, interfere. This is the case on the lower left hand side of Plate 7(a)–(c), see black rectangles.

Further theoretical work to improve the agreement between experiment and theory is in progress, both in our group as well as in several others (also see the contribution of S. Hughes in this book). Two aspects seem very important in this context. First, it is likely that, in the experiment, the pulses within the sample are modified with respect to the incident pulses. This might explain why the splitting between the two maxima in the third-harmonic spectra of the experiment is roughly a factor of two smaller than expected from the above modeling. Second, theory has to explain, why the damping of states far above the band gap becomes as short as one femtosecond and below.

2.3 Conclusions

In conclusion, our experiments on carrier-wave Rabi flopping have, for the first time, given access to semiconductor material dynamics on a timescale comparable to only one cycle of light. Preliminary analysis shows that the optical transitions which are several 100 meV above the band gap acquire dephasing times as short as 1 fs and below (which is shorter than the transition period itself). This would lead to a partial collapse of the band structure and, as a consequence of this, also to induced absorption at photon energies below the gap – as is observed in the experiment. Also, the extremely large electrical fields associated with the light field open new perspectives for studies on light-matter interaction in a highly unusual regime in which the potential drop over a single lattice constant is comparable to the band gap and the widths of the bands.

Acknowledgments

The work of M.W. is supported by the DFG Leibniz Award 2000. We thank W. Stolz for the high quality GaAs sample and U. Morgner and F.X. Kärtner for intense support in the initial phase of the experiment and for stimulating discussions. The theoretical part of this work has largely benefited from discussions with the group of H. Haug who shared their theoretical results concerning state-dependent damping in carrier-wave Rabi flopping with us prior to publication of their work.

References

1. Meystre, P., and Sargent III, M. (1991). *Elements of Quantum Optics*, second edition, Springer Verlag.
2. Schäfer, W., and Wegener, M. (2002). *Semiconductor Optics and Transport Phenomena, From Fundamentals to Current Topics*, Advanced Texts in Physics, Springer Verlag.
3. Hughes, S. (1998). Breakdown of the Area Theorem: Carrier-Wave Rabi Flopping of Femtosecond Optical Pulses, *Phys. Rev. Lett.* **81**, 3363–3366.
4. Mücke, O.D., Tritschler, T., Wegener, M., Morgner, U., and Kärtner, F.X. (2001). Signatures of Carrier-Wave Rabi Flopping in GaAs, *Phys. Rev. Lett.* **87**, 057401, 1–4.

5. Zener, C. (1934). A Theory of the Electrical Breakdown of Solid Dielectrics, *Proc. R. Soc. London* A **145**, 523–529.
6. Wegener, M., Bar-Joseph, I., Sucha, G., Islam, M.N., Sauer, N., Chang, T.Y., and Chemla, D.S. (1989). Femtosecond Dynamics of Excitonic Absorption in the Infrared In$_x$Ga$_{1-x}$As Quantum Wells, *Phys. Rev. B* **39**, 12794–12801.
7. Islam, M.N., Sunderman, E.R., Soccolich, C.E, Bar-Joseph, I., Sauer, N., Chang, T.Y., and Miller, B.I. (1990). Color Center Lasers Passively Mode Locked by Quantum Wells, *IEEE J. Quantum Electron.* **25**, 2454–2463.
8. Thoen, E.R., Koontz, E.M., Joschko, M., Langlois, P., Schibli, T.R., Kärtner, F.X., Ippen, E.P., and Kolodziejski, L.A. (1999). Two-photon Absorption in Semiconductor Saturable Absorber Mirrors, *Appl. Phys. Lett.* **74**, 3927–3929.
9. Binder, R., Koch, S.W., Lindberg, M., Peyghambarian, N., and Schäfer, W. (1990). Ultrafast Adiabatic Following in Semiconductors, *Phys. Rev. Lett.* **65**, 899–902.
10. Leo, K., Wegener, M., Shah, J., Chemla, D.S., Göbel, E.O., Damen, T.C., Schmitt-Rink, S., and Schäfer, W. (1990). Effects of Coherent Polarization Interactions on Time-Resolved Degenerate Four-Wave Mixing, *Phys. Rev. Lett.* **65**, 1340–1343.
11. Rappen, T., Peter, U., Wegener, M., and Schäfer, W. (1993). Coherent Dynamics of Continuum and Exciton States Studied by Spectrally Resolved fs Four-wave Mixing, *Phys. Rev. B* **48**, 4879–4882.
12. Rappen, T., Peter, U.-G., Wegener, M., and Schäfer, W. (1994). Polarization Dependence of Dephasing Processes: A Probe for Many-body Effects, *Phys. Rev. B* **R 49**, 10774–10777.
13. Bányai, L., Tran Thoai, D.B., Reitsamer, E., Haug, H., Steinbach, D., Wehner, M.U., Wegener, M., Marschner, T., and Stolz, W. (1995). Exciton-LO-Phonon Quantum Kinetics: Evidence of Memory Effects in Bulk GaAs, *Phys. Rev. Lett.* **75**, 2188–2191.
14. Fürst, C., Leitenstorfer, A., Laubereau, A., and Zimmermann, R. (1997). Quantum Kinetic Electron-Phonon Interaction in GaAs: Energy Nonconserving Scattering Events and Memory Effects, *Phys. Rev. Lett.* **78**, 3733–3736.
15. Wehner, M.U., Ulm, M.H., Chemla, D.S., and Wegener, M. (1998). Coherent Control of Electron-LO-Phonon Scattering in Bulk GaAs, *Phys. Rev. Lett.* **80**, 1992–1995.
16. Hügel, W.A., Heinrich, M.F., Wegener, M., Vu, Q.T., Bányai, L., and Haug, H. (1999). Photon Echoes from Semiconductor Band-to-Band Continuum Transitions in the Regime of Coulomb Quantum Kinetics, *Phys. Rev. Lett.* **83**, 3313–3316.
17. Vu, Q.T., Haug, H., Hügel, W.A., Chatterjee, S., and Wegener, M. (2000). Signature of Electron-Plasmon Quantum Kinetics in GaAs, *Phys. Rev. Lett.* **85**, 3508–3511.
18. Cundiff, S.T., Knorr, A., Feldmann, J., Koch, S.W., Göbel, E.O., and Nickel, H. (1994). Rabi Flopping in Semiconductors, *Phys. Rev. Lett.* **73**, 1178–1181.
19. Giessen, H., Knorr, A., Haas, S., Koch, S.W., Linden, S., Kuhl, J., Hetterich, M., Grün, M., and Klingshirn, C. (1998). Self-Induced Transmission on a Free Exciton Resonance in a Semiconductor, *Phys. Rev. Lett.* **81**, 4260–4263.
20. Schülzgen, A., Binder, R., Donovan, M.E., Lindberg, M., Wundke, K., Gibbs, H.M., Khitrova, G., and Peyghambarian, N. (1999). Direct Observation of Excitonic Rabi Oscillations in Semiconductors, *Phys. Rev. Lett.* **82**, 2346–2349.

21. Quochi, F., Bongiovanni, G., Mura, A., Staehli, J.L., Deveaud, B., Stanley, R.P., Oesterle, U., and Houdré, R. (1998). Strongly Driven Semiconductor Microcavities: From the Polariton Doublet to an ac Stark Triplet, *Phys. Rev. Lett.* **80**, 4733–4736.
22. Bányai, L., Vu, Q.T., Mieck, B., and Haug, H., (1998). Ultrafast Quantum Kinetics of Time-Dependent RPA-Screened Coulomb Scattering, *Phys. Rev. Lett.* **81**, 882–885.
23. Fürst, C., Leitenstorfer, A., Nutsch, A., Tränkle, G., and Zrenner, A. (1997). Ultrafast Rabi Oscillations of Free-Carrier Transitions in InP, *Phys. Stat. Sol.* (B) **204**, 20–22.
24. Takagahara, T. unpublished.
25. Kalosha V.P., and Herrmann, J. (1999). Formation of Optical Subcycle Pulses and Full Maxwell-Bloch Solitary Waves by Coherent Propagation Effects, *Phys. Rev. Lett.* **83**, 544–547.
26. Ziolkowski, R.W., Arnold, J.M., and Gogny, D.M. (1995). Ultrafast Pulse Interactions with Two-level Atoms, *Phys. Rev. A* **52**, 3082–3094.
27. Morgner, U., Kärtner, F.X., Cho, S.H., Chen, Y., Haus, H.A., Fujimoto, J.G., Ippen, E.P., Scheuer, V., Angelow, G., and Tschudi, T. (1999). Sub-two-cycle Pulses from a Kerr-lens Mode-locked Ti:sapphire Laser, *Opt. Lett.* **24**, 411–413.
28. Wehner, M.U., Ulm, M.H., and Wegener, M. (1997). Scanning Interferometer Stabilized by Use of Pancharatnam's Phase, *Opt. Lett.* **22**, 1455–1457.
29. The first microscope objective has a focal length of $f = 5.41$ mm and a numerical aperture of NA = 0.5 *Coherent 25–0522*), the second one $f = 13.41$ mm and NA = 0.5 (*Coherent 25-0555*).
30. Peyghambarian, N., Koch, S.W., and Mysyrowicz, A. (1993). *Introduction to Semiconductor Optics* (Prentice Hall, Englewood Cliffs, NJ).
31. Sheik-Bahae, M., Said, A.A., Wei, T.-H., Hagan, D.J., and van Stryland, E.W. (1990). Sensitive Measurement of Optical Nonlinearities Using a Single Beam, *IEEE J. Quantum Electron.* **26**, 760–769.
32. Private communication with Haug H. (June 2001).

Quantum Coherence, Correlation and Decoherence in
Semiconductor Nanostructures
T. Takagahara (Ed.)
Copyright © 2003 Elsevier Science (USA). All rights reserved.

Chapter 3
High-field effects in semiconductor nanostructures

S. Hughes*

Department of Physics, University of Surrey, Guildford, Surrey GU2 7XH, UK

Abstract

In this work we present a theoretical investigation of high electromagnetic field interactions in semiconductor nanostructures, qualitatively recovering several recent experimental observations, and predicting a few new ones. In the linear optical regime, a space–time method for modeling nonperturbative electron–hole wave packets in semiconductor quantum wells and wires is introduced. The technique is computationally efficient, physically intuitive, and can straightforwardly incorporate Coulomb, static, terahertz, and magnetic fields to all orders. Various electro-optical and electro-magneto-optical observables are obtained and a connection is made with recent measurements using free-electron and MIR lasers. For the high-intensity nonlinear optical regime, solutions of the semiconductor Bloch equations are shown including the relevant scattering mechanisms. Several nonlinear optical phenomena are predicted including excitonic trapping and adiabatic population transfer. Finally, we solve Maxwell's curl equations outside the slowly-varying envelope approximation to demonstrate carrier-wave Rabi flopping, an effect that was recently observed in GaAs using several-cycle extremely intense optical pulses.

3.1 Introduction

Driven by the pursuit to understand the differences and similarities between the semiconductor and atomic systems resonantly excited by laser pulses, the coherent nonlinear dynamics of direct-gap semiconductors has been vigorously investigated over the years. Moreover, with the ongoing advancement of short-pulse laser techniques and excellent semiconductors samples, new classes of coherent dynamic phenomena have

---

* Present address: Galian Photonics, 300-1727 West Broadway, Vancouver BC, Canada V6J 4W6.

been discovered recently including, for example, coherent exciton control [1,2], self-induced transmission [3], and carrier-wave Rabi flopping [4]. Several of these effects are discussed in other chapters of this book. Running in parallel have been new discoveries in atomic optics such as population trapping using frequency modulated fields [5], above threshold ionization and high-field harmonic generation (HHG) [6]. Collectively, these phenomena may be classed as high or extreme electromagnetic (EM) field effects in matter.

As the development of lasers continues to produce stronger and stronger fields, we can now study high-field-matter interactions in the laboratory, even with air as the matter, where the common theoretical techniques of nonlinear optics and perturbative field expansions break down. Indeed, it is now well established that the extremely-high-field physics of atomic ensembles presents many fascinating phenomena whereby the response of matter to high fields cannot be described within perturbation theory. For example, the process of HHG due to an intense atom-field interaction has received substantial attention in recent years [6], and harmonically-generated X-ray transients as short as 100 attoseconds have been predicted. For Rydberg atoms, higher frequency harmonics are produced from continuum-state to bound-state transitions, in which electrons release the energy absorbed from the field during its journey in the continuum. The theoretical problem of HHG is most tactfully treated nonperturbatively by exploring the wave packet motion by essentially-exact numerical methods.

In certain limits there is a one-to-one mapping between the theoretical description of atoms and excitons–Coulombically bound electron–hole pairs in solids analagous to Hydrogen; of course, the binding energies are substantially different. While the bound-state to continuum-state transitions in atoms are typically in the eV regime, in semiconductors these transitions are usually in the meV (THz) regime. Nevertheless, with an increase in scientific research that utilizes free-electron lasers as well as THz solid-state emitters, high-field THz and MIR spectroscopy is now timely entering similar *extreme* regimes for semiconductors [7,8]. However, the frequencies and field strengths (and thus ponderomotive energies) required are several orders of magnitude apart. For this reason, high-field effects in semiconductors can be observed at field intensities many orders of magnitude below what is required for the atoms. We present a new real-space-time method to calculate dynamic electron–hole wave packets in semiconductors, allowing the theoretical study of nonperturbative field regimes in a simplistic way. The technique can incorporate a variety of fields, such as Coulomb, magnetic, THz, static, all included exactly, in the two-particle low density limit. Experimental observables are extracted and qualitatively compared to measurements reported in the literature, where appropriate. Specific examples are presented for both quantum wells and quantum-well wires.

With regard to high-intensity ultrashort optical fields, Rabi oscillations of the population between two states can be seen in the temporal evolution of a two-level atom (TLA) [9]. Additionally, in the case of a sinusoidal frequency-modulated excitation, square-wave oscillations of the population – periodic state trapping – as well as more complicated and phase-dependent structures may appear [5]. In combination with other well-known concepts such as adiabatic rapid passage, multiphoton resonances, and Landau–Zener transitions, trapping was experimentally demonstrated in a TLA [10]. For two-band semiconductors, however, trapping and Rabi flopping are scarcely expected due to Coulomb many-body complications and the valence/conduction band-continua of free-carriers. At least this was the general consensus a few years ago. In semiconductors, the *two-level* model (as a first approximation to a two-band description) is considered inappropriate because Coulomb many-body interactions result in a renormalized Rabi energy and bandedge, and excitation-induced dephasing. Nevertheless, as mentioned before, multiple Rabi flopping on a semiconductor free exciton resonance has been recently reported for both bulk [3] and quantum wells (QWs) [11]; these measurements were successfully explained within the framework of the semiconductor Bloch equations (see also Chapter 1). We exploit this effect further to show that one can achieve population trapping dynamics in semiconductors using frequency-modulated or suitably-chirped broadband optical pulses. Finally, we will solve Maxwell's curl equations without incorporating the slowly-varying envelope approximations to study carrier-wave Rabi flopping of femtosecond optical pulses of only several carrier-cycles time duration [12]. Experimental evidence for this latter phenomenon was only just reported using thin film GaAs and extremely short optical pulses [4], and is also discussed in Chapter 2 of this book.

In the remainder of this chapter, we describe, in more detail, several high EM-field effects in semiconductor nanostructures. First, in section 3.2 we will present the general theoretical framework. In section 3.3 we investigate electron–hole wave packets in the presence of large static, THz, and magnetic fields. In section 3.4 we solve the semiconductor Bloch equations and quantum Bolzmann equations in the presence of a frequency-modulated optical pulse, demonstrating exciton trapping and quasi-adiabatic population transfer in a semiconductor quantum well. Section 3.5 deals with carrier-wave Rabi flopping and ultrafast coherent effects outside the rotating-wave approximation. In section 3.6 we give our conclusions and closing discussions.

3.2 General theory

The role of this theory is to calculate the propagated electromagnetic fields or the corresponding field correlation functions, and to make a connection to microscopic processes, such as many-body and excitonic Coulomb interactions between electrons and holes. In optical investigations of semiconductor structures, externally controlled EM fields are applied to the semiconductor material. The fields which contain frequency components close to the band edge of the semiconductor induce transitions of electrons from the populated valence bands into the empty conduction bands, thus creating electron–hole (e–h) excitations. On ultra-short time scales, e–h pairs decay by stimulated radiative recombination, emitting optical radiation that interferes with the applied electromagnetic fields. The EM field can subsequently be measured after its propagation through the entire sample. Experimentally, only the changes of the electromagnetic field after its propagation through the sample can be considered as a detector for microscopic processes in the material. The measurements are usually performed in the reflection or transmission geometry.

Therefore, our task is to solve Maxwell's equations for the propagating total EM fields with the respective initial conditions. The boundary conditions are dictated by the geometry of the semiconductor structure, whereas the initial conditions are fixed by the state of the sample before the arrival of the pulse (EM field). After the arrival of the field, the semiconductor experiences electromagnetic sources caused by the e–h excitations. These sources enter as macroscopic averages of the microscopic current, j, and charge density, ρ, into Maxwell's equations. These latter sources are calculated from the material Bloch equations, discussed below.

In this work, we use excitonic units throughout; $\hbar = m_r = e^2/\varepsilon_\infty = 1$ where m_r is the reduced mass of the e–h pair. The speed of light in the crystal is related to the background dielectric constant by $c = c_v/\sqrt{\varepsilon_\infty} = \sqrt{\varepsilon_\infty}/\vartheta$ where ϑ is the fine structure constant. The theory presented in this section will concentrate on semiconductor quantum wells, since they will form the main focus of our chapter. For more theoretical details, we refer the interested reader to the excellent textbook and review paper given in References [13] and [14], respectively.

We assume a familiarity with second quantization for particles and begin our dynamical description of optically excited carriers from the following Hamiltonian:

$$\mathbf{H} = \mathbf{H}_{fp} + \mathbf{H}_{cc} + \mathbf{H}_{cl}, \qquad (3.1)$$

where

$$\mathbf{H}_{fp} = \sum_{ki} \varepsilon^e_{i,\mathbf{k}} c^\dagger_{i,\mathbf{k}} c_{i,\mathbf{k}} + \sum_{kj} \varepsilon^h_{j,-\mathbf{k}} d^\dagger_{j,-\mathbf{k}} d_{j,-\mathbf{k}} \qquad (3.2)$$

is the free-particle (fp) Hamiltonian in the e–h picture. The terminology above is as follows: $\mathbf{k}$ is the in-plane wavevector; $i(j)$ refer to the electron (hole) subband index; $\varepsilon_{i,\mathbf{k}}^{e(h)}$ is the electron (hole) single-particle energy (band structure); and $c_{i,\mathbf{k}}^\dagger/c_{i,\mathbf{k}}$ and $d_{i,-\mathbf{k}}^\dagger/d_{i,-\mathbf{k}}$ are the creation/annihilation operators for electrons and holes.

The second contribution in Equation (3.1) describes the carrier–carrier (cc) interaction via the Coulomb potential V

$$\mathbf{H}_{cc} = \frac{1}{2} \sum_{ij,\mathbf{k}_1\mathbf{k}_2\mathbf{k}_3} V_{\mathbf{k}_3}^{ij} c_{i,\mathbf{k}_1+\mathbf{k}_3}^\dagger c_{j,\mathbf{k}_2-\mathbf{k}_3}^\dagger c_{j,\mathbf{k}_2} c_{i,\mathbf{k}_1}$$

$$+ \frac{1}{2} \sum_{ij,\mathbf{k}_1\mathbf{k}_2\mathbf{k}_3} V_{\mathbf{k}_3}^{ij} d_{i,\mathbf{k}_1+\mathbf{k}_3}^\dagger d_{j,\mathbf{k}_2-\mathbf{k}_3}^\dagger d_{j,\mathbf{k}_2} d_{i,\mathbf{k}_1}$$

$$- \sum_{ij,\mathbf{k}_1\mathbf{k}_2\mathbf{k}_3} V_{\mathbf{k}_3}^{ij} c_{i,\mathbf{k}_1+\mathbf{k}_3}^\dagger d_{j,\mathbf{k}_2-\mathbf{k}_3}^\dagger d_{j,\mathbf{k}_2} c_{i,\mathbf{k}_1}. \quad (3.3)$$

The final term accounts for the carrier–light (cl) coupling that we describe here within the dipole approximation

$$\mathbf{H}_{cl} = -\sum_{\mathbf{k},i} [\mu_{\mathbf{k}} E_i(t) c_{i,\mathbf{k}}^\dagger d_{i,-\mathbf{k}}^\dagger + \mu_{\mathbf{k}}^* E_i(t) d_{i,-\mathbf{k}} c_{i,\mathbf{k}}], \quad (3.4)$$

where all intraband terms have been neglected. Equation (3.4) describes the interaction in the z-direction (growth direction) localized dipolar plane with a classical optical laser field at the position of the plane, which is given by the matrix element $E_i(t) = \langle i | \mathbf{E}(\mathbf{x}, t) | i \rangle$, where $| i \rangle = \varphi_i(z)$ labels the confinement functions in the wells and $\mu_{\mathbf{k}}$ is the optical dipole matrix element between electron and hole states.

In the rotating wave approximation, it is advantageous to work with complex fields. Outside the rotating wave approximation, we will use real fields (see later). The particle current can be written in terms of the transition probability between different e–h states (polarization functions)

$$j = \frac{\partial}{\partial t} \sum_{\mathbf{k},i} \mu_{\mathbf{k}} \varphi_i^e(z) \varphi_i^h(z) \langle d_{i,-\mathbf{k}} c_{i,\mathbf{k}} \rangle, \quad (3.5)$$

where that the polarization and the material current are related through $j = \dot{P}$. To calculate the particle current, we use the electron and hole distribution functions, f^e and f^h, and the interband transition amplitude p as dynamic variables for the description of the material system. These quantities are defined as

$$f_{i,\mathbf{k}}^e = \langle c_{i,\mathbf{k}}^\dagger c_{i,\mathbf{k}} \rangle, \quad f_{i,\mathbf{k}}^h = \langle d_{i,\mathbf{k}}^\dagger d_{i,\mathbf{k}} \rangle, \quad p_{i,\mathbf{k}} = \langle d_{i,-\mathbf{k}} c_{i,\mathbf{k}} \rangle. \quad (3.6)$$

The equations of motion can be subsequently obtained from the Heisenberg equations, e.g.

$$\frac{d}{dt} f_{i,\mathbf{k}}^e = \frac{1}{i} \langle [c_{i,\mathbf{k}}^\dagger c_{i,\mathbf{k}}, H] \rangle \quad (3.7)$$

and correspondingly for the other quantities.

In the time-dependent Hartree–Fock approximation [13], we obtain the following semiconductor equations:

$$\frac{\partial}{\partial t} p_{i,\mathbf{k}} = -i(\omega^e_{i,\mathbf{k}} + \omega^h_{i,\mathbf{k}}) p_{i,\mathbf{k}} - i\Omega_{i,\mathbf{k}}(f^e_{i,\mathbf{k}} + f^h_{i,-\mathbf{k}} - 1) + \left.\frac{\partial p_{i,\mathbf{k}}}{\partial t}\right|_{\text{corr}} \qquad (3.8)$$

$$\frac{\partial}{\partial t} f^e_{i,\mathbf{k}} = -i(\Omega^*_{i,\mathbf{k}} p_{i,\mathbf{k}} - \Omega_{i,\mathbf{k}} p^*_{i,\mathbf{k}}) + \left.\frac{\partial f^e_{i,\mathbf{k}}}{\partial t}\right|_{\text{corr}} \qquad (3.9)$$

$$\frac{\partial}{\partial t} f^h_{i,-\mathbf{k}} = -i(\Omega^*_{i,\mathbf{k}} p_{i,\mathbf{k}} - \Omega_{i,\mathbf{k}} p^*_{i,\mathbf{k}}) + \left.\frac{\partial f^h_{i,-\mathbf{k}}}{\partial t}\right|_{\text{corr}}, \qquad (3.10)$$

where

$$\Omega_{i,\mathbf{k}} = \mu_{cv,\mathbf{k}} E_i(t) + \sum_{\mathbf{k}'} V^{ii}_{\mathbf{k}'-\mathbf{k}} p_{i,\mathbf{k}'}, \qquad (3.11)$$

$$\omega^e_{i,\mathbf{k}} = \varepsilon^e_{i,\mathbf{k}} - \sum_{\mathbf{k}'} V^{ii}_{\mathbf{k}-\mathbf{k}'} f^e_{i,\mathbf{k}'} \qquad (3.12)$$

and

$$\omega^h_{i,\mathbf{k}} = \varepsilon^h_{i,\mathbf{k}} - \sum_{\mathbf{k}'} V^{ii}_{\mathbf{k}-\mathbf{k}'} f^h_{i,\mathbf{k}'}, \qquad (3.13)$$

are the renormalized Rabi-frequencies and electron and hole energies, respectively, and $\mu_{cv,\mathbf{k}}$ is the transition dipole matrix element between the conduction (c) and valence (v) bands. We refer to the above as the semiconductor Bloch equations (SBE) in analogy with the well known atomic Bloch equations for the two level atom [9]. The correlation terms ($\dot{p}|_{\text{corr}}$, $\dot{f}|_{\text{corr}}$) will be discussed later. By obtaining a solution of the SBE, the carrier density can be calculated from the electron (or hole) population through $N = 2\sum_{\mathbf{k}} f_{\mathbf{k}}$, and similarly for the optical polarization, $P = 2\sum_{\mathbf{k}} \mu_{cv,\mathbf{k}} p_{\mathbf{k}}$. The factor of two accounts for spin and we have ignored some obvious normalization constants.

3.3 High-field electro-optics in quantum wells and wires

The aim of this section is to introduce a space-time method and apply it to study semiconductor wave packet (WP) dynamics in the presence of extreme fields – electric and (or) magnetic (dynamic and static). We will focus on e–h WPs that are dynamically created by resonantly exciting a semiconductor with a short (but finite) optical pulse. The pulse intensity is weak and thus we work in the linear regime for the optical field, but all other fields are treated nonperturbatively. This enables us to work in the low density limit where many-body effects do not come into play. In the presence of large THz, static, or (and) magnetic fields, one can explore WP motion and experimental observables that are distinct to the semiconductor system. Additionally, semiconductor studies in the dynamic extreme-field regime, although relatively new, offer several advantages over atomic environments including the ability to change the masses and dimensionalities of the system. At the outset, we emphasize that electro-absorption studies in semiconductors are certainly not new. The influence of a strong constant electric field **F** on the optical and electronic properties of semiconductors was brought to the fore over four decades ago, beginning with the Franz–Keldysh effect in bulk crystals [15]. For 2d semiconductors, such as quantum wells (QWs) [16], similar effects occur for a field polarized in the plane, while the quantum-confined Stark effect occurs for fields polarized in the growth direction. In 1d semiconductors, the Franz–Keldysh effect was recently theoretically investigated [17,18] using the real-space technique described below.

In the last few years there has been much interest in extending the semiconductor electro-optical studies into the dynamic (i.e. THz) regime. Besides being of fundamental interest, as highlighted in the introduction there are some very close analogies with high-field effects in atoms, though this connection is sometimes rarely made in the literature. An intuitive WP approach for modelling HHG in semiconductors was presented in Reference [19]. From a nonperturbative theoretical perspective, several techniques have been introduced, namely Green's function [20] and non-equilibrium Green's function techniques [21] (both restricted their handling of excitons and dynamics), as well as full scale numerical solution of the SBE in momentum space [22]. With regard to modelling high magnetic field interactions, usually one has to resort to a specialized basis set or (and) treat the Coulomb interaction perturbatively. In this work we set about solving the problem using a real space and real time technique, specializing in the low density regime (as in the above cases). The method (i) includes the fields (THz and magnetic) and the Coulomb interactions nonperturbatively, (ii) is an order of magnitude faster than the **K**-space SBE even without the magnetic field, and (iii) provides full spatial information thus providing a nice link with transport and optical properties. We will first concentrate on QWs, where the technique can be applied to recent free-electron laser studies and manipulation of excitonic states–rovering the dynamic Franz–Keldysh effect [7,22]. We will also model magneto-excitons and electro-magneto-excitons and explore their WP motion. The present theoretical understanding of semiconductors in large magnetic fields is limited since implementing the effects exactly with the Coulomb interaction is very difficult if one works in energy space; besides side-stepping this problem, moreover, we show that in the presence of crossed magnetic and THz fields, WP stabilization can occur. Additionally, we will explore extreme high-frequency induced sidebands that show good trends with recent MIR experiments [8]. Lastly, we investigate the static and dynamic Franz–Keldysh effect in semiconductor quantum-well wires.

Real space theoretical approach to electon–hole wave packets

High-field effects in atoms can be investigated nonperturbatively by numerically solving an effective Schödinger equation

$$i\frac{\partial \Psi(\mathbf{r},t)}{\partial t} = [-\nabla_r^2 - V(\mathbf{r}) + \mathbf{F}_{THz}(t) \cdot \mathbf{r}]\Psi(\mathbf{r},t), \quad (3.14)$$

where $\mathbf{r}$ is the spatial position, $\Psi(\mathbf{r}, t)$ is the WP, $\mathbf{F}_{THz}(t)$ is the oscillating (THz) electric field, and V is the Coulomb potential. A fully quantum method for integrating numerically the above equation without using a restrictive basis expansion can be implemented by employing the split-step method [23,24]. The essence of this method is to carry out the action of the kinetic operator efficiently in Fourier space, while the action of the potential operator is carried out in real space. (We will, however, take a different approach as the split-step method cannot handle magnetic fields.) On the other hand, semiconductor optical problems are usually tackled within the framework of the SBE as discussed earlier [see Equations (3.8–3.10)]. In the low density regime, for a two subband model, this leads to the following set of equations for the polarization only,

$$\frac{\partial p_\mathbf{k}(t)}{\partial t} = -\mathbf{F}_{THz}(t) \cdot \nabla_\mathbf{k} p_\mathbf{k}(t) - i\Delta_\mathbf{k} p_\mathbf{k}(t) + i\Omega_\mathbf{k}(t) - \Gamma p_\mathbf{k}(t), \quad (3.15)$$

where $\Delta_\mathbf{k} = E_\mathbf{k} - \omega_l + E_g$, E_g is the band gap, ω_l is the carrier frequency of the optical pulse, and $\mathbf{F}_{THz}(t)$ is the applied field that can be polarized in any direction (see below). Of course static effects can be described simply by dropping the time dependence of the dynamic field. We work here within the slowly-varying envelope approximations where the generalized Rabi frequency is $\Omega_\mathbf{k}(t) = \mu_{cv,\mathbf{k}}\tilde{\varepsilon}_{Opt}(t) + \sum_\mathbf{q} V_{\mathbf{k}-\mathbf{q}} p_\mathbf{q}(t)$, with $\tilde{\varepsilon}_{Opt}(t)$ the slowly-varying optical field polarized in the QW plane. To account qualitatively for Coulomb correlations, we assume the total dephasing rate of the optical polarization $\Gamma^{-1} = 500$ fs, unless stated otherwise. This is a reasonable assumption provided the input optical pulse is weak, which it is for the subsequent studies. The solution of the SBE can then be solved in $\mathbf{k}$ space, and the total optical polarization is $\mathbf{P}_{Opt}(t) = 2\sum_\mathbf{k} \mu_{cv,\mathbf{k}} p_\mathbf{k}(t)$. However we do not learn anything about the spatial dynamics; also, the $\mathbf{k}$ equations are extremely difficult to solve with the THz field since the dynamical components become anisotropic resulting in a major computational effort. Indeed, the solution alone with the THz field was only reported recently [22], and takes many days to simulate even with fast computer processors. With the possible addition of a magnetic field, the numerical problem becomes intractable, requiring a switch to a large number of Landau basis states with some perturbation approximations [13]. This is essentially because semiconductor physicists usually tackle most theoretical problems in energy space because of translational invariance. On the contrary, we will show in this work that the dynamics of semiconductor nanostructures excited hybridly by a weak sub-picosecond laser and a strong EM field is ideally studied in terms of WPs. We highlight that the pioneering work of Schmitt-Rink and co-workers into the semiconductor Stark effect also approached the electro-optical problem from the real-space perspective, essentially solving the stationary Schrödinger equation [16].

Assuming on-resonance excitation (zero detuning) we can rewrite the above equation in real space,

$$i\frac{\partial P(\mathbf{r},t)}{\partial t} = [-\nabla_\mathbf{r}^2 + \mathbf{r}\cdot\mathbf{F}_{THz}(t) - i\Gamma]P(\mathbf{r},t) - V(\mathbf{r})P(\mathbf{r},t) + \Omega_{cv}(t)\delta(\mathbf{r}), \quad (3.16)$$

where $P(\mathbf{r}, t)$ is the e–h WP, $\Omega_{cv}(t)$ is the Rabi frequency, and $\mathbf{r} = \mathbf{r}_e - \mathbf{r}_h$. The THz driving field will form the source for an oscillating dipole; a subsequent displacement of the WP from $\mathbf{r} = 0$ means it is polarized. For simplicity we neglect the $\mathbf{k}$-dependence of the interband dipole moment, though it is straightforward to include. The optical polarization is simply $\mathbf{P}_{Opt}(t) = 2\mu_{cv}P(\mathbf{r} = 0, t)$, while the THz-induced intraband dipole moment

$$\mathbf{P}_{THz}(t) = e\int dr\, P^*(\mathbf{r},t)\mathbf{r}P(\mathbf{r},t). \quad (3.17)$$

Therefore, for the emitted THz electric field (assuming a point source)

$$\mathbf{E}_{THz}(t) = -\mu_0 \ddot{\mathbf{P}}_{THz}(t)/4\pi r. \quad (3.18)$$

The structure of the Equation (3.16) is very similar to that for the atomic problem but we have dephasing and an optical pulse; therefore the e–h WP is created dynamically and relaxes dynamically (dephases). In the presence of a magnetic field, $\mathbf{B}$, terms like, for example, the momentum operator $p_x^2[i\partial/\partial x]^2$ are replaced by $[p_x + eA_x/c]^2$ [13] (where A_x is the vector potential) and do not pose any further refinements on the computational technique; relativistic correction terms can also be added. Thus, whether one studies free carriers, excitons, THz field ionization, or magnetic field effects, the computational technique is the same. This is in stark contrast to the $\mathbf{k}$-space approach where the numerics become increasingly more difficult with the various fields. We do point out that the $\mathbf{k}$-space approach is, however, much better suited to high-density and nonlinear optical studies, as demonstrated in many pioneering publications and in the next sections, but intractable with very large electric and magnetic fields.

Employing weak input optical fields, one can obtain the material properties via Maxwell's equations by Fourier transforming the WP at $\mathbf{r} = \mathbf{0}$ to the frequency domain and dividing by the spectrum of the input pulse. Thus absorption and the refractive index are, respectively, proportional to

$$\text{Im}[P_{\text{Opt}}(\omega)]/|\tilde{\varepsilon}_{\text{Opt}}(\omega)|^2 \text{ and } \text{Re}[P_{\text{Opt}}(\omega)]/|\tilde{\varepsilon}_{\text{Opt}}(\omega)|^2.$$

Our numerical strategy is based on an exploitation of the finite-difference time-domain [FDTD] [25] method and will be described in detail elsewhere; suffice to say here that FDTD has recently became the state-of-the-art computational method for solving Maxwell's equations exactly. A similar computational technique is also given in a recent textbook [26]. Essentially we replace the polarization equation by finite difference approximations, and implement them on a computational cell like the Yee cell [25] used in electrodynamics; in addition, the delta function is approximated as a narrow Gaussian and its numerical accuracy is verified to be in excellent agreement with other techniques in obtaining the excitonic and continuum properties for 1d, 2d, and 3d semiconductors. We work in the symmetric gauge, and take material parameters typical of GaAs throughout with a bulk exciton binding energy $E_0 = 4.2$ meV, and a Bohr radius $a_0 = 140$ Å.

Electro-magneto-optical simulations in quantum wells

Here we focus attention to a two-subband semiconductor QW excited by a short optical pulse where the center of the pulse corresponds to time $t = 0$ fs. For the electric field we choose a weak Gaussian pulse with a 40 fs FWe^{-2}M irradiance. To explore several applications of the real space method, we will investigate four separate semiconductor excitations. (a) A short optical pulse only. (b) As in (a) but also with a static magnetic field $\mathbf{B} = B_0 \hat{\mathbf{n}}_z$ with a corresponding Landau frequency of 4 meV [13] ($\approx$ 1 THz); $\hat{\mathbf{n}}_z$ is a unit vector perpendicular to the QW plane (growth direction). (c) As in (a) but with the addition of a THz electric field $\mathbf{F}_{\text{THz}}(t) = \hat{\mathbf{n}}_x F_0 \sin(\Omega t + \phi)$ with $\Omega = 4$ meV and phase ϕ (at the center of the optical pulse $t = 0$); F_0 the magnitude of the THz field is taken to be 5 kV/cm. And (d), as in (a) but with a mixed **B**-THz field applied, with identical parameters as to the above.

In Fig. 3.1 we show a snapshot of the corresponding WPs at $t = 1.8$ ps after the short pulse has gone. $|\mathbf{P}(\mathbf{r}, t)|$ is a measure of the probability of finding an electron and hole at position $\mathbf{r}$ at time t. Initially the WP is created at $\mathbf{r} = \mathbf{0}$ to satisfy energy and momentum requirements of a direct gap semiconductor. (Note that much larger spatial regions are accounted for computationally.) In Fig. 3.1(a) the WP is concentrated near the center due to the Coulomb interaction, since there is a high probability of finding the electron and hole at the same relative position (excitons). Figure 3.1(b) depicts the WP undergoing circular motion about the growth direction, and small Landau structures begin to appear at larger spatial positions Fig. 3.1(c) highlights a pronounced WP distortion and interference because of broken symmetry, and side lobes can be seen in the WP (these are formed by the combination of slow transverse spreading, the relatively fast field driven motion in the polarization direction, and Coulombic rescattering) Fig. 3.1(d) shows a unique spiralling interfering WP due to the combined THz and **B** fields.

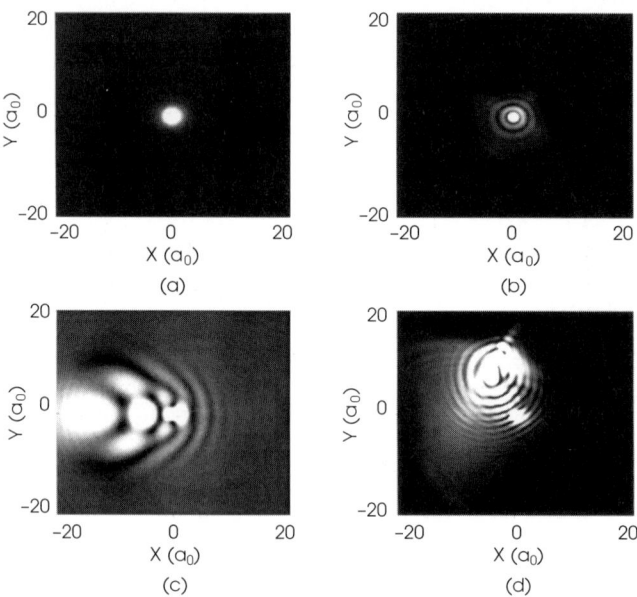

Fig. 3.1 Wave packet, | **P**(**r**, t) |, at time $t = 1.8$ ps for various excitation regimes (a) optical pulse only, (b) also with a **B** field, (c) with a THz field, and (d) with a mixed **B**-THz field. The spatial dimensions are given in units of the exciton Bohr radius.

To link the WP motion to familiar optical properties, in Fig. 3.2(a) we calculate the familiar field-free absorption spectrum for a QW, namely a strong 1s exciton resonance and a Coulombically-enhanced continuum; in the presence of a THz field we obtain the dynamic Franz–Keldysh effect with oscillations above the band gap [7]. Note that we have calculated the phase averaged WP results and thus the graphs represent the true absorption. Different phases, in principle, can coherently control the WP's motion. In Fig. 3.2(b) we obtain Landau oscillations and an enhancement of the lowest lying exciton. For clarity, (**B**-induced) diamagnetic shift terms of the band gap are not shown, though will depend on the helicity of the optical light pulse. To further highlight our technique with larger fields, we also employ a 12 meV Landau frequency, polarized, as before, in the growth direction. Substantial magneto-excitons are seen in the inset, with peaks separated by the Landau energy. These recover the known solution that employs many basis states in energy space [13], but without restrictions of perturbation theory and working with complicated basis states. We next obtain the mixed **B**-THz-field scenario resulting in interference between magneto-excitons (Landau levels) and nonlinear THz interactions, shown in Fig. 3.2(c). In Fig. 3.2(d) we plot the emitted THz field (from the oscillating dipole) in the presence of the THz driving field; a variety of harmonics similar to HHG in atoms appears. In the presence of a THz field the extra peaks correspond to the absorption of one optical photon and the absorption (or emission) of one or more THz photons. In the mixed field case we obtain a novel magnetic-THz

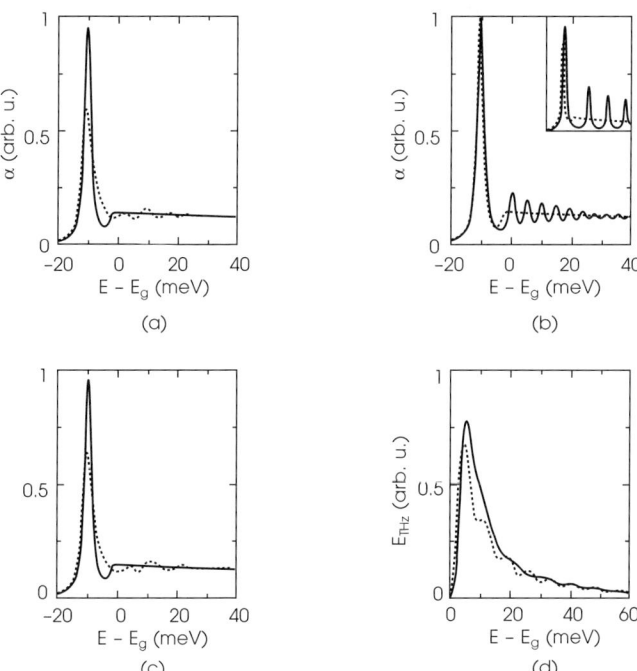

Fig. 3.2 Optical polarization versus energy corresponding to the following: (a) with (dotted line) and without a THz field; (b) with (solid line) and without a **B** field; (c) with (dotted line) and without a mixed **B**-THz field; (d) emitted dipole field with a THz driving field only and a mixed **B**-THz field (solid line).

coupling regime; interestingly the harmonics are suppressed when we add a magnetic field since WP stabilization occurs for longer times. This effect is consistent with the Green's function prediction [27] and all the physics can be explained from the WP's internal motion.

Next, we remove the time dependence of the oscillating THz field, e.g. $\mathbf{F} = F_0 \hat{\mathbf{n}}_x$, thus working in the regime of the static Franz–Keldysh effect. However, in addition we add a magnetic field in the growth direction. In Fig. 3.3(a) we show an example of the WP at time $t = 1$ ps. In this case, there is no longer a forced oscillating dipole though the WP is undergoing **B**-field spiralling and e–h ionization through the applied dc field. The corresponding absorption is shown in Fig. 3.3(b), demonstrating significant exciton ionization and mixed Franz–Keldysh magneto-exciton features. We note that the emitted THz field (see insert) shows no clear peaks beyond the dc signal which is due to the fact that the static electric field is dominating the WP motion.

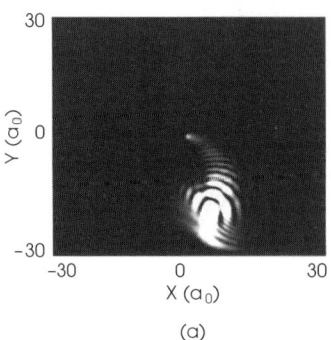
(a)

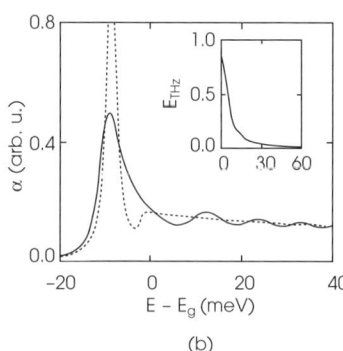
(b)

Fig. 3.3 (a) Wave packet at time $t = 1$ ps for a static **F-B** field excitation. (b) Absorption spectrum with (solid line) and without the mixed field. As an inset we show the emitted dipole field.

Before closing this subsection, we present an example of extreme high-frequency THz optics by using a 5 THz frequency field with a 100 kV/cm peak amplitude. We have also tried such a simulation using a momentum space approach and it is intractable, yet the present method solved the problem in about 10–20 CPU time minutes on a standard 500 MHz UNIX workstation. In Fig. 3.4 we depict the WP at $t = 0.5$ ps as well as the optical absorption and THz emission. Changes in the absorption 80 meV above and below the band gap can be seen, and several broadband harmonics appear; both these effects show similar trends with recent experiments studying extreme MIR interactions in semiconductors [8]. We note that although the field is almost 10 times the QW exciton binding energy, the Coulomb interaction still dominates the spatial interference patterns. We remark that at these field strengths the role of LO-phonons may become important for a quantitative explanation of experimental measurements.

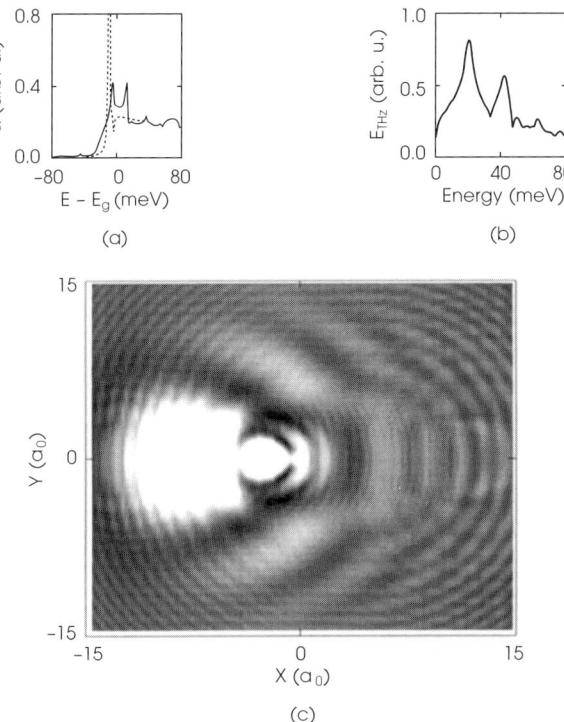

Fig. 3.4 (a) Absorption spectrum with (solid line) and without a large 5 THz-frequency field. (b) Emitted dipole field. (c) Wave packet at time $t = 0.5$ ps.

Static Franz–Keldysh effect in quantum wires

Semiconductor quantum wires offer a unique playground for investigating how electrons and holes move in one dimension [28]. A surprising feature of the 1d system is the inverse-square-root divergence of the joint density of states (DOS) at the band gap. However, as pointed out in Reference [29], the Sommerfeld factor, which is the intensity ratio of the optical density associated with excitonic scattering states to the free e–h pair above the band gap, removes this divergence. Moreover, in contrast to the 2d and 3d cases, the Sommerfeld factor is <1 for all frequencies above the band gap. Consequently, the singular 1d-DOS does not show up at all in the linear absorption spectrum (even with no dephasing).

Although Franz–Keldysh effects in bulk crystals and quantum wells have been actively investigated for a number of years, with regards to the theory of the Franz–Keldysh effect in quantum wires, besides a study for field effects for currents applied in the growth direction [30], no solution existed to the best of our knowledge until recently [17], which is described below. However, Coulomb ionization effects have been and remain to be intensely investigated for 1d atomic systems [31]. Although the 1d semiconductor was originally a model system, both experimental [32] and theoretical [33,34] studies on semiconductor quantum-well wires (QWWs) are receiving renewed interest as a result of pronounced strides in growth technology resulting in high quality samples with well defined characteristics.

Again we employ the real space approach. However, because we are dealing with quantum wires, some modification of the polarization equation is required. Assuming only the 1e subband (a) and various h subbands (b) (see below for more details):

$$i\frac{\partial}{\partial t} P_{ab}(x,t) = \left[E_g - \frac{\partial^2}{\partial x^2} + x\mathbf{F} - i\Gamma \right] P_{ab}(x,t)$$
$$- \sum_\beta V_{ab}^{\alpha\beta}(x) P_{\alpha\beta}(x,t) + \Omega_{ab}(t)\delta(x), \tag{3.19}$$

where P_{ab} is the induced polarization between two subbands a (electrons) and b (holes) and $\mathbf{F}$ is the static field polarized along the wire axis; in general the total polarization contains an infinite summation over all subbands, although accurate simplifications can be made. As before, we choose an optical pulse that is excited resonantly with the band gap.

For a tractable model for the QWW potential, we will assume that the electrons and holes are confined laterally by a harmonic oscillator potential with a subband spacing of $\Omega_{j=e,h}$. Furthermore, we assume essentially perfect confinement in the growth direction (i.e. only one associated subband). The Coulomb interaction between charge carriers thus becomes [30]

$$V_{ab}^{\alpha\beta}(x) = 2 \int dy \int dy' \frac{\psi_a(y)\psi_b^*(y')\psi_\alpha(y)\psi_\beta^*(y')}{\sqrt{x^2+(y-y')^2}}, \tag{3.20}$$

where $\psi_j(y)$ are the eigenfunctions [34] associated with the lateral carrier motion

$$\psi_n^{j=e,h}(y) = \left(\frac{1}{2^n n!} \sqrt{\frac{\mu\Omega_j}{m_j 2\pi}} \right)^{\frac{1}{2}} \exp[-(\mu\Omega_j/m_j)y^2] H_n\left(\sqrt{\frac{\mu\Omega_j}{2m_j}} y \right), \tag{3.21}$$

with μ the reduced mass, m_j the mass of an electron or hole, and H_n the Hermite polynomial.

For the moment we concentrate on a two-subband semiconductor model (ab = 1e-1hh) but will later discuss the role of multi-subbands as well as Coulomb mixing effects. For the eigenfunction calculations we choose subband spacings of $3E_0$ for the holes and $7E_0$ for the electrons, and $m_h = 5m_e$. This results in holes that are more strongly confined than the electrons which is typical of real QWWs. Since we are only dealing with 1d, the computational technique is extremely efficient, requiring a few seconds CPU time per run on desktop computers. We first calculate an absorption spectrum with a dephasing time of 500 fs in the absence of an external field **F**. In Fig. 3.5(a) is shown various snapshots of the polarization WP ($|P(x, t)|$), while Fig. 3.5(b) displays the corresponding absorption spectrum. In addition to a strong 1s exciton peak we also obtain the 2s peak for larger dephasing times (see below) and verify that the Coulomb interaction removes the singularity at the band gap associated with the 1d

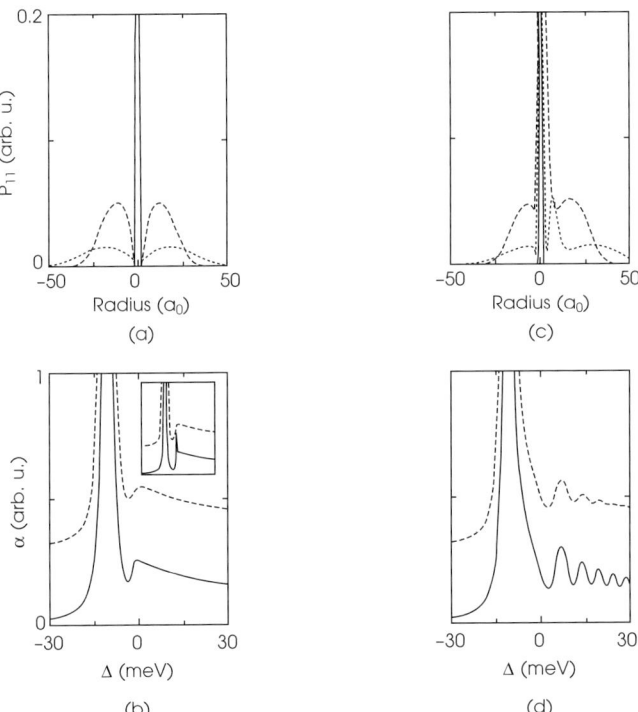

Fig. 3.5 (a) Wave packet at several times (solid curve: t = 0 fs, dashed curve: t = 400 fs, and dotted curve: t = 600 fs). (b) Optical absorption spectrum versus detuning from the band gap without (solid line) and with inhomogeneous broadening (dashed line). The dotted curve has been shifted vertically for clarity. The inset shows identical calculations but with a dephasing time of 2 ps. (c) As in (a) but with a field energy (eFa_0) of $0.8E_0$. (d) As in (b) but with a peak field energy of $0.8E_0$.

DOS [13,29]. The 1s binding energy is about 11 meV, 2–3 times the bulk value. We also show the absorption spectrum in the presence of 2-meV inhomogeneous broadening [35] (dashed line). For this calculation we carried out a standard convolution with a normalized Gaussian distribution after first obtaining the original absorption spectrum. As an additional simulation, we also calculate an absorption spectrum with a long dephasing time of 2 ps in the absence of an external field **F**, shown as an inset to Fig. 3.5(b); the 2s exciton is clearly resolved though inhomogeneous broadening apparently destroys any signatures of this resonance. The presence of inhomogeneous broadening again suppresses the above gap oscillations. The WP snapshots again highlight the physics; shortly after the pulse arrives there is a strong probability for the electrons and holes to be Coulombically bound, hence the sharp peak near the center of the relative motion space. At later times the WP spreads and quantum beating occurs between the continuum states and the excitonic states. Ultimately, the WP speads out and dephases.

Next we add an electric field aligned along the QWW axis (x-direction). In Fig. 3.5(c) and 3.5(d) we respectively display WP snapshots and the resulting absorption spectrum for a field energy of $0.8E_0$ (corresponding field strength of 2.4 kV/cm). These energies are quoted in terms of eFa_0, where E_0 and a_0 are the 1s Bohr radius and binding energies, defined earlier. In addition to a reduction of the 1s oscillator strength (of about 50%) and complete ionization of the 2s exciton, the free carrier–continuum–portion of spectrum exhibits pronounced oscillations that decrease in amplitude for higher energies. The WP is now strongly distorted, asymmetric, and propagates off to the right.

Now we employ large static fields and compare to the free carrier results, with energies greater than the quantum wire 1s binding energy. Specifically we adopt the field energy of $4E_0$ (12 kV/cm). In this regime, complete exciton ionization is expected. How do exciton effects modify the spectrum above the band gap since one expects signatures of a singularity for the free carrier results? Figure 3.6(a) shows snapshots of the WP at $t = 0$ fs and $t = 400$ fs; substantial interference effects are now evident for the propagating WP. In Fig. 3.6(b) is shown the corresponding absorption spectrum. Very large Franz–Keldysh oscillations now occur even in the presence of inhomogeneous broadening. Figures 3.6(c) and 3.6(d) show identical calculations but without the Coulomb interaction. Firstly a sharp discontinuity is not obtained as a result of field-induced tunneling, although the spectrum does tend to rise more and more for frequencies approaching the band gap. The oscillations have a slightly different frequency separation and larger amplitude in the free carrier case since the Sommerfeld factor tends to inhibit free particle absorption. The free-carrier WP smoothly propagates to the right, spreads, and dephases. Further numerical investigation (see the inset to Fig. 3.6(d)) at even higher fields results in a deeper modulation of the oscillations and an increase in the width of the oscillations. Moreover, field-induced transparancy can be achieved at particular spectral positions; there is also a continuing shift of the free carrier spectrum to lower energies (not shown). We note that the tunability of the peaks and the deep modulations are unique for QWWs and cannot be obtained for 2d and 3d semiconductors because of their relatively large background absorption (dependent on the DOS) and lower tunneling rates. Indeed for QWs and bulk semiconductors the Franz–Keldysh oscillations are typically rather small even for very large field strengths [13]. The reason for the present, substantial oscillations is because ionization tuneling in 1d systems is apparently much easier. Further, the results obtained with a dephasing time of 2 ps or in the limit of no background dephasing are almost identical to those above indicating that the broadening is primarily due to field-induced tunneling. The broadening can be understood as an

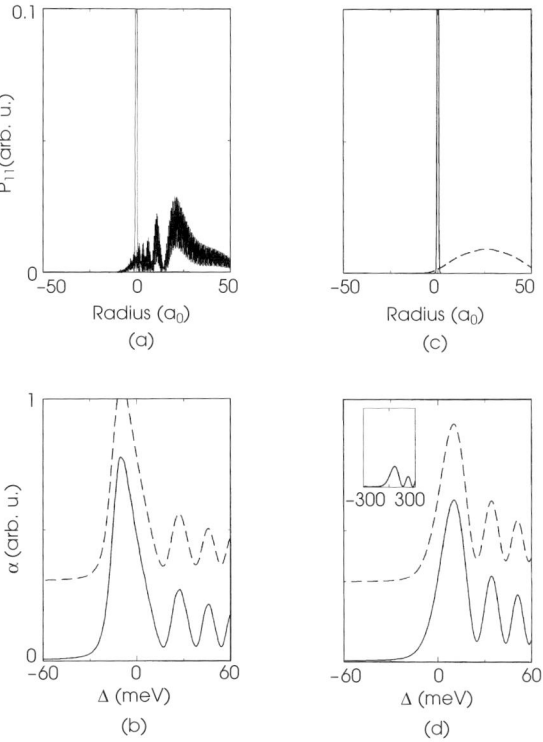

Fig. 3.6 (a) Wave packet at several times (solid curve: $t = 0$ fs, and dashed curve: $t = 400$ fs). (b) Optical absorption spectrum (with and without inhomogeneous broadening). (c) As in (a) but without excitons. (d) As in (b) but without excitons; the inset shows the absorption spectra obtained for the higher field strength of $16E_0$.

uncertainty principle effect; soon after the excitons are created, they are destroyed by field ionization, and hence the energy resonance is broadened by lifetime broadening [16].

Finally we employ a multi-subband model with 1 conduction subband (1e) and 4 valence subbands (1hh, 2hh, 3hh, 4hh). Since we are dealing with symmetric wires for this study only two WPs contribute to the optical properties at low density, P_{11} and P_{13}. Consequently we now have to solve two polarization equations that are coupled through the Coulomb matrix elements. Full Coulombic coupling is taken into account self-consistently by calculating the appropriate Coulomb and interband matrix elements. Figure 3.7(a) depicts the WP snapshots at $t = 0$ fs and $t = 400$ fs for both P_{11} and P_{13}; note that the scales are adjusted accordingly to account for the different dipole matrix elements. The corresponding absorption spectra (Fig. 3.7(b)) now exhibit 2 peaks separated by $6E_0$ (2 times the subband spacing). Figures 3.7(c) and 3.7(d) depict identical calculations with the field strength of $4E_0$. Surprisingly the structure of the second allowed transition

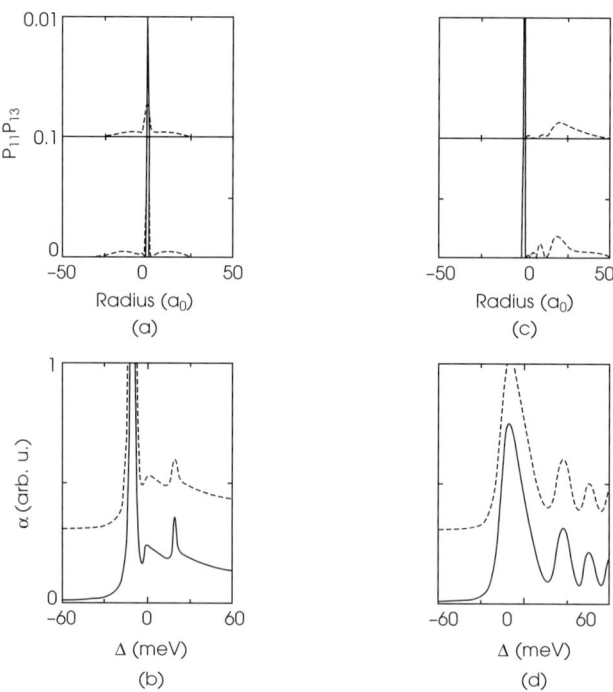

Fig. 3.7 (a) Multi-subband WPs P_{11} and P_{13} at several times (solid line: $t = 0$ fs, and dashed line: $t = 400$ fs. The dephasing time is 500 fs). (b) Optical absorption spectrum (with and without inhomogeneous broadening). (c) As in (a) but with a peak field energy of $4E_0$. (d) As in (b) but with a peak field energy of $4E_0$.

is completely washed out and the results are almost identical to those shown in Fig. 3.6(d). Finally, in relation to high-field transport experiments [36,37], effects such as transient velocity overshoot will not show up in our present study since we are dealing with the optical properties – where the only contributing spatial dynamics is at $x = 0$.

Dynamic Franz–Keldysh effect in quantum wires

In this subsection we investigate effects that can occur when the applied driving field becomes time dependent, that is when the QWW is subjected to intense THz-frequency electric fields $\mathbf{F}_{\text{THz}}(t)$, similar to the QW case described earlier. The theoretical treatment is identical to above, the only difference is that the applied electric field now oscillates at THz frequencies, e.g. $\mathbf{F}_{\text{THz}}(t) = \hat{\mathbf{n}}_x F_0 \sin(\Omega t + \phi)$, with $\hat{\mathbf{n}}_x$ a unit vector in the QWW axis, and F_0 is taken to be 8 kV/cm with a corresponding field energy $E = 3E_0$.

In what follows, we use the multi-subband model with 1 conduction subband and 4 valence subbands (see above). We mention, however, that essentially all the important dynamics can be obtained by assuming a two-subband semiconductor model (*ab* = le-lhh) since P_{11} is by far the predominant WP for this particular study; the additional subbands are included for completeness. The dephasing time is again 500 fs.

We first calculate absorption properties in the absence of an external field $\mathbf{F}(t)$. Figure 3.8(a) shows the input pulse (dashed curve), transmitted pulse (solid curve), and their ratio (dotted curve) which gives a measure of the spectral absorption. We note that a proper cw absorption spectrum is not well defined in the presence of a fixed phase THz field, so we choose the above approach to highlight phase effects. Although we could do phase averaging as in the QW case we explicitly want to domonstrate coherent control. Next we add a THz-frequency electric field aligned along the QWW axis (x-direction). In Figs 3.8(b), 3.8(c) and 3.8(d) we show the input and transmitted irradiances for $\Omega = 2$ THz, 2.5 THz, and 3 THz, respectively; with the phase $\phi = 0$. In Fig. 3.8(b), not only do we see spectral gain (which is discussed in more detail below) but, in addition to a reduction of the 1s oscillator strength, there is a clear absorption splitting of the fundamental exciton resonance. A single THz photon resonance from this splitting lies just below the band edge, that is the splitting energy plus a THz photon energy lies just below the band gap. Therefore what we are obtaining here is analogous to the well known Autler-Townes splitting for three-level atoms. The basic physics behind this phenomenon is to introduce a third level to a two-level system whereby the third level coherently couples to one of the other levels. Thus when two appropriately tuned fields are applied, the usual absorption spectrum has a large dip in its absorption resonance due to the coherent coupling to the third level. In the present case however the third level coherence is achieved via intraband coupling of the lowest (1s) and higher exciton states lying just below the band edge. In essence the higher lying exciton states are now coherently coupled to the lowest state by the THz field. To clarify this mechanism we see that the region of induced transmission (local resonance minimum) gradually red shifts proportionally with an increasing THz photon energy and eventually disappears when the THz photon energy becomes greater than the 1s exciton binding energy; to highlight the red shift, for clarity an arrow is shown in Fig. 3.8(d). This is expected as there is no dipole coupling between exciton and continuum states, and the effect of the higher subband does not play any significant role for our chosen QWW.

Now we investigate the observation of two-photon gain which results in a small gain in the absorption spectrum. The resulting effects will be quite distinct from the

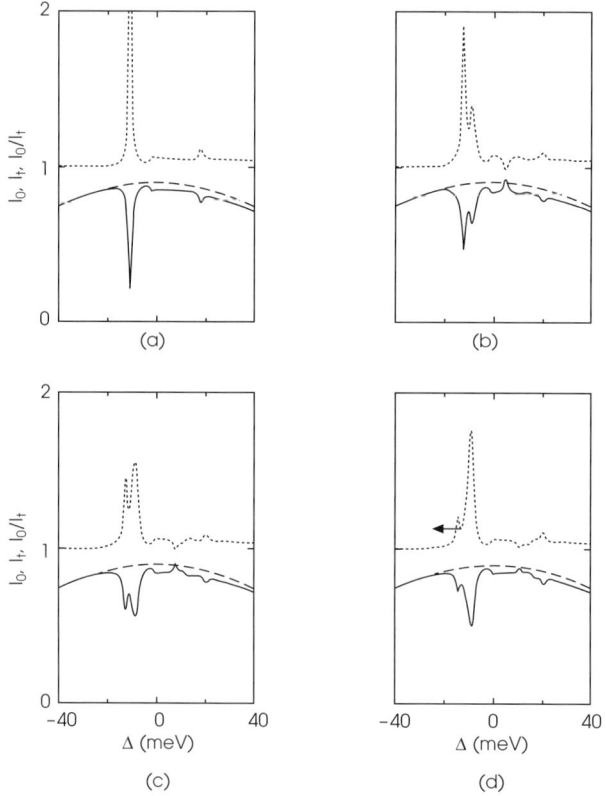

Fig. 3.8 (a) In the absence of $\mathbf{F}(t)$: Input pulse (dashed curve), transmitted pulse (solid curve), and their ratio (dotted curve) which gives a measure of the absorption. (b) As in (a) with $\mathbf{F}(t)$ applied and $\Omega = 2$ THz. (c) As in (a) with $\mathbf{F}(t)$ applied and $\Omega = 2.5$ THz. (d) As in (a) with $\mathbf{F}(t)$ applied and $\Omega = 3$ THz; the arrow shows the red shift of the local minimum (see text).

above one-photon scenario as they will appear as large changes in the spectrum in the region 2Ω above the $1s$ resonance in the continuum. Figures 3.9(a) and 3.9(b) respectively show the $\Omega = 2$ THz simulations with a phase of $\phi = 0$ and $\phi = \pi/2$. As can be recognized the two-photon replica can be one of induced gain or induced absorption, depending on the initial phase of the WPs.

In Figs 3.9(c) and 3.9(d) are shown snapshots of the polarization WP ($|P_{11}(x, t)|$) at the respective times of 200 fs (solid curve) and 600 fs (dashed curve). The snapshot at time $t = 200$ fs depicts the WPs shortly after the optical pulse has gone where there is a strong probability for the electrons and holes to be Coulombically bound. At later times the WP spreads and quantum beating occurs between the continuum states and the excitonic states, as in the QW case. Once again, since the WPs are highly anisotropic, there is a net dipole moment which results in the emission of THz radiation, is discussed below.

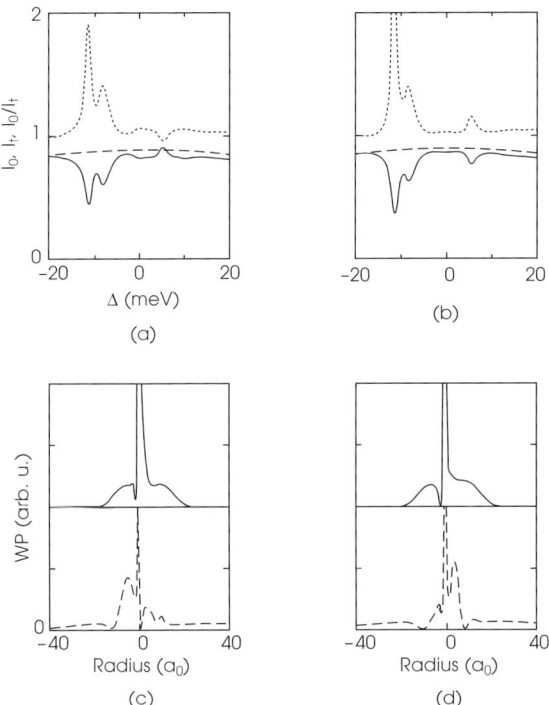

Fig. 3.9 (a) Input pulse (dashed curve), transmitted pulse (solid curve), and their ratio (dotted curve) which gives a measure of the absorption. The THz frequency is $\Omega = 2$ THz with $\phi = 0$. (b) As in (a) with $\phi = \pi/2$. (c) Wave packet (P_{11}) at several times (solid curve: $t = 200$ fs, and dashed curve: $t = 600$ fs) corresponding to (a). (d) WP (P_{11}) at several times (solid curve: $t = 200$ fs, and dashed curve: $t = 600$ fs) corresponding to (b).

In Fig. 3.10 we show the emitted THz field versus t for the phases (at $t = 0$, center of the optical pulse) of $\phi = 0$ (a) and $\pi/2$ (b), for the 2 THz field. Each transient is approximately 2 ps in duration reflecting the combined effect of WP spreading and dephasing. We note that the time $t = 0$ corresponds to the center of the optical pulse and hence has a finite contribution. In the corresponding EM spectra (Figs 3.10(c) and 3.10(d)), a series of harmonics separated in frequency by approximately 2Ω appear in the THz regime. (Of course they are not exactly spaced by 2Ω as here we are dealing with broadband pulses and non-perturbative effects.) Once more, the spectra do not seem to form any plateaus [38] (an extensive region of similar spectral intensity in frequency extending well above the fundamental excitonic binding energy), as in the QW study. But the range of additional frequency components is sufficiently large to temper experimental investigations.

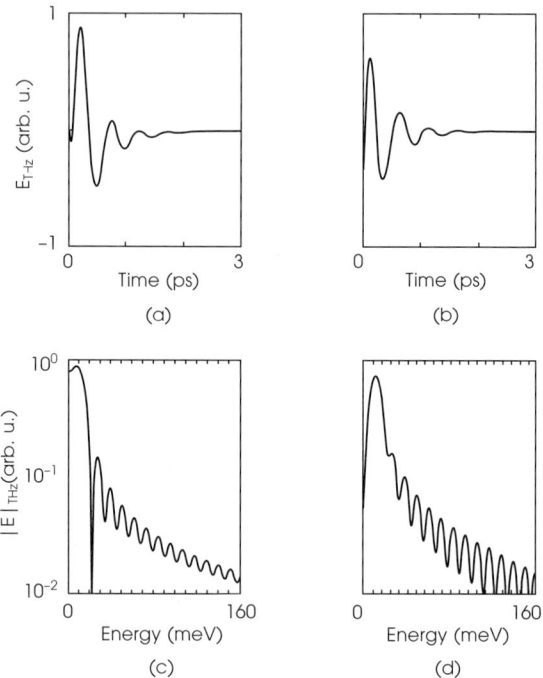

Fig. 3.10 (a) Emitted THz field as a function of time with $\phi = 0$ and $\Omega = 2$ THz. (b) As in (a) with $\phi = \pi/2$. (c) Corresponding spectrum to (a) [A log scale is used for clarity]. (d) Corresponding spectrum to (b).

We now briefly summarize the THz-frequency driving results: for Rydberg atoms [39], higher frequency harmonics are produced from continuum-state to bound-state transitions, in which electrons release the energy absorbed from the field during its excursion in the continuum. A similar mechanism is at work here when the ponderomotive energy of the e–h pair becomes comparable or greater than the bound states' energy. Infact, our value of $F_0 = 8$ kV/cm was chosen to produce ponderomotive energies comparable to the $1s$ binding energy.

3.4 Excitonic trapping, ultrafast population transfer, and Rabi flopping

Progress in the generation and exploitation of ultrashort sub-ps lasers in various regimes of the EM spectrum has given rise to innovative methods for research on materials, modern optoelectronics, and time-domain transient electrodynamics. In the nonlinear optical regime, the continued development of high-intensity ultrashort optical pulses combined with outstanding crystal growth techniques has led to a plethora of experimental techniques to probe the coherent carrier dynamics of low dimensional semiconductors.

In this section we will explore Rabi flopping, excitonic-state trapping (EST), and quasi-adiabatic population transfer (QAPT) dynamics. We discuss in detail conditions to achieve EST and QAPT even when a broadband pulse excites both excitons and free e–h pairs [40]. These effects are well known in atomic media [5], and coherent population trapping effect has been widely used in adiabatic population transfer, electromagnetically induced transparency, and velocity selective cooling. For solids, as pointed out recently by using adiabatic population transfer between heavy- and light-hole bands (to mimic a three-level atom) [41], the exact analog to the coherent-population-trapped state (dark state) used in atomic physics is not possible in semiconductors, but similar physics can indeed take place.

Theory of high optical field effects in quantum wells

In the following, the high-density regime comes into play because of intense optically-excited e–h pairs. Therefore, the Coulomb scattering between the carriers must be taken into account. In addition to the terms that result from the time-dependent Hartree–Fock approximation, the carrier–carrier and carrier–phonon collisions drive the nonequilibrium carrier distribution functions towards quasi-equilibrium Fermi functions and yield optical dephasing. Subsequently, the carrier–carrier collisions (Coulomb correlation terms) are calculated from the e–h quantum Boltzmann equations including also a correlation field, nondiagonal dephasing, and polarization scattering [42,43]. The influence of carrier–phonon interactions occurs over much longer time scales than those studied in this work and can be safely neglected. The required contributions can be calculated by factorizing the relevant 6-particle expectation values [44] or within a Greens function method [45]. For our theoretical approach we again assume the validity of the rotating-wave approximation and thus neglect the possibility of carrier-wave Rabi flopping [12] – a subject discussed in the next section. We further assume a two-subband QW where each e–h state within a certain band-structure with a wave number **k** contributes to the total optical polarization, as described earlier.

Henceforth we work in **k**-space for obvious reasons. To this end, we solve self-consistently the coherent SBE and the carrier–carrier correlations using the second-order Born and Markovian approximation for carrier in- and out-scattering. These contributions are well documented in the literature and essentially form the microscopic, parameter-free counterpart to the relaxation-time approximation [42,45]. The correlation contributions to the polarization function equations

$$\dot{p}|_{\text{corr}} = -\Gamma^{d}_{\text{corr}}(\mathbf{k},f)p_{\mathbf{k}} + \sum_{\mathbf{q}} \Gamma^{\text{nd}}_{\text{corr}}(\mathbf{k},\mathbf{q},f)p_{\mathbf{k}+\mathbf{q}}, \qquad (3.22)$$

where the diagonal dephasing rate Γ^{d}_{corr}, which accounts for loss of coherence of $p_{\mathbf{k}}$, can be written

$$\Gamma^{d}_{\text{corr}}(\mathbf{k},f) = 2 \sum_{\substack{\mathbf{k}',\mathbf{q} \\ b,a=e,h}} |W(\mathbf{q})|^2 [f^{a}_{\mathbf{k}+\mathbf{q}}(1-f^{b}_{\mathbf{k}'})f^{b}_{\mathbf{k}'-\mathbf{q}} + f \leftrightarrow (1-f) - p_{\mathbf{k}'}p_{\mathbf{k}'-\mathbf{q}*}]$$

$$\times \zeta \, [\varepsilon_{a}(\mathbf{k}) + \varepsilon_{b}(\mathbf{k}') - \varepsilon_{a}(|\mathbf{k}+\mathbf{q}|) - \varepsilon_{b}(|\mathbf{k}'-\mathbf{q}|)]. \qquad (3.23)$$

While the non-diagonal scattering rate is given by

$$\Gamma^{\text{nd}}_{\text{corr}}(\mathbf{k},\mathbf{q},f) = 2 \sum_{\substack{\mathbf{k}' \\ b,a=e,h}} |W(\mathbf{q})|^2 [f^{a}_{\mathbf{k}}(1-f^{b}_{\mathbf{k}'-\mathbf{q}})f^{b}_{\mathbf{k}'} + f \leftrightarrow (1-f) - p^{*}_{\mathbf{k}'}p_{\mathbf{k}'-\mathbf{q}}]$$

$$\times \zeta \, [\varepsilon_{a}(\mathbf{k}) + \varepsilon_{b}(\mathbf{k}') - \varepsilon_{a}(|\mathbf{k}+\mathbf{q}|) - \varepsilon_{b}(|\mathbf{k}'-\mathbf{q}|)), \qquad (3.24)$$

where $\zeta(x) = \pi\delta(x) + iP/x$, (P denotes the Principal value), $W_{\mathbf{q}}$ is the screened Coulomb potential (treated here within a quasi-static approximation), ε_{a} is the energy dispersion of the electrons or holes, and the pp terms account for polarization scattering. The essential point is that the nondiagonal dephasing reduces the dephasing rates in such a way that coherent transient effects become important for light propagation studies.

Similarly, the correlation contributions to the density equations read

$$f^{a}_{\mathbf{k}}|^{\text{in}}_{\text{corr}} = \Gamma^{\text{in}}_{\text{corr}}(\mathbf{k},f)(1-f^{a}_{\mathbf{k}}) - \Gamma^{\text{out}}_{\text{corr}}(\mathbf{k},f)f^{a}_{\mathbf{k}}$$

$$+ (p_{\mathbf{k}}\Omega^{*}_{\mathbf{k}}|_{\text{corr}} + c.c.) + (Q^{*}_{\mathbf{k}} + c.c.), \qquad (3.25)$$

where $\Gamma^{\text{in}}_{\text{corr}}$ and $\Gamma^{\text{out}}_{\text{corr}}$ ($a = e, h$) are the expressions for *in* and *out* scattering, for example

$$\Gamma_{\text{corr}}^{\text{in}}(\mathbf{k}, f) = 4 \sum_{\substack{\mathbf{k}',\mathbf{q} \\ b=e,h}} |W(\mathbf{q})|^2 f_{\mathbf{k}+\mathbf{q}}^a [(1 - f_{\mathbf{k}'}^b) f_{\mathbf{k}'-\mathbf{q}}^b]$$

$$\times \text{Re} \{\zeta[\varepsilon_a(\mathbf{k}) + \varepsilon_b(\mathbf{k}') - \varepsilon_a(|\mathbf{k}+\mathbf{q}|) - \varepsilon_b(|\mathbf{k}'-\mathbf{q}|)]\}, \quad (3.26)$$

$$\Omega_{\mathbf{k}}^*|_{\text{corr}} = 2 \sum_{\substack{\mathbf{k}',\mathbf{q} \\ b=e,h}} |W(\mathbf{q})|^2 (f_{\mathbf{k}'-\mathbf{q}}^b - f_{\mathbf{k}'}^b) p_{\mathbf{k}+\mathbf{q}}^*$$

$$\times \zeta^*[\varepsilon_{\tilde{a}}(\mathbf{k}) + \varepsilon_b(\mathbf{k}') - \varepsilon_{\tilde{a}}(|\mathbf{k}+\mathbf{q}|) - \varepsilon_b(|\mathbf{k}'-\mathbf{q}|)], \quad (3.27)$$

$$\Omega_{\mathbf{k}}^* = 2 \sum_{\substack{\mathbf{k}',\mathbf{q} \\ b=e,h}} |W(\mathbf{q})|^2 (f_{\mathbf{k}}^a - f_{\mathbf{k}+\mathbf{q}}^a) p_{\mathbf{k}'} p_{\mathbf{k}'-\mathbf{q}}^*$$

$$\times \zeta^*[\varepsilon_a(\mathbf{k}) + \varepsilon_b(\mathbf{k}') - \varepsilon_a(|\mathbf{k}+\mathbf{q}|) - \varepsilon_b(|\mathbf{k}'-\mathbf{q}|)], \quad (3.28)$$

(if $a = h$, then $\tilde{a} = e$ and vice versa) and $\Gamma_{\text{corr}}^{\text{out}}$ is obtained by replacing f by $1 - f$.

Besides optical absorption studies, nondiagonal dephasing has been shown to remove the well known theoretical artifact of predicting absorption below the bandgap in a semiconductor gain spectrum, that arises from a Lorentzian lineshape assumption [43]. Previous simulations that incorporate the above scattering contributions have also been successful, for example, in predicting four-wave mixing polarization times [45–47]. This type of dephasing is also an important factor in explaining ultrafast nonlinearities in semiconductor optical amplifiers and their microscopic origins [42,48,49]. It is equally important to descibe Rabi flopping and the like that we address below.

Excitonic trapping and ultrafast population transfer

We assume input optical pulses of the form $\mathbf{E}(\mathbf{r}, t) = \tilde{\varepsilon}_{\text{Opt}}(t) e^{-i[\omega_0 t + k_0 z + \phi(t)]}$ + c.c., polarized in the plane of the QW. Experimentally, $\text{In}_x\text{Ga}_{1-x}\text{As/GaAs}$ QWs are advantageous since with compressive strain one can increase the splitting of the heavy- and light-hole exciton and thus neglect the light-hole states, therefore validating the two-band model. In analogy with the atomic systems we modulate the input field by the phase factor $\phi(t) = M \sin(\Omega_m t)$, with M and Ω_m the index and frequency of modulation. For a TLA (two-level atom) the resonant part of the interaction Hamiltonian vanishes if M is chosen such that the Bessel function $J_0(M) = 0$, and one is left with the nonresonant rapidly oscillating terms [5]. The solid-state community may recall that a somewhat similar criterion has been derived to realize dynamic localization in infinite lattices driven by a harmonic time-dependent electric field [50,51].

In the TLA it is further known that the two bare levels cross at the times $t_n = n\pi/2\Omega_m$ (n = integer). The crossing of energy levels occurs quite naturally in the instantaneous frame of the modulated field. The population distribution at the crossings can be understood semiquantitatively by integrating numerically the time-dependent Schrödinger equation or by employing Landau–Zener theory [5]. For semiconductors the situation is much more complex, whereby the optical field provides coupling between two bands, and a modulated near-resonant field leads to coupling of higher spectral components of the field to larger **k**-states in the band, and the field components below the resonance provide a nonresonant coupling to the excitonic transition. Moreover, the Coulomb interaction yields exciton and plasma induced many-body effects. These conditions prohibit us from having a simple set of criteria for trapping. Thus we propose here a novel spectral distribution for the field which has its central frequency detuned far below the excitonic resonance, and, by strategic sweeping of the instantaneous carrier frequency, dynamical population trapping can be realized. As in the TLA case we do not necessarily have to choose M to be strictly a zero of the Bessel function.

Concerning the modulation frequency, we choose $\Omega_m = 8$ meV unless stated otherwise. For the *unmodulated* optical field, $\tilde{\varepsilon}_{Opt}(t)$, we employ a 150-fs pulse excited at the 1s exciton peak. In Fig. 3.11 we depict the spectral irradiance of the unmodulated

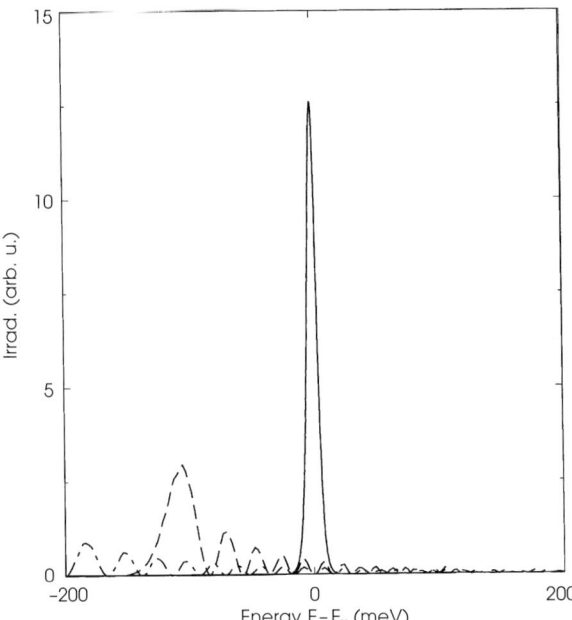

Fig. 3.11 Irradiance spectra for a frequency-modulated and unmodulated 150 fs optical pulse. The solid, dashed, and chain lines correspond to $M = 0$, $M = 14.9$, and $M = 30.6$, respectively. E_x is the bulk 1s binding energy.

and frequency-modulated fields. With modulation, a much larger bandwidth can be obtained with a series of larger peaks extending far below the exciton resonance; the injected pulse profile is sufficiently broad to couple many different excitation modes. One recognizes that the dominant spectral components are well below the band edge. By increasing the index of modulation M, the spectral width of the pulse increases. In contrast, without any modulation the spectrum of the excitation pulse is significantly narrower and peaked at the central frequency as shown later. Although such a modulation is difficult to obtain experimentally, one only requires about one approximate oscillation to obtain the appropriate energy level crossings. Indeed one can also achieve trapping using suitably chosen linear chirps (discussed below).

For an optical pulse with a Rabi energy of 30 meV, the area (integral of its Rabi frequency over time) of the pulse is approximately 6π (irradiance, $\sim$ 1 GW/cm^2). Figure 3.12(a) shows the corresponding excitation-induced density for the QW optically excited at the 1s exciton resonance using (i) an unmodulated input pulse [the solid curve depicts the carrier-density showing the familiar Rabi flopping [9,52] of the density (complete inversion is not possible due to Coulomb processes)]; and (ii) a frequency-modulation of the input pulse for two different values of the index of modulation M: the dashed and dotted curves are markedly different from the unmodulated case and display strong EST and ultrafast QAPT. The values of M were chosen in accord with the original work in atoms, while the pulse area was chosen to be large enough to drive the population into trapping (it is also close to typical experimental values). During the time interval from

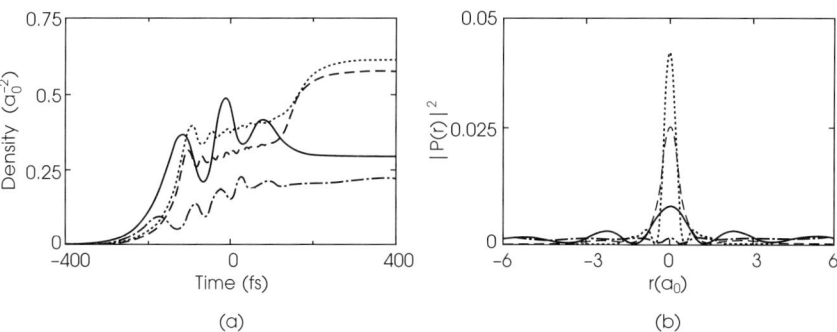

Fig. 3.12 (a) Pulse-induced carrier density (in units of inverse Bohr radius squared) as a function of time with [dashed (M = 30.6), dotted (M = 14.9), and chain curves (M = 14.9, but with a frequency modulation of $\Omega_m/2$) (see text)] and without frequency-modulation (solid curve). Excitonic-state trapping and ultrafast population transfer is clearly discerned. (b) The polarization density corresponding to the dashed case in (a) at several snapshots: t = -260 fs (solid curve), -80 fs (dashed curve), 100 fs (dotted curve), and 280 fs (chain curve).

about −160 fs to 160 fs, the population remains trapped in the exciton level, eventually jumping out in the wake of the exciton-continuum quasi-adiabatic crossing. For the case of $\Omega_m/2$ and $M = 14.9$ (chain curve), the trapping efficiency is much less as expected and QAPT does not occur, since the crossing occurs just once and the on-resonance field components are not sufficient to create the trapping-like feature; however, even in a regime where the pulse irradiance is negligible there is some evidence for small density changes at around 360 fs. In all cases there are signatures of phase intereference due to coherent carrier evolution along different interfering pathways and excitation-induced dephasing. We would like to point out that to obtain such good Rabi flopping one must include dephasing at a microscopic level that includes both nondiagonal dephasing and polarization scattering [42]; these contributions from carrier-carrier scattering reduce the interband optical linewidths of the higher $\mathbf{k}$-states and thus limit the higher-energy continuum occupations in comparison to the pure dephasing [13]. Physically, this is important since the leading edge of the pulse, which is detuned below the exciton resonance, prepares the system for the excitation of real population. A state and energy-independent dephasing time would result in erroneous large dephasing of the initially virtual excitation and therefore suppress the QAPT and EST.

To highlight the difference between the Rabi flopping and trapping we depict in Fig. 3.12(b) the polarization density (WP), $|P(\mathbf{r}, t)|^2$ with $\mathbf{r} = \mathbf{r}_e - \mathbf{r}_h$ at various temporal snapshots. During the period −260 fs to 100 fs the population becomes strongly trapped in the exciton state indicated in the figure by a high probability of finding the electron and hole at the same relative position. However at the later time of 280 fs, after the density increases rapidly (we mention that this increase is almost step-like in the absence of dephasing), the excitonic probability *decreases* substantially and the WP spreads out significantly, demonstrating that the population is no longer trapped. This dramatic spreading of the WP arises due to the modulation and the resulting crossing of carriers into the continuum. The above picture sheds much more light on the trapping scenario than, for example, $f_\mathbf{k}^{e,h}$ or $p_\mathbf{k}$, which do not clearly distinguish between Coulomb-bound excitons and free carriers in the continuum. The spatial polarization dynamics at $\mathbf{r} = 0$ may in fact be probed experimentally using conventional four-wave mixing techniques. We observe here a method to control, coherently, the excitonic WP by tailoring the conditions of the energy level crossing. To probe the entire spatial dynamics one would need to, e.g. couple a THz field with the optical field; this would involve a significantly more complex analysis. We also mention that to directly detect density changes, experimentally pulse-propagation studies are very difficult and one should, for example, use the technique reported recently in Reference [11]. This allows for the detection of density changes via simple differential transmission changes of a probe pulse.

We now attempt to clarify the physics using a simple energy crossing model: Adding a frequency-modulation to the excitation field is equivalent to modulating the energy separation between the ground and exciton/continuum states. In Fig. 3.13 we show, schematically, the various crossings of energy levels that lead to quasi-adiabatic transfer of population at these crossings. The crossings "B–D" transform into anti-crossings (solid lines) due to the coupling with the effective Rabi field. In contrast the crossing at "A" is unaffected due to the weak coupling of the $v \leftrightarrow c$ transition on the leading edge of the pulse. The modulation of the dressed energy levels results from transforming into a frame corresponding to the instantaneous field frequency, and are given as $\Sigma_c = -[\Delta_{\mathbf{k}}^2 - \Delta_x^1]$, $\Sigma_x = 0$ and $\Sigma_v = -[\Delta_x^2 + M\Omega_m \cos(\Omega_m t)]$, for the electrons in the conduction band (c), excitonic state (x), and electrons in the valence band (v), respectively. One can of course depict the diagram in reverse for the holes. Here Δ_x^1 ($\Delta_{\mathbf{k}}^2$) denote the detuning of the central laser field frequency from the excitonic (continuum) energy levels. The coupling of the field to the $v \leftrightarrow x$ and $v \leftrightarrow c$ transitions transforms these crossings into *avoided crossings*. Initially the system will be off resonant and far from any crossing; as the modulation changes, the population is swept through resonance and it evolves quasi-adiabatically into the excitonic state where it displays the trapping feature. The system further encounters two closely spaced crossings, resulting in an enhanced step-like transfer of population into the continuum. The arrows in Fig. 3.13 indicate one possible temporal path (high probability) along which the electrons may evolve ($\Delta_{\mathbf{k}}^2$ is fixed for simplicity). The detuning term $\Delta_{\mathbf{k}}^2 - \Delta_x^1$ is the difference

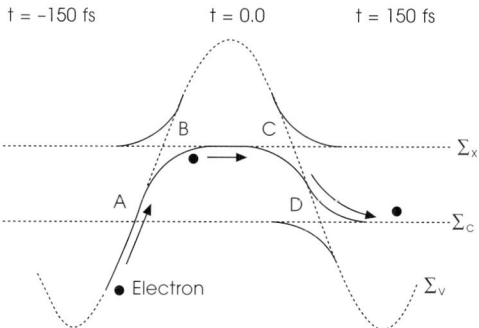

Fig. 3.13 Schematic of the energy level crossing that results from modulation of the field, where "A–D" represent the various crossings. The crossings "B–D" transform into anti-crossings (solid lines) due to the coupling with the field. The energy degeneracy is lifted due to the coupling and is proportional to the coupling $\Omega_{\mathbf{k}}$. In contrast the crossing at "A" is unaffected due to the weak coupling of the $v \leftrightarrow c$ transition on the leading edge of the pulse. The arrows depict the evolution of the population (large black dots).

between the continuum energy levels and frequency of the 1s exciton peak. One should keep in mind, however, that the above model is grossly simplified and many-body effects, included in our numerical results, will complicate things substantially; however our quantitative theoretical study is in fairly good agreement with the above level crossing model.

The essential difference in the trapping criterion between the atomic case and semiconductor case lies in the frequency content of the exciting field. In the atomic case the trapping-like phenomenon with nearly complete inversion results from *correlated sideband excitation* of the atom [53]. The frequency components of the modulated field excite the atom symmetrically about the atomic resonance, resulting in trapping. In the semiconductor, a symmetric frequency content of the excitation does not lead to the desired trapping due to its non-symmetric coupling to the band structure of the semiconductor; effectively the high frequency components of the modulated pulse selectively excite the continuum of states resulting in large carrier generation, thus washing out the trapping. We circumvent this problem by using a modulated pulse with its predominant frequency content away from the band edge.

Next we employ linear-chirped input pulses to discern if some of the same qualitative trapping features (EST and QAPT) can also be obtained using suitably chosen chirp parameters. To best immitate the frequency-modulated spectrum, we choose a frequency chirp of the form: $\omega \rightarrow \omega_x - 120$ meV $+ a_c t$ with a_c being ± 0.5 meV/fs. $\tilde{\varepsilon}_{Opt}(t)$ is identical to before. The chirped-pulse spectrum along with the unchirped spectrum is displayed in Fig. 3.14(b). The pulse-induced density with positive a_c is shown by the dashed line in Fig. 3.14(a) which, although it does not show signs of trapping, does show fast oscillations in the density as well as a rapid increase in the density near the crossing time. For the negative a_c (dotted curve) strong excitonic trapping is again achieved and seems to maintain its trapped state since no further crossing takes place. This is again clear from the simplified zero-order energy-level picture with appropriate crossing of the 1s exciton state, ground crystal state, and continuum states. The slope and sign of the chirp (a_c) gives us a handle to selectively suppress or enhance the excitonic transition. The trapping and population transfer are highlighted in Figs 3.14(c) and 3.14(d) which show the wave-packets at the times −80 fs (solid curve), 100 fs (dashed curve), and 280 fs (dotted curve) for both the negative and positive chirp pulse-excitation. Population transfer is not expected for the negative-chirp case as the zero-order energy-levels never cross.

Before finishing this section, we mention that coherent density oscillations, such as Rabi flopping or trapping, may also be probed or exploited through a novel THz emission scheme in dc-biased QWs [54,55].

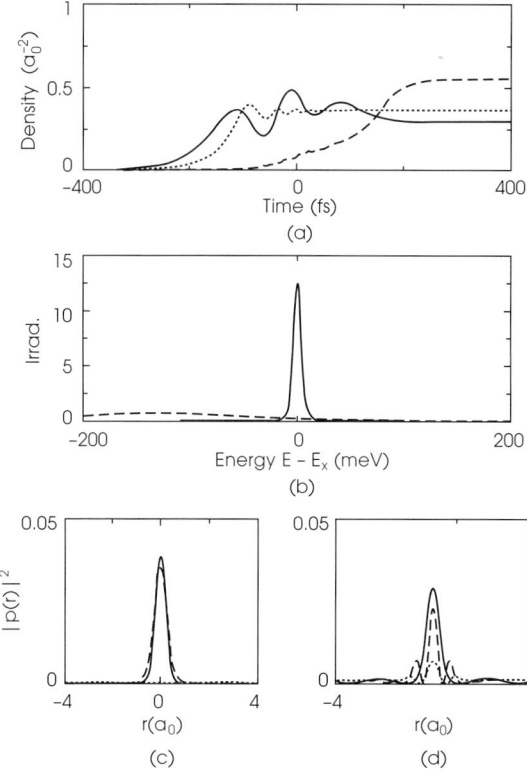

Fig. 3.14 (a) Pulse-induced carrier density as a function of time with [dashed (positive chirp) and dotted (negative chirp) curves] and without (solid line) linear chirp. (b) Corresponding input pulse spectra for (a), with the solid and dashed curves respectively representing unchirped and chirped pulses. (c) The polarization density for the positive chirp corresponding to the dashed case in (a) at several snapshots: −80 fs, 100 fs, and 280 fs. (d) The polarization density for the negative chirp corresponding to the dotted case in (a) at several snapshots: −80 fs, 100 fs, and 280 fs.

3.5 Carrier-wave Rabi flopping

Ultrashort optical pulses have allowed us to directly observe many fascinating nonlinear optical propagation studies including Rabi flopping [9], self-induced transparency [56], and photon echo [57]. From a theoretical viewpoint, all of the aforesaid can be described quite adequately by employing the appropriate coupled matter-Maxwell equations within the slowly-varying-envelope approximation (SVEA), e.g. the envelopes of the electromagnetic field and polarization are assumed to vary little over an optical period and wavelength. However, as optical pulses continue to get shorter and shorter, and materials become smaller and smaller, the SVEA that adopts slowly-varying phase and amplitude components may no longer be a good approximation.

As a matter of fact even with longer sub-ps pulses the theoretical description of Rabi flopping in semiconductors within the SVEA becomes questionable since the linear absorption lengths are typically comparable to the wavelength of light in the material (at least when excitonic processes dominate the response, at low temperature). Beyond that, for several-cycle optical pulses resonantly excited with atoms or semiconductors, one may also need to account for the dynamics on a time scale of the carriers themselves. The traditional analysis of Rabi flopping [52,58,59] assumes that the optical-frequency components of the energy density do not contribute to the nonlinearity. In this section, we return to the problem of Rabi flopping in a nonlinear optical material and present a Maxwell–Bloch analysis beyond the SVEA to model pulses of only a few optical-cycles time duration. In short, sub-carrier effects become important when the light intensity is so large that the period of one Rabi oscillation becomes comparable with the period of the excitation field. Apparent new features that arise in the full Maxwell solution may be termed: *carrier-wave Rabi flopping* (CWRF) [12]. Although this phenomenon was predicted for the TLA a few years ago, experimental signatures have only recently been reported in thin film GaAs [4] (see also Chapter 2 of this book). First, we describe some of the theory and simulation techniques required to account for CWRF and describe sub-carrier nonlinear optical phenomena – a regime long thought to be inaccessible. Then we investigate the simplified TLA and model pulse propagation in optically thick media to demonstrate the clear breakdown of the area theorem, leading to unstable temporal solitons. Finally, we solve Maxwell's curl equations self-consistently with the coherent semiconductor Bloch equations with dephasing, for semiconductor QWs, and make a connection to the experimental measurements and possible future work.

Theory and computation of sub-optical-carrier pulse propagation

We employ a finite-difference time-domain [FDTD] [26,60,61] approach for solving the full-wave Maxwell equations in one dimension, and a fourth-order Runge-Kutta method to solve the optical Bloch or semiconductor Bloch equations. Although we employed an FDTD approach earlier to model nonperturbative wave packets, we were still working within the rotating-wave approximation for the polarization. We begin by describing the simple atomic case to clarify the physics. A plane-wave pulse normally incident upon a TLA material that is unbounded in the transverse direction is considered. Assuming linear and nonlinear polarization, Maxwell's curl equations can be written

$$\frac{\partial B_y}{\partial t} = -\frac{\partial E_x}{\partial z} \tag{3.29}$$

$$\frac{\partial D_x}{\partial t} = \frac{\partial H_y}{\partial z}, \tag{3.30}$$

with $B_y = \mu_0 H_y$, $D_x = \varepsilon_0 E_x + P_x$, and $P_x = \varepsilon_0 (n_0^2 - 1) E_y + P_{nl}$ (n_0 is the background refractive index). The nonlinear polarization $P_{nl} = 2Nd \, \text{Re}[\rho_{12}]$, where N is the density of TLAs, and ρ_{12} is the off-diagonal density matrix element obtained from the optical Bloch equations $\dot{\rho}_{12} = i\Omega n - (\Gamma_2 + i\omega_{12})\rho_{12}$ and $\dot{n} = i2\Omega(\rho_{12} - \rho_{12}^*) - \Gamma_1(n - 1)$. These symbols have their usual meaning: $\Omega = dE$ is the Rabi frequency, $n = (\rho_{11} - \rho_{22})$ the population difference between the lower and upper states, ω_{12} the transition frequency, and d the field-direction dipole moment. The phenomenological population and polarization relaxation rates are given by Γ_1 and Γ_2, respectively. The initial field

$$E_x(z = 0, t) = \hat{\mathbf{n}}_x E_0 \, \text{sech}\left(-\frac{t - \tau_{\text{off}}}{\tau_0}\right) \sin(\omega t), \tag{3.31}$$

where E_0 is the peak input electric field, τ_{off} is the offset position of the pulse center (at $t = 0$), $\tau_p = 2 \, \text{arcosh}(1/\sqrt{0.5}) \tau_0$ is the FWHM of the pulse irradiance profile, $\hat{\mathbf{n}}_x$ is a unit vector perpendicular to the direction of propagation, and ω is the central pulse frequency.

For the semiconductor case, the approach is identical only we must solve the SBE and replace P_{nl} above by $P_{\text{Opt}} = 2 \sum_k \mu_{cv,k} \, \text{Re}[p_k]$. Since we must solve the equations with at least 30 points per carrier wavelength of interest, the semiconductor problem becomes a major computational *tour-de-force*. However, for fields propagating perpendicular to a semiconductor QW the analysis becomes much easier and the qualitative physics behind carrier-wave Rabi flopping can be explained by analyzing the reflection or transmission of the excitation field, in combination with the material properties of the medium.

Breakdown of the area theorem in a two-level atom

For this subsection, we model pulse propagation of various $2l\pi$ optical pulses (where l is an integer) of 18 fs time duration (FWHM irradiance). We closely follow the details in Reference [12]. For a pulse area of 2π the standard area-theorem results are recovered in agreement with the work presented in Reference [60]. However, the standard results for higher area pulses do not hold because of a strong reshaping of the individual optical carriers. We predict that electric field time-derivative effects will lead to CWRF and subsequently to the formation of higher spectral components on the propagating pulse.

Within the SVEA, it is well established that when the envelope of the pulse has an area that is an integer number of 2π, then lossless propagation is possible. For a 2π-pulse, the rotating dipoles are exactly returned to their initial state while maintaining the shape of the excitation pulse, and the hyperbolic secant solution propagates without change at a velocity which can be substantially slower than the speed of light. This will occur when $E_0^{2\pi} = 2/d\tau_0$. Moreover, by virtue of the area theorem [9], $2l\pi$ pulses are also asymptotic solutions to the coherent propagation problem, although such pulses are not stable and will split up into multiple 2π–sech pulses, as a consequence of multiple Rabi flopping. One other feature of coherent flopping is that coherence must be maintained in the system. Hence we choose relaxation times much longer than the input pulse duration, and adopt the following material and laser parameters: $\omega = \omega_{12} = 0.6 \times 10^{15}$ rads^{-1}, $\tau_0 = 10$ fs, $\Gamma_1^{-1} = \Gamma_2^{-1} = 1$ ns, $n_0 = 1$, $d = 2.65$ eÅ, $N = 2 \times 10^{18}$ cm^{-3}, and t_{off} is chosen appropriately to propagate the pulse in time from outside the computational domain (see below).

The peak amplitude of the required pulse to achieve a 2π envelope area is approximately 0.5 GV/m. Figure 3.15(a) shows an example of a propagating 2π-pulse in the TLA-medium. The pulse initially propagates in the free-space region, and thereafter enters the two-level medium at 20 µm; the pulse subsequently propagates the nonlinear medium and exits into the free-space region again at 80 µm. The 2π-pulse simulation approximately recovers the well known analytic results in agreement with Reference [60]: the excitation drives a complete transition of the TLA from its ground state to its excited state and back to its ground state while maintaining its shape. For comparison, Fig. 3.15(b) shows the corresponding temporal development of the inversion n (solid curve), electric field E_x (chain curve), and the polarization component Re[ρ_{12}] (dotted curve), at the fixed position of $z = 21$ µm that is near the input surface of the nonlinear medium. Although the medium is completely inverted and returned to its initial state, oscillation features at the zero points of the pulse arise due to the time-derivative behavior of the input field (see also Reference [60]). For longer propagation distances, these transient features cause local carrier modification though the envelope of the input pulse is essentially unchanged. Maxwell's curl equations also account for backwards propagating fields which do not occur in the present study since the linear refractive index is unity, and also because the absorption length is too large (thus the SVEA in space is valid here).

Next we investigate 4π-pulse excitation, and to emphasize the effects of propagation we have used a nonlinear medium length of 140 µm and $N = 4 \times 10^{18}$ cm^{-3}. Figure

High-field effects in semiconductor nanostructures 75

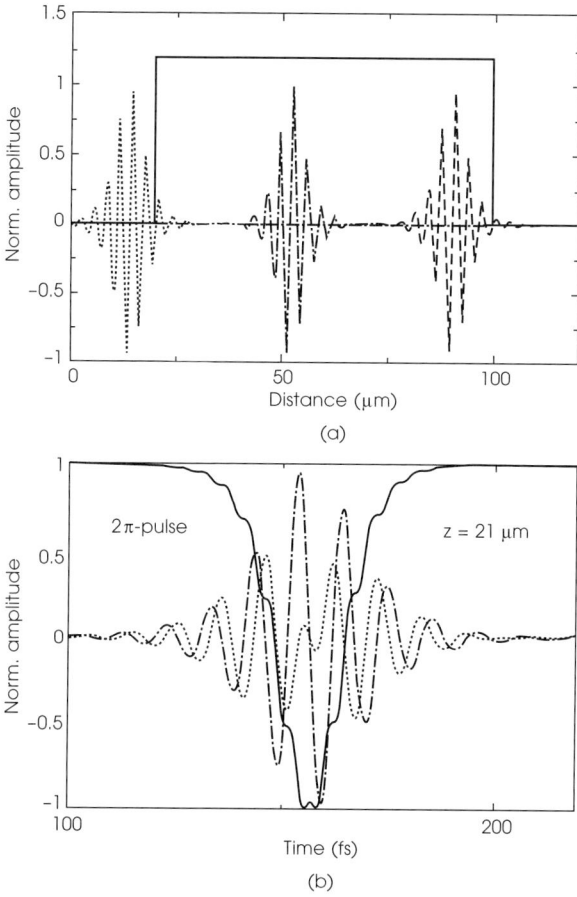

Fig. 3.15 (a) 2π–pulse propagation through the two-level system. The normalized electric field is shown at the respective times of 140 fs (dotted curve), 275 fs (chain curve), and 415 fs (dashed curve). (b) Normalized field (chain curve), inversion n (solid curve), and $\text{Re}[\rho_{12}]$ (dotted curve), near the front face of the two-level material ($z = 21$ µm).

3.16(a) depicts the electric field profile at the respective propagation times of 180 fs, 350 fs, and 525 fs. The driven density shows the expected two symmetric transversals between the ground and excited state. As a consequence of driving 2 complete Rabi flops (see Fig. 3.16(b)), the propagating pulse evolves into two separate pulses with differing spectral profiles. Again this is in agreement with the standard SVEA results. However, sub-carrier transient features arise once more due to the time-derivative nature of the input field.

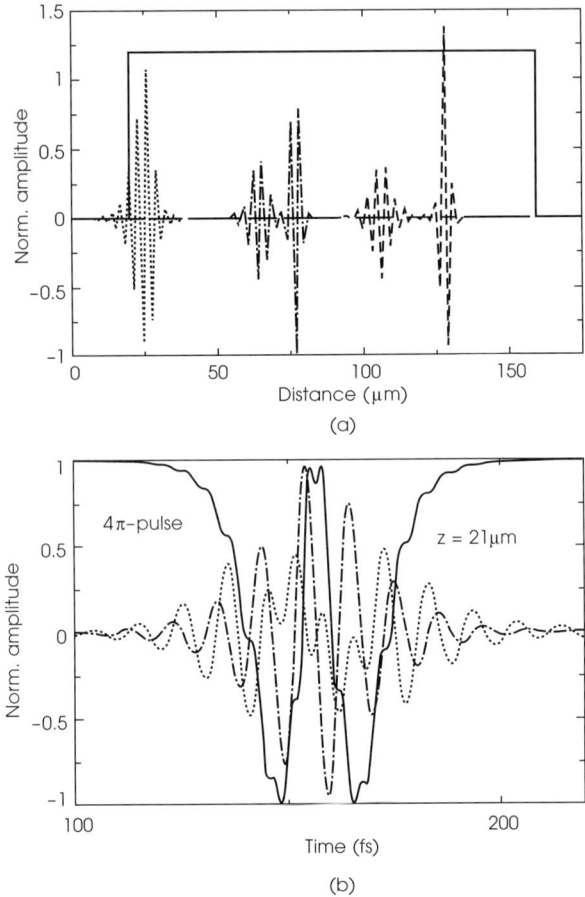

Fig. 3.16 (a) As in Fig. 3.15(a) but for 4π-pulse propagation at the respective times of 180 fs, 350 fs, and 525 fs. (b) As in Fig. 3.15(b) but for 4π-pulse propagation.

What now happens when larger input areas are injected into the material so that the area under the individual carriers may themselves cause Rabi flopping? The left-hand side of Fig. 3.17 shows the time-dependent inversion, E_x, at $z = 21$ µm, for pulse areas of 6π, 8π, 10π, 12π and 14π. The individual carriers now have a profound effect. First, it is noted that incomplete Rabi flops occurs instead of the anticipated integer number. For example, for the 10π-pulse case, 4.5 Rabi flops occur instead of 5. Complete flopping is very difficult to achieve because of the transient features in the Bloch equations beyond the rotating wave approximation. Second, local CWRF is clearly discerned. One finds that Re[ρ_{12}] follows E_x instantaneously so that its peak occurs at the peak in the E_x time derivative.

One consequence of CWRF is that carrier-wave reshaping is expected. To investigate this, the right-hand side of Fig. 3.17 displays the pulse time-profile at the sample exit face ($z = 180$ μm). Strong carrier-wave reshaping is indeed found in addition to a significant interaction with the free-induction decay of the material. For the chosen pulse and material parameters, it is not possible to achieve complete symmetric inversion for areas of and above 6π–pulse excitation, and less than 100% inversion is obtained in

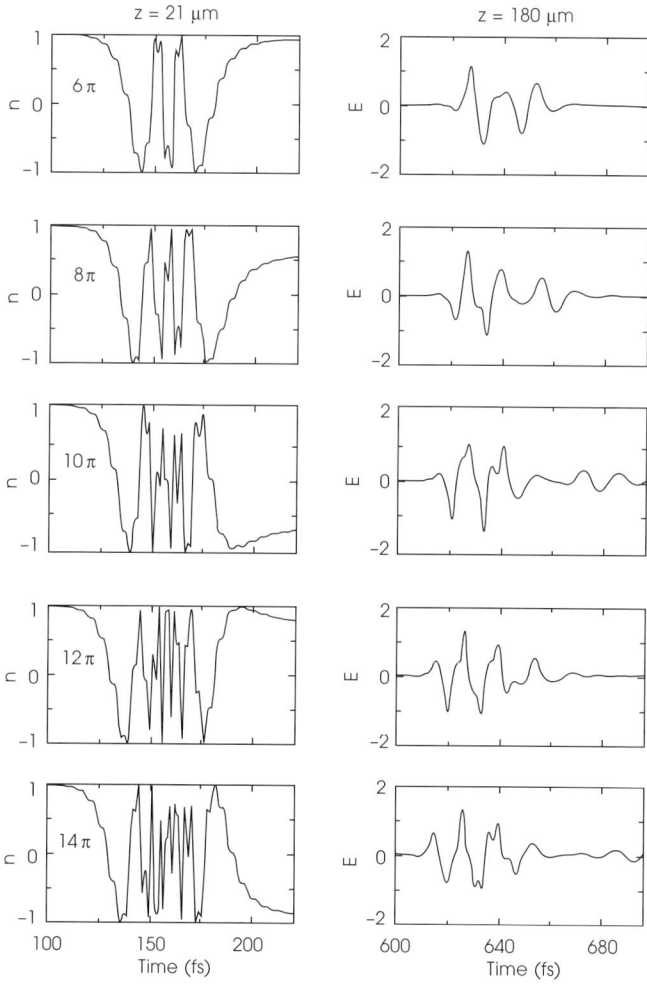

Fig. 3.17 The left graphs show 6–14π pulse-induced population difference near the front face of the two-level material ($z = 21$ μm). The right-side graphs display the corresponding propagated pulse (normalized electric field) at $z = 180$ μm.

each swing of the pulse since the medium is responding rapidly to the variations in the pulse shape and its time derivative. For the 10π–pulse, numerically it is found that an 11.4π–pulse approximately returns the inversion to the ground state while driving 5 density flops; however, these flops are far from symmetrical. The source for the CWRP is due to fast oscillations in the polarization equations outside the rotating wave approximation.

An experimental signature of such an effect could be seen, for example, on the output spectrum of the propagated pulse. Figures 3.18(a) and 3.18(b) show, in comparison to the input spectrum, the output irradiance of 0π– and 10π–pulses, respectively. Figure 3.18(a) reproduces the familiar 0π–pulse scenario in the resonant medium, showing the characteristic spectral hole resulting from a beating structure in the time-dependence;

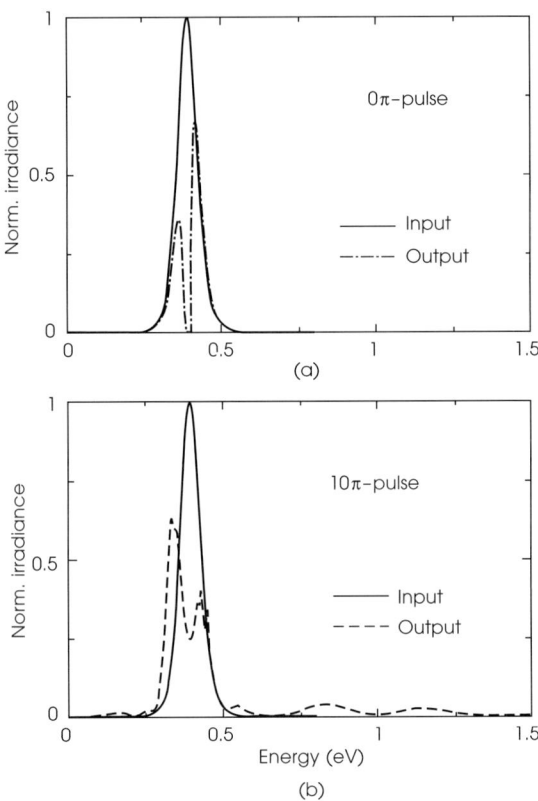

Fig. 3.18 (a) Input and output pulse irradiance-spectra for 0π–pulse propagation. (b) As in (a) but for 10π–pulse propagation.

the asymmetry is due to dephasing. (Note for numerical convenience in calculating the Fourier transform of the rapidly-varying time-dependent field, Γ_2 was increased to 1 ps^{-1} for this simulation.) However, Fig. 3.18(b) clearly shows the formation of higher (and some lower) spectral components due to CWRF. We have verified computationally that, if the input carrier frequency is increased by a factor of 10, then five symmetric transversals occur between the ground and excited state and there is no evidence for time-derivative features and no generation of these higher spectral features. This, of course, is also true if one increases the time duration of the exciting pulses since sufficient optical carriers will again be present to allow the validity of the SVEA in time. That said, for increasing irradiances the area theorem will again breakdown. Further, all the results obtained for the 18 fs pulse scale to much longer pulses, and spectrally more narrow two-level systems if the irradiances increase or the carrier frequency changes accordingly [62]. (This of course ignores exciton-induced dephasing.)

Carrier-wave Rabi flopping in semiconductors

Armed with the propagation results for the TLA, the problem we now face is what happens to a semiconductor material when one excites the system resonantly with extremely intense and extremely short optical pulses, in a regime where CWRF is expected? Fortunately Reference [4] has already addressed this question, experimentally, using 5 fs optical pulses with an area of up to 4π exciting thin film GaAs; see also Chapter 2 in this book. Although these pulses are on the lower area limit to observe the effects shown above, the experimentalists came up with an excellent idea to probe the first signs of sub-carrier coherent density oscillations, by carefully measuring the 3rd-harmonic signal. Additionally, their experimental measurements were qualitatively explained within a similar model to the above, applied to thin sample lengths ($l = 0.6$ μm).

Theoretically, we investigate the many-body system by exploring what happens to the several-cycle optical pulses when they propagate perpendicularly to a QW. This simplifies the numerics considerably, by analysing a different excitation geometry, but still allows us to simultaneously solve the coherent SBE with dephasing, in addition to Maxwell's curl equations. Additionally, our approach has not become a major computational nightmare (which sometimes obscures the physics), but solvable on desktop computers, and allows us to explore the important physical effects expected from the coherent SBE without employing a rotating-wave approximation. Quantum kinetics will almost certainly play an important role on these ultrashort time scales, but is ignored here for our simple semiconductor model study; the role of quantum kinetics will no doubt be reported in the literature in the near future.

We take the following parameters for the laser and QW system: $\omega = E_g - E_0$, where E_g is 1.42 eV (2.123×15^{15} rads^{-1}), $E_0 = 16$ meV is the (2d) QW 1s binding energy for GaAs, $\mu_{cv,\mathbf{k}=0} = 5$ eÅ (the **k**-dependence is modelled with a Kane dipole matrix element), the QW thickness is 100 Å, $\tau_0 = 5$ fs, and $\Gamma^{-1} = 100$ fs. The other material parameters for GaAs are defined earlier in the chapter. A SVEA 2-π pulse is then obtained with a peak electric field of 5×10^8 V/m. Outside the SVEA we note that the pulse described by Eq. (3.31) actually has an area of zero.

As a comparison, we first study the results for the simple TLA discussed earlier, but with the parameters above. In Fig. 3.19(a) we show the reflected pulse profile with an area of π and 4π, shown by the solid and dashed curves, respectively. For the lower

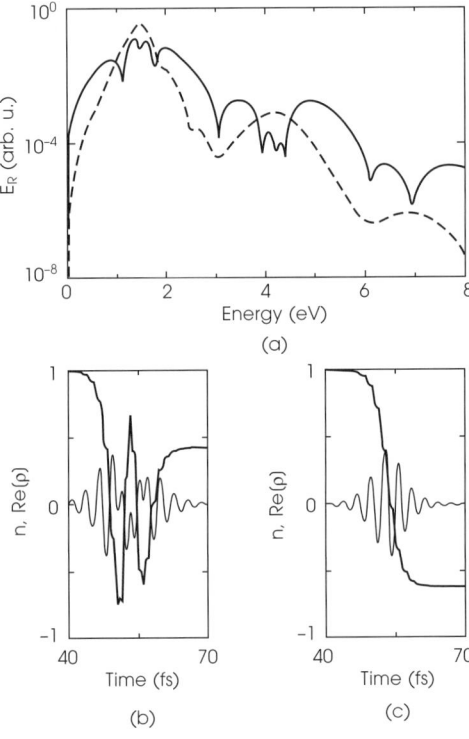

Fig. 3.19 (a) The reflected optical pulse exciting a two-level atom with a π (solid curve) and 4-π pulse (dashed curve). (b) Population (thick solid curve) and Re[ρ_{12}] versus time for the 4-π pulse. (c) As in (b) but for the π pulse.

field case, we obtain the familiar 3rd-order and 5th-order harmonic of the fundamental carrier frequency. However, for the higher intensity field we clearly obtain a splitting of the 3rd and 5th harmonic, that is characteristic of CWRF [4]; the same trend is also captured by solving the Maxwell–Bloch system for a thin film and looking at the emitted harmonic in the forward direction [4]. We also note that the oscillations of the fundamental is an excellent probe for the carrier-wave Rabi flops. Moreover, we have numerically verified that if the fundamental carrier (and transition) frequency is doubled, then this spectral split-up of the harmonics is not obtained; hence we are at the threshold to observe CWRF. We show the correspondong density changes and Re[ρ_{12}] in Figs 3.19(b) and 3.19(c). While the lower field polarization simply follows the input field, the higher field drives higher harmonics during the Rabi flopping within time scales that are shorter than the optical carrier cycle. We note that complete Rabi flopping is not obtained since we have employed a much larger dephasing rate. We also note the appearance of fine structure within the 3rd harmonic split-up region, verifying that a simple reflection analysis from a single QW is very sensitive to sub-carrier polarization oscillations; however, we mention that the fine structure disappears if the dephasing time is reduced to 50 fs or less.

Finally, we investigate the semiconductor case. In Fig. 3.20(a) we show the reflected pulse obtained with (solid curve) and without (dashed curve) a Coulomb interaction with a 4-π input pulse; in the latter case we immediately recognize a strong suppression of the harmonic split up since the many-body system has lost its simple two-level (homogeneously broadened) behavior and corresponds to that of an inhomogeneously broadened system. Since the pulse is so broad, many momentum states are excited and the higher ones do not exhibit clear Rabi flopping at all. Furthermore, even though the occupation of the higher momentum states is quite small, their influence is large because of the semiconductor density of states. Consequently, polarization components in the higher momentum states have a pronounced contribution to the polarization. The same arguments apply for the contributions to the total carrier density. With the inclusion of the coherent Coulomb interactions, there is a renormalized Rabi frequency and transition energy that we discuss below. The dotted curve also shows the reflected pulse for an 8-π pulse, demonstrating that significant splitting occurs in the fundamental and harmonics, making them somewhat indistinguishable; this demonstrates that (i) 4-π pulses are on the threshold to observe CWRF, and (ii) perturbation theory break downs anyway for very high intensity pulses such as those with areas greater than 4-π or so.

Although it has been shown elsewhere that, for much longer pulses, a typical 2-π input pulse has an effective (renormalized) area of around 4π in a semiconductor; this depends very much on the excitonic properties of the system and also on the spectral

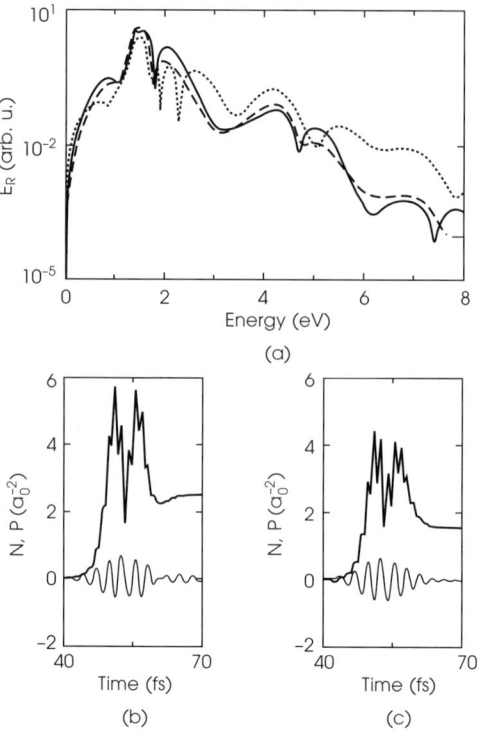

Fig. 3.20 (a) The reflected optical pulse exciting a semiconductor quantum well with a 4-π pulse, with (solid curve) and without (dashed curve) coherent Coulomb interactions. The reflected pulse for an 8-π pulse is also shown by the dashed curve. (b) Carrier density (thick solid curve) and optical polarization with Coulomb interactions (4-π pulse only). The scaling for the density and polarization is in units of the 2d inverse Bohr radius squared, and the latter has been multiplied by a factor of 4 for clarity. (c) As in (b) but without Coulomb interactions.

content of the excitation pulse. In the present case, although the lower momentum states cause Rabi flopping akin to an area of 3–4 π, the higher states exhibit flopping similar to that of a 2-π pulse; and the very high momentum states undergo adiabatic following of the pulse which shows up in the total density through pronounced oscillations. Of course, the pulse intensity in the QW is also reduced due to a finite reflection (though this reduction is rather weak). The corresponding density and optical polarization are shown in Figs 3.20(a) and 3.20(b), with and without the Coulomb interaction, respectively. In the former case the Coulomb interaction firstly creates more carriers than in the later case (through the renormalized Rabi energies), and also the polarization shows beating behavior around 60 fs. With regard to band-gap renormalization we can estimate that at the highest densities ($\approx 10^{13}$ cm^{-2}) a renormalization of about 100 meV occurs, though this does not seem to strongly affect the main spectral peaks in the reflected pulse.

The above analysis merely forms a starting point for solving the SBE outside the slowly-varying envelope approximation for ultrashort, extremely intense pulses. As highlighted earlier, clearly on these time scales one must account for quantum kinetics. Additionally, for such short pulses, one should also take care to include the proper band structure of the semiconductor, particularly at the higher momentum states. All this together becomes a fascinating theoretical playground into which to model these new excitation regimes using several-cycle optical pulses.

3.6 Conclusions

In summary, we have investigated several high-field excitation regimes in semiconductor nanostructures. First, we have applied a real-space-time method to calculate electron–hole wave packets in semiconductor quantum nanostructures, allowing the theoretical study of nonperturbative field regimes in a very intuitive way. The technique can incorporate a variety of fields, such as Coulomb, magnetic, THz, and static, in the low density limit. Experimental observables were extracted and in qualitative agreement with recent measurements. For quantum wells we studied a selection of high-field electro-optical effects, including the dynamic Frenz–Keldsyh effect, THz harmonic generation, and electro-magneto excitons. For quantum-well wires, we investigated the static and dynamic Franz–Keldysh effects. In the static case, exciton ionization effects and large Franz–Keldysh oscillations were highlighted by probing the electron-hole wave packet motion. For reasonable electric field strengths substantial oscillations appear above the band gap. Additionally, the Sommerfeld factor and field-induced tunneling significantly affect the continuum portion of the absorption spectrum and continue to remove the well-known divergence problem associated with the 1d DOS at all field stengths that we employ. For very large fields, tunneling-induced transparancy occurs at certain spectral frequencies. With a time-dependent field, several novel effects were obtained. For a THz photon just below the $1s$ binding energy, a strong Autler–Townes splitting of the exciton was obtained. By using coherent control, two-photon induced gain (or absorption) can be achieved in the continuum. Also, the emission of phase-dependent spectra in the THz regime was demonstrated, analogous to the high-field physics of atomic ensembles.

Second, we explored several nonlinear optical excitation regimes for semiconductor quantum wells. We predict the possibility of exciton trapping and quasi-adiabatic population transfer in a two-band semiconductor using frequency-modulated optical pulses. It is proposed that the trapping feature arises out of an interplay of the dominant excitonic resonance and the excitation by a sufficiently strong, broadband off-resonant pulse, with a weak yet broad spectral content to excite the continuum of states. The population redistribution results from the crossing of energy levels. Besides being an intriguing theoretical study, our results are timely with recent advances in frequency-modulated spectroscopy techniques and the observation of multiple Rabi flopping on free exciton transitions (see Chapter 1).

Third, an FDTD approach that utilizes the optical Maxwell–Bloch system, coupled to Maxwell's curl equations with a two-level atomic model for the polarization, was employed to describe carrier-wave Rabi flopping. It is shown that the nonlinear behavior is dependent not only on the electric field envelope but also on its propagating time-derivative effects. Standard *slowly-varying* results for 2π– and 4π–pulses are essentially reproduced with minor modifications due to local carrier effects. However, for higher pulse areas, it is found that carrier-wave effects become predominant, and a new, novel type of sub-carrier Rabi flopping is demonstrated. These features are absent in the standard area theorem and envelope-type models. Finally, we studied the semiconductor quantum well excited perpendicularly with extremely short, intense optical pulses and obtain clear spectral split-up of the first two harmonics for 4π–pulses that depends sensitively on the coherent Coulomb interactions. The latter simulations show qualitative trends with recent experiments (see Chapter 2), though a proper quantum kinetics approach is required for a future detailed comparison between theories and experiments.

Collectively, these investigations highlight some very fascinating light–matter interactions that are made possible by tremendous developments in short-pulse laser techniques, semiconductor growth and spectroscopy. They also point out the many similarities that continue to exist between high-field effects in semiconductors and atoms. As theoreticians, we congratulate all the experimentalists working in both fields, pushing the limits of light-matter excitation into ever shorter, ever intense, and hopefully more exciting regimes.

Acknowledgments

The author is indebted to David Citrin and Andreas Knorr for many valuable and insightful contributions thoughout this work. Many thanks also go to Martin Wegener for supplying Reference [4] prior to publication and to Dennis Sullivan for a free copy of his excellent FDTD book [26]. Useful discussions with Harald Geissen, Harshwardhan Wanare, Ben Murdin, Jeremy Allam, and Peter Knight are gratefully acknowledged. This work was supported by the Engineering and Physical Sciences Research Council (EPSRC), United Kingdom.

References

1. Heberle, A.P., Baumberg, J.J., and Köhler, K. (1995). Ultrafast coherent control and destruction of excitons in quantum-wells, *Phys. Rev. Lett.* **77**, 2598–2601.
2. Citrin, D.S., (1995). Self-pulse-shaping coherent control of excitons in a semiconductor microcavity, *Phys. Rev. Lett.* **77**, 4597–4599.
3. Giessen, H., Knorr, A., Haas, S., Koch, S.W., Linden, S., Kuhl, J., Hetterich, M., Grün, M., and Klingshirn, C. (1998). Self-induced transmission on a free exciton resonance in a semiconductor, J. Kuhl and M. Hetterich, *Phys. Rev. Lett.* **81**, 4260–4263.
4. Mücke, O.D., Tritschler, T., Wegener, M., Morgner, U., and Kärtner, F.X. (2001). Signatures of carier-wave Rabi flopping in GaAs, *Phys. Rev. Lett.*, **87**, 57401–57404.
5. Agarwal, G.A., and Harshawardhan, W. (1994). Realization of trapping in a two level system with frequency modulated fields, *Phys. Rev. A* **50**, 4465–4467.
6. Protopapus, M., Lappas, D.G., and Knight, P.L. (1997). Strong field ionization in arbitrary laser polarizations, *Phys. Rev. Lett.* **79**, 4550–4553; Paulus, G.G., Zacher, F., Walther, H., Lohr, A., Becker, W., and Kleber, M. (1998). Above-threshold ionization by an elliptically polarized field: Quantum tunneling interferences and classical dodging, *Phys. Rev. Lett.* **80**, 484–487.
7. Nordstrom, K.B., Johnsen, K., Allen, S.J., Jauho, A.P., Birnir, B., Kono, J., Noda, T., Akiyama, H., and Sakaki, H. (1998). Excitonic dynamical Franz–Keldysh effect, *Phys. Rev. Lett.*, **81**, 457–460; Kono, J., Su, M.Y., Inoshita, T., Noda, T., Sherwin, M.S., Allen, S.J., Jr., and Sakaki, H. (1997). Resonant terahertz optical sideband generation from confined magnetoexcitons, *Phys. Rev. Lett.* **79**, 1758–1761.
8. Chin, A.H., Calderon, O.G., and Kono, J. (2001). Extreme midinfrared nonlinear optics in semiconductors, *Phys. Rev. Lett.* **86**, 3292–3295.
9. See, for example, Allen, L., and Eberly, J.H. (1995). *Optical Resonance and Two-Level Atoms*, Wiley, New York.
10. Noel, M.W., Griffith, W.M., and Gallagher, T.F. (1998). Frequency-modulated excitation of a two-level atom, *Phys. Rev. A* **58**, 2265–2276.
11. Schülzgen, A., Binder, R., Donovan, M.E., Lindberg, M., Wundke, K., Gibbs, H.M., Ghitrova, G., and Peyghambarian, N. (1999). Direct observation of excitonic Rabi oscillations in semiconductors, *Phys. Rev. Lett.* **11**, 2346–2349.
12. Hughes, S. (1998). Breakdown of the area theorem: Carrier-wave Rabi flopping of femtosecond optical pulses, *Phys. Rev. Lett.* **81**, 3363–3366.
13. See Haug, H., and Koch, S.W. (1994). *Quantum Theory of the Optical and Electronic Properties of Semiconductors*, World Scientific, Singapore, 3rd edn, and references therein.
14. Zimmermann, R. (1997). *Spectroscopy and Dynamics of Collective Excitations in Solids*, Ed. Di Bartoli, Plenum Press, New York, 126.
15. Franz, W. (1958). Einfluss eines elektrischen Feldes auf eine optische Absorptionskante. *Z. Naturforsch.* **A 13**, 484–489; Keldysh, L.V. (1958). The effect of a strong electric field on the optical properties of insulating crystals. *Soviet Physics JETP* **34**, 788–790.
16. Schmitt-Rink, S., Chemla, D.S., and Miller, D.A.B. (1989). Linear and nonlinear optical properties of semiconductor quantum wells, *Advances in Physics* **38**, 89–188.

17. Hughes, S., and Citrin, D.S. (2000). High-field Franz–Keldysh effect and exciton ionization in semiconductor quantum wires, *Phys. Rev. Lett.* **84**, 4228–4231.
18. Hughes, S. (2001). Nonperturbative terahertz-field interactions in semiconductor quantum wires, *Phys. Rev. B* **63**, 15308–15311.
19. Citrin, D. (1997). Generation of 10-THz transients from a subpicosecond optical pulse and a 1-THz field in quantum wells, *Appl. Phys. Lett.* **70**, 1187–1191.
20. Citrin D.S., and Hashawardhan, W. (1999). Terahertz sideband generation in quantum wells viewed as resonant photon tunneling through a time-dependent barrier: An exactly solvable model, *Phys. Rev. B*, **60**, 1759–1763.
21. Jauho A.-P., and Johnsen, K. (1996). Dynamical Franz–Keldysh effect, *Phys. Rev. Lett.* **76**, 4576–4579.
22. Hughes S., and Citrin, D.S. (1999). Creation of highly anisotropic wave packets in quantum wells: Dynamical Franz–Keldysh effect in the optical and terahertz regimes, *Phys. Rev. B* **59**, R5288–R5291.
23. Grobe R., and Eberly, J.H. (1993). One-dimensional model of a negative-ion and its interaction with laser fields, *Phys. Rev. A* **48**, 4664–4681.
24. Protopapas, M., Lappas, D.G., and Knight, P.L. (1997). Strong field ionization in arbitrary laser polarizations, *Phys. Rev. Lett.* **79**, 4550–4553.
25. See Taflove, A. (1995). *Computational Electrodynamics: The Finite-Difference Time-Domain Method*, Artech House, Boston, London.
26. Sullivan, D. (2000). *Electromagnetic Simulation Using the FDTD Method*, IEEE Press Series on RF and Microwave Technology, IEEE Press, New York.
27. Hughes, S., and Citrin, D.S. (2000). Dynamic Franz–Keldysh effect on magnetoexcitons in quantum wells: beyond perturbation theory, *Solid State Commun.* **113**, 11–15.
28. Elliott, R.J., and Loudon, R. (1959). Theory of fine structure on the absorption edge in semiconductors. *J. Phys. Chem. Solids* **8**, 382–388; Elliott, R.J., and Loudon, R. (1960). Theory of the absorption edge in semiconductors in a high magnetic field. *J. Phys. Chem. Solids* **15**, 196–207.
29. Ogawa, T., and Takagahara, T. (1991). Optical absorption and Sommerfeld factors of one-dimensional semiconductors – an exact treatment of exciton effects, *Phys. Rev. B* **44**, 8138–8156.
30. Benner, S., and Haug, H. (1993). Influence of external electric and magnetic fields on the excitonic absorption spectra of quantum-well wires, *Phys. Rev. B* **47**, 15750–15754.
31. Schwengelbeck, U., and Faisal, F.H.M. (1994). Ionization of the one dimensional Coulomb atom in an intense laser field, *Phys. Rev. A* **50**, 632–640.
32. See, for example, Rinaldi, R., Cingolani, R., Lepore, M., Ferrara, M., Catalano, I.M., Rossi, F., Rota, L., Molinari, E., Lugli, P., Marti, U., Martin, D., Morier-Gemoud, F., Ruterana, P., and Reinhart, F.K. (1994). Exciton binding energy in v-shaped quantum wires, *Phys. Rev. Lett.* **73**, 2899; Constantin, C., Martinet, E., Rudra, A., and Kapon, E. (1999). Observation of combined electron and photon confinement in planar microcavities incorporating quantum wires. *Phys. Rev. B* **59**, R7809-R7812; Wang, K.H., Bayer, M., Forchel, A., Ils, P., Benner, S., Haug, H., Pagnod-Rossiaux, Ph., and Goldstein, L., (1996). Subband renormalization in dense electron-hole plasmas in In0.53Ga0.47As/InP quantum wires, *Phys. Rev. B* **53**, R10505–R10509.

33. See, for example, Tassone, F., and Piermarocchi, C. (1999). Electron-hole correlation effects in the emission of light from quantum wires, *Phys. Rev. Lett.* **82**, 843–846; Mauritz, O., Goldoni, G., Rossi, F., and Molinari, E. (1999). Local optical spectroscopy in quantum confined systems: A theoretical description, *Phys. Rev. Lett.* **82**, 847–850.
34. Tran Thoai, D.B., and Thien Cao, H. (1999). Subband renormalization of highly excited quantum-well wires, *Solid State Commun.* **111**, 67–72.
35. Rossi, F., and Molinari, E. (1996). Coulomb-induced suppression of band-edge singularities in the optical spectra of realistic quantum-wire structures, *Phys. Rev. B* **53**, 3642–3645.
36. Son, J.-H., Norris, T.B., and Whitaker, J.F. (1994). Terahertz EM pulses as probes for transient velocity overshoot in GasAs and SI, *J. Opt. Soc. Am. B* **11**, 2519–2527.
37. Leitenstorfer, A., Hunsche, S., Shah, J., Nuss, M.C., and Knox, W.H. (1999). Femtosecond charge transport in polar semiconductors, *Phys. Rev. Lett.* **82**, 5140–5143.
38. Moreno, P., Plaja, L., and Roso, L. (1996). High-order harmonic generation in a partially ionized medium, *J. Opt. Soc. Am. B* **13**, 430–435.
39. Krause, J.L., Schafer, K.J., Ben-Nun, M., and Wilson, K.R. (1998). Creating and detecting shaped Rydberg wave packets, *Phys. Rev. Lett.* **79**, 4978–4981.
40. Hughes, S., Harshawardhan, W., and Citrin, D.S., (1999). Excitonic-state trapping and quasi-adiabatic population transfer in a two-band semiconductor, *Phys. Rev. B* **60**, 15523–15526.
41. Binder, R., and Lindberg, M. (1998). Ultrafast adiabatic population transfer in p-doped semiconductor quantum wells, *Phys. Rev. Lett.* **81**, 1477–1480.
42. Knorr, A., Hughes, S., Stoucken, T., and Koch, S.W. (1996). Theory of ultrafast spatio-temporal dynamics in semiconductor heterostructures, *J. Chem. Phys.* **210**, 27–47, special issue; Hughes, S., Knorr, A., and Koch, S.W. (1997). Interplay of optical dephasing and pulse propagation in semiconductors, *J. Opt. Soc. Am. B* **49**, 754–760.
43. Hughes, S., Knorr, A., and Koch, S.W. (1996). The influence of electron–hole-scattering on the gain spectra of highly excited semiconductors, *Solid State Commun.* **100**, 555–559.
44. Lindberg, M., and Koch, S.W. (1988). Effective Bloch equations for semiconductors. *Phys. Rev. B* **38**, 3342–3350.
45. Rappen, T., Peter, U.-G., Wegener M., and Schäfer, W. (1994). Polarization dependence of dephasing processes – a probe for many-body effects, *Phys. Rev. B* **49**, 10774–10776.
46. Borri, P., Langbein, W., Mørk, J., and Hvam, J.M. (1999). Heterodyne pump-probe and four-wave mixing in semiconductor optical amplifiers using balanced lock-in detection, *Opt. Commun.* **169**, 317–324.
47. Hofmann, M., Brorson, S.D., and Mecozzi, A. (1996). Time resolved four-wave mixing technique to measure the ultrafast coherent dynamics in semiconductor optical amplifiers, *Appl. Phys. Lett.* **68**, 3236–3238.
48. Knorr A., and Hughes, S. (2001). Microscopic theory of ultrashort pulse compression and break-up in a semiconductor optical amplifier, *Photon. Tech. Lett.*, **13**, 782–784.
49. Hughes, S., (1998). Carrier–carrier interaction and ultrashort pulse propagation in a highly excited semiconductor laser amplifier beyond the rate equation, *Phys. Rev. A* **58**, 2567–2576.
50. Dunlap, D.H., and Kenkre, V.M. (1986). Dynamic localization of a charged particle moving under the influence of an electric field. *Phys. Rev. B* **34**, 3625–3633.
51. Holthaus, M. (1992). Collapse of minibands in far-infrared irradiated superlattices, *Phys. Rev. Lett.* **69**, 351–354.

52. Rabi, I.I. (1937). Space quantization in a gyrating magnetic field. *Phys. Rev.* **51**, 652–654.
53. Lam, P.K., and Savage, C.M. (1994). Complete atomic population-inversion using correlated side-bands, *Phys. Rev. A* **50**, 3500–3516.
54. Hughes, S., and Citrin, D.S. (1999). Tunable terahertz emission through multiple Rabi flopping in a dc-biased quantum well: a new strategy, *Opt. Lett.* **24**, 1242–1244; ibid, *Optics and Photonics News*, Dec., (1999), special issue on Highlights of Optics in 1999.
55. Hughes, S., and Citrin, D.S. (2001). Broadband terahertz emission through exciton trapping in a semiconductor quantum well, *Opt. Lett.*, **26**, 1–3.
56. McCall, S.L., and Hahn, E.L. (1967). Self-induced transparency by pulsed coherent light. *Phys. Rev. Lett.* **18**, 908–911; Lamb, G.L. Jr. (1971). Analytical descriptions of ultrashort optical pulse propagation in a resonant medium. *Rev. Mod. Phys.* **43**, 99–124.
57. Abella, I.D., Kurnit, N.A., and Hartmann, S.R. (1966). Photon echoes. *Phys. Rev.* **141**, 391–406.
58. Rabi, I.I., Millman, S., Kusch, P., and Zacharias, J.R. (1938). The magnetic moments of $_3\text{Li}^6$, $_3\text{L}^7$ and $_9\text{F}^{19}$. *Phys. Rev.* **53**, 495.
59. Cundiff, S.T., Knorr, A., Feldmann, J., Koch, S.W., Göbel, E.O., and Nickel, H. (1994). Rabi flopping in semiconductors, *Phys. Rev. Lett.* **73**, 1178–1181.
60. Ziolkowski, R.W., Arnold, J.M., and Gogny, D.M. (1995). Ultrafast pulse interactions with two-level atoms, *Phys. Rev. A.* **52**, 3082–3094.
61. Basinger, S.A., and Brady, D.J. (1994). Finite-difference time-domain modeling of dispersive nonlinear Fabry–Ferot cavities, *J. Opt. Soc. Am. B.* **11**, 1504–1511.
62. Hughes, S. (2000). Subfemtosecond soft-x-ray generation from a two-level atom: Extreme carrier-wave Rabi flopping, *Phys. Rev. A* **62**, 55401–55404.

Quantum Coherence, Correlation and Decoherence in
Semiconductor Nanostructures
T. Takagahara (Ed.)
Copyright © 2003 Elsevier Science (USA). All rights reserved.

Chapter 4
Theory of resonant secondary emission: Rayleigh scattering versus luminescence

R. Zimmermann,[1] E. Runge,[1,2] and V. Savona[1]

[1] Humboldt University Berlin, Hausvogteiplatz 5-7, D-10117 Berlin, Germany

[2] Max Planck Institute for the Physics of Complex Systems, Nöthnitzer Str. 38, D-01187 Dresden, Germany

Abstract

The light emission after optical excitation in a semiconductor quantum structure is dominated by exciton effects. For understanding the interplay between homogeneous and inhomogeneous broadening of lines, excitons in quantum wells with interface roughness are investigated theoretically. Using the disorder eigenstates as basis, a density matrix theory is derived for excitons in interaction with the light field and with acoustic phonons. The secondary emission is decomposed into the coherent part (Rayleigh scattering) and the incoherent part (photoluminescence). This distinction is based on the speckle analysis. Quantum mechanical features such as level repulsion and enhanced resonant backscattering are discussed. The polarization dependence of the emission is related to the exchange (spin) splitting of anisotropic exciton states. The interplay of disorder and polaritonic effects is exemplified for the time-dependent emission.

4.1 Introduction

Excitons, i.e. Coulomb bound states of electron and hole, determine the optical properties of semiconductors near the fundamental band edge. They show up as distinct lines in absorption, reflection, and photoluminescence (PL). Their study formed for decades a major part of optics in bulk semiconductors. With the arrival of semiconductor nanostructures such as quantum wells (QW), quantum wires, and quantum dots, excitonic effects became even more important, since the exciton binding energy is strongly enhanced. As a handwaving argument, the kinetic motion is hindered by the confinement, while the Coulomb attraction remains nearly unchanged [1]. Thus, undoped semiconductor nanostructures are an ideal playground for exciton physics. However, the definition of the interfaces on an atomic scale is never as ideal as physicists would like it to be. Even with the highly sophisticated molecular beam epitaxy (MBE), interface fluctuations of one or a few monolayers can hardly be avoided. A further source of disorder is alloy fluctuation if a ternary compound is used as barrier or well material.

These disorder effects determine the inhomogeneous broadening of the exciton line seen in optical measurements, and even tend to dominate their linewidth in narrow quantum structures. Rather than disregarding this unwanted feature, the exciton linewidth in PL is usually taken as a quality measure of the growth process. However, a distinction between inhomogeneous linewidth (disorder-induced) and homogeneous linewidth (induced by inelastic scattering) is no possible from PL alone. A further indicator of disorder is the Stokes shift between the peaks in PL and absorption (or rather photoluminescence excitation, PLE). Another quite important effect brought in by disorder is the breaking of the in-plane translational symmetry. Therefore, the center-of-mass momentum (COM) of the exciton is no longer a good quantum number. Exciting the sample with an electromagnetic plane wave, there is not only transmission and reflection (specular optics), but coherent emission in other directions as well. This will be called resonant Rayleigh scattering (RRS) in the following, and is due to elastic scattering of excitons into different COM directions. However, also inelastic processes as phonon emission and absorption or exciton-exciton interaction will redistribute the excitons over COM momenta, and thus over emission angle. Here, luminescence is the appropriate notion. For the sake of simplicity we will consider only low excitation where phonon scattering dominates over Coulomb scattering.

In order to avoid any preconception, the notion of *secondary emission* has been introduced to describe any emission outside the specular directions. This chapter is devoted to elucidate carefully both scattering channels: elastic or Rayleigh, and inelastic or luminescence. As we will show, this distinction is completely in line with a separation of the secondary emission into *coherent* (with the exciting beam) and *incoherent* parts. For few-level atoms, this is standard textbook knowledge [2]. However, in semiconductors the microscopic processes are much more complex and rich.

While photoluminescence experiments form one of the backbones of semiconductor optics, Rayleigh scattering measurements are relatively rare. Pioneering experiments RRS in the frequency domain are due to Hegarty [3], showing clearly a sharp spike in resonance with the exciting laser on top of a broad PL band. New interest evolved with time-resolved light scattering experiments. Stolz [4] raised the question on how RRS depends on polarization of excitation and emission. A new era was opened by experiments with femtosecond time resolution in the Shah group [5,6]. Quite as a surprise, the secondary emission was seen to appear with a definite time delay compared to the excitation pulse. Further, a beating of the signal with the heavy- to light-hole splitting clearly indicated a coherent process. In Fig. 4.1, we display a measurement from [6] together with a fit using an early version of theory [7]. The mechanisms and, in particular, the delayed onset of secondary emission after resonant femtosecond excitation was the subject of an intense debate in the last years [8–14]. For a more general review on the experimental situation, see [15] and [16]. Right at the heart of any ultra-fast optical experiment is the question of coherence and coherence loss (or dephasing). The quantum coherence of excitons (or any other elementary excitation) in semiconductors is most naturally studied with optical methods, because it is directly related to the coherence of the emitted light. A specific example is the well established and widely used technique of four-wave mixing, in particular the photon echo. An advantage is that the signal is separated from the excitation direction. However, it is insensitive to the incoherent emission, and intrinsically a nonlinear optical process. Therefore, experimental excitation powers are quite large, and processes like exciton-exciton scattering contribute substantially to, e.g., dephasing rates. An extrapolation to vanishing excitation densities is not an easy task. With interferometric methods [17], the very existence of coherent emission (having a fixed phase relation to the exciting wave) can be detected. However, again quantitative information on the degree of coherence is hard to obtain.

Light scattering into non-specular directions is background-free, too. Only such excitons which have experienced at least one real scattering process can be detected. For non-resonant excitation, this implies energy relaxation, thus destroying the coherence

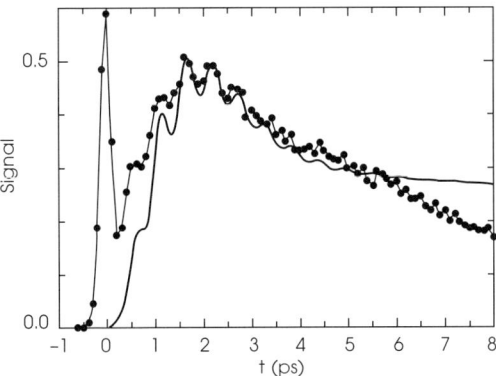

Fig. 4.1 Time-resolved secondary emission after pulsed excitation (120 fs FWHM), tuned 6 meV below the heavy-hole exciton of a 13 nm wide AlGaAs quantum well (dots, from [6]). The full curve is calculated using a kinetic equation for excitons in momentum space, and treating the disorder in the weak-memory approach; from [7].

within the exciton system. Therefore, the emitted radiation is intrinsically incoherent. For resonant excitation and in particular for short-pulse excitation, elastic Rayleigh scattering and inelastic relaxation are superimposed. However, the coherently scattered light is characterized by a fixed, but irregular pattern of destructive and constructive interference, both with itself and with a phase-coherent reference pulse. Fluctuations occur both in time and in space (direction) and will henceforth be referred to as *speckles* [18]. It was the idea of Langbein [19,20] to introduce and establish the speckle analysis in the secondary emission from semiconductors. For the first time, a clearcut and quantitative determination of coherence became possible.

Let us compare the specific way the coherence is preserved or destroyed in the exciton system. Scattering events by emission/absorption of phonons (or other excitons) can happen at any intermediate time between generation and ultimate radiative decay. Therefore, the related phase will differ from one single laser pulse to the next, disabling any interference pattern. This leads to a speckle-free background intensity. In contrast, disorder scattering is identical for all pulses, giving rise to an irregular but fixed phase pattern. In fact it is this average over tens of thousands of individual femtosecond pulses which makes the difference between inelastic and elastic scattering processes in view of coherence. This pulse repetition is present in any ultra-fast (and also interferometric) experiment – and is almost never mentioned explicitly. It is the basis for the method of statistical speckle analysis [21] which will be presented in detail in section 4.6.

The theory starts with a description of excitons in a quantum structure with interface disorder in section 4.2. For a summary of own work, see [22]. Starting with the electron-hole Schrödinger equation in effective mass approximation, several well-justified steps lead to a single-particle Schrödinger equation for the exciton center-of-mass (COM) motion. It contains a random potential which is spatially correlated at least over distances of the exciton Bohr radius. Compared to the monolayer fluctuation energy, the strength of the potential is strongly reduced in energy. Both effects (correlation and reduction) are related to the averaging effect of the electron–hole relative motion within the exciton, which is displayed schematically in Fig. 4.2. Solving the Schrödinger equation in this disordered potential landscape [23,24] allows to construct immediately the optical density (OD) of the $1s$ exciton as seen directly in absorption or approximately in PLE.

While up to this point the exciton Schrödinger equation with disorder is completely sufficient, the description of the emitted radiation and in particular the inclusion of inelastic scattering effects calls for a deeper foundation. In section 4.3, we construct a model Hamiltonian for excitons interacting with photons and phonons. In doing so we use consequently the basis of disorder eigenstates. Thus, the disorder is included from the very beginning, and in a non-perturbative fashion. Then, standard methods of density

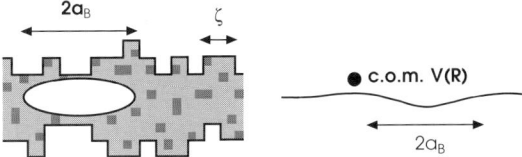

Fig. 4.2 Schematic illustration of a quantum well with rough interfaces (island size ζ) and alloy disorder. The averaging with the exciton relative wave function (radius a_B) leads to an effective single-particle problem with a smooth potential $V(\mathbf{R})$ for the center of mass motion.

matrix theory are applied. The coupling to acoustic phonons is considered as major source of dephasing, and treated in second Born approximation (section 4.4). The interaction with the (quantized) light field allows to describe the emission in general. Disregarding for the moment light propagation (polariton) effects, the secondary emission follows straightforwardly from the exciton density operator. Its expectation value contains a factorized part (product of polarizations) which is shown to be coherent with the incoming light, and produces Rayleigh scattering (section 4.5). For the incoherent part, a kinetic equation is derived and analyzed. The occupation of disorder eigenstates for the exciton can be followed in time, or calculated under steady state conditions (section 4.7). This goes well beyond any kinetic equation formulated in terms of exciton distribution as function of momentum, which we have tried in the beginning [7,25]. Some rather intricate experimental findings can be explained: The non-monotonic Stokes shift with temperature [26], the existence of a relaxation mobility edge [27], the deviation between PLE and absorption, to mention a few. The price to be paid is of a technical nature: To generate the eigenstates, huge matrix diagonalizations on a two-dimensional spatial grid have to be performed. On top comes the solution of the kinetic equation. This program has to be repeated a large number of times for different disorder realizations, until smooth curves are obtained which can be compared to experiments with a macroscopic excitation and detection spot.

The coherent part of the secondary emission is characterized by speckles. In section 4.6, we derive the expressions for the speckle statistics and quote a simple relation for the coherence degree in terms of intensity average and fluctuations [28]. The relation to interferometric setups is clarified. The issue of level repulsion in near-field exciton spectra is addressed in section 4.8. According to random matrix theory, a dip in the autocorrelation function (here of the optical spectrum) is expected. However, the interplay with the potential correlation and the finite spectral resolution reshape the feature into a shoulder. The agreement between simulation and recent experimental data gives striking evidence for the quantum-mechanical level repulsion of localized excitons

[29]. The relation to oscillating features in the time-resolved Rayleigh scattering is clarified [30]. The wave aspect of the exciton COM states is furthermore responsible for the phenomenon of enhanced resonant backscattering as predicted in [31] (section 4.9). However, the enhancement in the backscattered direction is found to be much less than the factor of two often quoted in connection with (off-resonant) light scattering in inhomogeneous media. Apart from full simulations in two dimensions (2D), standard many-particle theory is applied to sum up ladder and maximally crossed diagrams, and checked with a 1D calculation. At small dephasing rates, however, this well-known diagrammatic approach misses the important feature of enhanced forward scattering.

The intimate relation between spin degree of freedom, light polarization, and wave function anisotropy of excitons is investigated in section 4.10. Exchange (or spin) splittings in the ensemble of localized exciton states are calculated. The secondary emission shows a transfer of polarization into the counterpolarized channel, which, however, is not related to spin relaxation. The correct description is spin beating, preserving a high degree of coherence in both channels. These findings are fully substantiated by time-resolved experiments [32]. In section 4.11, the quantum well containing localized excitons is embedded into a dielectric medium, and the full vector Maxwell equations are solved in the basis of disorder eigenstates. Spin splitting and polarization transfer are now supplemented by radiative losses, which produces a clear non-exponential decay in the time-resolved RRS signal.

4.2 Disorder eigenstates of excitons

Let us consider the exciton Schrödinger equation for a two-band model in effective mass approximation,

$$\left(-\frac{\hbar^2}{2m_e}\Delta_{\mathbf{r}_e} - \frac{\hbar^2}{2m_h}\Delta_{\mathbf{r}_h} - \frac{e^2/\varepsilon_0}{|\mathbf{r}_e - \mathbf{r}_h|} + W_e(\mathbf{r}_e) + W_h(\mathbf{r}_h) - \varepsilon_\alpha\right)\Psi_\alpha(\mathbf{r}_e, \mathbf{r}_h) = 0. \quad (4.1)$$

Apart from the standard kinetic term and the Coulomb attraction, we have introduced the confinement potentials $W_a(\mathbf{r}_a)$ ($a = e, h$) which describe the spatial variation of the local band edges. The z-axis is taken along the growth direction. If interface roughness dominates the disorder, it depends on the band edge difference between barrier and well material Δ_a and the local well width $L_z(\boldsymbol{\rho}) = \bar{L}_z + \Delta L_z(\boldsymbol{\rho})$ (see Appendix A).

If the exciton binding energy is small compared to the (energetic) sublevel distance, but also well above the disorder-induced broadening, only the lowest bound state $1s$ at the fundamental sublevel transition has to be considered. Consequently, the total wave function can be factorized into

$$\Psi_\alpha(\mathbf{r}_e, \mathbf{r}_h) = u_e(z_e)\, u_h(z_h)\, \phi_{1s}(\boldsymbol{\rho}_e - \boldsymbol{\rho}_h)\, \psi_\alpha(\mathbf{R}), \qquad (4.2)$$

introducing the 2D center-of-mass coordinate $\mathbf{R} = (m_e \boldsymbol{\rho}_e + m_h \boldsymbol{\rho}_h)/M$ with the exciton kinetic mass $M = m_e + m_h$. Both the confinement wave functions $u_a(z_a)$ and the relative wave function $\phi_{1s}(\boldsymbol{\rho})$ obey Schrödinger equations of the QW structure with (average) thickness $\overline{L}_z$ (see [1] for an introduction into the single-particle confinement). Finally one is left with the COM equation

$$\left(-\frac{\hbar^2}{2M}\Delta_\mathbf{R} + V(\mathbf{R})\right)\psi_\alpha(\mathbf{R}) = \varepsilon_\alpha\, \psi_\alpha(\mathbf{R}). \qquad (4.3)$$

Note that the 1s exciton energy $\hbar\omega_x$ of the averaged QW is taken as zero of energy here and in what follows. The normalization is taken over the in-plane area A, with orthogonality relation

$$\int_A d\mathbf{R}\, \psi_\alpha(\mathbf{R})\, \psi_\beta(\mathbf{R}) = \delta_{\alpha,\beta}. \qquad (4.4)$$

The random COM potential resulting from well-width fluctuations is derived in Appendix A and reads

$$V(\mathbf{R}) = \int d\mathbf{R}' \sum_{a=e,h} \eta_a^2\, \phi_{1s}^2(\eta_a(\mathbf{R} - \mathbf{R}'))\, \frac{dE_a}{dL_z}\, \Delta L_z(\mathbf{R}'). \qquad (4.5)$$

Due to the mass factors $\eta_e = M/m_h$, $\eta_h = M/m_e$, different weight is given to the electron and the hole contribution. Even if the band offset for the hole is less than the electron one – as it is the case for the GaAs/AlGaAs system – the hole may dominate the potential fluctuations: Due to its larger mass, the hole part in the exciton visits a smaller volume, and averages not as efficient as the electron. Similar relations for alloy disorder have been derived elsewhere [33].

A reasonable assumption for the well width fluctuations is

$$\overline{\Delta L_z(\mathbf{R})\, \Delta L_z(\mathbf{R}')} = h^2 \exp(-|\mathbf{R}-\mathbf{R}'|^2/2\zeta^2) \qquad (4.6)$$

introducing a typical thickness fluctuation h and a correlation length ζ (typical island size). The overline denotes the average over different disorder realizations within the statistical ensemble. The convolution with the 1s wave function gives a final potential correlation length ξ which depends on the (mass-weighted) exciton Bohr radius a_B as well. If $a_B > \zeta$ holds, the potential Eq. (4.5) is constructed as a large sum of random independent contributions, and the central limit theorem holds: The potential values are Gauss distributed with variance $\sigma^2 = \overline{V^2(\mathbf{R})}$. Evaluated for an exciton $1s$ wave function of exponential type with Bohr radius $a_B > \zeta$, the result is (see Eq. (4.169))

$$\sigma^2 = \frac{2h^2\zeta^2}{a_B^2}((E'_e\eta_e)^2 + 8E'_eE'_h + (E'_h\eta_h)^2). \tag{4.7}$$

The energetic fluctuations of the COM potential are reduced by the ratio between the statistically independent areas of size ζ^2 and the exciton averaging area a_B^2. Therefore, a direct assignment of σ or of the exciton linewidth to the energy fluctuations on an atomic scale is not possible. In general, the confinement energies fluctuate on a much larger energy scale than the resulting inhomogeneous exciton linewidth seems to predict. Upon growth interruption in the MBE deposition, h is expected to decrease, whereas ζ gets larger. Since the product $h \cdot \zeta$ enters Eq. (4.7), it is not clear *a priori* if growth interruption leads to a reduction of the linewidth or not [34].

In the opposite limit $a_B < \zeta$, the potential values are no longer Gauss distributed, but attain discrete values. Correspondingly, the exciton line splits into a multiplet related to different monolayer energies [35]. This may be achieved in high-quality samples with growth interruption. Having in mind samples with dominant disorder effects, we do not consider monolayer splitting in this review.

Equations (4.3) and (4.5) are the starting point for calculations on simulated structures. One could use a model for the MBE growth to generate realistic interfaces by Monte Carlo simulation [36,37]. If a specific roughness model is not of interest, it suffices to use an artificially constructed random potential. The only ingredient is the potential correlation function

$$g(\mathbf{R} - \mathbf{R}') = \overline{V(\mathbf{R})\,V(\mathbf{R}')}. \tag{4.8}$$

In the numerical calculation, the choice is between exponential type correlation (resembling the 2D exciton wave function) and a Gaussian shape, which emphasizes more the random island structure. For the latter case, $g(\mathbf{R}) = \sigma^2 \exp(-R^2/(2\xi^2))$ is used. Apart from the chosen type, only two parameters characterize the potential: variance σ and correlation length ξ. According to what has been said above, ξ cannot be smaller than the (mass-weighted) exciton Bohr radius.

The first quantity of interest is the density of states (DOS) of COM excitons,

$$\rho(\omega) = \frac{1}{A}\sum_\alpha \delta(\omega - \omega_\alpha), \tag{4.9}$$

which gives the unbiased distribution of energy levels ($\varepsilon_\alpha \equiv \hbar\omega_\alpha$).

Introducing the Fourier-transformed wave function

$$\psi_{\alpha\mathbf{k}} = \int_A d\mathbf{R} e^{i\mathbf{k}\cdot\mathbf{R}} \psi_\alpha(\mathbf{R}), \tag{4.10}$$

we can construct the optical density (OD) by weighting with the wave function squared at $\mathbf{k} = 0$,

$$D(\omega) = \frac{1}{A} \sum_\alpha \psi^2_{\alpha\mathbf{k}=0} \delta(\omega - \omega_\alpha). \tag{4.11}$$

In the context of the (effective) one-particle problem Eq. (4.3), this is the spectral function at zero momentum, $D(\omega) = \text{Im } G_{\mathbf{k}=0}(z = \omega - i0^+)$, where $G_{\mathbf{k}}(z)$ is the one-particle (disorder-averaged) Green's function. More important, $D(\omega)$ is proportional to the absorption lineshape of $1s$ excitons at normal incidence, as shown in more detail in section 4.3. In a given potential realization, both expressions Eqs (4.9) and (4.11) consist of a series of delta lines (or narrow peaks if a broadening is artificially introduced). Only after performing a large number of simulation runs with different potential realizations, a smooth curve results. For the absorption, this average corresponds to a transmission experiment with large focus and a finite aperture. We will see in section 4.5 that other quantities like the angle-resolved scattered intensity do not contain this kind of ensemble average, and display an irregular spiky behavior (speckles).

Numerical results are shown in Fig. 4.3. We use $E_c = \hbar^2 \xi^{-2}/2M$ as a measure of the

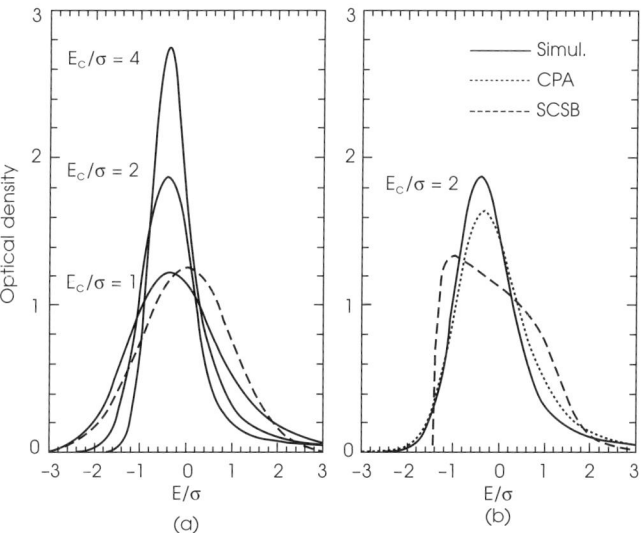

Fig. 4.3 Optical density $D(\omega)$ for exponentially correlated disorder in 2D. (a) Different correlation lengths ξ, expressed as $E_c = \hbar^2\xi^{-2}/2M$ over potential variance σ. Dashed curve – Gauss potential distribution. (b) Comparison between simulation, coherent potential approximation, and selfconsistent second Born approximation (for $E_c/\sigma = 2$ only).

confinement energy of the lowest COM state in a typical potential minimum. The Schrödinger equation can be rescaled by using appropriate units for energy and length [30]. Then, it is only the ratio E_c/σ which determines the optical lineshape, Fig. 4.3(a). $E_c/\sigma = 0$ describes classical excitons, where the OD coincides with the Gaussian potential distribution function (dashed curve). For finite values of this ratio, the lineshape develops a distinct asymmetry towards higher energies and gets narrower. This kind of motional narrowing has been reported often in the literature (e.g. for 1D excitons in [38]). The asymmetry can be understood by a simple perturbation theory argument: Starting with exciton states in the plane-wave basis (having kinetic energy $\hbar^2 k^2/2M$), the disorder can be considered as elastic scattering with the amplitude $g_{\mathbf{k}-\mathbf{k}'}$, the Fourier transform of the potential correlation function Eq. (4.8). Within second order perturbation theory, the density of final states is regulating the strength of tails in the spectral function. The (ideal) DOS vanishes below the zero-momentum exciton energy, and this argument yields a broadening towards higher energies only. In reality, the DOS acquires a low-energy tail, but still an asymmetry remains. Along these lines, we have started with a phenomenological model to describe an asymmetric OD found in experiment [39]. A distinct improvement is to implement a selfconsistency loop into second order perturbation theory, which is the selfconsistent second Born approximation (SCSB). Introducing the self-energy $\Sigma_{\mathbf{k}}(z)$ via $G_{\mathbf{k}}(z) = 1/(z - \hbar k^2/2M - \Sigma_{\mathbf{k}}(z))$, one has to solve

$$\Sigma_{\mathbf{k}}^{\text{SCSB}}(z) = \frac{1}{\hbar^2 A} \sum_{\mathbf{k}'} g_{\mathbf{k}-\mathbf{k}'} G_{\mathbf{k}'}(z). \tag{4.12}$$

Although used several times for the determination of the OD [40], the results are rather poor, Fig. 4.3(b). In particular, the SCSB is not able to describe the low-energy tail reliably (it gives zero below a threshold energy). On the high-energy side, the agreement is better, since perturbation theory is expected to work in this limit. We note in passing that the momentum dependence of the self-energy is not very strong, and putting $\Sigma_{\mathbf{k}}(z) \approx \Sigma_0(z)$ does not deteriorate the SCSB further. Another standard method for disorder problems is the coherent potential approximation (CPA) [41]. We have modified the CPA for including the potential correlation and got reasonable results both in 1D and 2D [23], Fig. 4.3(b).

An explicit ensemble average can be carried out for the moments of the optical density, which are defined as

$$M_n = \int d\omega\, (\hbar\omega)^n\, \overline{D(\omega)} = \frac{1}{A} \int_A d\mathbf{R}\, \overline{H^n(\mathbf{R})}, \tag{4.13}$$

expressed through powers of the Hamilton operator $H(\mathbf{R})$ of the COM Schrödinger equation (4.3). The first three moments are $M_0 = 1$ (normalization), $M_1 = 0$, $M_2 = \sigma^2$, and coincide with those of the underlying potential distribution. The first non-classical moment is M_3 (for Gauss-type correlation: $M_3 = 3E_c\sigma^2$). Its positive value is another hint on the asymmetry towards higher energies. Using computer algebra, we have automatically generated up to 20 moments and tried to reconstruct the OD numerically. Even after applying a resummation in the exponent (cumulant expansion), the results do not converge, in particular for $E_c/\sigma > 1$. In [42], this has been tried including the term M_3 only, with pathological behavior such as negative portions in the OD.

It is only in 1D and for an uncorrelated potential, that exact (non-simulation) results are available, using a method due to Halperin [43]. We have used these results [44] as a benchmark for any of the other approximations just mentioned. All these methods allow to calculate the OD, but if we are aiming at modeling luminescence or Rayleigh scattering, a numerical generation of the eigenstates in a given realization is indispensable.

For the numerics, the Laplacian in the 2D real-space Schrödinger equation is discretized on a square grid $\mathbf{R}_j$ (step size Δ), which maps the problem to a 2D tight-binding version of the Anderson model with diagonal (correlated) disorder. To generate the disorder landscape, a real number U_j is attached to each grid point, with the U_j being drawn randomly from a Gauss distribution with unity variance. The potential is then generated by summing

$$V_j = \sum_k W(\mathbf{R}_j - \mathbf{R}_k) U_k \tag{4.14}$$

with a smoothing function $W(\mathbf{R})$. For Gauss and exponential correlation type, this reads

Gauss: $W(\mathbf{R}) = \dfrac{\Delta\sigma}{\xi} \sqrt{2/\pi}\, e^{-R^2/\xi^2}$, (4.15)

Expon.: $W(\mathbf{R}) = \dfrac{\Delta\sigma}{\xi} \sqrt{2/\pi}\, e^{-R/\xi}$. (4.16)

Fast Fourier transform speeds up the generation of the potential, as well as the conversion of wave functions into $\mathbf{k}$ space. For more details on the numerical procedures, the reader is referred to [22,33,45,46].

The distribution of oscillator strengths, $\sim \psi_{\alpha 0}^2$ is displayed in Fig. 4.4, c.f. [47], and forms a special application of wavefunction statistics [48]. At energies well above the line center (panels 5, 6, 7), the distribution falls monotonously, as large oscillator strengths are rare. This can be attributed to the strong spatial oscillations of high-energy wave functions. Remember that in an ideal ordered system, these states would carry no oscillator strength at all (momentum selection rule $\mathbf{k} = 0$). Using statistical arguments, it can be shown that the oscillator strength distribution follows here a Porter–Thomas law,

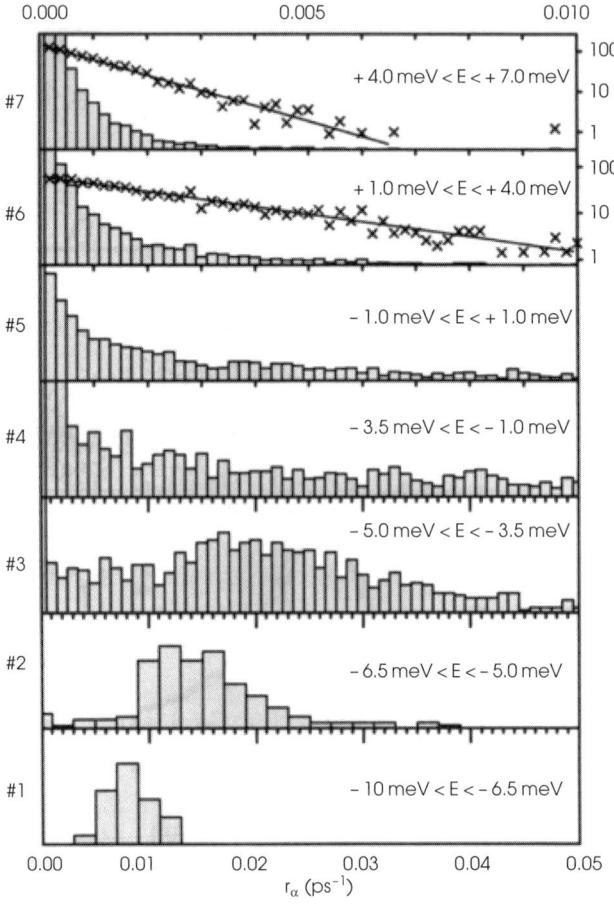

Fig. 4.4 Distribution of optical oscillator strengths $r_\alpha \propto \psi_{\alpha,0}^2$ in different energy windows. Symbols in panels 6 and 7 are rescaled in order to show the Porter–Thomas law Eq. (4.17) as a straight line (solid). Based on simulation data for a 5 nm AlGaAs QW with $\sigma = 8$ meV; after [47].

$$\mathscr{P}(r, E) = \overline{\delta(r - \psi_{\alpha,\mathbf{k}=0}^2)\, \delta(E - \varepsilon_\alpha)} \propto \frac{1}{\sqrt{r}}\, e^{-r/r_0(E)}. \tag{4.17}$$

Below the line center, exciton states are getting sparse and strongly localized. Going deeper into the tail, the oscillator strength distribution narrows appreciably, panels 1, 2 in Fig. 4.4. This can be easily explained by optimum fluctuation theory [49,50]: At a given energy in the tail, there are only a few specific disorder arrangements which give the major contribution to the OD within exponential accuracy. To get the dominant state $\psi_{OF}(\mathbf{R})$, the potential in Eq. (4.3) has to be replaced by

$$V(\mathbf{R}) \Rightarrow -\kappa \int d\mathbf{R}\, g(\mathbf{R} - \mathbf{R}')\, \psi_{OF}^2(\mathbf{R}'), \tag{4.18}$$

where κ is a Lagrange parameter. Thus, a nonlinear Schrödinger equation has to be solved.

Note that around the line center (Fig. 4.4, panels 3, 4), the distribution is rather broad and neither of the limiting cases discussed above applies.

4.3 Exciton Hamiltonian and density-matrix approach

For the narrow energy window around the $1s$ exciton, it suffices to construct an exciton Hamiltonian from the underlying electron-hole picture. Traditionally, this is done by defining creation and annihilation operators of excitons in plane-wave COM states. Then, the disorder enters as an elastic scattering process, commonly treated in second Born approximation. Our emphasis, however, is on a proper description of disorder effects. Therefore, we prefer to work in the representation of disorder eigenstates and assign to each of the states α an exciton creation operator $B_\alpha^\dagger$. Since the analysis is here restricted to low optical excitation, Boson commutation rules for the $B_\alpha^\dagger$ and B_α can be assumed.

The Heisenberg exciton operator $B_\alpha(t)$ describes a local oscillating dipole within the quantum well. Therefore, it is the source of an emitted electromagnetic wave. We use the terminology by Stolz [15] to express the field operator (positive rotating part $\mathbf{E}_\alpha^{(+)} \equiv \mathcal{E}_\alpha$) at a position $\mathbf{r}_1$ outside the sample as

$$\mathcal{E}_\alpha(\mathbf{r}_1, t) = \int d\mathbf{r}\, \frac{\omega_x^2 \mu_{cv}}{c^2 |\mathbf{r}_1 - \mathbf{r}|} \, \Psi_\alpha(\mathbf{r}_e = \mathbf{r}, \mathbf{r}_h = \mathbf{r})\, B_\alpha(t'). \tag{4.19}$$

This holds for dipole-allowed optical transitions, where the exciton-light coupling is proportional to the probability to find electron and hole at the same position. The retarded time $t' = t - |\mathbf{r}_1 - \mathbf{r}|/c$ accounts for the free propagation from the source at $\mathbf{r}$ to the observation point at $\mathbf{r}_1$. In the far-field limit, the expansion $|\mathbf{r}_1 - \mathbf{r}| \approx r_1 - \mathbf{r}_1 \cdot \mathbf{r}/r_1$ holds and reduces the retarded time to $t' = t - r_1/c + \mathbf{k} \cdot \mathbf{r}/\omega_x$, where the emission wave vector is introduced as $\mathbf{k} = (\mathbf{r}_1/r_1)(\omega_x/c)$. The retardation along the sample extension is tiny (below 1 fs), and consequently the leading time dependence of the operator $B_\alpha(t')$ $\sim \exp(i\,\omega_x t')$ is sufficient to replace

$$B_\alpha(t') \approx B_\alpha(t - r_1/c)\, e^{i\mathbf{k} \cdot \mathbf{r}}, \tag{4.20}$$

thus converting the retardation into a spatial interference pattern. Altogether, we have

$$\mathcal{E}_\alpha(\mathbf{r}_1, t) = \frac{k^2 \mu_{cv}}{r_1} \int d\mathbf{r}\, \Psi_\alpha(\mathbf{r},\mathbf{r})\, e^{i\mathbf{k}\cdot\mathbf{r}}\, B_\alpha(t - r_1/c). \tag{4.21}$$

Summing over all states, we obtain

$$\mathcal{E}_\mathbf{k}(t) = \frac{k^2}{r_1} \sum_\alpha M_\alpha(\mathbf{k})\, B_\alpha(t - r_1/c) \tag{4.22}$$

with the definition of the state-dependent optical matrix element as

$$M_\alpha(\mathbf{k}) = \mu_{cv} \int d\mathbf{r}\, e^{i\mathbf{k}\cdot\mathbf{r}}\, \Psi_\alpha(\mathbf{r},\mathbf{r}). \tag{4.23}$$

Due to the factorization Eq. (4.2), the matrix element can be simplified as

$$M_\alpha(\mathbf{k}) = \mu_{cv}\, \phi_{1s}(0)\, O_{eh}\, \psi_{\alpha\mathbf{k}}, \quad \psi_{\alpha\mathbf{k}} = \int d\mathbf{R}\, e^{i\mathbf{k}\cdot\mathbf{R}}\, \psi_\alpha(\mathbf{R}), \tag{4.24}$$

where $\mathbf{R}$ is as before the two-dimensional COM coordinate in the quantum well. Therefore, in the Fourier-transformed wave function, only the in-plane component $\mathbf{k}_\parallel$ of the light wave vector $\mathbf{k}$ enters. Due to $\lambda \gg L_z$, the confinement overlap $O_{eh} = \int dz\, u_e(z)\, e^{ik_z z}\, u_h(z)$ is nearly independent of k_z.

Exciting the system with a coherent light field, we expect a nonvanishing expectation value of the induced field, i.e. polarization

$$P_\mathbf{k}(t) = \langle \mathcal{E}_\mathbf{k}^\dagger(t)\rangle = \sum_\alpha M_\alpha^*(\mathbf{k})\, \langle B_\alpha^\dagger(t)\rangle. \tag{4.25}$$

The brackets are understood as the quantum mechanical expectation value (over bath variables etc.). This has to be distinguished from the disorder average which will be denoted by overlining throughout this work. Another quantity of interest is the time-resolved intensity

$$I_\mathbf{k}(t) = \langle \mathcal{E}_\mathbf{k}^\dagger(t) \mathcal{E}_\mathbf{k}(t)\rangle = \sum_{\alpha\beta} M_\alpha^*(\mathbf{k})\, M_\beta(\mathbf{k})\, \langle B_\alpha^\dagger(t) B_\beta(t)\rangle, \tag{4.26}$$

which will be investigated in great detail in the following sections. The light propagation from the sample to the detector placed at $\mathbf{r}_1$ is not of interest here. Therefore, we have omitted in Eqs (4.25) and (4.26) the k^2/r_1 prefactor and the overall retardation $-r_1/c$. This is consistent with putting into the following Hamiltonian the light coupling as $\mathcal{E}_\mathbf{k} M_\alpha^*\, B_\alpha^\dagger$ plus Hermitian conjugate,

$$\mathcal{H} = \sum_\alpha \varepsilon_\alpha B_\alpha^\dagger B_\alpha + \sum_{\alpha\mathbf{k}} [\mathcal{E}_\mathbf{k}\, M_\alpha^*(\mathbf{k})\, B_\alpha^\dagger + \mathcal{E}_\mathbf{k}^\dagger\, M_\alpha(\mathbf{k})\, B_\alpha]$$
$$+ \sum_{\alpha\beta\mathbf{q}} t_{\alpha\beta}^\mathbf{q} (a_\mathbf{q}^\dagger + a_{-\mathbf{q}}) B_\alpha^\dagger B_\beta + \sum_\mathbf{q} \hbar w_\mathbf{q}\, a_\mathbf{q}^\dagger a_\mathbf{q}. \tag{4.27}$$

Apart from the disorder effects which are implemented exactly into Eq. (4.27) using the disorder eigenstates of excitons, there are other (inelastic) scattering processes. They give rise to dephasing, and contribute to the linewidth as homogeneous broadening. In general, they can be described by a linear coupling to Bose fields $a_\mathbf{q}^\dagger$ with dispersion $\hbar w_\mathbf{q}$. For the lowest exciton states under consideration here, scattering with acoustic phonons is in most cases the dominant process. More details on the corresponding exciton-phonon matrix elements $t_{\alpha\beta}^\mathbf{q}$ are given in section 4.4. The following derivation applies quite generally to a Hamiltonian of type Eq. (4.27).

Equation (4.27) has to be supplemented by the Hamiltonian of the free light field with photon dispersion

$$\omega_\mathbf{k} = kc/n, \quad n = \varepsilon_b^2. \tag{4.28}$$

Here, n is the refractive index of the medium, and ε_b the background dielectric constant in the gap region, excluding the 1s exciton contribution. We have chosen to drop this part of the total Hamiltonian at the moment, since specific light field properties such as polarization degree of freedom and spontaneous emission will be treated later in sections 4.10, 4.11, and Appendix C.

The central quantities within the density matrix approach are the (equal-time) expectation values for the exciton density matrix

$$N_{\alpha\beta}(t) = \langle B_\alpha^\dagger(t) B_\beta(t) \rangle \tag{4.29}$$

and the polarization

$$P_\alpha(t) = \langle B_\alpha^\dagger(t) \rangle. \tag{4.30}$$

Their equations of motion are obtained from the time-dependence of the Heisenberg operators $-i\hbar\partial_t B_\alpha^\dagger = [\mathcal{H}, B_\alpha^\dagger]$ with the Hamiltonian Eq. (4.27):

$$(-i\hbar\partial_t - \varepsilon_\alpha + \varepsilon_\beta) N_{\alpha\beta} = \sum_{\rho\mathbf{q}} (t_{\rho\alpha}^\mathbf{q} T_{\rho\beta,\mathbf{q}} + t_{\rho\alpha}^\mathbf{q} T_{\beta\rho,-\mathbf{q}}^* - T_{\alpha\rho,\mathbf{q}} t_{\beta\rho}^\mathbf{q} - T_{\rho\alpha,-\mathbf{q}}^* t_{\beta\rho}^\mathbf{q})$$
$$+ \sum_\mathbf{k} (M_\alpha(\mathbf{k}) \langle \mathscr{E}_\mathbf{k}^\dagger B_\beta \rangle - M_\beta^*(\mathbf{k}) \langle B_\alpha^\dagger \mathscr{E}_\mathbf{k} \rangle) \tag{4.31}$$

$$(-i\hbar\partial_t - \varepsilon_\alpha) P_\alpha = \sum_{\rho\mathbf{q}} (t_{\rho\alpha}^\mathbf{q} \tilde{T}_{\rho,\mathbf{q}} + t_{\rho\alpha}^\mathbf{q} \hat{T}_{\rho,-\mathbf{q}}) + \sum_\mathbf{k} M_\alpha(\mathbf{k}) \langle \mathscr{E}_\mathbf{k}^\dagger \rangle. \tag{4.32}$$

This is the first step of an infinite hierarchy: The two- and one-operator expectation values $N_{\alpha\beta}$ and P_α couple to the so-called phonon-assisted density matrices [51],

$$T_{\rho\beta,\mathbf{q}} \equiv \langle a_\mathbf{q}^\dagger B_\rho^\dagger B_\beta \rangle, \quad \hat{T}_{\rho,\mathbf{q}} \equiv \langle a_\mathbf{q}^\dagger B_\rho^\dagger \rangle, \tag{4.33}$$

which involve three and two operators, respectively. Their equations of motion in turn contain expectation values involving two phonon operators. In order to truncate the hierarchy, we factorize into exciton and phonon contributions [51–55], neglect the small phonon distortion $\langle a_q^\dagger a_{-q}^\dagger \rangle$ [56], and use a Bose distribution for the phonons,

$$\langle a_q^\dagger a_{q'} \rangle = \delta_{qq'}\, n_q. \tag{4.34}$$

The rationale for assuming the phonons to be in equilibrium is our restriction to the low-excitation case and the fact that bulk-like phonons will easily dissipate energy into regions far from the quantum structure under consideration. The resulting equations for the phonon-assisted density matrices contain cross terms describing phonon-assisted optical transitions. These are dropped, and we are left with

$$(-i\hbar\partial_t - \hbar w_q - \varepsilon_\rho + \varepsilon_\beta)\, T_{\rho\beta,q} = \sum_\eta ((n_q + 1)\, t_{\eta\rho}^{-q} N_{\eta\beta} - n_q\, t_{\beta\eta}^{-q} N_{\rho\eta}) \tag{4.35}$$

$$(-i\hbar\partial_t - \hbar w_q - \varepsilon_\rho)\, \tilde{T}_{\rho,q} = \sum_\eta (n_q + 1)\, t_{\eta\rho}^{-q} P_\eta. \tag{4.36}$$

This equation of motion will be solved within the Markov approximation: The crucial step is to observe that most expectation values involve rapidly varying phase factors ($\varepsilon_\alpha \equiv \hbar\omega_\alpha$),

$$\langle B_\alpha^\dagger \rangle = P_\alpha \sim e^{+i\omega_\alpha t},\; N_{\alpha\beta} \sim e^{+i(\omega_\alpha - \omega_\beta)t},\; T_{\rho\beta,q} \sim e^{+i(w_q + \omega_\rho - \omega_\beta)t}. \tag{4.37}$$

As shown in Appendix B, this can be used to obtain

$$T_{\rho\beta,q} = i\pi \sum_\eta \delta(\hbar w_q + \varepsilon_\rho - \varepsilon_\eta)(n_q + 1)\, t_{\eta\rho}^{-q} N_{\eta\beta}$$
$$- i\pi \sum_\eta \delta(\hbar w_q + \varepsilon_\eta - \varepsilon_\beta)\, n_q\, t_{\beta\eta}^{-q} N_{\rho\eta} \tag{4.38}$$

$$\tilde{T}_{\rho,q} = i\pi \sum_\eta \delta(\hbar w_q + \varepsilon_\rho - \varepsilon_\eta)(n_q + 1)\, t_{\eta\rho}^{-q} P_\eta. \tag{4.39}$$

These terms will contribute significantly to the dynamics of $N_{\alpha\beta}(t)$ and $P_\alpha(t)$ only if the leading frequencies coincide. For the first term on the r.h.s. of Eq. (4.38) and for Eq. (4.39), the resonance condition is: $\varepsilon_\eta \approx \varepsilon_\alpha$. For non-degenerate exciton states, it is reasonable to assume that this implies $\eta = \alpha$. This argument yields the following contributions to Eq. (4.31) and (4.32) (with $t_{\alpha\beta}^{-q} = t_{\beta\alpha}^{q*}$)

$$t_{\rho\alpha}^q T_{\rho\beta,q} \to i\pi\delta(\hbar w_q + \varepsilon_\rho - \varepsilon_\alpha)(n_q + 1)\, |t_{\rho\alpha}^q|^2 N_{\alpha\beta}(t) \tag{4.40}$$

$$t_{\rho\alpha}^q \tilde{T}_{\rho,q} \to i\pi\delta(\hbar w_q + \varepsilon_\rho - \varepsilon_\alpha)(n_q + 1)\, |t_{\rho\alpha}^q|^2 P_\alpha(t). \tag{4.41}$$

We combine these phonon emission processes with the corresponding contributions from phonon absorption ($\hat{T}$) and obtain

$$\sum_{\rho q} (t_{\rho\alpha}^q T_{\rho\beta,q} + t_{\rho\alpha}^q T_{\beta\rho,-q}^*) \to \frac{i\hbar}{2} \sum_\rho \gamma_{\rho \leftarrow \alpha} N_{\alpha\beta}(t) \tag{4.42}$$

$$\sum_{\rho q} (t_{\rho\alpha}^q \tilde{T}_{\rho,q} + t_{\rho\alpha}^q \hat{\tilde{T}}_{\rho,-q}) \to \frac{i\hbar}{2} \sum_\rho \gamma_{\rho \leftarrow \alpha} P_\alpha(t), \tag{4.43}$$

where we have introduced the phonon scattering rates as

$$\gamma_{\rho \leftarrow \alpha} = \frac{2\pi}{\hbar} \sum_{\mathbf{q}} ((n_{\mathbf{q}} + 1)\, \delta(\varepsilon_\rho + \hbar w_{\mathbf{q}} - \varepsilon_\alpha) + n_{\mathbf{q}}\, \delta(\varepsilon_\rho - \hbar w_{\mathbf{q}} - \varepsilon_\alpha))\, |t^{\mathbf{q}}_{\rho\alpha}|^2. \quad (4.44)$$

Qualitatively different is the second contribution to the r.h.s. of Eq. (4.38), because its rapidly oscillating factors do not depend on α or β. The resonance condition is $\varepsilon_\alpha - \varepsilon_\beta \approx \varepsilon_\rho - \varepsilon_\eta$. The case $\{\rho, \eta\} = \{\alpha, \beta\}$ is excluded for $w_{\mathbf{q}} \neq 0$ by the delta-function. Thus, we expect this term to contribute only if $\varepsilon_\alpha = \varepsilon_\beta$ and furthermore $\varepsilon_\rho = \varepsilon_\eta$. Again, we assume that this implies $\alpha = \beta$ and $\rho = \eta$:

$$t^{\mathbf{q}}_{\rho\alpha} T_{\rho\beta,\mathbf{q}} \to -\delta_{\alpha\beta}\, i\pi\, \delta(\hbar w_{\mathbf{q}} + \varepsilon_\rho - \varepsilon_\alpha)\, n_{\mathbf{q}}\, |t^{\mathbf{q}}_{\rho\alpha}|^2\, N_{\rho\rho}(t) \quad (4.45)$$

$$\sum_{\rho\mathbf{q}} (t^{\mathbf{q}}_{\rho\alpha} T_{\rho\beta,\mathbf{q}} + t^{\mathbf{q}}_{\rho\alpha} T^*_{\beta\rho,-\mathbf{q}}) \to -\delta_{\alpha\beta}\, \frac{i\hbar}{2} \sum_{\rho} \gamma_{\alpha \leftarrow \rho}\, N_{\rho\rho}(t). \quad (4.46)$$

Thus, this contribution turns out to be of in-scattering form. A corresponding term is not present in the polarization equation (4.39).

Next, we discuss the mixed photon-exciton expectation values. We separate the coherent polarization $\langle B^\dagger_\alpha \rangle = P_\alpha$, which is driven by the macroscopic classical field $\langle \mathscr{E}_\mathbf{k} \rangle = E_\mathbf{k}(t)$, from a fluctuation term. The latter is evaluated analogously to the phonon-assisted matrices in the Markov approximation, with photon distribution set equal to zero, as we consider emission into the field vacuum. At this point, we have to express the field operators via creation- and destruction operators of Bose type. Due to the rotating wave approximation, this is a one-to-one correspondence, and gives a modified commutator

$$[\mathscr{E}_\mathbf{k}, \mathscr{E}^\dagger_{\mathbf{k}'}] = \delta_{\mathbf{k},\mathbf{k}'}\, g_\mathbf{k} \quad (4.47)$$

with the coupling function $g_\mathbf{k}$ (see Appendix C for details). Assuming the exciting field to be directed along $\mathbf{k}_0$, $E_\mathbf{k}(t) = \delta_{\mathbf{k}\mathbf{k}_0} E_0(t)$, the result is

$$\langle B^\dagger_\alpha \mathscr{E}_\mathbf{k} \rangle = P_\alpha(t)\, \delta_{\mathbf{k}\mathbf{k}_0}\, E_0(t) + i\pi\, g_\mathbf{k} \sum_\eta \delta(\varepsilon_\eta - \hbar\omega_\mathbf{k})\, M_\eta(\mathbf{k})\, N_{\alpha\eta}(t). \quad (4.48)$$

This contributes to the equation of motion for $N_{\alpha\beta}$, Eq. (4.31), only if $\varepsilon_\eta = \varepsilon_\beta$ or $\eta = \beta$. Thus,

$$M^*_\beta(\mathbf{k})\, \langle B^\dagger_\alpha \mathscr{E}_\mathbf{k} \rangle \to P_\alpha(t)\, \delta_{\mathbf{k}\mathbf{k}_0}\, E_0(t)\, M^*_\beta(\mathbf{k}_0) + i\pi g_\mathbf{k} \delta(\varepsilon_\beta - \hbar\omega_\mathbf{k})\, |M_\beta(\mathbf{k})|^2\, N_{\alpha\beta}(t) \quad (4.49)$$

and

$$\sum_\mathbf{k} M^*_\beta(\mathbf{k})\, \langle B^\dagger_\alpha \mathscr{E}_\mathbf{k} \rangle \to P_\alpha(t)\, E_0(t)\, M^*_\beta(\mathbf{k}_0) + \frac{i\hbar}{2}\, r_\beta\, N_{\alpha\beta}(t). \quad (4.50)$$

Here, we have introduced the rate for spontaneous radiative decay

$$r_\beta = \frac{2\pi}{\hbar} \sum_{\mathbf{k}} g_{\mathbf{k}} \, \delta(\varepsilon_\beta - \hbar\omega_{\mathbf{k}}) \, |M_\beta(\mathbf{k})|^2, \qquad (4.51)$$

which is further evaluated in Appendix C.

With the total out-scattering rate defined as

$$2\Gamma_\alpha = r_\alpha + \sum_\beta \gamma_{\beta \leftarrow \alpha}, \qquad (4.52)$$

the final result takes the form

$$\partial_t N_{\alpha\beta}(t) = (i(\omega_\alpha - \omega_\beta) - (\Gamma_\alpha + \Gamma_\beta)) \, N_{\alpha\beta}(t) + \delta_{\alpha\beta} \sum_\rho \gamma_{\alpha \leftarrow \rho} N_{\rho\rho}(t)$$

$$+ \frac{i}{\hbar} P_\beta^*(t) \, M_\alpha(\mathbf{k}_0) \, E_0^*(t) - \frac{i}{\hbar} P_\alpha(t) \, M_\beta^*(\mathbf{k}_0) \, E_0(t) \qquad (4.53)$$

$$\partial_t P_\alpha(t) = (i\omega_\alpha - \Gamma_\alpha) \, P_\alpha(t) + \frac{i}{\hbar} M_\alpha(\mathbf{k}_0) \, E_0^*(t). \qquad (4.54)$$

The polarization equation can be solved without problems,

$$P_\alpha(t) = \int_{-\infty}^{t} dt' \, e^{(i\omega_\alpha - \Gamma_\alpha)(t-t')} \, \frac{i}{\hbar} M_\alpha(\mathbf{k}_0) \, E_0^*(t'). \qquad (4.55)$$

Obviously, the polarization decays with Γ_α, the dephasing rate (this was the reason for introducing into Eq. (4.52) the factor of two).

For the nondiagonal density matrix $N_{\alpha\beta}$ with $\alpha \neq \beta$, there is no inscattering term in Eq. (4.53), and a careful inspection shows that the factorized ansatz

$$N_{\alpha\beta}(t) = P_\alpha(t) P_\beta^*(t), \quad \alpha \neq \beta. \qquad (4.56)$$

solves the equation of motion directly. Therefore, it is only the kinetic equation of the diagonal density $N_{\alpha\alpha}(t)$ which needs to be solved numerically,

$$\partial_t N_{\alpha\alpha}(t) = S_\alpha(t) + \sum_\rho \gamma_{\alpha \leftarrow \rho} N_{\rho\rho}(t) - 2\Gamma_\alpha N_{\alpha\alpha}(t). \qquad (4.57)$$

Besides the phonon scattering rates (4.44) and radiative rates (4.51), we have introduced the source term $S_\alpha(t)$ as

$$S_\alpha(t) = \operatorname{Im} \frac{2}{\hbar} P_\alpha(t) \, M_\alpha^*(\mathbf{k}_0) \, E_0(t). \qquad (4.58)$$

In standard time-resolved light-scattering experiments, a very short excitation pulse is used ($t_P \sim 10 - 500$ fs). In most cases, $\sigma t_P \ll 1$ holds, and the excitation is close to a *delta-like* excitation $E_0(t) \approx \hbar\Omega_0 \, \delta(t)$ on the scale of the inhomogeneous exciton line ($\hbar\Omega_0$ is the pulse area). Then, the polarization is simply

$$P_\alpha(t) = i\Omega_0 \, M_\alpha(\mathbf{k}_0) \, e^{(i\omega_\alpha - \Gamma_\alpha)t} \, \Theta(t), \qquad (4.59)$$

and the corresponding source term reads

$$S_\alpha(t) = \Omega_0^2 \, |M_\alpha(\mathbf{k}_0)|^2 \, \delta(t). \tag{4.60}$$

The expressions for stationary excitation are given in section 4.7. Numerical results for different cases are presented in sections 4.4 and 4.6.

Some remarks are in place here. First, the exciton population $N_{\alpha\alpha} = \langle B_\alpha^\dagger B_\alpha \rangle$ is refilled from other states. Thus, it decays in general much slower than the off-diagonal elements $N_{\alpha\beta}$, $\alpha \neq \beta$. Phonon scattering destroys the excitonic polarization, but, apart from radiative recombination, the total number of excitons is conserved.

Second, it is important to notice that the kinetic equation (4.57) involves only state-diagonal terms. Therefore, this kind of standard rate equation is not sufficient for the initial phase of coherent decay, where the off-diagonal terms (i.e. the polarization) are of central relevance.

Third, while the density matrix formulation is independent of the chosen basis in principle, a state-diagonal kinetics can only be an approximation. We shall briefly discuss the approximations involved in the derivation. Using the disorder eigenstates as basis is quite plausible. This eliminates off-diagonal components due to disorder scattering, which would be present in a plane-wave basis. The weakest point in the derivation above is to replace (approximate) energy conservation $\varepsilon_\eta = \varepsilon_\alpha$ by the condition $\eta = \alpha$. For infinite systems, states are arbitrarily close in energy, and this simple identification by itself is questionable. However, in Eqs (4.38), (4.39) and (4.45), the energy-dependent factor is weighted with phonon matrix elements $t_{\rho\alpha}^q t_{\rho\eta}^{q*}$. This factor assures that ψ_α^2 and ψ_η^2 both have a substantial overlap with ψ_ρ^2, which, in turn, implies that the states α and η are localized in the same spatial region. It will be discussed in detail in section 4.8 that two states which are localized in the same spatial region and which have nearly the same energies are almost certainly the same states (level repulsion).

Fourth, even with the diagonal assumption accepted, the present kinetic equation (4.57) for $N_{\alpha\alpha}$ is much more rich compared to a rate equation formulated in terms of an energy- (or momentum-) dependent exciton occupation, as treated in [57]. It is just the local spatial surrounding of a given state which determines its character: There are states being able to emit phonons easily, and others where the decay rests upon the radiative rate, and this may happen within the same energy window. A splitting of the exciton states into two groups according to this idea has been tried in [58], and some puzzling features seen in experiment could be explained.

Fifth, the specific way of selecting fully resonant terms in the phonon scattering and the subsequent Markov approximation has produced only strictly energy conserving processes. In a complete treatment, each exciton level is not only polaron shifted but gets phonon satellites with spectral weight on the expense of the sharp main line. Taking fully into account these subtle quasiparticle effects opens the way to understand a partial phase loss – even without a real phonon scattering to be happen. This has been called *pure dephasing* in the literature [59,60]. The coupling strength of acoustic phonons with excitons in GaAs, however, is not sufficient to influence the Rayleigh scattering markedly in this way, not to speak on luminescence which is much less sensitive. This is quite different in the strongly localized situation of quantum dots [61].

4.4 Exciton kinetics with acoustic phonon scattering

The dominant dephasing mechanism in excitonic resonant secondary emission from QW is acoustic phonon scattering with deformation-potential coupling [62]. Interaction with optical phonons is efficient at excitations above the band gap only. At higher excitation intensities, exciton-exciton scattering and the interaction of excitons with free carriers will take over [60].

First, we discuss the differences of QW and bulk material qualitatively. In the simple case of a single isotropic bulk band, the deformation-potential coupling is

$$H_{ph} = \sum_{a=e,h} \sum_{\mathbf{q}} \sqrt{\frac{\hbar q}{2\Omega \rho_m u_s}} (D_a a_{\mathbf{q}}^\dagger e^{i\mathbf{q}\cdot\mathbf{r}_a} + h.c.), \qquad (4.61)$$

in terms of the particle position $\mathbf{r}_a$ ($a = e, h$), and second-quantized generation and annihilation operators $a_{\mathbf{q}}^\dagger$ and $a_{\mathbf{q}}$ for longitudinal acoustic phonons with linear dispersion, $w_{\mathbf{q}} = u_s q$. Note that in the electron-hole language used throughout, the deformation potentials D_a are negative quantities ($D_e = D_c$ and $D_h = -D_v$ in the standard terminology). The sample volume, the mass density, and the speed of sound are Ω, ρ_m, and u_s, respectively [63–66].

For free carriers having a kinetic energy of the order of $k_B T$, the scattering efficiency of thermal acoustic phonons and the phonon-emission probability are both given by the squared matrix element (proportional to q and thus to $k_B T$ via the phonon dispersion), and therefore vanishingly small in the low-temperature limit. Thus, other scattering processes, in particular piezo-electric scattering with squared matrix elements $\sim 1/q$, dominate for $T \to 0$.

This argument applies to excitons in bulk material as well. Phonons interact with the constituents separately. Their momenta can be rather large ($\sim 1/a_B$) even for low-energy excitons. However, the momentum transfer is supplied by the center-of-mass motion, which, again, limits it to $k \sim \sqrt{k_B T}$.

The situation is qualitatively different for quantum structures such as quantum wells. Only the in-plane momentum is a conserved quantity. The vertical confinement in a well of width L_z provides at no cost in energy momentum differences up to about $q_z \approx 2\pi/L_z$. The deformation potential scattering of free carriers and excitons is thus enhanced relative to the bulk case [67].

For the following derivation of the kinetic equation for the exciton population, we neglect the anisotropy and multi-band character of the valence-band maximum and simply sum the interaction (4.61) for both constituents. On the one hand, this simplifies the notation. On the other hand, it reflects the fact that for many materials only the isotropic total deformation potential is known. The deformation potential matrix element for scattering between disorder eigenstates is [67]

$$t_{\alpha\beta}^q = \sqrt{\frac{\hbar w_q}{2u_s^2 \rho_m \Omega}} \iint d\mathbf{r}_e d\mathbf{r}_h \, \Psi_\beta^*(\mathbf{r}_e \mathbf{r}_h)(D_e e^{i\mathbf{q}\cdot\mathbf{r}_e} + D_h e^{i\mathbf{q}\cdot\mathbf{r}_h})\Psi_\alpha(\mathbf{r}_e \mathbf{r}_h). \tag{4.62}$$

Within the factorization approximation, Eq. (4.2), this takes the form [68]

$$t_{\alpha\beta}^q = \sqrt{\frac{\hbar w_q}{2u_s^2 \rho_m \Omega}} (D_e K_e(q_z) \chi(\mathbf{q}_\parallel/\eta_e) + D_h K_h(q_z) \chi(\mathbf{q}_\parallel/\eta_h)) \cdot (\psi_\alpha \psi_\beta)_{\mathbf{q}_\parallel}. \tag{4.63}$$

The mass factors η_a have been defined in Eq. (4.5). The dependence on the disorder COM eigenstates is via the Fourier-transformed overlap

$$(\psi_\alpha \psi_\beta)_{\mathbf{q}_\parallel} = \int d\mathbf{R} \, \psi_\alpha^*(\mathbf{R}) \, e^{-i\mathbf{q}_\parallel \cdot \mathbf{R}} \, \psi_\beta(\mathbf{R}). \tag{4.64}$$

Note that $(\psi_\alpha \psi_\beta)_{\mathbf{q}_\parallel} \to \delta_{\alpha\beta}$ for $q_\parallel \to 0$. Further, the Fourier transforms K_a and χ of the squared confinement and relative wavefunctions $u_{e,h}^2$ and ϕ_{1s}^2, respectively, enter:

$$K_a(q_z) = \int dz \, u_a^2(z) \, e^{-iq_z z}, \quad \chi(\mathbf{q}_\parallel) = \int d^2r \, \phi_{1s}^2(r) \, e^{-i\mathbf{q}_\parallel \cdot \mathbf{r}}. \tag{4.65}$$

In order to illustrate the dependence on geometry and material parameters, we give explicit expressions using the infinite-barrier limit for the confinement function, and an exponential for the exciton relative motion, respectively,

$$u_a(z) = \sqrt{\frac{2}{L_z}} \cos\left(\frac{\pi z}{L_z}\right), \quad K_a(q_z) = \frac{(2/q_z L_z) \sin(q_z L_z/2)}{1 - (q_z L_z/2\pi)^2};$$

$$\phi_{1s}(\mathbf{r}) = \sqrt{\frac{2}{\pi}} \frac{1}{a_B} e^{-r/a_B}, \quad \chi(\mathbf{q}_\parallel) = (1 + (q_\parallel a_B/2)^2)^{-3/2}. \tag{4.66}$$

Decay rates for a typical QW system are given in Fig. 4.5. The top panel shows the phonon scattering rates at $T = 4$ K, while the middle panel displays the radiative rates, Eq. (4.176). Their largest values are found near the line center [47]. The phonon rates increase rapidly with increasing energy. This reflects primarily the increasing final-state density. A comparison of the absolute values shows that only for low-energy excitons the radiative decay can be faster than phonon scattering. In particular at elevated temperatures, where phonon absorption and stimulated emission dominate, the total dephasing rate is mainly due to phonon scattering. The optical density is shown in the bottom panel. The low-energy part results from few strongly absorbing states and exhibits some structure, whereas the high-energy part is the sum of very many small contributions. In order to illustrate the broadening due to dephasing, a dashed curve obtained with a uniform Gauss broadening is shown for comparison. Concerning the numerics, a single realization consisted of 128×128 sites with step size $\Delta = 1.65$ nm, covering an area of $(211 \text{ nm})^2$. Altogether, 186 realizations have been added for the bottom panel. Covering thus a total area of $(3 \text{ μm})^2$, this is close to a wide-focus spectrum. Note that for the sake of clarity, data points from 10 realizations only are displayed in the two upper panels.

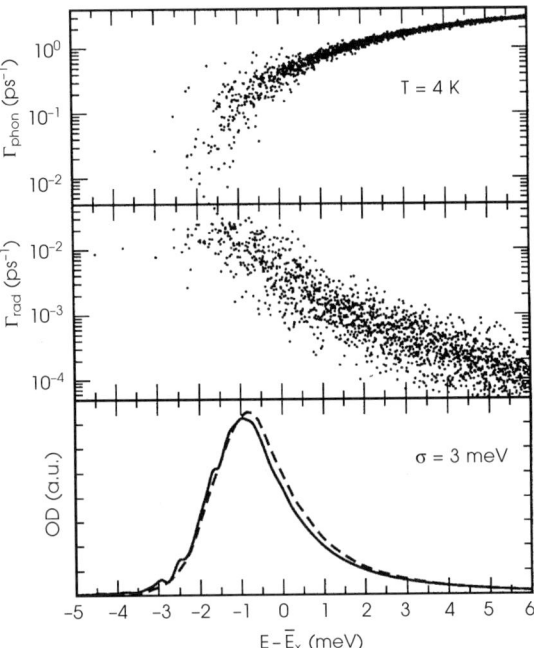

Fig. 4.5 Distribution of phonon rates (top panel) and radiative rates (middle panel) for a 5 nm wide AlGaAs quantum well at $T = 4$ K. Each dot refers to a given COM eigenstate. The bottom panel shows the corresponding optical density. Full curve – each state contributes with a Lorentzian of half width $\hbar\Gamma_\alpha$. Dashed curve – with uniform Gauss broadening of variance 0.2 meV. (Model calculation by courtesy of G. Mannarini).

The kinetic Eq. (4.57) has been applied to many questions related to the energy relaxation and thermalization of an exciton population. In most cases, only the dynamics of the diagonal density matrix was considered. It is quite plausible that the spectrally resolved secondary emission at time t is given by

$$I(\omega, t) = \sum_\alpha M_\alpha^2 \, \delta(\omega - \omega_\alpha) \, N_\alpha(t), \qquad (4.67)$$

which might be decomposed into a Rayleigh (coherent) and photoluminescence (incoherent) contribution, as detailed in section 4.5. More precise expressions for the frequency resolved spectra are derived in section 4.7. Within a similar reasoning, the optical density (absorption) is given by

$$D(\omega) = \sum_\alpha M_\alpha^2 \, \delta(\omega - \omega_\alpha). \qquad (4.68)$$

Depending on the experiment under consideration, different source terms resp. initial occupations of the exciton states have to be considered when solving the kinetic equation: (i) Hot excitation into continuum states with rapid energy relaxation by LO phonon emission and exciton formation within less than 20 ps. All exciton states are filled with approximately equal weight. Here, the entire emission is incoherent luminescence (PL). (ii) Resonant but broad-band excitation gives a population of optically active excitons with initial weight given by their squared optical matrix element. (iii) Monochromatic resonant excitation is described by an initial population at a single energy.

Photoluminescence excitation experiments (PLE) combine a resonant excitation at frequency ω_0 with a possible spectral resolution in the detection (frequency ω). In the present framework,

$$I^{\text{PLE}}(\omega_0, \omega, t) = \sum_\alpha M_\alpha^2 \, \delta(\omega - \omega_\alpha) \, N_\alpha(t), \qquad (4.69)$$

which has to be calculated with the source term $S_\alpha = E_0^2 \, M_\alpha^2 \, \delta(\omega_0 - \omega_\alpha)$. Finally, the resonant secondary emission is obtained from the PLE spectrum at $\omega = \omega_0$. An example of such calculations is given in Fig. 4.6. The red-shift of luminescence relative to absorption is called Stokes shift in analogy to Raman scattering. It depends on temperature and vanishes at elevated temperatures. The Stokes shift is not always a monotonous function of temperature: With decreasing temperature, a blue-shift of the PL follows sometimes the usual red-shift. This anomaly has been seen experimentally and analyzed theoretically [26] using the exciton kinetics (4.57). In contrast, the peak of the PLE signal is often shifted to energies above the absorption peak, as seen in Fig. 4.7. Qualitatively, the temperature dependence of the PLE line can be described as disappearance of the low-energy wing with decreasing temperature. This has been explained in terms of an effective mobility edge for exciton relaxation [27,69]. Excitons generated below the effective mobility edge are strongly localized in local potential minima. During their finite lifetime, they are unable to reach other minima lower in energy which would contribute to low-energy detection. Obviously, the position of the effective mobility edge depends on temperature and exciton lifetime.

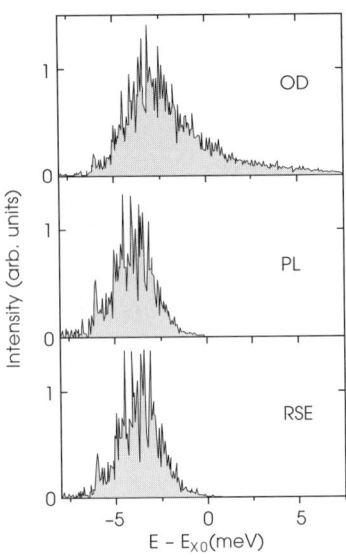

Fig. 4.6 Results from simulations based on Eq. (4.57) for various steady-state spectra of a GaAs/Al$_x$Ga$_{1-x}$As QW system. From top to bottom: Optical density or absorption (OD), photoluminescence (PL) for broad resonant excitation, and resonant secondary emission (RSE), all in arbitrary units. The spikes due to individual eigenstates can be seen since the simulation area was restricted to (2 μm)^2; from [70].

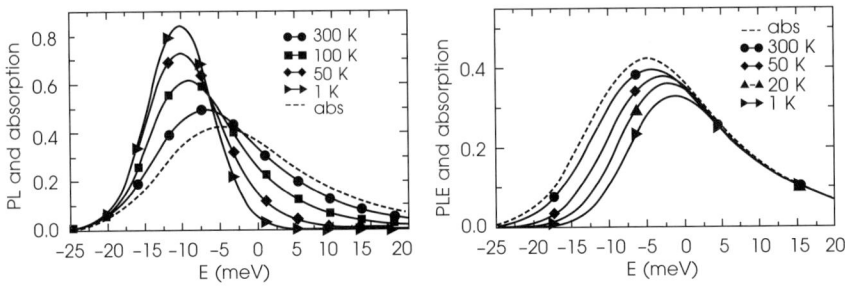

Fig. 4.7 Calculated PL (left) and PLE (right) for a strained In$_x$Ga$_{1-x}$As/GaAs sample with strong disorder. A strong Stokes shift of PL and a moderate blue-shift of PLE are seen; from [26].

4.5 Coherent and incoherent emission in the time domain

The distinction between coherent and incoherent secondary emission and the related subject of quantum coherence are hotly debated issues of both fundamental and, possibly, practical importance. *Losing phase coherence* is generally associated with processes changing the energy of the particle under consideration. For the present case, these are phonon-scattering events. We do not want to discuss fundamental questions of decoherence in quantum systems nor present sample-specific results, but rather focus on the experimental possibilities to measure the degree of coherence quantitatively, and the opportunities which this opens. Some examples will show that the separation of coherent and incoherent secondary emission yields information which otherwise would not be accessible.

A good starting point for a quantitative theory of the coherent and incoherent secondary emission is a detailed discussion of the intensity $I_\mathbf{k}(t)$ in direction $\mathbf{k}$ at time t after a short-pulse excitation at $t = 0$ [21,30]. Initially, the sample is excited coherently: All excitonic oscillators are in phase with each other and with the exciting light pulse. Here, we consider a single quantum well sitting in a uniform dielectric background. This allows to neglect reabsorption or propagation effects (see section 4.11 for relaxing this restriction). As shown explicitly in section 4.3, the electric field $\mathcal{E}_\mathbf{k}$ at the detector is directly proportional to the polarization in the sample at the same wave vector. In terms of the disorder eigenstates, the quantum mechanical expectation value is

$$P_\mathbf{k}(t) = \langle \mathcal{E}_\mathbf{k}^* \rangle_t = \mu_{cv}\, \phi_{1s}(0) O_{eh} \sum_\alpha \int d\mathbf{R}\, f(\mathbf{R})\, e^{-i\mathbf{k}\cdot\mathbf{R}}\, \phi_\alpha(\mathbf{R})\, P_\alpha(t). \tag{4.70}$$

We include a function $f(\mathbf{R})$ describing the spot whose emission is collected.

For a discussion of the dependence of P on $\mathbf{k}$ and t, we assign to each COM eigenstate an exciton position $\mathbf{R}_\alpha$, e.g., by the expectation value of the position operator. The precise definition does not matter, as we will only consider functions of $\mathbf{R}_\alpha$ which vary smoothly with a characteristic length of the order of the wavelength of the light λ. The latter is assumed to be large compared to the scale of the disorder and to the exciton COM extension, $\lambda \gg \Lambda$. For typical single QW with Λ below 100 nm, this condition is easily fulfilled.

Separating rapidly and slowly varying factors, we obtain

$$P_\mathbf{k}(t) = \sum_\alpha e^{-i\mathbf{k}\cdot\mathbf{R}_\alpha} f(\mathbf{R}_\alpha)\, M_\alpha^*\, P_\alpha(t) \tag{4.71}$$

with the optical matrix element at zero momentum, $M_\alpha \equiv M_\alpha(\mathbf{k}=0)$. With the explicit solution Eq. (4.59) for delta-like excitation and assuming normal incidence for the excitation, $\mathbf{k}_0 = 0$, we have

$$P_{\mathbf{k}}(t) = i\Omega_0 \sum_{\alpha} f(\mathbf{R}_\alpha) e^{-i\mathbf{k}\cdot\mathbf{R}_\alpha} |M_\alpha|^2 e^{i\omega_\alpha t} e^{-\Gamma_\alpha t} \qquad (4.72)$$

($t \geq 0$ is understood within the remainder of this section.) The sum involves many complex numbers. Therefore, the central limit theorem is applicable, and the fields are Gaussian distributed. For each disorder realization, the expectation value $P_{\mathbf{k}}(t)$ varies strongly with $\mathbf{k}$ and t. It also varies from one potential realization to the next. We have shown in [30] that an angular average over $\mathbf{k}$ directions (speckle average) is equivalent to an ensemble average over disorder realizations. At small times $t \approx 0$ and observation along the normal direction, $\mathbf{k} \approx 0$, all contributions are in phase and add up to the strong reflected/transmitted beam. Outside this region, however, the average of the expectation value (4.72) vanishes. It is the fluctuation which remains and shows up as finite intensity. This behavior for the Rayleigh part has been called *non-ergodic* by Shah and coworkers [17].

As an illustration, we consider the following toy model: The dephasing rates Γ_α are replaced by an average value Γ, and the optical density is assumed to have Gaussian shape with spectral variance σ_1. This yields for the ensemble average

$$\overline{P_{\mathbf{k}}(t)} \sim i\Omega_0 \tilde{f}_{\mathbf{k}} e^{-\sigma_1^2 t^2} e^{-\Gamma t} \qquad (4.73)$$

with the Fourier transform $\tilde{f}_{\mathbf{k}}$ of the focus $f(\mathbf{R})$. Clearly, this describes the reflected/transmitted beam centered at $\mathbf{k} = 0$ with width given by the inverse focus size. The ensemble-averaged polarization disappears with a characteristic time given by the inverse width of the optical density, i.e. $1/\sigma_1$ in this example. For the ensemble average of the squared expectation value, however, we find in this model a finite result

$$\overline{|P_{\mathbf{k}}(t)|^2} \sim \Omega_0^2 e^{-2\Gamma t}, \qquad (4.74)$$

which depends neither on the direction $\mathbf{k}$ nor shape or width of the optical density.

Next, we discuss the time-dependent intensity

$$I_{\mathbf{k}}(t) = \sum_{\alpha\beta} f(\mathbf{R}_\alpha) f(\mathbf{R}_\beta) e^{-i\mathbf{k}\cdot(\mathbf{R}_\alpha-\mathbf{R}_\beta)} M_\alpha^* M_\beta \langle B_\alpha^\dagger(t) B_\beta(t) \rangle. \qquad (4.75)$$

The frequency-resolved counterpart is derived in section 4.7. The expectation value $N_{\alpha\beta} = \langle B_\alpha^\dagger B_\beta \rangle$ will factorize for two spatially well separated eigenstates: $N_{\alpha\beta} \approx \langle B_\alpha^\dagger \rangle \langle B_\beta \rangle = P_\alpha P_\beta^*$. When we add and subtract the factorized expectation value, the double sum splits naturally into two contributions,

$$I_{\mathbf{k}}(t) = I_{\mathbf{k}}^{\text{coh}}(t) + I^{\text{inc}}(t):$$

$$I_{\mathbf{k}}^{\text{coh}}(t) = |\sum_{\alpha} f(\mathbf{R}_\alpha) e^{-i\mathbf{k}\cdot\mathbf{R}_\alpha} M_\alpha^* P_\alpha(t)|^2 = |P_{\mathbf{k}}(t)|^2, \qquad (4.76)$$

$$I^{\text{inc}}(t) = \sum_{\alpha\beta, \mathbf{R}_\beta \neq \mathbf{R}_\alpha} f(\mathbf{R}_\alpha) f(\mathbf{R}_\beta) M_\alpha^* M_\beta (N_{\alpha\beta}(t) - P_\alpha(t) P_\beta^*(t)). \qquad (4.77)$$

We have written the restriction $\mathbf{R}_\beta \approx \mathbf{R}_\alpha$ explicitly, even though it is fulfilled automatically: As shown in Eq. (4.56), the off-diagonal elements of the density matrix factorize, $N_{\alpha\beta}(t) = P_\alpha(t) P_\beta^*(t)$ for $\alpha \neq \beta$. Consequently, only the diagonal terms contribute, and we may define an incoherent exciton density via

$$N_\alpha^{\text{inc}}(t) = N_{\alpha\alpha}(t) - |P_\alpha(t)|^2. \tag{4.78}$$

Thus, the incoherent contribution to the emission takes the form

$$I^{\text{inc}}(t) = \sum_\alpha f^2(\mathbf{R}_\alpha) |M_\alpha|^2 N_\alpha^{\text{inc}}(t). \tag{4.79}$$

It is quite instructive to formulate the kinetic equation for the incoherent density directly. The source Eq. (4.58) is replaced by a single out-scattering term from the coherent part $|P_\alpha(t)|^2$,

$$\partial_t N_\alpha^{\text{inc}} = \sum_\beta \gamma_{\alpha \leftarrow \beta} |P_\beta(t)|^2 + \sum_\beta \gamma_{\alpha \leftarrow \beta} N_\beta^{\text{inc}}(t) - 2\Gamma_\alpha N_\alpha^{\text{inc}}(t). \tag{4.80}$$

After short-pulse excitation, first the polarization is excited, followed by an increase of the incoherent density on the expense of phonon scattering, as seen in Fig. 4.8. Obviously, the incoherent contribution is independent of the detection direction, whereas the coherent contribution varies strongly with $\mathbf{k}$. These strong fluctuations (speckle) reflect the interference between the field components originating from different eigenstates which is a coherent process [18,71,72]. In contrast, the incoherent emission comes from summing

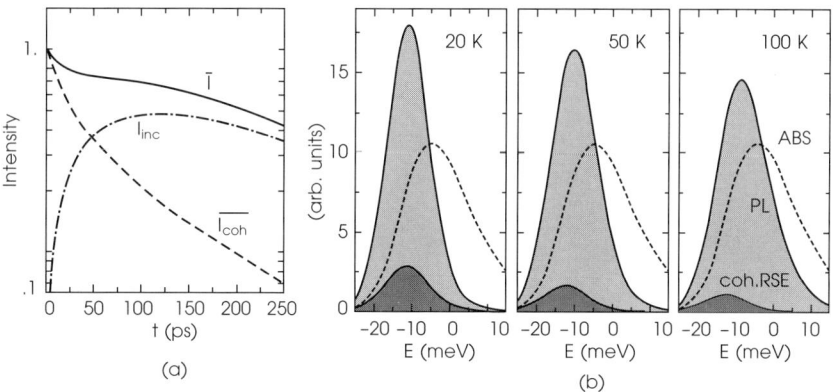

Fig. 4.8 (a) Calculated coherent ($\overline{I^{\text{coh}}}$, dashed) and incoherent (I^{inc}, dot-dashed) contributions to the speckle-averaged emission ($\overline{I}$, solid line) following resonant excitation of a 1.2 nm wide In$_{0.5}$Ga$_{0.5}$As/GaAs quantum well with σ = 8.5 meV at 5 K on a logarithmic scale; from [21]. (b) Relative weight of coherent (dark gray) and incoherent (light gray, PL) emission at various temperatures. Optical density (ABS) for comparison; from [70].

up the contributions from the individual exciton states independently, since the phase information got lost during the phonon scattering process.

Restoring the full **k**-dependence in the matrix elements, we rewrite the time-resolved coherent emission Eq. (4.76) after short-pulse excitation as

$$I^{coh}_{\mathbf{k}\mathbf{k}_0}(t) = \Omega_0^2 \left| \sum_\alpha M_\alpha^*(\mathbf{k}) M_\alpha(\mathbf{k}_0) e^{(i\omega_\alpha - \Gamma_\alpha)t} \right|^2. \tag{4.81}$$

If needed, the focus function $f(\mathbf{R})$ can be included in the integral defining the optical matrix element, Eq. (4.24). As before, the momentum $\mathbf{k}_0$ denotes the direction of the incoming light. It is instructive to write Eq. (4.81) as a double sum over states,

$$I^{coh}_{\mathbf{k}\mathbf{k}_0}(t) = \Omega_0^2 \sum_{\alpha\beta} M_\alpha^*(\mathbf{k}) M_\alpha(\mathbf{k}_0) M_\beta(\mathbf{k}) M_\beta^*(\mathbf{k}_0) e^{i(\omega_\alpha - \omega_\beta)t} e^{-(\Gamma_\alpha + \Gamma_\beta)t}. \tag{4.82}$$

Treating for the moment the dephasing rates as state-independent (which we know is a poor approximation), we can recast Eq. (4.82) into a compact shape using the completeness of states $\psi_\alpha(\mathbf{R})$,

$$I^{coh}_{\mathbf{k}\mathbf{k}_0}(t) = \Omega_0^2 \, e^{-2\Gamma t} \iint d\mathbf{R}\, d\mathbf{R}' \, e^{-i\mathbf{k}\cdot(\mathbf{R}-\mathbf{R}')} \, e^{it(H(\mathbf{R})-H(\mathbf{R}'))} \, e^{i\mathbf{k}_0\cdot(\mathbf{R}-\mathbf{R}')}. \tag{4.83}$$

Again, the COM Schrödinger Hamilton operator $H(\mathbf{R})$ has been invoked, Eq. (4.3). The initial behavior ($t \to 0$) is governed by moment-like expressions. In contrast to Eq. (4.13), however, we have two different spatial arguments: The coherent emission falls into the realm of two-particle Green's functions, while the optical density is related to the one-particle Green's function. For $t = 0$, a Kronecker symbol in momentum follows easily, $I^{coh}_{\mathbf{k}\mathbf{k}_0}(0) = \Omega_0^2 A^2 \delta_{\mathbf{k}\mathbf{k}_0}$, which is the specular emission. This argument relies on the completeness of states. If by purpose the excitation pulse does not cover spectrally all optically active states, a nearly instantaneous appearance of the RRS is expected. This has been confirmed in a sample with monolayer-split excitons, where only one sub-ensemble has been excited [14].

Continuing with Eq. (4.83), there is no first-order contribution in time due to $\overline{V(\mathbf{R})} = 0$, and the emission in non-specular direction starts with

$$\overline{I^{coh}_{\mathbf{k}\neq\mathbf{k}_0}(t \to 0)} = \Omega_0^2 \, A \, e^{-2\Gamma t} \cdot t^2 \int d\mathbf{R}\, g(\mathbf{R}). \tag{4.84}$$

This is the famous quadratic rise of the coherent emission (see Fig. 4.1). Therefore, it may happen that the incoherent emission dominates at very early times, since it rises linearly in time (at least within the Markov approximation for phonon scattering). It is rather cumbersome to proceed with the series expansion of $I^{coh}_{\mathbf{k}\mathbf{k}_0}(t)$. However, in the classical limit where the Hamilton operator reduces to the potential $V(\mathbf{R})$, all averages can be performed directly [7], yielding

$$\overline{I^{\text{coh}}_{\text{class}}(t)} = \Omega_0^2\, A\, e^{-2\Gamma t}\, e^{-\sigma^2 t^2} \cdot \int d\mathbf{R}\, (e^{+g(\mathbf{R})t^2} - 1). \tag{4.85}$$

Here, for simplicity, the momentum dependence has been dropped *after* observing $\mathbf{k} \neq \mathbf{k}_0$.

The speckle average in the coherent intensity singles out the diagonal part $\alpha = \beta$ in Eq. (4.82), since all other terms have random phases (strictly speaking, this holds only for times $t\sigma > 1$). Therefore, at later times we are left with the simple form,

$$\overline{I^{\text{coh}}_{\mathbf{kk}_0}(t)} = \Omega_0^2\, \sum_\alpha |M_\alpha(\mathbf{k}) M_\alpha(\mathbf{k}_0)|^2\, e^{-2\Gamma_\alpha t}, \tag{4.86}$$

which exhibits a non-exponential decay. In the subsequent sections, several of the expressions listed here will be taken as starting point.

4.6 Speckle measurement and interferometry

We saw in section 4.5 that the $\mathbf{k}$ dependence allows to separate the coherent from the incoherent scattering intensity. The latter is $\mathbf{k}$-independent, whereas the former fluctuates strongly with $\mathbf{k}$. Both real and imaginary part of the emitted field are Gaussian distributed (central limit theorem) and uncorrelated. This implies an exponential distribution for the absolute squares (intensity). Note that the Jacobian $\partial I_\mathbf{k}/\partial |E_\mathbf{k}|$ cancels the two-dimensional volume element in the complex plane, and we consider detection of a single polarization direction. Adding up the non-fluctuating incoherent intensity, the total intensity exhibits a displaced exponential distribution [20],

$$\mathscr{P}(I) = \theta(I - I^{\text{inc}})\, \frac{e^{-(I - I^{\text{inc}})/I^{\text{coh}}}}{I^{\text{coh}}}. \tag{4.87}$$

A full analysis using the variation of $I_\mathbf{k}(t)$ with either observation direction $\mathbf{k}$ or time t for constructing the intensity distribution (histogram) has been performed [21]. However, Langbein *et al.* [20] noted that there is a much easier way to extract the degree of coherence

$$c = \frac{\overline{I^{\text{coh}}}}{\overline{I^{\text{coh}}} + I^{\text{inc}}} \tag{4.88}$$

from the data: It suffices to evaluate the averages of intensity and intensity fluctuations which leads for ideal detectors to the remarkably simple relation

$$c = \frac{1}{\overline{I}}\sqrt{\overline{I^2} - \overline{I}^2}. \tag{4.89}$$

Again, the clue is the exponential distribution of the coherent intensity. In contrast, the incoherent contribution I^{inc} is not fluctuating, since an average over many laser shots is involved. This is the meaning of the quantum mechanical expectation value (bath variables), which is denoted by brackets in the present text. We reiterate that overlining marks an average over ensembles or speckles (direction **k**). A decomposition of the measured time-resolved emission into the coherent and incoherent parts along this line is displayed in Fig. 4.9.

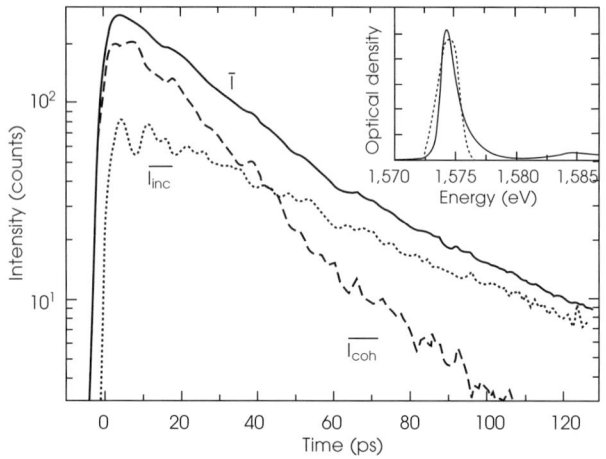

Fig. 4.9 Coherent ($\overline{I_{coh}}$, dashed) and incoherent ($\overline{I_{inc}}$, dotted) contributions to the speckle-averaged resonant secondary emission ($\overline{I}$, solid line) from an 8 nm wide GaAs/AlGaAs single quantum well with a small inhomogeneous broadening of 0.85 meV, see inset. The separation is obtained from the experimentally determined coherence degree as a function of time; from [20].

On a semilogarithmic plot, the distribution (4.87) forms a triangle, horizontally displaced from the origin by I^{inc}. Due to finite temporal and spatial resolution, the width of the experimentally obtained speckle distributions is less than ideal. A triangular shape with a rounded vertex is obtained, Fig. 4.10, and the expression (4.89) would lead to an underestimation of the degree of coherence. This can be avoided by a more detailed analysis of the resolution-modified speckle distribution [28]. From its very shape, the actual resolution parameter can be revealed, and the underlying full coherence degree can be reconstructed.

Quantitative speckle analysis has been used to extract the coherence degree of the secondary emission and to study the energy dependence of the phonon scattering rate [20]. Another application is the study of the loss of spin polarization [73]. Additional insight into the anisotropy of disorder localized wave functions can be obtained from

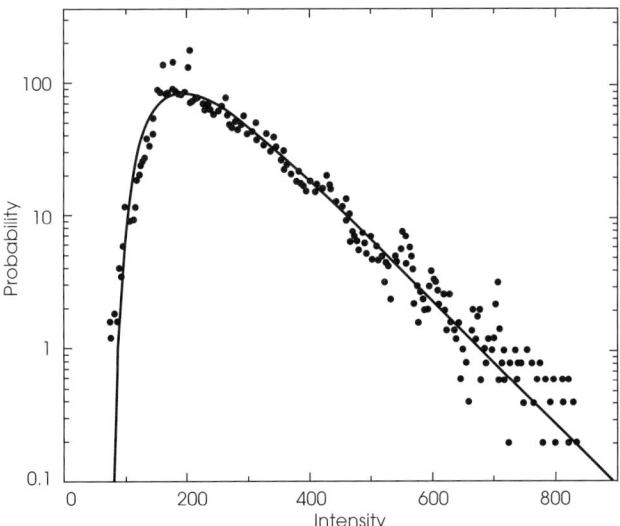

Fig. 4.10 Intensity distribution of the emission at a delay time of 5 ps from Fig. 4.9, and analytical fit (full curve); from [28].

cross-correlations of speckle intensities [74], which show up also in the (speckle-averaged) polarization degree (see section 4.10).

Alternatively, the degree of coherence can be determined using interferometric (IF) measurements, which actually predated the speckle analysis. In a nutshell, interferometric experiments study a superposition of the excitation pulse $E_0(t) = \sqrt{I_0} e^{i\omega_0 t}$ with the secondary emission. The former is attenuated by a factor α and phase-shifted by a time-delay τ,

$$I_{\mathbf{k}}^{\text{IF}}(t) = \langle (\alpha E_0(t-\tau) + \mathscr{E}_{\mathbf{k}}) \times (h.c.) \rangle$$
$$= \alpha^2 I_0 + 2\alpha \sqrt{I_0 I_{\mathbf{k}}^{\text{coh}}(t)} \cos(\omega_0(t-\tau) + \varphi_0) + I_{\mathbf{k}}^{\text{coh}}(t) + I^{\text{inc}}(t). \quad (4.90)$$

The phase φ_0 summarizes constant phase factors.

The time-dependent intensity can be detected e.g. via frequency up-conversion [16]. An oscillatory behavior with respect to the delay τ proves the presence of a coherent component. A quantification of the latter is obtained if the attenuation α is adjusted to maximize the fringe contrast between maxima and minima,

$$F_{\mathbf{k}} = \frac{I_{\mathbf{k}}^{\text{IF}}|_{\max} - I_{\mathbf{k}}^{\text{IF}}|_{\min}}{I_{\mathbf{k}}^{\text{IF}}|_{\max} + I_{\mathbf{k}}^{\text{IF}}|_{\min}}. \quad (4.91)$$

The maximal contrast is reached at $\alpha^2 I_0 = I^{\text{coh}} + I^{\text{inc}}$ and is given by

$$F_{\mathbf{k}}(t) = \sqrt{\frac{I_{\mathbf{k}}^{coh}(t)}{I_{\mathbf{k}}^{coh}(t) + I^{inc}(t)}}. \tag{4.92}$$

It shows the same speckle as the intensity itself [75]. In this sense, interferometric measurements and the method of speckle statistics are equivalent. The contrast $F_{\mathbf{k}}(t)$ can be anything between one (for an exceptionally strong speckle) and zero (in between speckles). The message is that interferometric methods are not conclusive if only one scattering direction is investigated. Only after singling out mean value and variance of the interferometric amplitude from an average over directions, the degree of coherence (4.88) can be extracted. To characterize the statistical ensemble of exciton states, statistical methods of evaluation are indispensable. Measuring a single speckling trace (over time) is not conclusive at all, and a comparison with simulations on a few disorder realization is not meaningful either [76]. Using two phase-locked excitation pulses gives additional information (coherent control) on the secondary emission, but is hampered by the speckle statistics, too [77].

While the analysis and understanding of the speckled nature of any resonant emission is of great practical importance, we will for the rest of this review concentrate on speckle-averaged quantities.

4.7   Frequency-resolved secondary emission

Up to now we were concerned with equal-time expectation values which is sufficient if no spectral resolution of the intensity is wanted. Now, if a spectrometer is placed into the outgoing direction $\mathbf{k}$, we need two different times for the field (or exciton) operators. With a spectrometer resolution of Δ_S we have for the spectrally resolved (Fourier-limited) light intensity at time t [15] (compare Eq. (4.75))

$$I_{\mathbf{k}}(\omega, t) = \Delta_S \int_{-\infty}^{t} dt_1 e^{(i\omega - \Delta_S)(t-t_1)} \int_{-\infty}^{t} dt_2 \, e^{(-i\omega - \Delta_S)(t-t_2)} \tag{4.93}$$
$$\sum_{\alpha\beta} M_{\alpha}^*(\mathbf{k}) M_{\beta}(\mathbf{k}) \langle B_{\alpha}^{\dagger}(t_1) B_{\beta}(t_2) \rangle.$$

The double-time expectation value is found from an equation of motion with respect to one time only. We apply $-i\partial_t B_{\alpha}^{\dagger} = (\omega_{\alpha} + i\Gamma_{\alpha}) B_{\alpha}^{\dagger} + E_0^*(f)(t) M_{\alpha}(\mathbf{k}_0)/\hbar$ to connect the temporal evolution to quantities with equal times. For the case $t_1 > t_2$ we find

$$\langle B_{\alpha}^{\dagger}(t_1) B_{\beta}(t_2) \rangle = P_{\alpha}(t_1) P_{\beta}^*(t_2) + e^{(i\omega_{\alpha} - \Gamma_{\alpha})(t_1 - t_2)} [N_{\alpha\beta}(t_2) - P_{\alpha}(t_2) P_{\beta}^*(t_2)] \tag{4.94}$$

and correspondingly for $t_1 < t_2$

$$\langle B_{\alpha}^{\dagger}(t_1) B_{\beta}(t_2) \rangle = P_{\alpha}(t_1) P_{\beta}^*(t_2) + e^{(-i\omega_{\beta} - \Gamma_{\beta})(t_2 - t_1)} [N_{\alpha\beta}(t_1) - P_{\alpha}(t_1) P_{\beta}^*(t_1)]. \tag{4.95}$$

The term within square brackets has been shown to be diagonal ($\alpha = \beta$, see Eq. (4.56)) and equal to the incoherent density, Eq. (4.78), which simplifies the two-time expectation value to

$$\langle B_\alpha^\dagger(t_1)B_\beta(t_2)\rangle = P_\alpha(t_1)P_\beta^*(t_2) + \delta_{\alpha\beta}e^{i\omega_\alpha(t_1-t_2)}e^{-\Gamma_\alpha|t_1-t_2|}N_\alpha^{inc}(\min(t_1,t_2)). \quad (4.96)$$

Obviously, the polarization product (first term) has a different time dependence compared to the remainder, and this splitting of terms coincides with the separation into coherent (or Rayleigh) part and incoherent (or luminescence) part.

We concentrate on the incoherent part first. Assuming a slow temporal variation of the incoherent density on the scale of both the dephasing rate Γ_α and the spectrometer resolution Δ_S, we may approximately take N_α^{inc} at the (measurement) time t when inserting Eq. (4.96) into Eq. (4.93). Carrying out both time integrations, we end up with

$$I_\mathbf{k}^{inc}(\omega,t) = \sum_\alpha M_\alpha^*(\mathbf{k})M_\alpha(\mathbf{k})N_\alpha^{inc}(t)\frac{\Gamma_\alpha+\Delta_S}{(\omega-\omega_\alpha)^2+(\Gamma_\alpha+\Delta_S)^2}. \quad (4.97)$$

Dephasing and spectrometer resolution contribute additively to the Lorentz broadening. Neglecting both quantities leads to the simple result quoted in section 4.4,

$$I_\mathbf{k}^{inc}(\omega,t) = \sum_\alpha |M_\alpha(\mathbf{k})|^2 N_\alpha^{inc}(t)\pi\delta(\omega-\omega_\alpha). \quad (4.98)$$

However, from the derivation it is quite clear that a perfect spectrometer ($\Delta_S = 0$ in Eq. (4.93)) would spoil any time resolution.

Only for stationary excitation, $E_0(t) = E_0\exp(-i\omega_0 t)$, Eq. (4.98) is fully consistent. To derive the source Eq. (4.58) in the kinetic equation, we first quote the polarization for stationary excitation,

$$P_\alpha(t) = \langle B_\alpha^\dagger(t)\rangle = \frac{1}{\hbar}E_0 e^{i\omega_0 t}\frac{M_\alpha(\mathbf{k}_0)}{\omega_0-\omega_\alpha-i\Gamma_\alpha}, \quad (4.99)$$

ending up with the time-independent source term

$$S_\alpha = \frac{2E_0^2}{\hbar^2}|M_\alpha(\mathbf{k}_0)|^2\frac{\Gamma_\alpha}{(\omega_0-\omega_\alpha)^2+\Gamma_\alpha^2}. \quad (4.100)$$

If dephasing is small, this simplifies to

$$S_\alpha = E_0^2\frac{2\pi}{\hbar}|M_\alpha(\mathbf{k}_0)|^2\delta(\hbar\omega_0-\varepsilon_\alpha). \quad (4.101)$$

For the coherent (or Rayleigh) part, we assume stationary excitation from the outset and use Eq. (4.99) to carry out the time integrations in Eq. (4.93) directly, with the result

$$I_{\mathbf{k}\mathbf{k}_0}^{coh}(\omega) = \frac{\Delta_S}{(\omega_0-\omega)^2+\Delta_S^2}\frac{E_0^2}{\hbar^2}\sum_{\alpha\beta}\frac{M_\alpha^*(\mathbf{k})M_\alpha(\mathbf{k}_0)M_\beta(\mathbf{k})M_\beta^*(\mathbf{k}_0)}{(\omega_0-\omega_\alpha-i\Gamma_\alpha)(\omega_0-\omega_\beta+i\Gamma_\beta)}. \quad (4.102)$$

Note the strict resonance with the incoming light field: Under steady-state illumination, the Rayleigh scattering has no broadening related to dephasing. The decay rates Γ_α enter only the sum which gives the weight of the sharp resonance peak. In contrast, the luminescence is spectrally broad due to both inhomogeneous (ω_α distribution) and homogeneous effects (Γ_α). The schematic sketch in Fig. 4.11 serves to illustrate these findings. This clear distinction in the spectral regime has been used first by Hegarty [3] to extract the Rayleigh spectrum from the total secondary emission. What is called Rayleigh intensity in the following is just the integrated strength of this emission as a function of excitation frequency,

$$I^{Ray}_{kk_0}(\omega_0) = \sum_{\alpha\beta} \frac{M^*_\alpha(\mathbf{k})M_\alpha(\mathbf{k}_0) M_\beta(\mathbf{k}) M^*_\beta(\mathbf{k}_0)}{(\omega_0 - \omega_\alpha - i\Gamma_\alpha)(\omega_0 - \omega_\beta + i\Gamma_\beta)}. \tag{4.103}$$

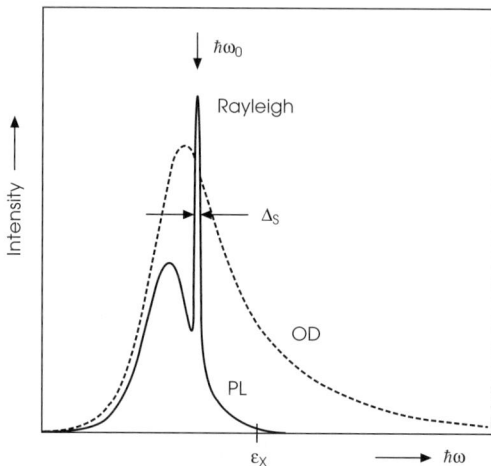

Fig. 4.11 Schematic illustration of the secondary emission spectrum under monochromatic excitation at $\hbar\omega_0$. Sharp spike – resonant Rayleigh scattering, broadened by the spectrometer resolution Δ_S. The optical density (dashed curve) is shown for comparison.

(The prefactor $\pi E_0^2/\hbar^2$ has been omitted for brevity.) The speckling behavior (dependence on scattered direction $\mathbf{k}$) resides in the factor $M^*_\alpha(\mathbf{k}) M_\beta(\mathbf{k}) \sim \psi^*_{\alpha k}\psi_{\beta k}$. Again, the angular average over the speckles can be replaced by an ensemble average. A careful analysis of Eq. (4.103) follows in section 4.9 on enhanced resonant backscattering.

A simplification is possible if the inhomogeneous broadening exceeds the dephasing, which resembles the long-time limit in the time-resolved emission, Eq. (4.86). Then, the diagonal terms $\alpha = \beta$ are expected to dominate the double sum, and we are left with

$$I^{\text{Ray}}(\omega_0) = \sum_\alpha \overline{\frac{M_\alpha^4(0)}{(\omega_0 - \omega_\alpha)^2 + \Gamma_\alpha^2}}. \tag{4.104}$$

As a rule, if only a speckle-averaged quantity is considered, we may safely neglect the momentum dependence of the optical matrix elements, putting $M_\alpha(\mathbf{k}) \approx M_\alpha(0)$. The strict small-damping limit reads

$$I^{\text{Ray}}(\omega_0) = \sum_\alpha \overline{\frac{M_\alpha^4(0)}{\Gamma_\alpha} \pi \delta(\omega_0 - \omega_\alpha)}. \tag{4.105}$$

The signal emitted in transmission direction $\mathbf{k}_0$ itself contains the coherent polarization only once, and allows to define the linear optical susceptibility from the ratio $P_\alpha(t)/E_0(t)$ as

$$\chi(\omega_0) = \sum_\alpha \frac{|M_\alpha(\mathbf{k}_0)|^2}{\omega_0 - \omega_\alpha - i\Gamma_\alpha}. \tag{4.106}$$

Its imaginary part gives the absorption (or optical density). Dropping the dephasing Γ_α and some nearly constant prefactors, we recover the simple expressions Eq. (4.11), resp. Eq. (4.68) for normal incidence.

The Rayleigh signal in comparison to the optical density shows two modifications (see Fig. 4.6): (i) The optical matrix element enters the Rayleigh signal in fourth power, which essentially reduces the high-energy tail in the OD and gives a narrowing of the spectrum. (ii) The dephasing rate in the denominator emphasizes states down in the tail – they need a longer time to dephase. This is the main reason for the distinct red shift of the RRS spectrum compared with the OD. Both peak shift and line narrowing have been found experimentally [15,78]. The enormous scatter in the dependence of Γ_α versus ε_α (see Fig. 4.5) prevents a simple interpretation as $\Gamma(E)$. In his seminal work on resonant Rayleigh scattering, Hegarty [3] did not take into account this statistical effect, but was still able to extract a consistent physical picture from the data. Based on our present experience, we claim that deducing dephasing rates from a comparison of frequency-resolved spectra (PLE vs. RRS) is nearly impossible without a detailed simulation.

4.8 Signatures of level repulsion

Optical near-field techniques which have been introduced recently allow a spectroscopy of excitons on a microscopic scale. With these methods, it became possible to 'zoom' into the inhomogeneous exciton spectrum and to resolve a multitude of single lines, each corresponding to an individual localized state. Knowing about the quantum-mechanical effect of level repulsion, we have proposed some years ago a spectral correlation technique which allows to get statistical information on the exciton eigenenergies, as fingerprint of their quantum-mechanical localization [79]. Only recently, this suggestion has been followed with specifically tailored experiments [29,80,81].

The energy-level statistics is a tool which was developed in the 1930s for the analysis of the spectra of highly excited nuclei. These spectra present many narrow resonances which seem to be distributed at random on the energy axis. For such complex systems a detailed modeling of the system Hamiltonian would be an extremely difficult task. However, an analysis of statistical properties, such as the level spacing distribution, bears very general information about the quantum-mechanical nature of the system. The level spacing distribution is defined as ensemble average

$$C(\Delta E) = \overline{\sum_{\alpha\beta} \delta(\Delta E - (\varepsilon_\alpha - \varepsilon_\beta))}. \tag{4.107}$$

The general finding was that this quantity displays a dip for $\Delta E \to 0$ suggesting that energy levels cannot be arbitrarily close to each other. This result reflects a general property of quantum mechanical systems: In the absence of special symmetries which induce degeneracy, the energy levels tend to *repel* each other. The simplest example of this general behavior is the *avoided level crossing* which takes place for example in the case of two interacting quantum states as the detuning is varied. In the case of nuclear spectra as well as in other complex systems, like for example in the optical spectra of large molecules, a very successful method is the random matrix theory (RMT) [82]. The RMT approach assumes that the elements of the Hamiltonian matrix are random numbers taken from a given ensemble. Because no further assumptions are made, the results have a universal character and apply to a large variety of complex systems. RMT with real symmetric matrices having Gauss distributed elements, the so-called *Gauss orthogonal ensemble*, has been very successful in reproducing the level statistics of, e.g., nuclear spectra and of other complex systems.

In this section, we consider the concept of level repulsion (LR) in the context of localized excitons. We will develop the basic LR formalism and present sample numerical results to illustrate the behavior for the case of the exciton COM. We then discuss recent experimental results where the numerical simulations have provided clear evidence for quantum mechanical LR.

In order to gain an intuitive picture of LR in the case of the Schrödinger problem with disorder, we must keep in mind that each wave function is characterized by a localization length Λ and therefore by a finite spatial extension. If two wave functions are at least partially overlapping, then the requirement of orthogonality implies that one of the two must be changing sign in space more frequently than the other. This implies a larger kinetic energy, and consequently a shift in the energy compared to the other state. On the other hand, wave functions which are well separated in space have an exponentially vanishing overlap and can therefore be quasi degenerate. This argument is illustrated in Fig. 4.12. It suggests that the level repulsion is essentially a local property of a disordered system.

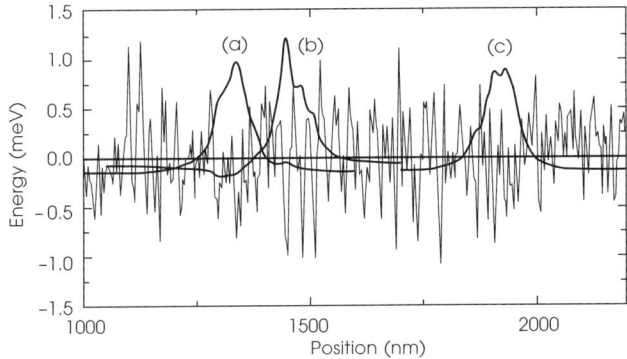

Fig. 4.12 Three localized states from the 1D Schrödinger problem are plotted together with the disorder potential. Each wave function is vertically displaced by its eigenenergy. Wave functions (a) and (b) have a non-negligible overlap and exhibit level repulsion, as opposite to wave functions (a) and (c) which are quasi degenerate.

For the present applications, it is only the statistics of the optically active states which matters. Therefore, we define the optically weighted level spacing distribution in analogy with Eq. (4.107) as

$$R(\Delta E) = \frac{1}{A} \overline{\sum_{\alpha\beta} M_\alpha^2 M_\beta^2 \, \delta(\Delta E - (\varepsilon_\alpha - \varepsilon_\beta))}, \quad (4.108)$$

where A denotes the quantum well area and M_α are the optical matrix elements as defined in Eq. (4.23). With the definition of the optical density without ensemble averaging, Eq. (4.68), the previous expression can be rewritten as a convolution of experimental (or theoretical) spectra,

$$R(\Delta E) = A \overline{\int dE' \, D(E') \, D(E' - \Delta E)}. \quad (4.109)$$

The statistics in Eq. (4.108) includes pairs of overlapping states as well as pairs which are far apart in space. We expect the statistical weight of the latter pairs to dominate if the system size is larger than the typical localization length of the optically active states.

Therefore, the quantum-mechanical level repulsion is expected to appear as a small feature superimposed to a large background originating from counting all the energy-uncorrelated pairs of eigenstates. In order to isolate the LR feature we subtract the uncorrelated counterpart of Eq. (4.109),

$$R_0(\Delta E) = \Omega \int dE' \, \overline{D(E')} \cdot \overline{D(E' - \Delta E)}, \tag{4.110}$$

and obtain the level autocorrelation function as the difference

$$R_c(\Delta E) = R(\Delta E) - R_0(\Delta E). \tag{4.111}$$

The contributions from pairs of non-overlapping states cancel out in the level autocorrelation $R_c(\Delta E)$. The autocorrelation is thus expected to provide information on the level repulsion.

A detailed analysis of the autocorrelation function depending on the amplitude and correlation length of the disorder potential has been presented in [30]. All the ensemble averaged results related to the COM Schrödinger equation depend on the dimensionless parameter E_c/σ upon scaling, as already pointed out in section 4.2. An analysis of the autocorrelation function $R_c(\Delta E)$ as a function of $\Delta E/\sigma$ is presented in Fig. 4.13. Numerical simulations have been performed for a 1D system of 2 μm size with $\sigma = 1$ meV for various values of E_c/σ. The classical limit $E_c/\sigma \to 0$ can be obtained analytically,

$$R^{cl}(\Delta E) = \int_A d\mathbf{R} \frac{1}{2\sqrt{\pi(\sigma^2 - g(\mathbf{R}))}} \exp\left(-\frac{\Delta E^2}{4(\sigma^2 - g(\mathbf{R}))}\right)$$

$$R_0^{cl}(\Delta E) = \frac{A}{2\sigma\sqrt{\pi}} \exp\left(-\frac{\Delta E^2}{4\sigma^2}\right). \tag{4.112}$$

The autocorrelation R_c has a maximum at $\Delta E = 0$ and negative wings for larger energy. This shape reflects the spatial correlation of the potential since in this limit, the energy levels are just the potential values. The positive values reflect that positions which are close in real space (distance $\leq \xi$) have very similar potential values. The negative values at large ΔE show that it is very unlikely for them to have vastly different potential energies. The curves are plotted by rescaling the vertical axis with the potential correlation length ξ. A feature common to the numerical simulations with $E_c/\sigma \neq 0$ is the delta peak at $\Delta E = 0$, corresponding to the correlation of each level with itself. The integral of the autocorrelation function, including this δ-contribution, equals zero by definition. The first case shown in Fig. 4.13(a) has $E_c/\sigma = 1$ and is close to the white-noise limit, in which the role of the spatial correlation is negligible. Strong level repulsion is seen: The autocorrelation function is negative for all values of the energy difference. An important feature displayed by the numerical result is the apparent sharp drop or even singularity at $\Delta E \to 0$. For larger values of ΔE the shape of the autocorrelation function bears similarity with the RMT result. At present, we have no analytical result which reproduces this behavior. The exponential decay of the localized wave functions at large distance and the orthogonality argument suggest that there might be a logarithmic singularity for vanishing energy spacing, induced by the level-repulsion effect between wave functions with an exponentially small overlap. This result represents a significant difference with respect to the RMT, where the autocorrelation function drops linearly for $\Delta E \to 0$. The discrepancy originates from the substantial difference between the present tight-binding matrix with disorder on the diagonal only, and a full random matrix. Obviously, the Anderson problem is not properly described within RMT.

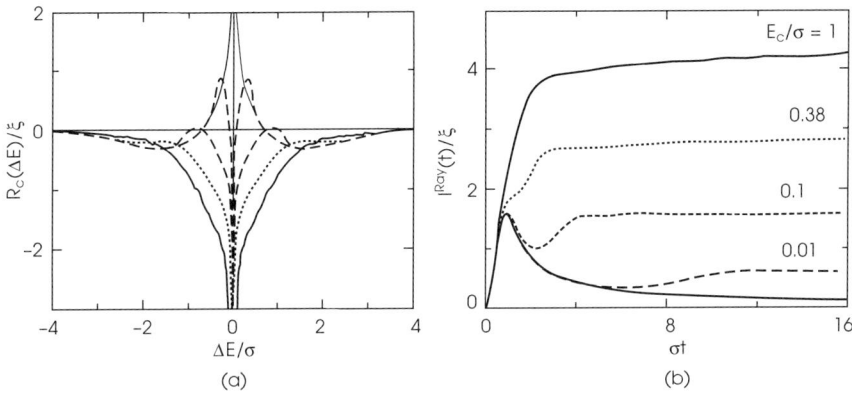

Fig. 4.13 (a) Autocorrelation function $R_c(\Delta E)$ computed for a 1D case for different values of E_c/σ. The thin line is the classical limit $E_c/\sigma \to 0$. (b) Computed Rayleigh scattering signal corresponding to the five cases depicted in panel (a); from [30].

The intermediate situations with $1 > E_c/\sigma > 0$ can be looked at as a quantum-mechanical LR effect superimposed onto the classical autocorrelation function. This produces a dip at $\Delta E \to 0$, as in the white-noise case, but closely matches the classical limit for larger energy spacing. This behavior results in a peak at $\Delta E \sim E_c$, which might be interpreted as an average energy spacing between successive excited states localized in the same local minimum of the spatially correlated potential. This intuitive picture must however be considered with caution, since the localization length Λ strongly increases as a function of energy, eventually resulting in localized states spanning several potential minima. We expect a spatial correlation length of 30–300 nm in realistic situations of excitons in QWs, see Fig. 4.15 below. Therefore, an intermediate situation like the ones depicted in Fig. 4.13(a), is expected to occur.

In the last decade, many experiments have been performed revealing the localized exciton structure on the sub-micron scale and providing at the same time spectral information [29,80,81,83–88]. Some of these works have dealt specifically with the concept of level repulsion. Recently, we were able to provide strong evidence for quantum-mechanical LR by comparing our numerical simulations to a statistical analysis of spectra measured by means of near-field scanning optical microscopy (NSOM) at low temperatures [29]. The investigated sample was a 3 nm thick GaAs/Al$_{0.5}$Ga$_{0.5}$As single QW with rather strong disorder. The near-field PL spectra at a sample temperature of 20 K were recorded with high spatial and spectral resolution (150 nm and 100 μeV, respectively). The spectra are dominated by a set of spectrally sharp and intense emission lines. In contrast, the spatially averaged PL spectrum is characterized by a 15 meV broad structureless emission band. The near-field spectra were measured on the same spot for different excitation powers coupled into the fiber ranging from 4 nW to 4 μW. The spectral structures are practically unchanged within this intensity range. This rules out biexcitonic effects which might show up as an (unwanted) feature at the biexciton binding energy in the spectral autocorrelation. We note in passing that level-correlation spectroscopy in the high-density regime could be of interest by itself. The data shown are taken at an excitation of 800 nW. A set of 432 photoluminescence spectra spanning a rectangular region of 3×5 μm^2 has been collected.

The simulation has been carried out on a 2D square domain with 130 nm lateral size, corresponding roughly to the experimental resolution. The spatial correlation of the potential has been simulated with an exponential smoothing function Eq. (4.16). This choice is justified assuming that the 1s exciton wave function is exponential. The potential correlation length ξ is expected to be of the order of the exciton Bohr radius. Using simulated spectra for the optical density for a large number of disorder realizations, the autocorrelation function has been generated according to Eqs (4.109), (4.110) and (4.111). The experimental spectra have been processed accordingly. In the numerical calculations, a Lorentz broadening of the spectra has been introduced, in order to account for the homogeneous exciton linewidth and for the finite energy resolution of the experiment. The parameters σ and ξ have been adjusted to fit the experimental autocorrelation curve, obtaining $\sigma = 5.3$ meV and $\xi = 17$ nm. The results are shown in Fig. 4.14. The agreement between theory and experiment is extremely good. Note that we have compared a calculation of the optical density with experimental results for the PL. This, however, is only a difference in the prefactors (exciton occupation), while the level structure is a persistent feature in all autocorrelation spectra.

The dashed curve shows the theoretical result without introducing a Lorentz broadening. Here, the level repulsion appears as a dip for $\Delta E \to 0$, in analogy with the

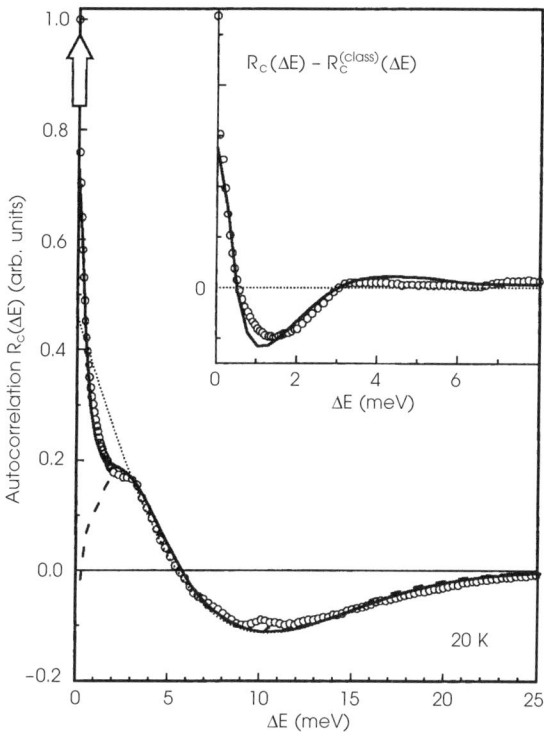

Fig. 4.14 The autocorrelation function $R_c(\Delta E)$ at $T = 20$ K is shown for the experimental data (circles), the Lorentz-convoluted numerical simulation (solid curve), the classical limit (dotted curve), and the raw numerical simulation (dashed curve). Level repulsion is evident from the shoulder around 3 meV (circles, solid curve). The arrow on the vertical axis denotes the δ-function part of the dashed curve. Inset: Difference between $R_c(\Delta E)$ and the classical limit $R_c^{cl}(\Delta E)$. Circles – experiment, solid curve – simulation; from [29].

intermediate case in Fig. 4.13. Due to the Lorentz broadening, the delta function and the level repulsion dip are smeared out, and only a shoulder remains in the autocorrelation function. The broadening used for the theoretical curve is consistent with the linewidth of the spikes as measured in the individual near-field spectra. In absence of quantum mechanical level repulsion, we would expect a classical result in which $R_c(\Delta E)$ is determined by the correlations of the potential only. This result is shown in Fig. 4.14 as a dotted line for comparison. The marked difference with the experiment, shown in detail in the inset, supports the interpretation in terms of quantum mechanical level repulsion.

The magnitude of the level repulsion dip is related implicitly to the spatial extension of the relevant COM wave functions. However, with the present model, and knowing the values of the parameters σ and ξ, it is possible to extract more detailed wave function informations. As an example, we evaluate the exciton localization length Λ_α of state ψ_α via the *participation ratio* [89]

$$\Lambda_\alpha^{-D} = \int d\mathbf{R}\, \psi_\alpha^4(\mathbf{R}). \tag{4.113}$$

(*D* denotes the system dimensionality.) Figure 4.15 shows the distribution of Λ_α across the exciton spectrum. The dots are computed from 10 potential realizations, while the line is an optically weighted average. Two features can be extracted from this plot. First, the exciton localization length is much larger than the potential correlation length ξ = 17 nm, proving that the wave functions are localized over several minima of the disorder potential. Second, the localization length Λ_α varies strongly as a function of energy, and even for a given energy it is broadly distributed. Obviously, the idea of a unique exciton localization length for a given system, often claimed in the exciton literature, is inappropriate.

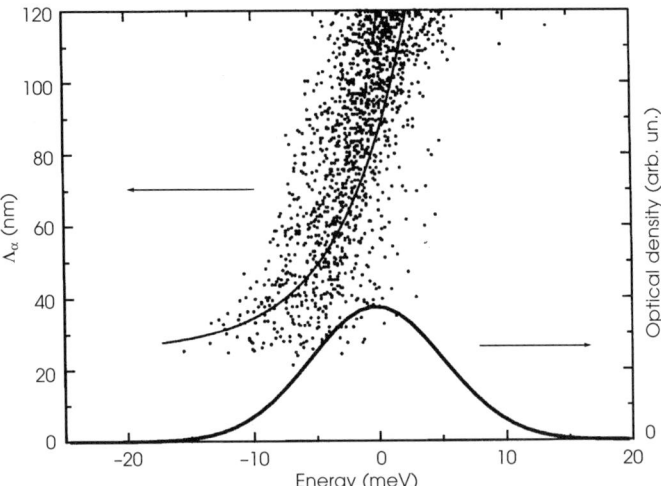

Fig. 4.15 Exciton localization length Λ_α (circles). The full curve is the optically-weighted average of Λ_α as a function of energy. Bold curve – averaged exciton optical density. All data refer to the simulation of Fig. 4.14.

Signatures of the LR can also be found in the time-resolved RRS signal (see Eq. (4.82) in section 4.5). Here we point out the relation between the time-dependent scattered intensity and the LR feature. It has been shown [30] that the speckle-averaged RRS can be expressed in terms of the autocorrelation function as Fourier integral,

$$I^{Ray}(t) = \int dE\, R_c(E) \cos(Et). \tag{4.114}$$

Figure 4.13(b) shows the time-dependent RRS signal calculated for different values of E_c/σ in rescaled units, in correspondence to the respective autocorrelation functions of Fig. 4.13(a). Since $\Gamma_\alpha = 0$ has been assumed, each curve approaches a finite value at long times,

$$I^{Ray}(t \to \infty) = \frac{1}{A} \overline{\sum_\alpha M_\alpha^4} \equiv L_{opt}^D. \tag{4.115}$$

The *optical length* L_{opt} can be interpreted as the average localization length of the optically active exciton states. The white-noise case, $E_c/\sigma = 1$, in Fig. 4.13(b) starts with the expected quadratic rise and increases monotonically up to the optical length. For large σt, the rise is very slow, as a consequence of the logarithmic feature at small energy spacing in $R_c(\Delta E)$ as seen in Fig. 4.13(a). This slow rise is therefore a first signature of LR. In contrast, a Gauss distribution of classical oscillators without any spatial correlation produces a RRS signal $I^{Ray}(t) \propto 1 - \exp(-\alpha^2 t^2)$, which approaches its long-time limit exponentially fast. The opposite situation, represented by the classical limit, displays a maximum at $\sigma t \sim 1$ and decays to zero as $(\sigma t)^{-D}$. The classical limit can be interpreted as a situation in which the particle has infinite mass and is infinitely localized, thus justifying a vanishing optical length. The intermediate situations display a non-monotonic behavior after the first maximum, which initially follows the classical result. This feature results from the Fourier transform of the maximum at finite ΔE present in $R_c(\Delta E)$ and can be considered a feature of LR in a situation with spatially correlated disorder potential, as we expect for excitons in QWs. Such a level-repulsion interpretation of time-resolved spectra has been suggested in two recent works [90,91]. It must be noted, however, that there are two basic problems: The first one is related to the difficulty of separating the coherent RRS signal from the total emitted intensity. In [90], it was assumed that at short times the RRS dominates the secondary emission, while in [91] a speckle interferometry was performed to select that part of the emitted intensity which is coherent with the incoming laser pulse. The other, more basic problem, is related to polariton effects [92,93], which are discussed in section 4.11. Here we just remark that multiple photon absorption and emission between different localized exciton states may give rise to multiple interference and thus to oscillations in the time-resolved signal. The radiative coupling constant Γ_0 in a single GaAs QW is of the order of a few tens of μeV, therefore considerably smaller than the energy range of LR features in $R_c(\Delta E)$. In this case we expect polariton effects to be negligible. They become more pronounced in multiple QWs [92,93]. An unambiguous demonstration of LR signatures in the time-resolved RRS signal thus requires measurements on a single QW in order to minimize polariton effects, and a robust method for sorting out the coherent RRS component from the secondary emission. Such an experimental evidence, to our knowledge, is still to come.

4.9 Enhanced resonant backscattering

One feature common to many cases of wave propagation in disordered systems is *enhanced backscattering* (EBS) [76]. If a monochromatic wave is scattered off an ideally static random medium, the angle-dependent scattered intensity displays speckles. In most experimental situations, however, a speckle average is introduced when measuring the angle-dependent intensity. This average can be due to the finite angular and frequency resolution in the experiment, and to a possible time evolution of the scattering medium during the measurement. As a result of this averaging, and for a non-resonant isotropic medium, the scattered intensity is nearly constant over the solid angle with, however, the remarkable exception of a narrow cone around the backscattering direction. Within this cone an increase of the scattered intensity is generally observed, peaking at the backscattering direction where a factor-of-two enhancement can be reached. The reason for this enhancement is usually explained within a perturbation picture [94–96]. The enhancement is then an interference effect in the coherent superposition of different scattering paths inside the medium. More precisely, for each multiple-scattering path the time-reversed path also contributes to the total scattered intensity and the two have a relative phase which tends to zero for angles close to the backscattering direction $\mathbf{k} = -\mathbf{k}_0$, independently of the disorder configuration. Thus the coherent summation of all pairs of time-reversed paths implies a factor-of-two enhancement in the backscattering direction. This ideal situation is never realized in practice, since single scattering events do not have a time-reversed equivalent, and contribute with an angle-independent intensity background. This very simple picture suggests also that the angular width of the EBS peak is proportional to the inverse of the average distance between the entrance- and exit-point into and from the disordered medium.

The EBS has been directly observed in a wide variety of systems. Light scattering experiments have revealed EBS in suspensions of microspheres in water [94–96], two-dimensional organic systems, and ultra-cold atom gases. Moreover, EBS was observed in the scattering of ultrasonic acoustic waves and of elastic waves in crystals. The concept of EBS is also relevant in the scaling behavior of the low-temperature conductivity of metals (weak localization) and, in three dimensions, in the metal-insulator transition.

Anderson localization can be interpreted as a result of EBS. In a system having infinite spatial extension, a wave originating at one point has higher probability of being scattered backwards than being diffused in other directions. This implies a vanishing probability of diffusing over long distances, which means localization. The difference to a system with boundaries is that for the latter the wave enters the system at a given time and eventually leaves it at a later time, giving rise to scattering paths rather than to localized modes. Excitons are produced by optical excitation and eventually recombine to give rise to secondary emission. In this sense, the light scattering on QWs is analogous to wave scattering from a random medium with boundaries. The multiply-scattered wave is the microscopic exciton polarization while the optical excitation and recombination play the role of the initial and final events of all multiple-scattering paths within the system. Thus we expect EBS to occur for excitons in disordered QWs, and the angular pattern of the scattered light to carry a corresponding signature.

The exciton, however, presents substantial differences with respect to the usual EBS systems. The perturbation picture mentioned above was developed for weak, isolated scatterers with mean free path much larger than the wavelength. In our disorder model, scattering is strong and each position acts as scattering site. Furthermore, dephasing and radiative damping introduce an 'external' characteristic time besides the scattering time.

In what follows, we will address these issues and characterize the excitonic EBS with the help of both numerical simulations and a perturbation-theory approach [31]. First, we present the result of a simulation based on Eq. (4.3) which shows EBS to occur for typical exciton parameters, in a simulated RRS experiment with resonant steady-state excitation. Then we make a connection with the perturbation theory of EBS in terms of two-particle diagrams which has been developed by Vollhardt and Wölfle [97] and is widely used to describe EBS.

Figure 4.16 shows the computed optical density and the RRS signal at normal incidence for parameters characteristic of a high-quality GaAs/AlGaAs QW. Panel (b) is a (k_x, k_y) contour plot of the scattered intensity at fixed frequency denoted by the dot in panel (a). Three relevant features can be extracted: First, the bright spot at the incoming wave vector represents the intensity in the reflected direction. Second, a bright ring at larger wave vectors is determined by the exciton dispersion relation $\varepsilon_x + \hbar^2 k^2/(2M) = \hbar\omega$ and simply reflects the resonant nature of the scattering process. The third and most important feature is a broad enhancement of the intensity centered at $\mathbf{k} = -\mathbf{k}_0$ constituting the EBS. Figures 4.16(c) and (d) show the RRS intensity as a function of k_x (k_y) for $k_y = 0$ ($k_x = 0$), respectively. The backscattering appears in Fig. 4.16(c) as a small peak opposite to the sharp reflected peak at $\mathbf{k}_0$, while the plot in Fig. 4.16(d) is symmetric around $k_y = 0$. A similar behavior is found for other values of the frequency of the incident field, with the exciton dispersion ring moving to different wave vectors according to the dispersion relation. The intensity enhancement is by far smaller than the factor of two expected from standard EBS theory. As shown below, this is partly due to the dominance of the single-scattering paths contributing a structureless background to the total scattered intensity.

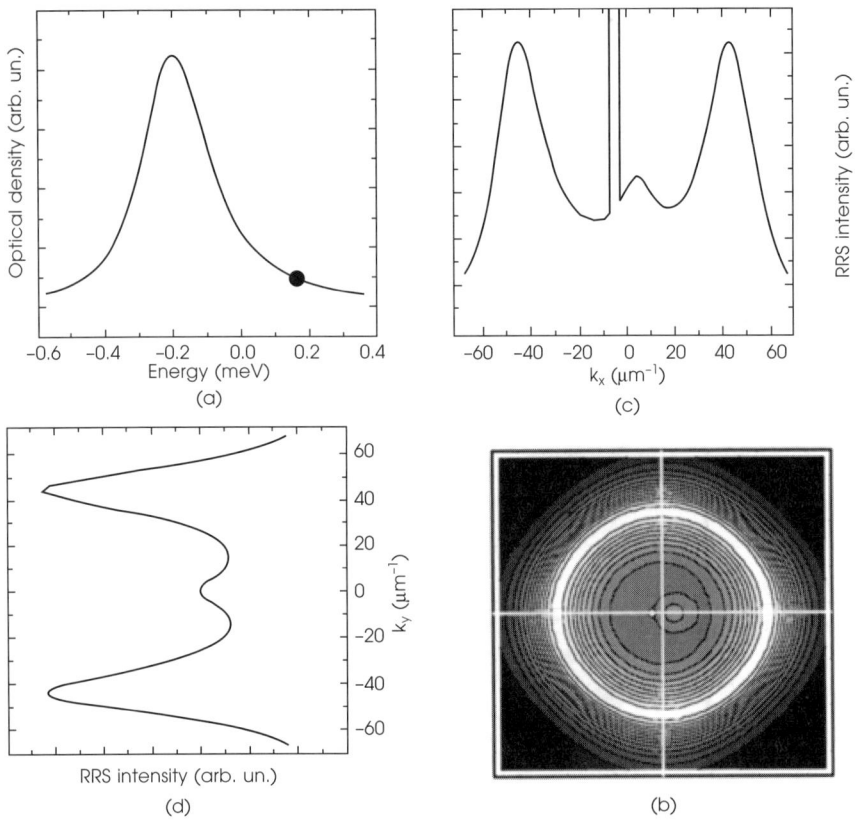

Fig. 4.16 (a) Computed exciton optical density. The dot refers to the energy value considered in the other panels. (b) Contour plot of the RRS intensity in the **k**-plane for fixed ω and for a two-dimensional white-noise disorder potential with $\sigma = 0.4$ meV. A square region of 2.6 µm lateral size is sampled with a grid of 128×128 points. The incident field is a plane wave at $\mathbf{k}_0 = (-4.4, 0)$ µm^{-1}. The scattered intensity has been averaged over 5×10^4 configurations which realizes the average over speckles. (c) and (d) are plots of the RRS intensity taken along the horizontal and vertical line on panel (b), respectively.

In a two-dimensional heterostructure, excitons interact with the external electromagnetic field if their momenta are within the *light cone* defined by $k \leq K_0 = \omega/c$, which corresponds to the whole external solid angle. For GaAs with fundamental gap of about 1.5 eV, $K_0 = 8$ μm^{-1}. The angular width of the EBS feature in Fig. 4.16 is larger than the actual light cone, in sharp contrast to e.g., diffuse light scattering in classical disordered media, where the EBS peak has an angular width of a few degrees or less. In the present case, it can be argued that the EBS feature is related to the average extension of the exciton wave function or localization length Λ_α. The width of the EBS peak in k space is then expected to be roughly the inverse of the localization length Λ_α of states in resonance with the incident field. Typically, in the center of the optical line, we have $\Lambda \cdot K_0 \gg 1$, thus explaining the very broad EBS peak. Facing such an unfavorable situation, a method to detect EBS in QWs is to measure an anisotropy of the speckle-averaged emission intensity around $k = 0$, for light incident at a fixed angle away from the normal.

Let us relate the present theory to the self-consistent perturbation approach in terms of two-particle diagrams, formulated by Vollhardt and Wölfle to describe the Anderson transition [97]. The resolvent expression for the coherent (and speckle-averaged) intensity Eq. (4.103) can be written as

$$I_{\mathbf{k},\mathbf{k}_0}(\omega) = \overline{\mathscr{G}^+_{\mathbf{k},\mathbf{k}_0}(\omega)\,\mathscr{G}^-_{\mathbf{k},\mathbf{k}_0}(\omega)}, \tag{4.116}$$

where $\mathscr{G}^\pm$ are the retarded and advanced propagators of the exciton COM motion in the disorder potential, and the overlining denotes the average over the statistical disorder ensemble. It is then possible to formally express the intensity (4.116) as a two-particle diagram expansion in terms of the disorder potential strength and the averaged exciton propagator

$$\overline{\mathscr{G}^\pm_{\mathbf{k},\mathbf{k}'}(\omega)} = G^\pm_{\mathbf{k}}(\omega)\delta_{\mathbf{k},\mathbf{k}'} \tag{4.117}$$

as well as the one-particle self-energy $\Sigma_{\mathbf{k}}(z)$, defined as usual via (see section 4.2)

$$G_{\mathbf{k}}(z) = \frac{1}{z - \hbar k^2/2M - \Sigma_{\mathbf{k}}(z)}, \tag{4.118}$$

where z is defined on the complex frequency plane. The retarded and advanced quantities are obtained by letting $z \to \omega \pm i0$, respectively.

The first- and second-order diagrams are displayed in Fig. 4.17(a). The total wave vector at each vertex is conserved, and each loop implies a sum over wave vectors. As an example, the first-order diagram in Fig. 4.17(a) is

$$I^{(1)}_{\mathbf{k},\mathbf{k}_0} = G^+_{\mathbf{k}} G^-_{\mathbf{k}}\, g_{\mathbf{k}-\mathbf{k}_0}\, G^+_{\mathbf{k}_0} G^+_{\mathbf{k}_0}. \tag{4.119}$$

From now on, within this section, the ω dependence is omitted in our notation, unless necessary. A diagram entering the scattering matrix $T_{\mathbf{k},\mathbf{k}_0}$ is *reducible* if cutting one

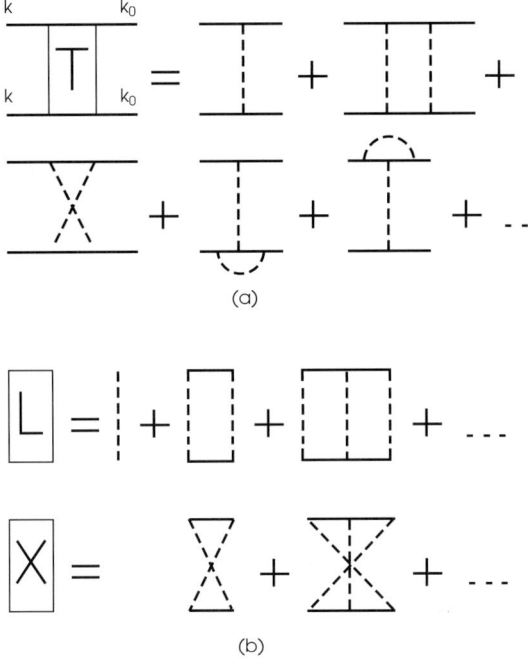

Fig. 4.17 (a) The first two orders of the diagram expansion of the RRS intensity. The upper (lower) full lines denote G^+ (G^-) propagators while the dashed lines stand for the Fourier transform g_q of the potential correlation function (4.8). (b) Diagram expansions of the ladder and maximally crossed contributions to the scattering matrix.

retarded and one advanced propagator line breaks it into two separate diagrams. Otherwise, it is called *irreducible*. The total scattering matrix is then expressed by means of a Bethe-Salpether equation in terms of the irreducible vertex T_{k,k_0}^{irred}, consisting of all possible irreducible diagrams:

$$T_{k,k_0} = T_{k,k_0}^{\text{irred}} + \sum_{k'} T_{k,k'}^{\text{irred}} G_{k'}^+ G_{k'}^- T_{k',k_0}. \tag{4.120}$$

In practice, only some classes of irreducible diagrams are taken into account in the irreducible vertex. Also, the $G_k^\pm$ are not the exact one-particle propagators, but should be determined in a self-consistent way by those diagrams only which enter the two-particle expansion. More specific, self-consistency is provided by satisfying the Ward identity

$$\Sigma_k^+ - \Sigma_k^- = \sum_{k'} T_{k,k'}^{\text{irred}} (G_{k'}^+ - G_{k'}^-). \tag{4.121}$$

In the present analysis, we will follow a simpler approach which consists in explicitly summing two classes of diagrams in the expansion for the full vertex T_{k,k_0}.

In addition, we introduce a phenomenological damping term γ_0 by replacing $\omega \pm i0$ with $\omega \pm i\gamma_0$ in the definition of retarded and advanced quantities. This damping accounts in a simplified way for the radiative recombination rate as well as for the dephasing processes and cannot be neglected if we wish to obtain quantitative information from the present theory. In order to fulfill the self-consistency between one- and two-particle propagators, the Ward identity can be rewritten in the following form

$$\Sigma^+_{\mathbf{k}_0} - \Sigma^-_{\mathbf{k}_0} = 2i\gamma_0 \sum_{\mathbf{k}} G^+_{\mathbf{k}} G^-_{\mathbf{k}} T_{\mathbf{k},\mathbf{k}_0}, \tag{4.122}$$

which has the advantage of being expressed in terms of the full vertex $T_{\mathbf{k},\mathbf{k}_0}$ instead of the irreducible one. We note in passing that Eq. (4.122) is an optical theorem, relating the forward scattering cross section to the total scattered intensity. Another important feature is that Eq. (4.122) admits nontrivial solutions only for $\gamma_0 > 0$, suggesting that a finite damping is an essential element of the present perturbation approach.

In order to compare the perturbation result to a full numerical calculation, we consider white-noise disorder in 1D. In this case, the potential correlation function is expressed in terms of the white-noise energy and length units E_1, l_1 as

$$g(x - x') = \overline{V(x)V(x')} = E_1^2 l_1 \delta(x - x'), \tag{4.123}$$

where $E_1 = \hbar^2/(2Ml_1^2)$. Then, the Fourier-transformed potential correlation $g_q = E_1^2 l_1$ is constant, simplifying the subsequent analysis. We take into account the *ladder* (L) diagrams and the *maximally crossed* (X) ones, which are depicted in Fig. 4.17(b). The ladder series, e.g., follows from solving the Bethe-Salpeter Eq. (4.120) with $T^{\text{irred}}_{\mathbf{k},\mathbf{k}'} \to g_{\mathbf{k}-\mathbf{k}'}$. These diagram series are expected to give the dominant contributions to the scattered intensity, as pointed out in [97]. Further, we neglect the k dependence in the exciton self-energy $\Sigma^\pm_k(\omega)$. Then, the summations for L_{k,k_0} and X_{k,k_0} can be performed analytically, resulting in

$$T_{k,k_0} = L_{k,k_0} + X_{k,k_0} = \frac{1 + A(\omega)}{B^2(\omega)} + \frac{A(\omega)}{(k_0 + k)^2 + B^2(\omega)}. \tag{4.124}$$

$A(\omega)$ and $B(\omega)$ are explicit analytical expressions depending on $\Sigma(\omega)$ and γ_0. The EBS stems from the second part (maximally crossed series) which is peaked at $k = -k_0$, whereas the ladder contribution (first part) is k-independent. The '1' appearing in the numerator of the ladder sum originates from the single scattering contribution, given by the first-order term in Fig. 4.17(b). Without that term, an intensity enhancement in the backscattering direction by exactly the well-known factor of two would result. In the light-scattering experiments mentioned above, the single-scattering contribution is either negligible or does not appear in the measurement due to polarization selection rules [94–96]. In the present case, the single-scattering contribution cannot be neglected and leads to a substantial reduction of the enhancement factor, thus explaining in part the result of our numerical simulation.

For a semi-quantitative description, we start with the self-energy $\Sigma(\omega)$ in the self-consistent second Born approximation Eq. (4.12),

$$\Sigma^{SCSB}(\omega) = -\frac{(E_1/\hbar)^{3/2}}{2\sqrt{\Sigma^{SCSB}(\omega) - \omega - i\gamma_0}}, \quad (4.125)$$

which is k-independent. We then evaluate the T matrix (4.124), while adjusting for each ω the imaginary part of the self-energy to satisfy Eq. (4.122). The real part of $\Sigma(\omega)$ is left unchanged. Results of the perturbation analysis are shown in Fig. 4.18 and compared with a full numerical simulation. The agreement is quite satisfactory, both for the optical density and for the Rayleigh signal. It must be said, however, that the present

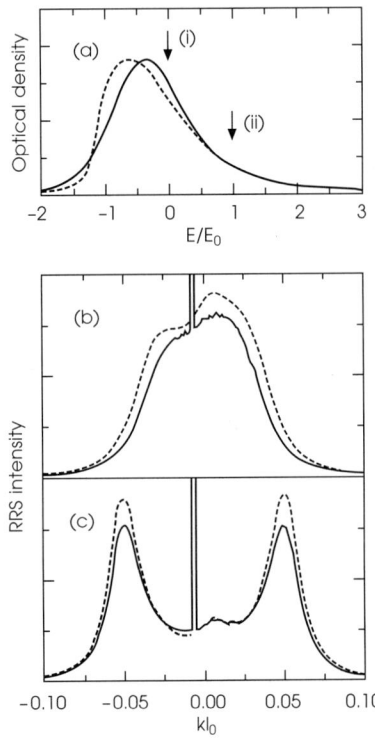

Fig. 4.18 Enhanced backscattering in 1D with white-noise disorder ($E_0 = 0.35$ meV, $\gamma_0 = 0.1 E_0$). Full curves – numerical simulation, dashed curves – perturbation approach (see text). (a) Optical density. (b) and (c) Rayleigh intensity as a function of k for two energies denoted by the arrows labeled (i) and (ii) in panel (a).

perturbation approach is successful only thanks to the finite damping γ_0. In the perturbation treatment, the EBS peak becomes infinitely narrow for $\gamma_0 \to 0$, whereas the corresponding exact result still displays a finite EBS width, related to the inverse of the average exciton localization length at the given energy. We argue that in presence of a finite damping, the EBS peak width is dominated by γ_0 (inverse exciton dissipation length) rather than by $1/\Lambda$ (localization). An even more striking discrepancy between the perturbation approach and the exact result is the following: In the limit $\gamma_0 \to 0$, the intensity as a function of $\mathbf{k}$ should be inversion symmetric around $\mathbf{k} = 0$, while Eq. (4.124) is obviously not. For the proof of the exact property, we write the scattered intensity as resolvent in terms of exciton COM eigenstates (compare Eq. (4.103)),

$$I_{\mathbf{k},\mathbf{k}_0}(\omega) = \sum_{\alpha\beta} \frac{\psi^*_{\alpha,\mathbf{k}} \psi_{\alpha,\mathbf{k}_0}}{\omega - \omega_\alpha - i\gamma_0} \frac{\psi_{\beta,\mathbf{k}} \psi^*_{\beta,\mathbf{k}_0}}{\omega - \omega_\beta + i\gamma_0}. \qquad (4.126)$$

For small values of γ_0, this expression is dominated by the diagonal contribution $\alpha = \beta$, thus

$$I^{\text{diag}}_{\mathbf{k},\mathbf{k}_0}(\omega) = \frac{\pi}{\gamma_0} \sum_\alpha |\psi_{\alpha,\mathbf{k}}|^2 |\psi_{\alpha,\mathbf{k}_0}|^2 \delta(\omega - \omega_\alpha). \qquad (4.127)$$

The mentioned symmetry $I_{\mathbf{k},\mathbf{k}_0}(\omega) = I_{-\mathbf{k},\mathbf{k}_0}(\omega)$ follows from the reality of the exciton COM wave functions in real space. As expected, the numerical simulations in the limit of small damping display this symmetry. This means that in addition to the EBS peak, a *forward* scattering enhancement around the reflected direction ($\mathbf{k} = \mathbf{k}_0$) appears, which has no counterpart in the perturbation approach.

We conclude that the perturbation expansion presented here, although providing a suggestive interpretation of the EBS mechanism, is only valid in presence of not too small damping, which means in the diffusive regime. In the strongly localized regime, however, it fails to describe certain essential properties of EBS. A quantitative information on the exciton localization length Λ can, at present, only be obtained by comparing experimental data with a full simulation result.

4.10 Spin- and polarization-dependent emission

Up to this point, the spin degree of freedom of electrons was of no importance, and has been neglected. In the present section, however, the spin content of the exciton states plays the decisive role. Since spin and orbital momentum are coupled in the valence band, there is a specific connection between spin and polarization, which will be detailed in the following. Without disorder, there are simple selection rules for the two optically active excitons: Circularly polarized light excites only one exciton state having a definite spin. Therefore, any cross-polarized emission seems to indicate a spin relaxation among the exciton states. Following this argument, the cross-polarized emission has been often considered to be fully incoherent [4]. The spin transfer between exciton states was measured and modelled using rate equations [5,98]. Within a momentum-state representation for the exitons, transition rates have been introduced phenomenologically, being understood as incoherent spin relaxation.

However, things change if disorder is included in a proper way: Due to the exchange interaction, the true exciton eigenstates refer to linear polarization, and the spin is no longer a good quantum number. Using the speckle analysis, we could show that the cross-polarized emission can have nearly the same coherence degree as the co-polarized one [32]. Then, the correct paradigm to explain the experimental findings is spin beating within a fully coherent exciton doublet, and not spin relaxation. In what follows we give a comprehensive outline of these effects which relate wave function anisotropy, spin content, and polarization direction in the secondary emission of localized excitons. This is in full agreement with results by Gammon et al. [80,85] who have analyzed exchange splittings of individual exciton lines using high spatial and spectral resolution.

In the spirit of the effective mass approximation in direct-gap semiconductors, it is sufficient to concentrate on the band edge states at the Γ point. The conduction band forms a Kramers doublet of s-like symmetry, $|+1/2\rangle = |S, \uparrow\rangle$ and $|-1/2\rangle = |S, \downarrow\rangle$. The labels $\pm 1/2$ refer to the total angular momentum. Due to the confinement in the quantum well, the p-like valence band states are split into heavy-hole and light-hole states (the spin-split band is even further away). Concentrating on the upper one in GaAs, the heavy-hole band, the basic Bloch states are $|+3/2\rangle = |X + iY, \uparrow\rangle/\sqrt{2}$ and $|-3/2\rangle = |X - iY, \downarrow\rangle/\sqrt{2}$. The electron-heavy-hole basis contains now four states which can be labelled with total momentum $L = +2, +1, -1, -2$. Since optical transition do conserve spin, only $L = \pm 1$ are radiative. This property is found in the exciton states as well: two states are bright, the other two are dark. The optically allowed states have the following dipole matrix elements,

$$L = +1: \quad \mathbf{\mu}_{cv} = \frac{1}{\sqrt{2}} \langle S, \uparrow| e\mathbf{r} | X + iY, \uparrow\rangle = \mu_{cv} \frac{1}{\sqrt{2}} (1, i, 0) \qquad (4.128)$$

$$L = -1: \quad \mathbf{\mu}_{cv} = \frac{1}{\sqrt{2}} \langle S, \downarrow| e\mathbf{r} | X - iY, \downarrow\rangle = \mu_{cv} \frac{1}{\sqrt{2}} (1, -i, 0)$$

where $\mu_{cv} = e \langle S | x | X \rangle$.

Now, the four-fold degeneracy ($L = \pm 1, \pm 2$) is lifted via the exchange interaction [99]. It is traditionally divided into a short-range part and a long-range part. The latter is nothing else than the (non-retarded) interaction with the radiation field and acts on the bright states alone. It falls off as $1/R^3$ in real space [67], which corresponds to k^{+1} in momentum space.

We concentrate on the dominant exchange splitting between degenerate states sharing the same COM wave function. The two bright states are coupled via a Hermitian 2×2 matrix with the exchange energy on the diagonal,

$$W_\alpha^{on} = \pi [\mu_{cv} \phi_{1s}(0) O_{eh}]^2 \int d\mathbf{k} |\psi_{\alpha k}|^2 k, \qquad (4.129)$$

and off-diagonal,

$$W_\alpha^{off} = \pi [\mu_{cv} \phi_{1s}(0) O_{eh}]^2 \int d\mathbf{k} |\psi_{\alpha k}|^2 (k_x + ik_y)^2/k \equiv e^{2i\theta_\alpha} \hbar\Delta_\alpha/2. \qquad (4.130)$$

The new eigenenergies are labeled with $(\pm)$,

$$\varepsilon_\alpha^{(\pm)} = \varepsilon_\alpha + W_\alpha^{on} \pm |W_\alpha^{off}|, \qquad (4.131)$$

and the corresponding exciton states read

$$\psi_\alpha^{(\pm)} = \psi_\alpha(\mathbf{R}) \frac{1}{\sqrt{2}} (e^{+i\theta_\alpha}|L=+1\rangle \pm e^{-i\theta_\alpha}|L=-1\rangle). \qquad (4.132)$$

Constructing the dipole vector of the new states using Eq. (4.128) and Eq. (4.132),

$$\boldsymbol{\mu}^{(+)} = \mu_{cv} (\cos\theta_\alpha, -\sin\theta_\alpha, 0),$$
$$\boldsymbol{\mu}^{(-)} = \mu_{cv} (i\sin\theta_\alpha, i\cos\theta_\alpha, 0), \qquad (4.133)$$

it is clear that $(-\theta_\alpha)$ gives the (in-plane) orientation of the exciton dipole. Going back to the definition Eq. (4.130), one can see that this orientation is directed along one of the main axis of the COM wave function. For an isotropic wave function, Eq. (4.130) vanishes identically, giving a zero spin splitting Δ_α. Thus, spin effects in the exciton are intimately related to anisotropies in the spatial COM wave function. Detailing all prefactors [98,100], one ends up with

$$\Delta_\alpha = \frac{3}{8} \left[\frac{\phi_{1s}^{QW}(0)}{\phi_{1s}^{bulk}(0)} \right]^2 \Delta_{LT}^{bulk} |I_\alpha|, \qquad (4.134)$$

with I_α being the integral in Eq. (4.130). Δ_{LT}^{bulk} is the longitudinal-transverse polariton splitting frequency in the bulk. If it comes to numbers, the diagonal energy (shift between dark and bright states) is of the order of 100 µeV, while the exchange splittings $\hbar\Delta_\alpha$ are much less, typically 10 µeV [85].

The corresponding exchange coupling between spatially separated exciton states leads in principle to another energy correction and wave function mixing, in analogy to the Foerster energy transfer. However, since the interaction falls off rapidly in space, and the energies are not degenerate, the induced energy changes are much less than the numbers given above, and may be neglected. This will be quantified in section 4.11 on polariton effects, where all couplings are included. In Ref. [67], a phonon-assisted Foerster transfer has been considered which might play a role in the (incoherent) relaxation of localized excitons.

Next we evaluate the optical matrix elements. For the strong localization under study in the present section, we are led to neglect the momentum dependence in the COM wave functions. However, direction and polarization of incoming/outgoing light beams need thorough analysis. We write the wave vector of the light wave in the farfield limit in spherical coordinates (z is the growth direction of the QW),

$$\mathbf{k} = k \, (\sin \vartheta \cos \phi, \, \sin \vartheta \sin \phi, \, \cos \vartheta). \tag{4.135}$$

The two possible polarization vectors are conveniently written as

$$\mathbf{e}_{TE} = (-\sin \phi, \, \cos \phi, \, 0),$$
$$\mathbf{e}_{TM} = (\cos \vartheta \cos \phi, \, \cos \vartheta \sin \phi, \, -\sin \vartheta) \tag{4.136}$$

and called transverse electric (TE, with electric field in the x-y-plane) and transverse magnetic (TM, with electric field in the plane of incidence). The scalar product with the dipole Eq. (4.133) leads to the following form of the optical matrix elements

$$M_{\alpha,+}^{TM} = -\cos(\theta_\alpha + \phi) \cos \vartheta \, M_\alpha, \quad M_{\alpha,+}^{TE} = -\sin(\theta_\alpha + \phi) \, M_\alpha,$$
$$M_{\alpha,-}^{TM} = -i \sin(\theta_\alpha + \phi) \cos \vartheta \, M_a, \quad M_{\alpha,-}^{TE} = +i \cos(\theta_\alpha + \phi) M_a, \tag{4.137}$$

The polarization-independent part M_α is the optical matrix element defined in Eq. (4.24), taken at $\mathbf{k} \approx 0$. In what follows we consider nearly normal incidence by putting $\cos \vartheta = 1$ (note that ϑ is the internal angle of incidence). The label TM should then be better called parallel (to the direction given by ϕ), and TE perpendicular.

Neglecting incoherent processes, the intensity of resonant Rayleigh scattering after a short-pulse excitation of unit strength is given by Eq. (4.82). In the present case, each state index α has to be supplemented by the label ($\pm$) of the bright states, according to Eq. (4.132). Writing the spin-split energies explicitly, we have

$$I_{\text{out,in}}(t) = \Omega_0^2 \sum_{\alpha\beta} \sum_{l,m=\pm 1} M_{\alpha l}^{\text{out}*} M_{\alpha l}^{\text{in}} M_{\beta m}^{\text{out}} M_{\beta m}^{\text{in}*} e^{i(\omega_\alpha + l\Delta_\alpha/2 - \omega_\beta - m\Delta_\beta/2)t}. \tag{4.138}$$

We are not interested in the speckling of the emitted intensity, but rather concentrate on speckle-averaged quantities. Correlations between speckles in different polarizations have been studied in [74]. As shown before, averaging over the angular speckles is equivalent to the ensemble average over disorder realizations. Here, we take advantage of the very different energy scales involved: The $\hbar\omega_\alpha \equiv \varepsilon_\alpha$ have a spread of σ, which corresponds to the inhomogeneous linewidth and is of the order of a few meV in standard samples. For times $\sigma t/\hbar > 1$, destructive interference will kill all terms $\alpha \neq \beta$ in Eq. (4.138). However, due to the much smaller spin splittings (10 μeV), different $l = \pm 1$ and $m = \pm 1$ are still active on a picosecond time scale and result in a damped oscillatory time dependence. We stress that this kind of spin beating is present already without any magnetic field.

Thus, keeping only $\alpha = \beta$ in Eq. (4.138) and inserting the matrix elements Eq. (4.137), we arrive at

$$\overline{I_{\parallel,\parallel}(t)} = \frac{\Omega_0^2}{2} \overline{\sum_\alpha M_\alpha^4 (1 + \cos^2(2(\theta_\alpha + \phi)) + \sin^2(2(\theta_\alpha + \phi))\cos(\Delta_\alpha t))}$$

$$\overline{I_{\perp,\parallel}(t)} = \frac{\Omega_0^2}{2} \overline{\sum_\alpha M_\alpha^4 \sin^2(2(\theta_\alpha + \phi))(1 - \cos(\Delta_\alpha t))}. \qquad (4.139)$$

The linear polarization degree is defined as

$$\mathscr{P}_\phi^{\text{lin}} = \frac{I_{\parallel,\parallel} - I_{\perp,\parallel}}{I_{\parallel,\parallel} + I_{\perp,\parallel}}. \qquad (4.140)$$

It can be written in compact form as

$$\mathscr{P}_\phi^{\text{lin}}(t) = \langle \cos^2(2(\theta_\alpha + \phi))\rangle + \langle \sin^2(2(\theta_\alpha + \phi))\cos(\Delta_\alpha t)\rangle, \qquad (4.141)$$

where we have introduced in this section a notation for the ensemble average of any quantity weighted with the matrix element in fourth power,

$$\frac{\sum_\alpha M_\alpha^4 \cdots}{\sum_\alpha M_\alpha^4} \stackrel{\text{def}}{=} \langle \cdots \rangle. \qquad (4.142)$$

In general, the polarization degree (4.140) depends on the direction ϕ which is given as a label. To be definite, we choose the crystallographic axis $[1\bar{1}0]$ as $\phi = 0$. For a random orientation of all wave functions, we get simply

Random: $\quad \mathscr{P}^{\text{lin}}(t) = \frac{1}{2} + \frac{1}{2}\langle \cos(\Delta_\alpha t)\rangle, \qquad (4.143)$

independent of the direction ϕ. In particular, at large times a constant polarization of exactly 50% survives.

In Fig. 4.19, the time-resolved polarization degree is displayed, measured on a 6 nm wide AlGaAs quantum well [32]. The sample was placed in a cryostat at a temperature of 5 K. The excitation pulse (2 ps duration) was centered at the lowest heavy-hole exciton transition. Note that the incoherent contribution from the intensity data has been subtracted before deducing $\mathscr{P}(t)$, applying the speckle analysis [20]. The inhomogeneous broadening gives a FWHM of 6 meV in the Rayleigh spectrum. Therefore, the limit $\sigma t \gg 1$ is reached after a few picoseconds. The experimental data show a distinct dependence of $\mathscr{P}_\phi^{\text{lin}}$ on observation angle, pointing to a non-random orientation of the COM wave functions. This preferential orientation may result from e.g. steps or elongated islands on the interfaces which have a preferential alignment. The data at large times exhibit a maximum polarization for $\phi = 0°$. Since only the first term in Eq. (4.141) survives at $t \to \infty$, we are led to the conclusion that the preferential orientation is along $[1\bar{1}0]$ ($\theta_\alpha = 0$). This is in accordance with experimental information on step orientation on the [100] GaAs surface [85]. Then, with $\langle \sin(\theta_\alpha) \rangle = 0$ we arrive at

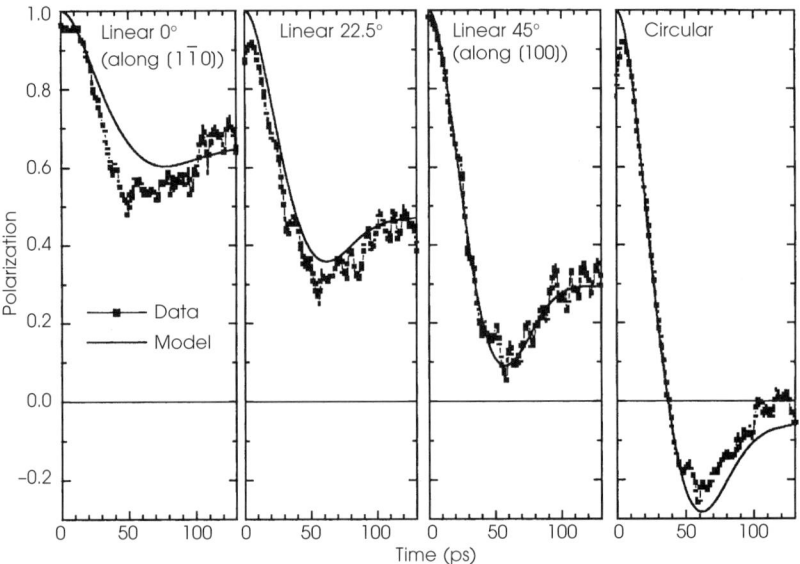

Fig. 4.19 Polarization degree as a function of time for linear and circular polarization. Experimental data (symbols) are compared with model calculations (curves), using the simulation results in Fig. 4.20 for an anisotropically correlated landscape; from [73].

$$\mathscr{P}_\phi^{\text{lin}}(t) = \frac{1}{2}(1 + \langle \cos(\Delta_\alpha t) \rangle) + \frac{1}{2}\cos(4\phi)\langle \cos(4\theta_\alpha)(1 - \cos(\Delta_\alpha t)) \rangle. \quad (4.144)$$

Two different averages determine the time dependence: The plain average with $\cos(\Delta_\alpha t)$, and one weighted with $\cos(4\theta_\alpha)$. The underlying d-like symmetry of the spin-splitting is obvious from the $\cos(4\phi)$ dependence. Introducing $C_4 = \langle \cos(4\theta_\alpha) \rangle$, the polarization at large times reduces to

$$\mathscr{P}_\phi^{\text{lin}}(t \to \infty) = \frac{1}{2}(1 + \cos(4\phi) C_4) \tag{4.145}$$

and can be used to determine C_4. The parameter C_4 quantifies the preferential orientation since $C_4 = 0$ holds for random orientation, and $C_4 = 1$ for a strictly uniaxial one. The measured data of Fig. 4.19 gave a value of $C_4 = 0.37$.

For circular polarization $\sigma+$, $\sigma-$, the polarization vectors are defined as $\mathbf{e}_{\sigma\pm} = (\mathbf{e}_{TM} \mp i\mathbf{e}_{TM})/\sqrt{2}$. According to this definition, $\sigma+$ excites the spin-up transition only (for normal incidence). The matrix elements can be evaluated with Eq. (4.137) as

$$M_{\alpha l}^{\sigma+} = e^{+i(\theta_\alpha + \phi)} M_\alpha / \sqrt{2}; \quad M_{\alpha l}^{\sigma-} = l\, e^{-i(\theta_\alpha + \phi)} M_\alpha / \sqrt{2}. \tag{4.146}$$

Putting again $\alpha = \beta$ in Eq. (4.138), all phase factors in Eq. (4.146) cancel, and we obtain

$$\overline{I_{++}(t)} = \frac{\Omega_0^2}{2} \overline{\sum_\alpha M_\alpha^4 (1 + \cos(\Delta_\alpha t))}$$

$$\overline{I_{-+}(t)} = \frac{\Omega_0^2}{2} \overline{\sum_\alpha M_\alpha^4 (1 - \cos(\Delta_\alpha t))} \tag{4.147}$$

for the co- and cross-circular polarized intensity, respectively. The degreee of circular polarization takes the simple form

$$\mathscr{P}^{\text{circ}} = \frac{I_{++} - I_{-+}}{I_{++} + I_{-+}} = \langle \cos(\Delta_\alpha t) \rangle. \tag{4.148}$$

Please note that for circular polarization, the orientation of the COM wave functions has no influence since both angles θ_α, ϕ dropped out completely. The polarization degree (rightmost panel in Fig. 4.19) starts again with unity at $t = 0$ and changes sign at a finite time, clearly indicating a spin beating. At times large compared to the inverse spin splitting, the polarization degree vanishes completely.

The essential link between experiment and theory (i.e. simulation) is the distribution of spin (or exchange) splittings,

$$p_0(\Delta) = \langle \delta(\Delta - \Delta_\alpha) \rangle. \tag{4.149}$$

In the case of spectral selective excitation, a proper window of eigenvalues has to be considered as well. Exchange splittings for excitons within artificially constructed anisotropic quantum boxes have been calculated by Ivchenko and coworkers [101]. In a disordered quantum well, we are faced with a full potential landscape, and must proceed to an ensemble language. Such a distribution of exchange splittings has been defined and evaluated by Nickolaus *et al.* [100] and Maialle [102], using the topological method of minimum counting in a disordered landscape introduced to quantum well physics by Wilkinson and coworkers [103]. One, at first glance, surprising finding is that almost all potential minima are anisotropic and lead to finite spin splittings – even within an isotropically correlated landscape (no preferential orientation). We have extended the minimum counting model to an anisotropically correlated landscape in [32]. Now, the potential correlation contains two different correlation lengths ξ, η. For simplicity, we consider a Gauss type correlation,

$$\overline{V(\mathbf{R}=(x,y))V(\mathbf{0})} = \sigma^2 \exp\left(-\frac{x^2}{2\xi^2} - \frac{y^2}{2\eta^2}\right). \tag{4.150}$$

While the Wilkinson model allows to obtain analytical expressions for the splitting distributions, a full quantum-mechanical calculation is superior. We present here results of corresponding simulations carried out on anisotropic potential landscapes of type Eq. (4.150) [73]. For the linear polarization, we need in addition to the plain splitting distribution Eq. (4.149) the cosine weighted variant

$$p_4(\Delta) = \langle \cos(4\theta_\alpha)\, \delta(\Delta - \Delta_\alpha) \rangle, \tag{4.151}$$

which contains the information on wave function orientation θ_α as well. If the orientation were independent of the spin splitting, the shape of $p_4(\Delta)$ would coincide with $p_0(\Delta)$. However, Fig. 4.20 demonstrates that this is not the case: The cosine weighting shifts the Δ-distribution to slightly larger values. This was to be expected since in the preferential direction, we expect larger wavefunction anisotropies and, therefore, a larger spread in exchange splittings. In general, the anisotropic correlation tends to enhance the spin splitting compared to the isotropic case with a comparable correlation length $l_{\text{corr}} = \sqrt{\xi\eta}$ (not shown).

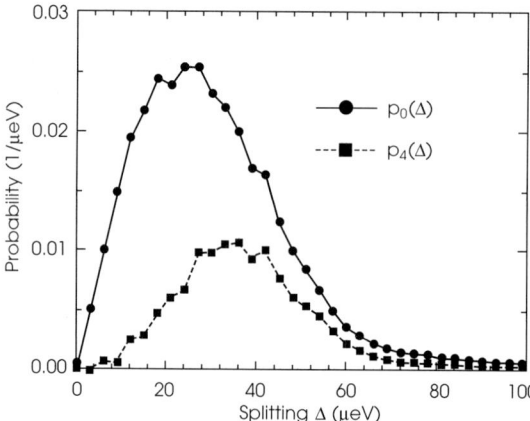

Fig. 4.20 Distribution $p_0(\Delta)$ of the exchange splittings, derived from a simulation with an anisotropically correlated disorder ($\eta/\xi = 0.5$). The distribution $p_4(\Delta)$ is weighted with $\cos(4\theta_\alpha)$.

Integrating $p_4(\Delta)$ over Δ gives the angular average $C_4 = \langle\cos(4\theta_\alpha)\rangle$. We have used the value of $C_4 = 0.37$ as extracted from the polarization data for adjusting the ratio $\rho = \eta/\xi$. Test runs for the simulation gave best results for $\rho = 0.50$. The zero-crossing of the circular polarization degree,

$$\mathscr{P}^{\mathrm{circ}}(t) = \langle\cos(\Delta_\alpha t)\rangle = \int_0^\infty d\Delta\, p_0(\Delta) \cos(\Delta t), \tag{4.152}$$

sets the scale for the average spin splitting. Taking from the data 37 ps for the zero crossing, we found reasonable agreement by adjusting the correlation lengths to $\xi = 30$ nm and $\eta = 15$ nm. A potential variance of $\sigma = 3$ meV has been taken to reproduce the spectral width. With these input data, the simulation produced an average spin splitting of $\langle\hbar\Delta_\alpha\rangle = 31$ μeV. The comparison of the calculated polarization degree (curves) in Fig. 4.19 with the measured data is not perfect but underlines the relevance of the model. While the correlation product $\xi\eta$ is only vaguely related to the (average) island size, its ratio ρ gives a clear indication of the aspect ratio of the elongated islands. The preferential orientation along $[1\bar{1}0]$ could be extracted unambiguously, which is corroborated by findings from scanning tunneling microscopy, as reported in [85].

4.11 Polariton effects in the secondary emission

The Rayleigh scattering model adopted so far treats the exciton-photon coupling in second Born approximation, Eq. (4.82). In physical terms, this corresponds to neglecting the multiple photon emission and reabsorption events that may occur up to infinite order before a photon is finally emitted out of the system. The scattered electric field at the QW boundary is assumed to be proportional to the macroscopic exciton polarization, which implies a direct relation between the emitted intensity and the exciton density matrix. A full treatment requires that the electric field and the macroscopic polarization are related to each other via Maxwell's equations, using a material susceptibility based on the exciton COM Schrödinger equation. The mixed exciton-radiation modes arising from the coupled Maxwell-Schrödinger system of equations are called *polaritons*, and the scheme is known in the literature as the polariton model.

The polariton concept is of central importance for exciton optics in semiconductors. A comprehensive review has been given by Andreani [99], covering all the relevant aspects for bulk semiconductors as well as for nanostructures in the absence of disorder. Polaritons have a dramatic effect in bulk semiconductors. As a consequence of the translational invariance, each polariton mixes a single exciton state and a single photon mode with the same wave vector, in close analogy to a pair of coupled harmonic oscillators. The polariton dispersion differs substantially from the original exciton and photon dispersions, displaying an avoided crossing. Bulk polaritons are steady-state modes with virtually infinite lifetime (were it not for interactions with phonons and/or impurities, and with the system boundaries).

For excitons in systems of lower dimensionality, polariton effects are less prominent. In a QW (without disorder) a single exciton is coupled to a one-dimensional continuum of photon modes, resulting in a polariton self energy whose imaginary part gives the radiative decay rate. It is nonzero only within the light cone $k \leq K_0 = (\omega/c)\sqrt{\varepsilon_b}$. For $k > K_0$ the z component of the polariton wave vector is imaginary, giving rise to an evanescent mode (or surface polariton) outside the QW. The exciton radiative broadening in typical GaAs QW structures amounts to a few tens of μeV. The energy shifts are of the same order, and are irrelevant in most experimental situations.

Since spatial symmetry is absent for localized excitons in a disordered QW, they can couple to all photon modes. A fundamental question with respect to polaritons arises: Can the optical response of the system be described still in terms of localized exciton states, perhaps with small corrections for radiative damping and shifts? Or is the exciton–photon coupling dominant, and the entity which is going to be localized is the polariton? In his review work [99], Andreani considers the relevance of the polariton concept in terms of temporal and spatial coherence. Temporal coherence requires the characteristic energy for the polariton coupling to be much larger than typical energies governing the time-evolution of the bare exciton. Spatial coherence, in a similar way, requires the exciton state to be coherent over a length exceeding the photon wavelength λ. If these two conditions are satisfied, polaritons are the basic entities to start with.

Aiming at the relevance of the polariton picture for exciton localization, we restrict ourselves to zero temperature and neglect all homogeneous broadening effects other than radiative broadening. Therefore, we have to identify the exciton coherence length with the COM localization length Λ. The characteristic polariton coupling energy in QWs is the radiative broadening of the free exciton at zero momentum Eq. (4.177),

$$\Gamma_0 = \frac{2\pi}{\hbar} \frac{K_0}{\varepsilon_b} \mu_{cv}^2 \phi_{1s}^2(0) O_{eh}^2. \qquad (4.153)$$

This should be compared to the inhomogeneous energy distribution of localized excitons, i.e. to σ (or better the OD linewidth). For typical GaAs QWs, $\hbar\Gamma_0$ ranges between 20 and 60 μeV, whereas the inhomogeneous exciton linewidth is in most cases larger than 100 μeV. Correspondingly, $\Lambda \ll \lambda$ holds, and localized exciton states should be practically unaffected by the polariton coupling. However, any speculation on temporal and spatial coherence should be supported by a quantitative estimate.

The spin-splitting and the resulting polarization dynamics of the scattered field were already discussed in section 4.10 in terms of the long-range part of the exchange interaction. A full polariton treatment, which we are going to present here, automatically includes this coupling. Furthermore, retardation effects in the propagation along the QW are properly accounted for [99,104].

A formal theory for exciton–polaritons in QWs including disorder has been first developed by Citrin [105], including a nonlocal exciton susceptibility (resolvent of the exciton COM Schrödinger equation) and spin degrees of freedom as essential ingredients. The Maxwell boundary conditions are implemented via the transfer-matrix technique. Left and right propagating fields for the two polarizations have to be calculated from equations which couple all values of the in-plane wave vector **k**. This is a very demanding task with huge matrices. Therefore, Citrin resorted to a simple disorder model of circular 'islands', which excludes, however, any spin splitting (being related to an anisotropic COM wave function). The theory has been extended recently by Grote et al. [106] to include (i) excited exciton states, (ii) a disorder model in complete analogy to section 4.2, and (iii) multiple quantum wells (MQWs). For the numerical calculations, however, the authors have decided to average the nonlocal exciton susceptibility over azimuthal angles in **k** space. This removes again the polarization mixing, and partially smoothes out the speckle features. Despite these simplifications, the measured RRS dynamics of MQWs with a large number of QWs can be successfully reproduced, presumably due to dominating polariton effects in that case.

In what follows, we will develop a full Maxwell-Schrödinger formalism for excitons in a disordered QW, using again a representation in disorder eigenstates. By means of numerical simulations for single QWs, we discuss the following points: (i) radiative lifetimes and shifts of localized exciton states and the RRS dynamics, (ii) polarization mixing in RRS, (iii) polariton coupling between different localized states in view of the persistence of individual localized exciton states.

The starting point for a polariton theory in QWs is the (nonlocal) linear exciton susceptibility which is a tensor for the three spatial directions x, y, and z. For the heavy-hole exciton in a single QW and using the disorder eigenstates $\psi_\alpha(\mathbf{R})$, we have

$$\hat{\chi}(\mathbf{R}, \mathbf{R}', z, z', \omega) = \mu_{cv}^2 \phi_{1s}(0)^2 \sum_\alpha \frac{\psi_\alpha(\mathbf{R}) \psi_\alpha(\mathbf{R}')}{\hbar\omega_\alpha - \hbar\omega - i0^+} \rho(z) \rho(z') \times \begin{pmatrix} 1 & 0 & 0 \\ 0 & 1 & 0 \\ 0 & 0 & 0 \end{pmatrix}.$$

(4.154)

Its diagonal form follows by summing the outer product of polarization vectors Eq. (4.128) over the two bright states. Note that the z component of the electric field is not coupled to the excitonic polarization for heavy-hole excitons in GaAs. The susceptibility (4.154) is nonlocal both in the z and in the in-plane coordinate $\mathbf{R}$, but factorizes as a consequence of the factorization ansatz Eq. (4.2), with $\rho(z) = u_e(z)\, u_h(z)$ for the given subband pair. Equation (4.154) is completely equivalent to the resolvent expression with the COM Hamiltonian as used in [105,106]. The generalization to a MQW structure is straightforward: Susceptibilities of type Eq. (4.154) with a QW index, centered at the respective z-position, are simply added. Assuming no vertical correlation between different QWs, one would construct each QW with a different potential realization. This point, which has been overlooked by previous authors, will be addressed at the end of this section. For simplicity, we assume a uniform background dielectric constant ε_b, thus neglecting the effect of the sample surface and of the dielectric mismatch between barrier and well material.

The Maxwell equation for the electric field $\mathscr{E}$ can be written as

$$\nabla \times \nabla \times \mathscr{E}(\mathbf{R}, z, \omega) - \frac{\omega^2}{c^2} \left[\varepsilon_b\, \mathscr{E}(\mathbf{R}, z, \omega) \right.$$

$$\left. + 4\pi \int d\mathbf{R}'\, dz'\, \hat{\chi}(\mathbf{R}, \mathbf{R}', z, z', \omega) \cdot \mathbf{E}(\mathbf{R}', z', \omega) \right] = 0. \qquad (4.155)$$

We define the in-plane and z-components of the electric field as $\mathscr{E} = (\mathbf{E}, E_z)$. Since E_z is not coupled to the exciton, it can be eliminated from Eq. (4.155). We also Fourier-transform to reciprocal space as $\mathbf{E}(\mathbf{R}, z) = \sum_k \mathbf{E}_\mathbf{k}(z) \exp(i\mathbf{k} \cdot \mathbf{R})$. After some algebra, the resulting equation for the in-plane component $\mathbf{E}$ reads

$$-\left(1 + \frac{1}{k_z^2}\frac{\partial^2}{\partial z^2}\right)\begin{pmatrix} K_0^2 - k_y^2 & k_x k_y \\ k_x k_y & K_0^2 - k_x^2 \end{pmatrix} \cdot \mathbf{E}_\mathbf{k}(z)$$

$$= 4\pi \frac{K_0^2}{\varepsilon_b} \sum_{\mathbf{k}'} \int dz'\, \hat{\chi}_{\mathbf{k},\mathbf{k}'}(z, z') \cdot \mathbf{E}_{\mathbf{k}'}(z'), \qquad (4.156)$$

where $k_z = \sqrt{K_0^2 - k^2}$ is the z-component of the wave vector of light. As the boundary condition, we assume a plane wave $\mathbf{E}_{\mathbf{k}_0}$ to enter from the left, $z < 0$. From now on we omit the ω-dependence in the notation, unless necessary. At this point, instead of solving Eq. (4.156) directly, we prefer to adopt the scattering approach proposed by Martin and Piller [107]. A background Green's function $\hat{\mathbf{G}}_\mathbf{k}(z)$ is defined as the solution of the lhs of Eq. (4.156) with an inhomogeneity $\hat{\mathbf{1}}\,\delta(z)$, and with outgoing boundary conditions. It can be obtained analytically,

$$\hat{\mathbf{G}}_\mathbf{k}(z) = \frac{i}{2K_0^2 k_z}\begin{pmatrix} K_0^2 - k_x^2 & -k_x k_y \\ -k_x k_y & K_0^2 - k_y^2 \end{pmatrix} \exp(ik_z |z|). \qquad (4.157)$$

Next, Eq. (4.156) is written as a Dyson equation,

$$\mathbf{E}_k(z) = \mathbf{E}_k^{(0)}(z) + 4\pi \frac{K_0^2}{\varepsilon_b} \sum_{k'} \int dz' \, dz'' \, \hat{\mathbf{G}}_k(z-z') \cdot \hat{\boldsymbol{\chi}}_{k,k'}(z',z'') \cdot \mathbf{E}_{k'}(z''), \quad (4.158)$$

where $\mathbf{E}_k^{(0)}(z)$ is the incoming field in the dielectric background. If the wavelength is much larger than the QW thickness, $\rho(z) = O_{eh}\,\delta(z)$ can be used to simplify the Dyson equation (4.158),

$$\mathbf{E}_k = \mathbf{E}_k^{(0)} + 2\Gamma_0 K_0 \sum_{k'\alpha} \frac{\psi_{\alpha k}^* \psi_{\alpha k'}}{\omega_\alpha - \omega - i0^+} \hat{\mathbf{G}}_k \cdot \mathbf{E}_{k'}, \quad (4.159)$$

where all the quantities are defined at the QW position $z = 0$. The prefactors have been combined into the rate Eq. (4.153).

Equation (4.159) describes the electric field at the QW position once the incoming field $\mathbf{E}_k^{(0)}$ is known. In absence of disorder, the susceptibility would be diagonal in $\mathbf{k}$, and for each $\mathbf{k}$ the 2×2 tensor $\hat{\mathbf{G}}_k$ could be diagonalized. Two polarization modes TE and TM of the electric field are obtained which would be conserved in the propagation. In the presence of disorder, however, different momenta in Eq. (4.159) are coupled, and a global diagonalization is prevented since TE and TM refer to the direction of $\mathbf{k}$. Consequently, polarization mixing arises for any initial orientation of the field, as already shown in section 4.10. If we neglect in the background Green's function (4.157) the retardation by letting $K_0 \to 0$ in $K_0^2 \hat{\mathbf{G}}_k$, the long-range exchange interaction matrix of section 4.10 appears. Then, however, the imaginary part is lost which is responsible for the finite radiative lifetime, as seen below.

In order to proceed, we project Eq. (4.159) on the exciton COM eigenstates using their completeness, obtaining

$$\mathbf{E}_\alpha = \mathbf{E}_\alpha^{(0)} + \sum_\beta \frac{\hat{\mathbf{G}}_{\alpha\beta}}{\omega_\beta - \omega - i0^+} \mathbf{E}_\beta, \quad (4.160)$$

where

$$\mathbf{E}_\alpha = \sum_k \psi_{\alpha k} \mathbf{E}_k, \quad \hat{\mathbf{G}}_{\alpha\beta} = 2\Gamma_0 K_0 \sum_k \psi_{\alpha k} \hat{\mathbf{G}}_k \psi_{\beta k}^*. \quad (4.161)$$

Equation (4.160) clarifies the concept of polariton coupling between localized exciton states. In absence of polariton coupling, the input field is scattered by each COM localized state directly into the outgoing field, independently of the other states. Polariton coupling is responsible for the reabsorption of the scattered photons by other exciton states, due to the nondiagonal terms in $\hat{\mathbf{G}}_{\alpha\beta}$. These multiple scattering processes influence the RRS dynamics and the polarization mixing. The polariton coupling might, in principle, introduce a long-range spatial coherence along the QW plane [92,106]. In a single QW, we expect this modification to become substantial only if the radiative coupling $\hbar\Gamma_0$ is at least comparable to the exciton inhomogeneous broadening, as already stated. In order to test this hypothesis we derive a solution of Eq. (4.160) based on a full-scale numerical simulation. Introducing $\mathbf{E}_\alpha = (\omega_\alpha - \omega - i0^+) \tilde{\mathbf{E}}_\alpha$, we can rearrange Eq. (4.160) as

$$[\omega_\alpha - \omega - \hat{\mathbf{G}}_{\alpha\alpha}] \cdot \tilde{\mathbf{E}}_\alpha = \mathbf{E}_\alpha^{(0)} + \sum_{\beta \neq \alpha} \hat{\mathbf{G}}_{\alpha\beta} \cdot \tilde{\mathbf{E}}_\beta. \qquad (4.162)$$

The polariton problem in the form of Eq. (4.162) is easier to handle. By diagonalizing the 2×2 matrices $\hat{\mathbf{G}}_{\alpha\alpha}$, two complex eigenenergies $\omega_\alpha^\pm = \omega_\alpha + \Sigma_\alpha^\pm$ are obtained, where the sign label refers to the two main axes of the COM wave function (section 4.10). The real and imaginary part are the radiative shift (exchange splitting) and the radiative recombination rate of each exciton doublet, $\Sigma_\alpha^\pm = \Delta_\alpha^\pm + i\Gamma_\alpha^\pm$.

For the simulation, we take a Gauss-correlated COM potential and parameters corresponding to a 15 nm wide GaAs QW with high-quality interfaces. The COM Schrödinger equation is solved on a grid in real space, and $\mathbf{E}^{(0)}$ and $\hat{\mathbf{G}}$ are generated in the basis of the eigenstates so obtained. Finally, Eq. (4.162) is solved as a linear inhomogeneous set of equations. We point out that the numerical results shown are the solution of a full 2D problem. An azimuthal average of the response function [106] is not justified. It would remove completely all polarization-mixing effects which are found experimentally [32].

Calculated radiative shifts and rates are shown in Fig. 4.21. As expected, the radiative rates (oscillator strengths) are largest at the exciton peak, reaching values comparable to $\hbar\Gamma_0$. The strong energy dependence of $\Gamma_\alpha^\pm$ implies in the time domain a RRS signal which decays nonexponentially (see Fig. 4.22), since the most radiative states recombine first, and the decay is slowed down. The radiative shifts are comparable to $\hbar\Gamma_0$ at the exciton resonance, while the spin splitting ($\Delta_\alpha^+ - \Delta_\alpha^-$) is smaller by a factor of ten. At higher energies, the shifts increase more or less as a square root of energy. To explain this behaviour we note that the corresponding disorder eigenstates are composed primarily of plane waves of given $k \propto \sqrt{\omega_\alpha}$. Further, in a perfect planar thin QW both the LT splitting and the bright-dark splitting increase linearly with k outside the light cone. Eq. (4.162) has been numerically solved for many disorder realizations. The input field was x-polarized and proportional to $\delta_{\mathbf{k},\mathbf{k}_0}$. The time-dependent scattered field has been obtained by Fourier transforming $\mathbf{E}_\alpha(\omega)$ and averaging over the realizations. Figure 4.22(a) shows the time-dependent scattered intensity for both polarizations (parallel and perpendicular relative to the input field). We compare these results with the zeroth-order solution used in section 4.10. Including here the light-cone correction, this reads in the present language

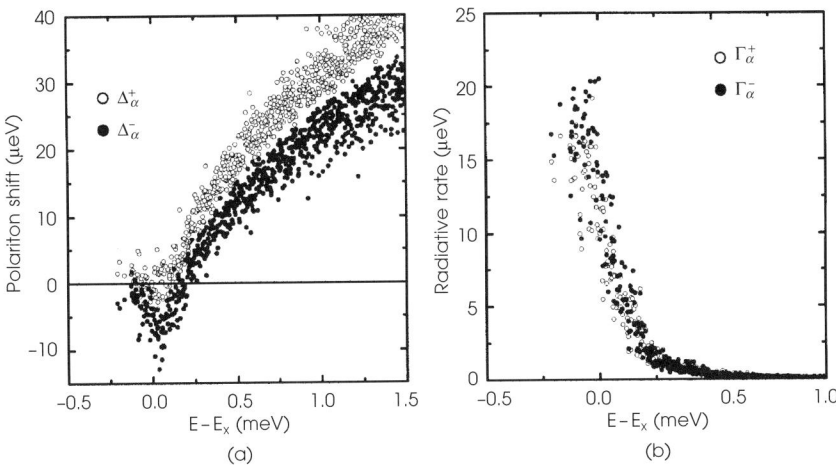

Fig. 4.21 (a) Computed polariton shifts $\Delta_\alpha^\pm$. (b) Computed radiative rates $\Gamma_\alpha^\pm$. The simulation refers to a Gauss-correlated disorder potential with $\sigma = 0.3$ meV and $\xi = 10$ nm. The polariton coupling is $\hbar\Gamma_0 = 0.02$ meV.

$$\mathbf{E}_\alpha = \mathbf{E}_\alpha^{(0)} + i\Gamma_0 \left[\omega_\alpha - \omega - \hat{\mathbf{G}}_{\alpha\alpha}\right]^{-1} \cdot \mathbf{E}_\alpha^{(0)}, \quad (4.163)$$

and includes the dominant diagonal spin splitting and radiative width. The full solution of Eq. (4.162) accounts additionally for the mixing of different COM wave functions, thus answering the question on the persistence of exciton localization in contrast to polariton localization or even delocalization. As the comparison between zeroth order and full results in Fig. 4.22 shows, the differences are not significant, and we conclude on the absence of strong polariton effects which would modify the wave functions. The nonexponential decay is clearly seen in both cases, and the perpendicularly polarized component rises slowly on a timescale of the inverse spin splitting. The linear polarization degree (Fig. 4.22(b)) starts with unity, falls below 0.5 before reaching the asymptotic value 0.5 from below, a clear signature of beating in the inhomogeneous ensemble. Note that in the present case, an isotropically correlated landscape has been used.

So far we have discussed the simple case of a single QW in a homogeneous dielectric medium. Polariton effects can be modified by the presence of dielectric interfaces. Both the sample surface [108] and the dielectric mismatch between well and barrier material [109] can influence the polariton dispersion. Since, however, the polariton effect is small for a single QW, these modifications could be accounted for by an effective renormalization of the coupling constant Γ_0. A completely different picture holds for QWs embedded in a multilayered dielectric structure which has by purpose strong resonances close to the exciton energy, as in a microcavity [110,111]. The other

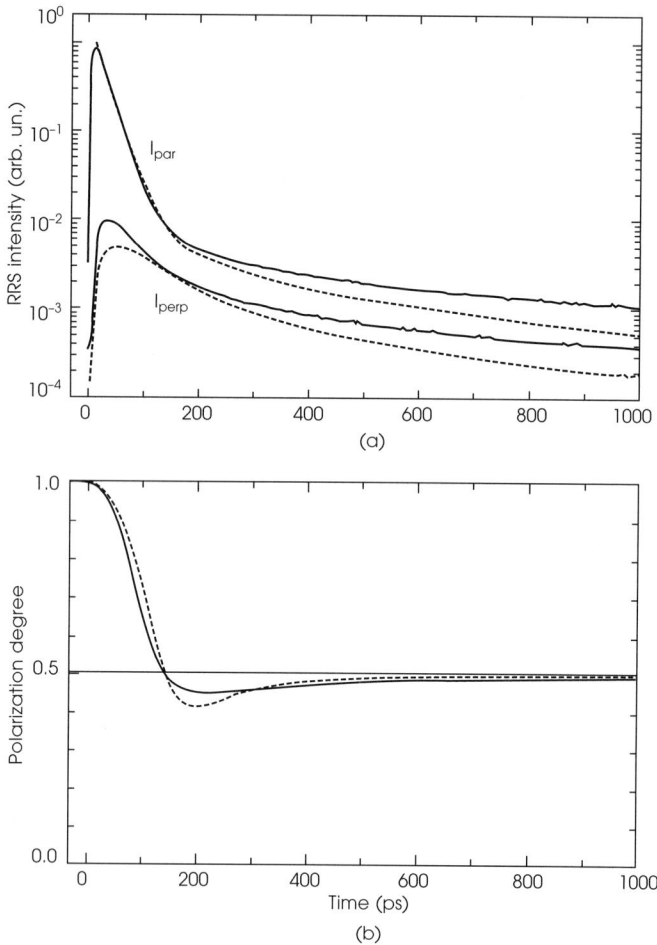

Fig. 4.22 (a) Speckle-averaged time-resolved RRS signal for a linearly-polarized input field in normal direction. Both parallel and perpendicular polarized scattered intensities are plotted. (b) Speckle-averaged polarization degree of the scattered field. In both plots, full and dashed curves refer to the full polariton result Eq. (4.162) and to the zeroth-order approximation Eq. (4.163), respectively. The following parameter values have been used in the simulation: $\sigma = 0.3$ meV, $\xi = 10$ nm, and $\hbar\Gamma_0 = 20$ μeV.

case where polariton effects are expected to be important is that of a multiple quantum well structure. Especially if the MQW period is small compared to the wavelength of light or close to a Bragg configuration, the electromagnetic field couples exciton states in different QWs into macroscopic superpositions, thus strongly enhancing the effective radiative coupling. Beatings are expected in the time-dependent RRS signal, corresponding to a multiple peak structure in the polariton spectrum. These effects have been experimentally observed and predicted in various fashions [92,93,106]. One important aspect, however, has always been overlooked here, namely the possibility of vertical disorder correlations within the MQW stack [112]. There is not much known on the correlation of the interface roughness between successive QWs as a result of the epitaxial growth process (sometimes called *heritage*). While some correlation is to be expected when growing thin barriers, the assumption of all QWs having the same disorder landscape is highly unrealistic. Depending on the amount of vertical correlation, polariton effects will change dramatically. For a strict correlation, each QW has identical COM localized states. The radiative modes are then formed by a symmetric linear superposition of all these identical states, giving rise to a coupling enhancement by roughly a factor of N^2 (N is the number of QWs). Already for medium vertical correlation, the localized states are expected to differ substantially between different QWs, and the enhancement effect ceases. While some theories [92] assume implicitly a full correlation between QWs, in other work the assumption made for the vertical correlation are not reported. To our knowledge, the influence of vertical disorder correlation on the secondary emission from MQWs has never been addressed theoretically. We believe that comparing the experimental data with a theory that includes such features would be of great help to understand growth-induced disorder correlations.

Appendix A: Potential variance

The quantum well with rough interfaces is characterized within the effective mass approximation by the confinement potential for electrons and holes ($a = e, h$), written as

$$W_a(\pmb{\rho}, z) = \Delta_a [\Theta(z - \overline{L_z} - h_2(\pmb{\rho})) - \Theta(z - h_1(\pmb{\rho}))]. \tag{4.164}$$

Here, Δ_a is the band edge difference between barrier and well material, and $h_1(\pmb{\rho})$, $h_2(\pmb{\rho})$ define the local variation of the lower and upper interface, respectively. The confinement wave functions are constructed from the in-plane averaged situation with well width $\overline{L_z}$,

$$\left[-\frac{\hbar^2}{2m_a}\frac{d^2}{dz^2} + \overline{W}_a(z)\right] u_a(z) = E_a u_a(z). \tag{4.165}$$

First-order perturbation theory of Eq. (4.1) with the factorized two-particle wave function Eq. (4.2) gives then a correction

$$V(\mathbf{R}) = \int d\mathbf{R}' \sum_{a=e,h} \eta_a^2 \phi_{1s}^2(\eta_a(\mathbf{R}-\mathbf{R}')) \int dz\, u_a^2(z) [W_a(\mathbf{R}', z) - \overline{W}_a(z)]. \tag{4.166}$$

For small deviations $h_j(\mathbf{R})$ from the average $\overline{h_j(\mathbf{R})} = 0$, we may expand

$$W_a(\mathbf{R}, z) - \overline{W}_a(z) = \Delta_a \left[\delta(z) h_1(\mathbf{R}) - \Delta(z - \overline{L_z}) h_2(\mathbf{R})\right]. \tag{4.167}$$

The roughness of the interfaces enters the potential with a weight proportional to $u_a^2 (z = 0)$ and $u_a^2 (z = \overline{L_z})$. Only for a symmetric quantum well, the result depends on the well width fluctuation $\Delta L(\mathbf{R}) = h_2(\mathbf{R}) - h_1(\mathbf{R})$ alone,

$$V(\mathbf{R}) = \int d\mathbf{R}' \sum_{a=e,h} \eta_a^2 \phi_{1s}^2 (\eta_a (\mathbf{R} - \mathbf{R}')) u_a^2 (0) \Delta L(\mathbf{R}'). \tag{4.168}$$

Introducing the first-order correction of Eq. (4.165),

$$E'_a \equiv \frac{dE_a}{dL_z} = -\Delta_a u_a^2 (0), \tag{4.169}$$

leads to the final result Eq. (4.5) in section 4.2.

Appendix B: Weak-memory and Markov approximation

As initial condition for the equations of motion (4.35) and (4.36), it is quite natural to assume $\langle a_\mathbf{q}^\dagger B_\alpha^\dagger B_\beta \rangle_{t_0} = 0$ at a time t_0 just before the excitation pulse arrived. The formal solutions are then

$$T_{\rho\beta,\mathbf{q}}(t) = \frac{i}{\hbar} \int_{t_0}^{t} dt' \sum_\eta e^{i(w_q + \omega_\rho - \omega_\beta)(t-t')} ((n_q + 1) t_{\eta\rho}^{-\mathbf{q}} N_{\eta\beta}(t') - n_q t_{\beta\eta}^{-\mathbf{q}} N_{\rho\eta}(t'))$$

$$\tilde{T}_{\rho,\mathbf{q}}(t) = \frac{i}{\hbar} \int_{t_0}^{t} dt' \sum_\eta e^{i(w_q + \omega_\rho)(t-t')} (n_q + 1) t_{\eta\rho}^{-\mathbf{q}} P_\eta(t'). \tag{4.170}$$

Plugging this into Eqs (4.31) and (4.32), we obtain quantum kinetic equations with memory: The right hand sides depend on all times in the past [113].

The weak-memory approximation starts with the observation that – in the absence of the exciton–phonon interaction – the density matrices $N_{\alpha\beta}(t)$ and $P_\alpha(t)$ were proportional to $\exp(i(\omega_\alpha - \omega_\beta)t)$ and $\exp(i\omega_\alpha t)$, respectively. Thus, terms $\exp(-i(\omega_\eta - \omega_\beta)t') N_{\eta\beta}(t')$ and $\exp(-i\omega_\eta t') P_\eta(t')$ in Eq. (4.170) are taken to be slowly varying in time. Corresponding factors are pulled in front of the integral. The remaining integrations can be done analytically,

$$\frac{1}{i} \int_{t_0}^{t} dt' \, e^{i(t-t')\Delta\omega} = \frac{1 - e^{i(t-t_0)\Delta\omega}}{\Delta\omega + i0^+}, \tag{4.171}$$

yielding, e.g.,

$$\tilde{T}_{\rho,q}(t) = \sum_{\eta}(n_q+1)\, t_{\eta\rho}^{-q} P_\eta(t)\,\frac{1-e^{i(t-t_0)(w_q+\omega_\rho-\omega_\eta)}}{(w_q+\omega_\rho-\omega_\eta)+i0^+}. \qquad (4.172)$$

This level represents the weak-memory approach [7,114], where the sudden switch-on of the excitation pulse is still remembered. In a last step, the Markov approximation is completed by letting $t_0 \to -\infty$,

$$\frac{1-e^{i(t-t_0)\Delta\omega}}{\Delta\omega+i0^+} \to \frac{1}{\Delta\omega+i0^+} = -i\pi\delta(\Delta\omega) + \mathscr{P}\!\left(\frac{1}{\Delta\omega}\right). \qquad (4.173)$$

Only now, the energy-conserving delta function evolved which is at the heart of Fermi's Golden Rule. The additional principle-value parts give rise to the polaron shifts of the exciton levels. For acoustic phonons, these are small and thought to be included into the eigenvalues $\varepsilon_\alpha = \hbar\omega_\alpha$. This yields Eq. (4.38) and Eq. (4.39).

Appendix C: Radiative rates

Here, explicit results for the radiative rate are derived following Andreani [115]. To do this properly, one has to include the polarization degree of freedom of the photons, together with the detailed spin degeneracy of the exciton states. For not overloading the general derivation, we did not pay attention to these refinements when starting with the Hamiltonian Eq. (4.27).

Rewriting Eq. (4.51) for the radiative rate,

$$r_\alpha = \frac{2\pi}{\hbar}\sum_{\mathbf{k},\nu} g_{\mathbf{k},\nu}\,\delta(\varepsilon_\alpha - \hbar\omega_\mathbf{k})\,|M_\alpha(\mathbf{k})|^2, \qquad (4.174)$$

we have summed over the polarization directions ν = TE, TM (transverse electric, transverse magnetic) with respect to the quantum well plane. For the heavy-hole exciton states under consideration, no contribution polarized along the growth direction z appears [116]. The coupling strengths for TE mode is given by

$$g_{\mathbf{k},\text{TE}} = \frac{\pi\hbar\omega_\mathbf{k}}{\varepsilon_\beta \Omega}, \qquad (4.175)$$

with Ω as the 3D normalization volume. For the TM mode, the analysis of the heavy-hole matrix element gives an additional factor of $\cos^2\theta = k_z^2/k^2$ (see section 4.10). The summation in Eq. (4.174) has to be performed over the full 3D photon momentum $\mathbf{k} = (\mathbf{k}_\parallel, k_z)$. The energy conserving delta function is used for the summation over k_z and produces a square root density-of-states, $K_0/\sqrt{K_0^2 - k_\parallel^2}$. Consequently, the in-plane momentum $k_\parallel$ is restricted to lie within the light cone $K_0 c/\sqrt{\varepsilon_b} = \omega_\alpha \approx \omega_x$:

$$r_\alpha = \frac{2\Gamma_0}{A} \sum_{\mathbf{k}_\|} \frac{\Theta(K_0^2 - k_\|^2)}{K_0 \sqrt{K_0^2 - k_\|^2}} (K_0^2 - k_\|^2/2) |\psi_{\alpha \mathbf{k}_\|}|^2. \tag{4.176}$$

A is the normalization area within the quantum well plane. The prefactor $2\Gamma_0$ is just the radiative rate of a plane-wave exciton state $\psi_{Q\mathbf{k}_\|} = \sqrt{A}\delta_{Q,\mathbf{k}_\|}$ at zero COM momentum, $Q = 0$. Combining the prefactors in Eqs (4.24) and (4.175), we have

$$\Gamma_0 = \frac{2\pi}{\hbar} \frac{K_0}{\varepsilon_b} \mu_{cv}^2 \phi_{1s}^2(0) O_{eh}^2. \tag{4.177}$$

Strictly speaking, Eq. (4.176) holds if the wave function is isotropic within the light cone. This is not a serious restriction, since usually the wave function extension in Fourier space ($\sim 1/\Lambda_\alpha$) is much larger than K_0. Note, however, that the anisotropy in the entire momentum space determines the exciton spin splitting, which is treated in section 4.10. As explained there, each COM eigenstate of the exciton has four nearly degenerate spin states, two of them are radiative (bright states). The expressions given here describe the radiative decay rate of the bright states alone. If the two dark states are in equilibrium with the bright ones, the total decay rate will decrease. There are good arguments and experimental results [5] that this transfer (true spin relaxation) is a relatively slow process, at least at the low excitation densities being considered here. A further refinement appears for a MQW structure, where a dramatic shortening of the decay times may occur (giant oscillator strength). Indeed, the radiative rate of excitons is not a function only of the exciton parameters, but depends on the specific photon mode structure in the sample as well (section 4.11).

For a well localized exciton state, $|\psi_{\alpha \mathbf{k}_\|}|^2$ is practically constant within the light cone and can be taken outside the integral, which then can be obtained analytically,

$$r_\alpha = C \cdot |\psi_{\alpha \mathbf{k}_\|=0}|^2; \quad C = \frac{2}{3\pi} \Gamma_0 K_0^2. \tag{4.178}$$

To give numerical values, Andreani derived $\hbar\Gamma_0 = 26$ μeV for a 10 nm wide GaAs quantum well (an error of 1/2 appearing in [115] has been corrected later [99]). With E_x = 1.58 eV and $n \equiv \sqrt{\varepsilon_b}$ = 3.6, we have $K_0 = 0.0288$ nm^{-1} and end up with $C = 7 \cdot 10^6$ s^{-1} nm^{-2}. The dependence on well width is mainly through the exciton relative wave function. If assumed to be of exponential type, we have $\phi_{1s}^2(0) = 2/(\pi a_B^2)$. Plugging in the exciton Bohr radius for a 5 nm GaAs quantum well, we obtain $\hbar\Gamma_0 = 34$ μeV and $C = 9.3 \cdot 10^6$ s^{-1} nm^{-2}, respectively.

In view of the parameter uncertainities and refinements in theory, calculating radiative exciton rates is not an easy task. E.g., the value $17 \cdot 10^6$ s^{-1} nm^{-2} quoted in our earlier work for the 5 nm QW [46,68] – applying a formula given by Bockelmann [117] – was an overestimation by a factor of two.

The radiative rates in Fig. 4.5 (middle panel) have been generated using the full expression (4.176). Compared with the approximate ones, Eq. (4.178), an increase of the values is obtained (not shown), in particular at higher energies where the wave functions are less localized.

References

1. Bastard, G. (1988). Wave Mechanics Applied to Semiconductor Heterostructures. Les Editions de Physique, Paris.
2. Loudon, R. (1983). The Quantum Theory of Light. 2nd ed., Clarendon Press, Oxford.
3. Hegarty, J., Sturge, M.D., Weisbuch, C., Gossard, A.C., and Wiegmann, W. (1982). Resonant Rayleigh scattering from an inhomogeneously broadened transition: A new probe of the homogeneous linewidth. *Phys. Rev. Lett.* **49**, 930.
4. Stolz, H., Schwarze, D., von der Osten, W., and Weimann, G. (1993). Transient resonance Rayleigh scattering from electronic states in disordered systems: Excitons in GaAs/Al$_x$Ga$_{1-x}$As multiple-quantum-well structures. *Phys. Rev. B* **47**, 9669.
5. Vinattieri, A., Shah, J., Damen, T.C., Kim, D.S., Pfeiffer, L.N., Maialle, M.Z., and Sham, L.J. (1994). Exciton dynamics in GaAs quantum wells under resonant excitation. *Phys. Rev. B* **50**, 10 868.
6. Wang, H., Shah, J., Damen, T.C., and Pfeiffer, L.M. (1995). Spontaneous emission of excitons in GaAs quantum wells: The role of momentum scattering. *Phys. Rev. Lett.* **74**, 3065.
7. Zimmermann, R. (1995). Theory of resonant Rayleigh scattering of excitons in semiconductor quantum wells. *Il Nuovo Cimento* **17 D**, 1801.
8. Garro, N., Pugh, L., Phillips, R.T., Drouot, V., Simmons, M.Y., Kardynal, B., and Ritchie, D.A. (1997). Resonant Rayleigh scattering by excitonic states laterally confined in the interface roughness of GaAs/Al$_x$Ga$_{1-x}$As single quantum wells. *Phys. Rev. B* **55**, 13 752.
9. Haacke, S., Taylor, R.A., Zimmermann, R., Bar-Joseph, I., and Deveaud, B. (1997). Resonant femtosecond emission from QW excitons: The role of Rayleigh scattering and luminescence. *Phys. Rev. Lett.* **78**, 2228.
10. Gurioli, M., Bogani, F., Ceccherini, S., and Colocci, M. (1997). Coherent vs. incoherent emission from semiconductor structures after resonant femtosecond excitation. *Phys. Rev. Lett.* **78**, 3205.
11. Hayes, G.R., Haacke, S., Kauer, M., Stanley, R.P., Houdre, R., Oesterle, U., and Deveaud, B. (1998). Resonant Rayleigh scattering versus incoherent luminescence in semiconductor microcavities. *Phys. Rev. B* **58**, R 10 175.
12. Shchegrov, A.V., Bloch, J., Birkedal, D., and Shah, J. (2000). Theory of resonant Rayleigh scattering from semiconductor microcavities: Signatures of disorder. *Phys. Rev. Lett.* **84**, 3478.
13. Hayes, G.R., Deveaud, B., Savona, V., and Haacke, S. (2000). Speckle-averaged resonant Rayleigh scattering from quantum-well excitons. *Phys. Rev. B* **62**, 6952.

14. Langbein, W., Leosson, K., Jensen, J.R., Hvam, J.M., and Zimmermann, R. (2000). Instantaneous Rayleigh scattering from excitons localized in monolayer islands. *Phys. Rev. B* **61**, R 10 555.
15. Stolz, H. (1994). Time-Resolved Light Scattering from Excitons. Springer Tracts in Modern Physics, Vol. **130**, Springer, Berlin.
16. Shah, J. (1996). Ultrafast Spectroscopy of Semiconductors and Semiconductor Nanostructures. Springer Series in Solid-State Sciences **115**, Springer, Berlin.
17. Birkedal, D., and Shah, J. (1998). Femtosecond spectral interferometry of resonant secondary emission from quantum wells: Resonant Rayleigh scattering in the nonergodic regime. *Phys. Rev. Lett.* **81**, 2372.
18. Dainty, J.C. (1984). Laser Speckle and Related Phenomena. Springer, New York.
19. Langbein, W., Hvam, J.M., and Zimmermann, R. (1999). Time-resolved speckle analysis: Probing the coherence of excitonic secondary emission. In: Proc. 24th Int. Conf. on the Physics of Semiconductors, Jerusalem (1998), (D. Gershoni, ed.), paper II-E-8. World Scientific, Singapore.
20. Langbein, W., Hvam, J.M., and Zimmermann, R. (1999). Time-resolved speckle analysis: A new approach to coherence and dephasing of optical excitations in solids. *Phys. Rev. Lett.* **82**, 1040.
21. Runge, E., and Zimmermann, R. (1999). Coherence properties of resonant secondary emission. In *Advances in Solid State Physics*, Vol. 39 (B. Kramer, ed.) p. 423, Vieweg, Braunschweig.
22. Runge, E., and Zimmermann, R. (1998). Optical properties of localized excitons in nanostructures: Theoretical aspects. In *Festkörperprobleme/Advances in Solid State Physics*, Vol. 38, (B. Kramer, ed.) 251. Vieweg, Braunschweig.
23. Zimmermann, R., and Runge, E. (1994). Exciton lineshape in semiconductor quantum structures with interface roughness. *J. Luminescence* **60/61**, 320.
24. Zimmermann, R. (1995). Excitonic spectra in semiconductor nanostructures. *Jap. J. App. Phys.* 34 Suppl. 1, 228.
25. Runge, E., Schülzgen, A., Henneberger, F., and Zimmermann, R. (1995). Relaxation kinetics and photoluminescence lineshape of excitons in II-VI semiconductor quantum wells. *Phys. Stat. Solidi (B)* **188**, 547.
26. Grassi Alessi, M., Fragano, F., Patane, A., Capizzi, M., Runge, E., and Zimmermann, R. (2000). Competition between radiative decay and energy relaxation in disordered InGaAs/GaAs quantum wells. *Phys. Rev. B* **61**, 10 985.
27. Jahn, U., Ramsteiner, M., Hey, R., Grahn, H.T., Runge, E., and Zimmermann, R. (1997). Effective exciton mobility edge in narrow quantum wells. *Phys. Rev. B* **56**, R 4387.
28. Runge, E., and Zimmermann, R. (2000) Statistical properties of speckle distribution in resonant secondary emission. *Phys. Rev. B* **61**, 4786.
29. Intonti, F., Emiliani, V., Lienau, Ch., Elsaesser, T., Savona, V., Runge, E., Zimmermann, R., Nötzel, R., and Ploog, K.H. (2001). Quantum mechanical repulsion of exciton levels in a disordered quantum well. *Phys. Rev. Lett.* **87**, 076801.
30. Savona, V., and Zimmermann, R. (1999). Time-resolved Rayleigh scattering of excitons: Evidence for level repulsion in a disordered system. *Phys. Rev. B* **60**, 4928.
31. Savona, V., Runge, E., and Zimmermann, R. (2000). Enhanced resonant backscattering of light from quantum well excitons. *Phys. Rev. B* **62**, R 4805.

32. Langbein, W., Zimmermann, R., Runge, E., and Hvam, J.M. (2000). Spin-relaxation without coherence loss: Fine-structure splitting of localized excitons. *Phys. Stat. Solidi (B)* **221**, 349.
33. Zimmermann, R., and Runge, E. (1995). Optical lineshape and radiative lifetime of excitons in quantum structures with interface roughness. In: Proc. 22nd Int. Conf. on the Physics of Semiconductors, Vancouver (1994) (D.J. Lookwood, ed.), p. 1424. World Scientific, Singapore.
34. Bimberg, D., Heinrichsdorff, F., Bauer, R.K, Gerthsen, D., Stenkamp, D., Mars, D.E., and Miller, J.N. (1992). Binary AlAs/GaAs versus ternary GaAlAs/GaAs interfaces: A dramatic difference of perfection. *J. Vac. Sci. Technol. B* **10**, 1793.
35. Castella, H., and Wilkins, J.W. (1998). Splitting of the excitonic peak in quantum wells with interfacial roughness. *Phys. Rev. B* **58**, 16 186.
36. Große, F., and Zimmermann, R. (1995). Growth simulation of ternary quantum structures and their optical properties. *Superlatt. and Microstruct.* **17**, 439.
37. Große, F., and Zimmermann, R. (2000). Monte Carlo growth simulation for AlGaAs heteroepitaxy. *J. Crystal Growth* **212**, 128.
38. Köhler, J., Jayannavar, A.M., and Reineker, P. (1989). Excitonic line shapes of disordered solids. *Z. Phys. B* **75**, 451.
39. Schnabel, R.F., Zimmermann, R., Bimberg, D., Nickel, H., Lösch, R., and Schlapp, W. (1992). Influence of exciton localization on recombination line shapes: InGaAs/GaAs quantum wells as a model. *Phys. Rev. B* **46**, 9873.
40. Stroucken, T., Anthony, T., Knorr, A., Thomas, P., and Koch, S.W. (1995). A Green's function approach to the description of optical properties of disordered semiconductors. *Phys. Stat. Solidi (B)* **188**, 539.
41. Velicki, B., Kirkpatrick, S., and Ehrenreich, H. (1968). Single-site approximation in the electronic theory of simple binary alloys. *Phys. Rev.* **175**, 747.
42. Glutsch, S., and Bechstedt, F. (1994). Theory of asymmetric broadening and shift of excitons in quantum structures with rough interfaces. *Phys. Rev. B* **50**, 7733.
43. Halperin, B.I. (1965). Green's functions for a particle in a one-dimensional random potential. *Phys. Rev.* **139**, A 104.
44. Zimmermann, R. (1992). Theory of dephasing in semiconductor optics. *Phys. Stat. Solidi (B)* **173**, 129.
45. Zimmermann, R., Runge, E., and Große, F. (1996). Optical spectra of quantum structures: Influence of interface and alloy disorder. In: Proc. 23rd Int. Conf. on the Physics of Semiconductors, Berlin (Germany) (1996) (M. Scheffler and R. Zimmermann, eds.), p. 1935. World Scientific, Singapore.
46. Zimmermann, R., Große, F., and Runge, E. (1997). Excitons in semiconductor nanostructures with disorder. *Pure and Applied Chemistry* **69**, 1179.
47. Runge, E., and Zimmermann, R. (2000). Porter-Thomas distribution of oscillator strengths of quantum well excitons. *Phys. Stat. Solidi (B)* **221**, 269.
48. Runge, E., and Zimmermann, R. (1999). Level repulsion and wavefunction statistics in semiconductor quantum systems with localized excitons. *Ann. Physik (Leipzig)* **8**, SI 229.
49. Halperin, B.I., and Lax, M. (1966). Impurity-band tails in the high-density limit. I. Minimum counting methods. *Phys. Rev.* **148**, 722.
50. Lifshitz, I.M., Gredeskul, S.A., and Pastur, L.A. (1988). Introduction to the Theory of Disordered Systems. Wiley, New York.

51. Zimmermann, R. (1990). Transverse relaxation and polarization specifics in the dynamical Stark effect of excitons. *Phys. Stat. Solidi (B)* **159**, 317.
52. Haug, H. (1992). Interband quantum kinetics with LO-phonon scattering in a laser-pulsed excited semiconductor: I. Theory.; (with Banyai, L., Tran Thoai, D.B., and Remling, C.:) II. Numerical studies. *Phys. Stat. Solidi (B)* **173**, 139 + 149.
53. Schilp, J., Kuhn, T., and Mahler, G. (1994). Electron-phonon quantum kinetics in pulse-excited semiconductors: Memory and renormalization effects. *Phys. Rev. B* **50**, 5435.
54. Axt, V.M., and Stahl, A. (1994). A dynamics-controlled truncation scheme for the hierarchy of density matrices in semiconductor optics. *Z. Phys. B* **93**, 195.
55. Victor, K., Axt, V.M., and Stahl, A. (1995). Hierarchy of density matrices in coherent semiconductor optics. *Phys. Rev. B* **51**, 14 164.
56. Zimmermann, R., Wauer, J., Leitenstorfer, A., and Fürst, C. (1998). Observation of memory effects in electron-phonon quantum kinetics. *J. Luminescence* **76&77**, 34.
57. Takagahara, T. (1985). Localization and homogeneous dephasing relaxation of quasi-two-dimensional excitons in quantum well structures. *Phys. Rev. B* **32**, 7013.
58. Neukirch, U., Weckendrup, D., Faschinger, W., Juza, P., and Sitter, H. (1994). Exciton relaxation dynamics in ultrathin CdSe/ZnSe single quantum wells. *J. Cryst. Growth* **138**, 849.
59. Takagahara, T. (1985). Dephasing of exciton polaritons. *Phys. Rev. B* **31**, 8172.
60. Takagahara, T. (1999). Theory of exciton dephasing in semiconductor quantum dots. *Phys. Rev. B* **60**, 2638.
61. Borri, P., Langbein, W., Schneider, S., Woggon, U., Sellin, R.L., Ouyang, D., and Bimberg, D. (2001). Ultralong dephasing time in InGaAs quantum dots. *Phys. Rev. Lett.* **87**, 157401.
62. Siantidis, K., Axt, V.M., and Kuhn, T. (2001). Dynamics of exciton formation for near band-gap excitation. *Phys. Rev. B* **65**, 035303.
63. Nag, B.R. (1980). Electron Transport in Compound Semiconductors. Springer Series in Solid-States Sciences **11**, Springer, Berlin.
64. Ridley, B.K. (1993). Quantum Processes in Semiconductors. Clarendon Press, Oxford.
65. Ridley, B.K. (1997). Electrons and Phonons in Semiconductor Multilayers. Cambridge University Press, Cambridge.
66. Yu, P.Y., and Cardona, M. (1996). Fundamentals of Semiconductors. Springer, Berlin.
67. Takagahara, T. (1985). Localization and energy transfer of quasi-two-dimensional excitons in GaAs–AlAs quantum-well heterostructures. *Phys. Rev. B* **31**, 6552.
68. Zimmermann, R., and Runge, E. (1997). Excitons in narrow quantum wells: Disorder localization and luminescence kinetics. *Phys. Stat. Solidi (A)* **164**, 511.
69. Runge, E., and Zimmermann, R. (1998). Spatially resolved spectra and effective mobility edge in narrow quantum wells. *Phys. Stat. Solidi (B)* **206**, 167.
70. Runge, E. (2002). Excitons in semiconductor nanostructures. To be published in the series Solid State Physics, Vol. 57, Academic Press, San Diego.
71. Chu, B. (1974). Laser Light Scattering. Academic Press, New York.
72. Mandel, L., and Wolf, E. (1995). Optical Coherence and Quantum Optics. Cambridge University Press, Cambridge.
73. Zimmermann, R., Langbein, W., Runge, E., and Hvam, J.M. (2001). Localized excitons in quantum wells show spin relaxation without coherence loss. *Physica E* **10**, 40.

74. Zimmermann, R., Langbein, W., Runge, E., and Hvam, J.M. (2001). Speckle correlation spectroscopy on localized spin-split exciton states in quantum wells. In: Proc. 25th Int. Conf. on the Physics of Semiconductors, Osaka (2000) (N. Miura and T. Ando, eds.), Springer Proceedings in Physics **87**, p. 523.
75. Haacke, S., Schaer, S., Deveaud, B., and Savona, V. (2000). Interferometric analysis of resonant Rayleigh scattering from 2D excitons. *Phys. Rev. B* **61**, R 5109.
76. Shchegrov, A.V., Birkedal, D., and Shah, J. (1999). Monte Carlo simulations of ultrafast resonant Rayleigh scattering from quantum well excitons: Beyond ensemble averaging. *Phys. Rev. Lett.* **83**, 1391.
77. Woerner, M., and Shah, J. (1998). Resonant secondary emission from two-dimensional excitons: Femtosecond time evolution of the coherence properties. *Phys. Rev. Lett.* **81**, 4208.
78. Gurioli, M., Bogani, F., Vinattieri, A., Colocci, M., Belitsky, V.I., Cantarero, A., and Pavlov, S.T. (1995). Resonant Rayleigh scattering in semiconductor structures. *Il Nuovo Cimento* **17 D**, 1487.
79. Runge, E., and Zimmermann, R. (1998). Level repulsion in excitonic spectra of disordered systems and local relaxation kinetics. *Annalen der Physik (Leipzig)* **7**, 417.
80. Guest, J.R., Stievater, T.H., Steel, D.G., Gammon, D., Katzer, D.S., and Park, D. (2000). Nonlinear near-field spectroscopy and microscopy of single excitons in a disordered quantum well. In: Quantum Electronics and Laser Science Conference, OSA Technical Digest, p. 6. Optical Soc. of America, Washington DC.
81. von Freymann, G., Kurtz, E., Klingshirn, C., and Wegener, M. (2000). Statistical analysis of near-field photoluminescence spectra of single ultrathin layers of CdSe/ZnSe. *Appl. Phys. Lett.* **77**, 394.
82. Mehta, M.L. (1990). Random Matrices. 2nd ed. Academic Press, San Diego.
83. Hess, H.F., Betzig, E., Harris, T.D., Pfeiffer, L.N., and West, K.W. (1994). Near-field spectroscopy of the quantum constituents of a luminescent system. *Science* **264**, 1740.
84 Brunner, K., Abstreiter, G., Böhm, G., Tränkle, G., and Weimann, G. (1994). Sharp-line photoluminescence of excitons localized at GaAs/AlGaAs quantum well inhomogeneities. *Appl. Phys. Lett.* **64**, 3320.
85. Gammon, D., Snow, E.S., Shanabrook, B.V., Katzer, D.S., and Park, D. (1996). Fine structure splitting in the optical spectra of single GaAs quantum dots. *Phys. Rev. Lett.* **76**, 3005.
86. Jahn, U., Kwok, S.H., Ramsteiner, M., Hey, R., Grahn, H.T., and Runge, E. (1996). Exciton localization, photoluminescence spectra, and interface roughness in thin quantum wells. *Phys. Rev. B* **54**, 2733.
87. Wu, Q., Grober, R.D., Gammon, D., and Katzer, D.S. (1999). Imaging spectroscopy of two-dimensional excitons in a narrow GaAs/AlGaAs quantum well. *Phys. Rev. Lett.* **83**, 2652.
88. Savona, V., Runge, E., Zimmermann, R., Intonti, F., Emiliani, V., Lienau, Ch., and Elsaesser, T. (2002). Level repulsion of localized excitons in disordered quantum wells. *Phys. Stat. Solidi (A)* **190**, 625.
89. Kramer, B., and MacKinnon, A. (1993). Localization: Theory and experiment. *Reports Prog. Phys.* **56**, 1469.
90. Savona, V., Haacke, S., and Deveaud, B. (2000). Optical signatures of energy-level statistics in disordered quantum systems. *Phys. Rev. Lett.* **84**, 183.

91. Birkedal, D., Shah, J., Shchegrov, A.V., and Pfeiffer, L.N. (2000). Experimental investigation of quantum effects in time-resolved resonant Rayleigh scattering from quantum well excitons. *Phys. Stat. Solidi (A)* **178**, 5.
92. Malpuech, G., Kavokin, A., Langbein, W., and Hvam, J.M. (2000). Resonant Rayleigh scattering of exciton–polaritons in multiple quantum wells. *Phys. Rev. Lett.* **85**, 650.
93. Prineas, J.P., Shah, J., Grote, B., Ell, C., Khitrova, G., Gibbs, H.M., and Koch, S.W. (2000). Dominance of radiative coupling over disorder in resonance Rayleigh scattering in semiconductor multiple quantum-well structures. *Phys. Rev. Lett.* **85**, 3041.
94. Van Albada, M.P., and Lagendijk, A. (1985). Observation of weak localization of light in a random medium. *Phys. Rev. Lett.* **55**, 2692.
95. Wolf, P.-E., and Maret, G. (1985). Weak localization and coherent backscattering of photons in disordered media. *Phys. Rev. Lett.* **55**, 2696.
96. Akkermans, E., Wolf, P.-E., and Maynard, R. (1986). Coherent backscattering of light by disordered media: Analysis of the peak line shape. *Phys. Rev. Lett.* **56**, 1471.
97. Vollhardt, D., and Wölfle, P. (1980). Diagrammatic, self-consistent treatment of the Anderson localization problem in $d \leq 2$ dimensions. *Phys. Rev. B* **22**, 4666.
98. Maialle, M.Z., de Andrada e Silva, E.A., and Sham, L.J. (1993). Exciton spin dynamics in quantum wells. *Phys. Rev. B* **47**, 15 776.
99. Andreani, L.C. (1995). Optical transitions, excitons, and polaritons in bulk and low-dimensional semiconductor structures. In: Confined Electrons and Photons. (E. Burstein and C. Weisbuch, eds.) p. 57. NATO ASI Series B **340**, Plenum Press, New York.
100. Nickolaus, H., Wünsche, H.-J., and Henneberger, F. (1998). Exciton spin relaxation in semiconductor quantum wells: The role of disorder. *Phys. Rev. Lett.* **81**, 2586.
101. Goupalov, S.V., Ivchenko, E.L., and Kavokin, A.V. (1998). Anisotropic exchange splitting of excitonic levels in small quantum systems. Superlatt. and Microstruct. 23, 1205.
102. Maialle, M.Z. (2000). Spin dynamics of localized excitons in semiconductor quantum wells in an applied magnetic field. *Phys. Rev. B* **61**, 10 877.
103. Wilkinson, M., Fang Yang, Austin, E.J., and O'Donell, K.P. (1992). A statistical model of exciton luminescence spectra. *J. Phys.: Condens. Matter* **4**, 8863.
104. Andreani, L.C., and Bassani, F. (1990). Exchange interaction and polariton effects in quantum-well excitons. *Phys. Rev. B* **41**, 7536.
105. Citrin, D.S. (1994). Time-domain theory of resonant Rayleigh scattering by quantum wells: Early-time evolution. *Phys. Rev. B* **54**, 14 572.
106. Grote, B., Ell, C., Koch, S.W., Gibbs, H.M., Khitrova, G., Prineas, J.P., and Shah, J. (2001). Resonance Rayleigh scattering from semiconductor heterostructures: The role of radiative coupling. *Phys. Rev. B* **64**, 045330.
107. Martin, O.J.F., and Piller, N.B. (1998). Electromagnetic scattering in polarizable backgrounds. *Phys. Rev. E* **58**, 3909.
108. Ammerlahn, D., Grote, B., Koch, S.W., Kuhl, J., Hübner, M., Hey, R., and Ploog, K. (2000). Influence of the dielectric environment on the radiative lifetime of quantum-well excitons. *Phys. Rev. B* **61**, 4801.
109. Jorda, S. (1994). Quantum theory of the interaction of quantum-well excitons with electromagnetic waveguide modes. *Phys. Rev. B* **50**, 2283.

110. Savona, V., Piermarocchi, C., Quattropani, A., Schwendimann, P., and Tassone, F. (1999). Optical properties of microcavity polaritons. *Phase Transitions* **68**, 169.
111. Khitrova, G., Gibbs, H.M., Jahnke, F., Kira, M., and Koch, S.W. (1999). Nonlinear optics of normal-mode-coupling semiconductor microcavities. *Rev. Mod. Phys.* **71**, 1591.
112. Menniger, J., Kostial, H., Jahn, U., Hey, R., and Grahn, H. (1995). Depth correlated lateral variations of layer thicknesses in GaAs-AlGaAs multiple quantum wells investigated by cathodoluminescence. *Appl. Phys. Lett.* **66**, 2349.
113. Haug, H., and Jauho, A.-P. (1996). Quantum Kinetics in Transport and Optics of Semiconductors. Springer Series in Solid-State Sciences **123**, Springer, Berlin.
114. Meden, V., Wöhler, C., Fricke, J., and Schönhammer, K. (1995). Hot-electron relaxation: An exactly solvable model and improved quantum kinetic equations. *Phys. Rev. B* **52**, 5624.
115. Andreani, L.C. (1991). Radiative lifetime of free excitons in quantum wells. *Solid State Comm.* **77**, 641.
116. Citrin, D.S. (1993). Radiative lifetime of excitons in quantum wells: Localization and phase-coherence effects. *Phys. Rev. B* **47**, 3832.
117. Bockelmann, U. (1993). Exciton relaxation and recombination in quantum dots. *Phys. Rev. B* **48**, 17637.

Quantum Coherence, Correlation and Decoherence in
Semiconductor Nanostructures
T. Takagahara (Ed.)
Copyright © 2003 Elsevier Science (USA). All rights reserved.

Chapter 5
Higher-order Coulomb correlation effects in semiconductors

S.R. Bolton

*Physics Department, Williams College, 33 Lab Campus Drive,
Williamstown, MA 01267, USA*

Abstract

Optically excited semiconductors provide an ideal model system for the study of many body interactions. Recent breakthroughs in theory and experiment have revealed unique signatures of individual Coulomb-correlated many-body interactions, up to the six-particle level. Experiments are performed using ultrafast spectroscopy, so that the Coulomb-correlated quasi-particles can be measured on time scales short compared with their dephasing times. The simultaneous progress in theory and experiment in this field has resulted in a burst of collaborative activity, yielding detailed understanding of the influence of each many-body term. In this work we present the new experimental and theoretical techniques which have allowed such rapid progress. We also summarize the results which have been achieved thus far, and present the outstanding questions which are the subject of current study.

5.1   Introduction

The interactions among quasi-particles underlie many of the important problems in condensed matter physics, including superconductivity, the quantum Hall-effect, magnetism, and nonlinear optics (NLO). Particularly at low temperatures, such interactions are poorly represented by classical approximations, and bear distinct signatures of quantum mechanical phase. Only a quantum system can exist in a coherent, correlated state, that is, in a superposition of eigenstates with well-defined relative phases. The entanglement in such a state gives rise to the properties necessary for quantum computation. In condensed matter, the Coulomb force is responsible for interparticle correlations which may extend over thousands of lattice sites. That same Coulomb force is also the cause of dephasing, by which initially coherent excitations evolve into incoherent superpositions. The dynamics of quantum coherence and dephasing are of central importance in condensed matter physics. However, the many-body interactions which underlie these processes are extremely complex and not easily modeled theoretically. Recently, simultaneous advances in theory and experiment have demonstrated that individual many-body interactions have specific spectral and temporal signatures in semiconductor nonlinear optical response. As a result, new ultrafast spectroscopic techniques combined with microscopic many-body theory have brought dramatic progress in the study of many-body Coulomb interactions both in bulk and in quantum confined systems. In particular, the role of Coulomb correlations among more than two particles, which was inaccessible to both theory and experiment before 1994, has been beautifully revealed in the last several years.

This chapter will be devoted to a review of recent research into such "high order" Coulomb correlations (HOCs). Section 5.2 outlines the basic physics of ultrafast excitation and the measurement of Coulomb correlations in semiconductors, and defines the state of the field before the advent of techniques specific to measurements of high-order correlations. Section 5.3 reviews the theoretical advances which have led to a new microscopic understanding of nonlinear optical response. Section 5.4 presents the results of series of recent experiments, from which the roles of four- and six-particle Coulomb correlations can be explicitly identified by comparison with microscopic theory. Finally, section 5.4 outlines possible directions for future work.

5.2 Ultrafast spectroscopy of semiconductor nanostructures as probes of Coulomb correlations

Optically excited semiconductors provide an ideal laboratory for investigating correlated many-body quantum kinetics. Almost perfect samples are available, including both very pure bulk materials and well-controlled quantum well heterostructures. The variety of high quality semiconductor heterostructures is constantly increasing, and now comprises samples with a wide range of band gaps, exciton binding energies, and degrees of Coulomb correlation. In addition to the most commonly used III-V materials (GaAs and $In_xGa_{1-x}As$) which have band gaps ranging from 0.8 to 1.5 eV, high purity heterostructures of the larger gap II-VI (CdSe, CdS, ZnSe) materials are now becoming available. II-VI materials allow the study of dynamics in systems with very strongly bound excitons, as well as spectrally resolved bi-excitons. Optically excited semiconductors are also excellent testing grounds from the theoretical point of view, because the low temperature ground state (full valence band, empty conduction band) is very well understood via effective mass and mean-field theories. Therefore, although the semiconductor in its ground state is a true correlated many-body system, the single-particle excitations near the band gap provide a well-defined and completely uncorrelated starting point for the dynamics. From this ground state, optical excitation creates electron–hole pairs which become correlated via the Coulomb interaction. These interparticle correlations are induced by the laser field, and are thus experimentally controllable, in contrast with the strong ground state correlations present in systems such as superconductors. The combination of high purity samples and external control of correlations has placed the field of time-resolved nonlinear optical spectroscopy of semiconductors in an exceptional situation. Very precise experimental data can be compared with first principles theory, providing the opportunity for insights into many-particle interactions which are extremely difficult to obtain in most condensed matter systems. In the following paragraphs we outline the physical processes taking place upon excitation of a semiconductor with a short optical pulse.

In order to obtain maximal information about correlated many-body states, it is necessary to perform spectroscopy on time scales short with respect to the dephasing of those correlations, typically on the fs or ps scale. The field of a laser pulse on this time scale is typically expressed as $\mathscr{E}(t) = E(t)e^{i\omega t} + c.c.$, where the time dependence of the field is separated into a "slow" envelope $E(t)$ and a fast oscillation at the carrier frequency, ω. Upon excitation of a semiconductor with an above-gap ultrafast pulse, a series of interactions take place by which the initially created polarization evolves toward equilibrium. The initial excitation of electron–hole pairs follows the carrier wave, creating a coherently oscillating macroscopic polarization. Immediately thereafter, exciton–exciton, electron–electron, and electron–phonon scattering begin to randomize the relative phases of the oscillations, gradually dephasing the macroscopic polarization. The dephasing time, T_2, is very much dependent on both the energy distribution of the photo-carriers and on the temperature of the sample. For excitons at low temperature T_2 can be several picoseconds, while for electrons high in the band T_2 is only some tens of femtoseconds. Once the phase coherence is destroyed, carriers continue to interact via both Coulomb and electron–phonon scattering, typically taking hundreds of femtoseconds to evolve from non-thermal energy distributions to quasi-equilibrium Fermi–Dirac distributions. Finally, e–h recombination occurs on a nano-second time scale, returning the semiconductor to its ground state. In the following section we outline the theoretical formulations which describe the creation and interaction of electron–hole pairs (details may be found in a number of excellent reviews, for example Haug and Koch [1]). In so doing we encounter the many-body problem in optically excited condensed matter.

Overview of the semiconductor equations of motion with optical excitation

Consider a two-band semiconductor, which has a Hamiltonian of the form

$$\hat{H}_{el} = \sum_{k,\lambda} E_{\lambda,k} \hat{a}^+_{\lambda,k} \hat{a}_{\lambda,k} + \sum_{\substack{k,k' \\ q \neq 0 \\ \lambda,\lambda'}} V_q \hat{a}^+_{\lambda,k+q} \hat{a}^+_{\lambda',k'-q} \hat{a}_{\lambda,k} \hat{a}_{\lambda',k'} \tag{5.1}$$

Here the "a"s are creation and annihilation operators in the bands indexed by λ, and V gives the Coulomb interaction. The first term in equation (5.1) describes the dispersion of the conduction and the valence bands, while the second contains both inter- and intra-band Coulomb terms. The driving optical field is most commonly treated in the rotating wave and dipole approximations, so that we can write the interaction of electrons and photons as

$$\hat{H}_I = -\sum_k [\mu_{cv,k} E(t) \hat{c}^+_k \hat{v}_k + \mu^*_{cv,k} E^*(t) \hat{v}^+_k \hat{c}_k] \tag{5.2}$$

where c and v represent the annihilation of particles in the conduction and valence bands, respectively. The total Hamiltonian, $\hat{H} = \hat{H}_{el} + \hat{H}_I$, has no known solutions, and approximations are necessary at this point.

The general approach for describing the optical properties of a many-particle system is to determine the expectation values of observables through the density matrix operator, ρ, which is written as a Fourier sum, $\hat{\rho}(t) = \sum_k \hat{\rho}_k(t)$. For optical measurements the observable of most importance is the macroscopic interband polarization, P_k, whose expectation value is given by $\langle P_k \rangle = Tr(\hat{\rho}_k P_k)$ (Shen, [2]). In the two parabolic band model, the matrix elements of $\hat{\rho}_k$ are the expectation values of two particle operators, thus

$$\hat{\rho}_k = \begin{bmatrix} \langle \hat{c}^+_k \hat{c}_k \rangle & \langle \hat{v}^+_k \hat{c}_k \rangle \\ \langle \hat{c}^+_k \hat{v}_k \rangle & \langle \hat{v}^+_k \hat{v}_k \rangle \end{bmatrix} = \begin{bmatrix} n_{e,k} & p^*_k \\ p_k & n_{h,k} \end{bmatrix} \tag{5.3}$$

Here n_{ek} and n_{hk} represent electron and hole populations, while the off-diagonal components, p, describe interband polarizations (also known as electron–hole transition amplitudes). In order to obtain the equation of motion for the macroscopic polarization, one must start with the Heisenberg equation of motion for the two-particle operators, i.e.,

$$-i\hbar \frac{\partial \hat{\rho}}{\partial t} = [\hat{H}, \hat{\rho}], \tag{5.4}$$

Even in the absence of driving fields, the equation of motion for any two-particle observable will contain four-particle operators due to the Coulomb terms in H. In the presence of a driving field, $E(t)$, the equation of motion for the kth component of polarization reads as follows.

$$\hbar\left[i\frac{d}{dt}-(\varepsilon_{e,k}+\varepsilon_{h,k})\right]p_k = (n_{e,k}+n_{h,k}-1)\mu_{c,v}E(t)$$
$$+ \sum_{k',q\neq 0} V_q[\langle c^+_{k'+q}v_{k-q}c_{k'}c_k\rangle + \langle v_{k'+q}v_{k-q}v^+_{k'}c_k\rangle$$
$$+ \langle v_{k'}c^+_{k'-q}c_{k'}c_{k-q}\rangle + \langle v_{k'}v_{k'-q}v^+_{k'}c_{k-q}\rangle] \quad (5.5)$$

Here ε_{ek} and ε_{hk} are the single-particle electron and hole energies. Equation (5.5) shows that the polarization is driven by three source terms – the dipole coupling to the electric field (μE), the Pauli blocking (terms proportional to $n\mu E$), and the (four-particle) Coulomb interactions among particles. Thus, to solve the equations of motion for the two-particle operators, one must write the equation of motion for the four-particle terms, which in turn contain six-particle operators, leading to an infinite hierarchy of coupled equations. This behavior is a fundamental problem of many-body systems – in all cases one must select appropriate approximations to truncate the hierarchy. Experimental tests of many-body approximation schemes are typically very difficult to obtain. However, ultrafast spectroscopy measures directly the time evolution of the macroscopic polarization, on time scales short compared with the correlation decoherence times, thus accessing directly the Coulomb correlations. Polarization selection rules and variation of excitation intensity allow significant experimental control of the types and densities of excited populations, and thus of the relative strength of Coulomb correlations. As a result, carefully controlled spectroscopy techniques measure quite directly the many-body Coulomb correlations, and can readily distinguish the contributions of two-particle correlations, four-particle correlations (4PC), and even higher-order contributions. The details of these techniques and their results will be presented in section 5.4. In the next section we review initial theoretical approaches to semiconductor NLO response, including the role of two-particle Coulomb correlations.

Non-interacting and Hartree–Fock approximations

The simplest approximation one could make to equation (5.5) would neglect Coulomb interactions altogether, treating the carriers as entirely noninteracting and driven only by the external electric field. For semiconductors this noninteracting electron approximation fails quite dramatically even in describing linear absorption, which is dominated by the presence of bound excitons below the gap, and which is strongly enhanced by the presence of Coulomb interactions even in the continuum. (The Coulomb-induced "Sommerfeld enhancement" of the linear absorption gives a 10% enhancement above the noninteracting absorption even at a photon energy which is 50 exciton Rydbergs above the gap (Elliott, [3]; Tanguy, [4]).) One important situation in which the noninteracting theory *does* apply is that of high density electron–hole plasmas, such as would be created by intense optical excitation well above the gap, or by phonon-induced ionization of excitons in samples at room temperature. In this case, the e–h plasma screens the Coulomb interaction so significantly that the non-interacting limit is nearly recovered.

The most straightforward approximation scheme including Coulomb interactions is the time-dependent Hartree–Fock/random phase approximation (HF-RPA). In this approximation all four-particle expectation values are *factorized into products of two-particle expectation* values. For example,

$$\langle c_{k'}^+ v_{k-q}^+ c_{k'} c_k \rangle \cong \langle c_k^+ c_k \rangle \langle v_k^+ c_k \rangle \delta_{k-q,k'} = n_{e,k} p_k^* \delta_{k-q,k'}.$$

The HF-RPA yields a closed set of equations of motion for the expectation values of polarization, $p_k = \langle c_k v_k^+ \rangle$, and population, $n_{e,k} = \langle c_k^+ c_k \rangle$, $n_{h,k} = \langle v_k^+ v_k \rangle$, and neglects all higher-order Coulomb correlations (Lindberg and Koch, [5]; Haug and Koch, [1]). The resulting equations of motion for the two particle observables are known as the semiconductor Bloch equations (SBE) by virtue of their similarity to the well-known Optical Bloch equations for two-level (atomic) systems. We reproduce the SBE here, as they are the starting point for analysis of the vast majority of optical experiments in semiconductors.

$$\frac{\partial p_k}{\partial t} = -i(e_{e,k} + e_{h,k}) p_k - i(n_{e,k} + n_{h,k} - 1)\omega_{R,k} \tag{5.6}$$

$$\frac{\partial n_{e,k}}{\partial t} = -2\,\text{Im}(\omega_{R,k} p_k^*) = \frac{\partial n_{h,k}}{\partial t} \tag{5.7}$$

The Coulomb interaction influences the SBE in two ways. First, the original single particle energies, ε_i, are renormalized to $e_{i,k} \equiv \varepsilon_{i,k} + \sum_q V_{|k-q|} n_{i,q}$, $i = e, h$. This shifting of the energies ultimately induces an alteration in both band gap and band curvature, known as band gap renormalization (BGR). In addition, the coupling of the electrons and holes to the driving electric field is altered, as indicated by the generalized Rabi frequency, ω_R, which replaces the bare Rabi frequency as follows:

$$\omega_{R,k} \equiv \frac{1}{\hbar}\left[\mu_{c,v} E + \sum_{q \neq k} V_{|k-q|} p_q\right].$$

Note that although the SBE appear superficially to be diagonal in the momentum index k, in reality the Coulomb terms in the generalized Rabi frequency and the renormalized particle energies lead to a coupling of all momentum states. The SBE describe quite well a number of experiments in ultrafast spectroscopy, and yield a straightforward interpretation of the physical origins of NLO signals. The first term in equation (5.6) describes the coupling of the polarization to the combined external electric field and the internal Coulomb fields from other dipoles. The second term includes the influence of e–h populations on absorption (Pauli blocking) – showing that electrons cannot make transitions into already occupied states. If the Coulomb interaction is set to zero in the SBE, one recovers the optical Bloch equations for free carrier transitions (non-interacting limit). The homogeneous parts of the SBE (no driving electric field) reproduce the famous Wannier equations describing the energies of electron–hole pairs, but with the e–h Coulomb interaction renormalized by the carrier occupation factors. Finally, equation (5.7) shows how the time evolution of the electron and hole populations is driven by the combination of external and internal fields.

Beyond the coherent SBE: screening and scattering

The physics of the coherent regime, in which the e–h pairs retain the phase of the laser, is often well described by the semiconductor Bloch equations without scattering terms. However, the processes of carrier–carrier scattering and carrier–phonon scattering, which lead to dephasing and energy rearrangement, are not included in the derivation of the SBE. In practice, these interactions can be included in the fit to experimental data at various levels of approximation. Most frequently, scattering terms are included in a relaxation time approximation, in which phenomenological terms dp/dt_{coll} and dn/dt_{coll} are added to equations (5.6) and (5.7).

Collisional interactions can be understood more completely by approximating the first-order corrections to the SBE. This may be done by evaluating the equation of motion of the correlated part of the four-particle Coulomb term which is factorized in the HF/RPA. In the last step of such an evaluation all six- and four-particle terms are factorized into products of two-particle terms. From this analysis one obtains the complete Boltzmann collision rates for electron–electron and electron–hole scattering. The rates describe the effective scattering into and out of each state k (Lindberg, [5]). If the initial carrier distribution is in the form of a Fermi–Dirac distribution, the terms of the Boltzmann equation cancel, giving no net scattering into or out of a given state. Because the full scattering rates derived in this manner are very complicated and highly nonlinear, it is far more common to treat both dephasing and energy relaxation by describing them with simple phenomenological rates, γ_p and γ_n. The dephasing rate, γ_p, is often taken to be independent of the distribution. For the population scattering processes, which tend toward Fermi distributions within the respective bands, a linear approximation of the scattering integrals can be extremely useful if the initial distribution is not too far away from a Fermi distribution. This gives a population scattering rate of the form:

$$\left.\frac{\partial n_{e,k}}{\partial t}\right|_{\text{collisional}} = -\gamma_n (n_{e,k} - f_{e,k}),$$

and neglects the population distribution dependence of the rate term, γ_n. It should be noted that any Boltzmann approach to scattering requires the use of the Markov and Second Born approximations, in which collisions are treated as instantaneous and memory effects are neglected. Although these are excellent approximations over long times, measurements on time scales short compared to the natural oscillation times of the bath (determined by the plasma frequency for carrier–carrier scattering, and the phonon frequency for carrier phonon interactions) reveal significant departures from the Boltzmann limit.

At high excitation density screening of the Coulomb interaction must be taken into account. Screening is an inherently four-particle process, and is thus not included at any level in the SBE. However, a number of studies have included screening phenomenologically by replacing the unscreened Coulomb potential with a screened interaction, V_s, in equations (5.6) and (5.7). The most commonly used screening model for this purpose is the Linhard dielectric function, treated quasistatically. This gives a screened Coulomb potential which is strongly dependent on the density of excited carriers, through both the screening length and the plasma frequency (Lindberg, [6]). The inclusion of screening alters the effects of all the Coulomb factors in the semiconductor response, including BGR, the Rabi frequency, and the exchange interaction.

Ultrafast optical measurement techniques

A number of ultrafast spectroscopic techniques have been developed which allow both tests of the validity of the HF/RPA and the study of higher-order correlations. Most prominent among these are four-wave mixing (FWM) and pump/probe (P/P) spectroscopy, both of which allow access to the time evolution of the macroscopic polarization, and thus give results which can be directly compared with the equations of motion (5.6) and (5.7). A schematic of the experimental arrangement for these techniques is shown in Fig. 5.1. For all ultrafast time-resolved experiments the original short pulse from the laser is split into two parts, with a precisely controllable delay. In P/P an intense pump beam excites the semiconductor, which is then probed by a much weaker probe pulse. The pump-induced changes in the probe absorption spectrum are measured as a function of pump–probe delay. Pump–probe spectroscopy is particularly useful for measuring the incoherent parts of the dynamics, such as the inelastic scattering which results in thermalization and cooling of an initially non-thermal carrier distribution. Four-wave mixing is the most common technique for measurements of coherent and dephasing dynamics. In FWM, the first pulse (along k_1) creates a coherent macroscopic polarization. The second pulse (k_2) then interferes with this polarization, creating a coherent population grating from which a second k_2 photon can scatter. The emitted signal in the momentum conserving $2k_2 - k_1$ direction is studied as a function of time delay between pulses. Initial work studied simply the total FWM intensity as a function of delay, however significantly more information can be extracted from the FWM signal by resolving its spectral and/or temporal profile. The signal emitted along $2k_2 - k_1$ is calculated by performing a perturbative expansion of the equations of motion (e.g. (5.6) and (5.7)) in the exciting fields. This takes the form:

$$P = \chi E + \chi^{(2)}E^2 + \chi^{(3)}E^3 + \ldots \equiv P^{(1)} + P^{(2)} + P^{(3)} + \ldots \qquad (5.8)$$

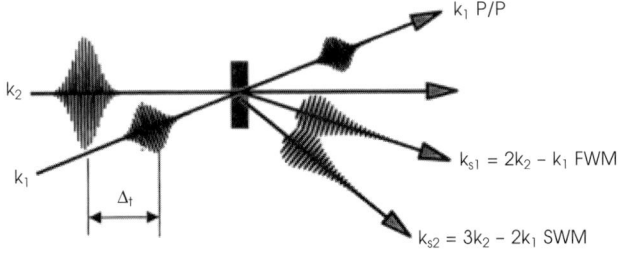

Fig. 5.1 Experimental arrangements for ultrafast spectroscopy. For P/P the probe beam k_1 is much weaker than the pump beam, k_2. For four-wave-mixing (FWM) and six-wave-mixing (SWM) k_1 and k_2 are of equal intensity. The P/P, FWM, and SWM signals may be spectrally resolved by sending them to a spectrometer with multichannel detection. FWM and SWM may be temporally resolved by upconverting the emitted signal with part of the original laser pulse.

The FWM signal is of leading order in E^3 (two photons from k_2, one photon from k_1) and can have as its origin any of the nonlinear sources in the equation of motion (5.5) – these include the Pauli blocking as well as the Coulomb coupling. Typically theoretical predictions are made in the leading order alone, however recent work demonstrates that even at low densities the fifth-order contributions to the FWM signal are non-negligible (Wegener, [7]; Haase, [8]).

The most straightforward interpretation of FWM is obtained in a non-interacting electron approximation (setting $V = 0$ in equations (5.6) and (5.7)), and reads the FWM signal as a direct measure of the dephasing time, T_2 (Yajima and Taira, [9]). The second pulse simply samples the remaining coherence present from the first excitation, and the total intensity in the FWM direction decays with a time constant equal to $T_2/2$. The noninteracting electron model predicts that the FWM signal should be emitted

Higher-order Coulomb correlation effects in semiconductors 175

instantaneously upon the arrival of pulse two, should have an spectral profile identical to that of pulse two, and should be completely absent if pulse two precedes pulse one (negative time delay). Although the noninteracting interpretation of FWM data is still quite commonly used, FWM signals in condensed matter are actually far more complicated and contain significantly more dynamical information. In particular, the polarization is driven by an effective local field, consisting of the sum of the external driving field and the internal Coulomb sources. Local field effects strongly alter the FWM, causing both a non-zero signal at negative time delay and a change in the real-time profile of the emission (Leo, [10]; Kim, [11]; Weiss, [12]; Chemla and Bigot, [13]). As shown in Figs 5.2 and 5.3, the local field contributions to the FWM signal *dominate* at low excitation densities, leading to two distinct components in the FWM emission, one instantaneous (due to Pauli blocking effects) and one delayed, due to the slow build up

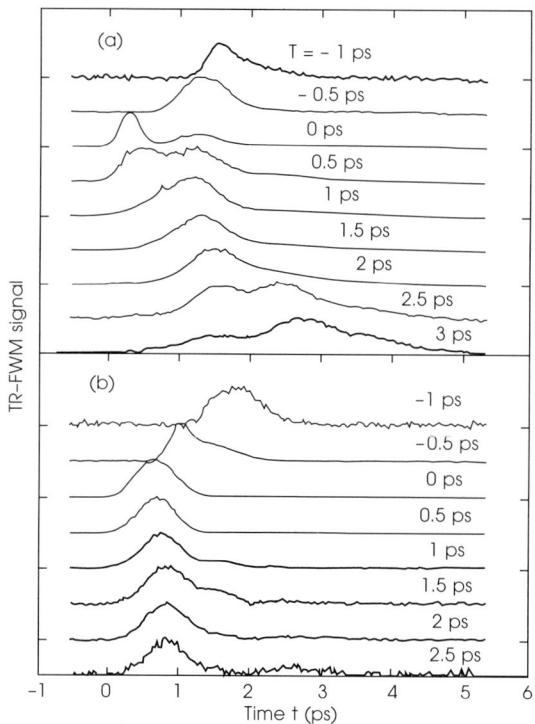

Fig. 5.2 Time-resolved FWM signal from a 17 nm GaAs quantum well at 10 K. (a) is measured with co-polarized excitation, (b) with cross-polarized excitation. Note that the emission often appears as two well-separated pulses – the first is due to Pauli blocking nonlinearities, while the second results from the Coulomb interaction. Adapted from reference [11].

of the internal polarization. Local field effects are included in the SBE at the two-particle level.

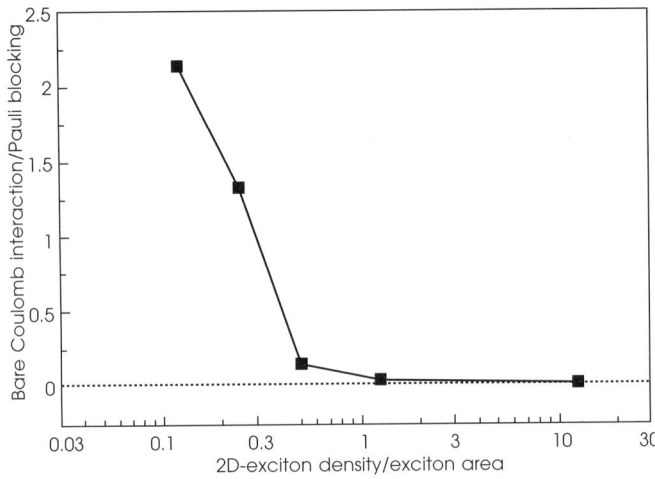

Fig. 5.3 Ratio of the contribution of the Coulomb interaction and Pauli blocking to the FWM emission, as a function of the density of carriers. Adapted from reference [13].

The role of screening in NLO response was demonstrated in 1993–1994 through studies of the polarization and density dependence of P/P and FWM signals (Lindberg, [14]). The polarization dependence of these signals stems from the selection rules indicated schematically in Fig. 5.4. Circular polarization allows selective excitation of

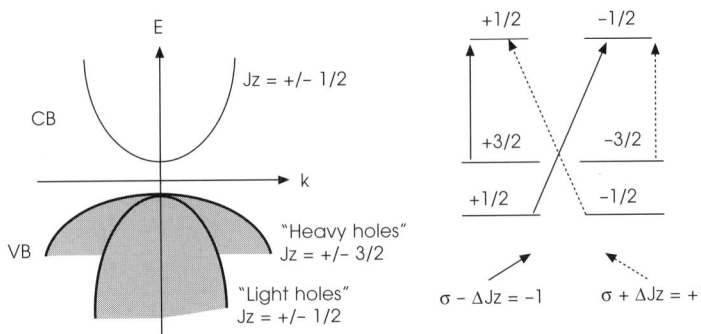

Fig. 5.4 Band structure and selection rules for zinc-blende semiconductors. Note that in a quantum well or mechanically strained bulk sample, the heavy hole and light hole bands split in energy. Thus, excitons of only one type may be excited if the binding energy and hh-lh splittings are larger than the laser bandwidth.

one spin sub-band, thus differentiating between those many-body processes which are spin dependent (such as Pauli blocking) and those which are spin independent, such as screening and other carrier–carrier scattering processes.

An important result of Coulomb screening is *density dependence* of the dephasing rates which arise in the screened HF approximation (Haug, [15]). P/P and FWM measurements demonstrate commensurate decreases in dephasing time and increases in exciton linewidth with increasing excitation density (Wang, [16]) (See Fig. 5.5). This "excitation induced dephasing" (EID) gives an additional driving nonlinearity in the equations of motion, leading to a delayed FWM signal from spatial modulation in the exciton dephasing rate, as well as to FWM signals at negative times (Wang, [17]).

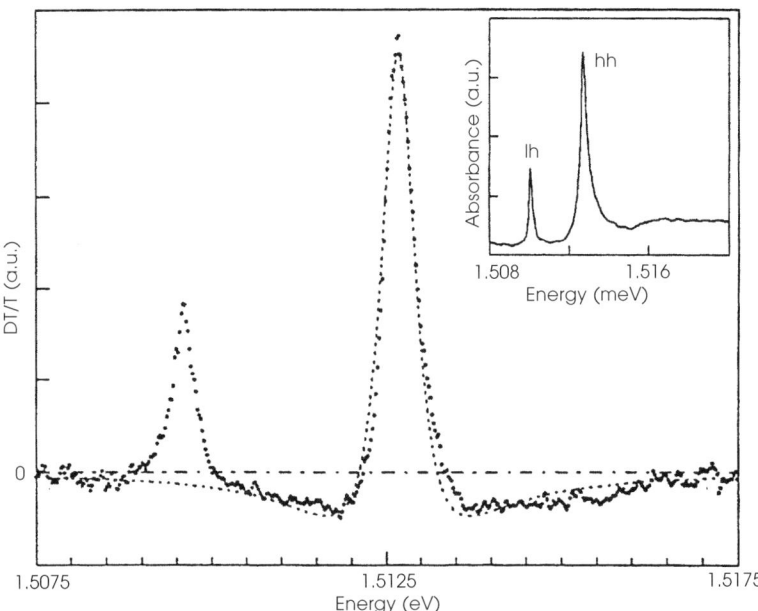

Fig. 5.5 Differential transmission signal from P/P spectroscopy in an 0.2 μm GaAs sample at low temperature. The DTS signal at the hh is very well modeled by the difference of two Lorentzians (dotted line) indicating that the pump excitation is causing a broadening of the exciton line, consistent with excitation induced dephasing. Adapted from reference [16].

Polarization resolution is critical in resolving the EID contribution to FWM, as this term cancels for perpendicular linear k_1 and k_2, while for parallel linear polarizations the EID signal is non-zero, and can even be dominant (Hu, [18]). EID-induced signals can thus be distinguished from local field contributions, which are polarization independent.

Screening effects for carriers high in the band have also been studied by FWM. Spectrally resolved experiments in bulk germanium revealed that co- and cross-linearly polarized excitation yield dramatically different FWM emission even for excitation in the continuum of states 7 Rydberg above the direct gap (Rappen, [19]). Careful theoretical analysis using nonequilibrium Green's functions to calculate the dephasing rates in the second-order Born approximation was performed. This analysis revealed that vertex and exchange contributions, which had been neglected in previous work, were very important in the balance of continuum and excitonic contributions to the emission.

5.3 Beyond the screened HF approximation – theoretical approaches to many-body correlations

Though very successful in many cases, the semiconductor Bloch equations represent an inherently mean field theory. The polarization of each exciton (or electron–hole pair) interacts with an effective electric field made up of the external applied field and the *average* polarization field created by other excitons. Thus, although the SBE provide a good description of the interaction of Coulomb correlated electron–hole pairs with photons, they do not include correlations between two or more electron–hole pairs. With the advent of higher quality samples and more precise experimental techniques, it became clear that such high-order correlations could cause important, or even dominant, contributions to the optical nonlinearity of semiconductors. The limitations of the SBEs are particularly apparent for excitation at the excitonic resonance, at low temperatures and low densities. Under these circumstances excitons do not interact with a sufficient fraction of their neighbors over the time scales measured by a short pulse for mean field theories to be appropriate, and screening of the Coulomb interaction is not sufficient to dampen correlations. Thus, a theory is required which does more than add relaxation or screening terms to the SBE – one must account for the Coulomb interaction consistently to arbitrary order.

Biexcitons and few-level theories

The most obvious example of dominant four-particle correlations is the bound biexciton – a singlet four-particle state analogous to H_2. Biexcitons have been well known in large gap semiconductors, such as the I-VII and II-VI series, for many years (Mysyrowicz, [20]; Hannamura, [21]). Such materials have relatively small dielectric constants, and thus exceptionally large exciton binding energies (Ry~200 meV in bulk CuCl, as opposed to only about 4.2 meV in bulk GaAs). The biexciton binding energies are similarly enhanced in large gap materials – in CuCl the biexciton binding energy is 28 meV – and thus the biexcitonic resonance is quite easily observed in two-photon absorption. Large gap materials excited near the excitonic resonance are often modeled as few-level systems, ignoring both continuum of unbound electron–hole pairs and the interactions among excitons. (The simplest such models treat three levels – ground state, excitonic state, and bound biexcitonic state.) This few-level approach has been quite successful in describing coherent NLO processes in CuCl on the ns time scale (Maruani, [22]; Chemla, [23]).

More recently it has become clear that biexcitonic states play an important role even in III-V semiconductors, where the biexcitonic binding energy is only ~1 meV – no larger than the typical exciton linewidth, and thus the biexciton cannot be spectrally resolved. Evidence of bound biexcitons in GaAs/AlGaAs quantum well structures was found nearly simultaneously in P/P (Bar-Ad, [24]) and FWM (Lovering, [25]) experiments. An example of the exciton/biexciton oscillations seen in P/P is shown in Fig. 5.6.

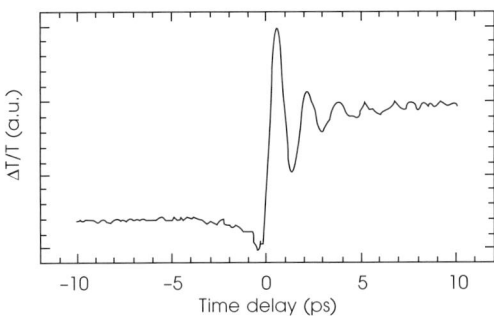

Fig. 5.6 Exciton–biexciton oscillation observed in a GaAs quantum well using the P/P technique, with cross-circular polarization of pump and probe. Adapted from reference [26].

Polarization-selective experiments are key to resolving the role of the bound biexciton, since it requires two electrons of opposite spin to form a singlet state, and thus can only be created when photons of both $\sigma+$ and $\sigma-$ polarization are present. (It is possible, in principle, to form a bound biexciton from an electron–hole population which has only one spin state, via spin flip processes. Spin flips are typically quite slow in semiconductors, however, and do not contribute to coherent signals on short time scales.) Although the NLO response of GaAs including biexcitonic states was initially modeled using the phenomenological few-level approach, this rapidly becomes extremely complicated due to the large number of states which are energetically close. (A description of the overall lineshapes and beat phases in polarization resolved FWM experiments required a model based on a ten-level system!) (Mayer, [27]; Denton, [28]). Ultimately, such a phenomenological model is clearly not adequate. As well as being very complex for systems with weakly bound excitons, it treats the bound biexciton as the only four-particle correlation, and ignores others that can be of equal importance. Full understanding of the nonlinear optical response requires a formalism which treats n-particle interactions in a consistent way.

The dynamics-controlled truncation scheme

Several such theoretical approaches have been developed. The first originated in the context of molecular systems (Spano, [29], [30]; Dubovsky, [31]; Leegwater, [32]; Mukamel, [33]), while others are based on diagrammatic techniques (Maille, [34]) or the development of correlation functions in the basis of n-exciton eigenstates (Ostereich, [35]). I will focus in this chapter on the formalism most commonly used at present, known as the dynamics-controlled truncation scheme (DCT). The DCT naturally extends the density-matrix approach of the SBEs, and provides an expansion in terms of the most easily controlled experimental variable – the driving optical field.

The semiconductor ground state is well understood, and Coulomb correlations among electrons and holes arise due to optical excitation of quasi-particles above the gap. It is therefore appealing to consider Coulomb correlations via an expansion in powers of the driving optical field. Axt and Stahl showed in 1994 that such an expansion allows a systematic decoupling of the infinite hierarchy of equations of motion generated by equation (5.5) (Axt, [36]). The essence of the theorem underlying the truncation is as follows. Let $\hat{A}_n$ be an n-point operator in normal order according to the electron–hole representation, and let n_e (n_h) denote the number of electron (hole) operators contained in $\hat{A}_n$, such that $n = n_e + n_h$. It then follows, for a system initially in the ground state and governed by the Hamiltonians in equations (5.1) and (5.2) that $\langle \hat{A}_n \rangle$ is of order E^m, with $m \geq \max \{n_e, n_h\}$. Thus, *the nonlinear response of a semiconductor material to a given order in the electric field is influenced by only a FINITE number of density matrices*. This theorem allows a systematic truncation of the infinite many-body hierarchy once the desired order in the electric field has been chosen. One arrives at a closed set of equations of motion which allow the calculation of the system response exactly (for a given order in the driving field). It is critical to note that the desired order in the electric field is used only to select those density matrices which contribute to the equations of motion. Once the equations of motion are determined, they are solved *to infinite order in the field*.

Consider, for example, the results of the DCT truncation at the $\chi^{(3)}$ level. These are critical as $\chi^{(3)}$ gives the leading order contribution to both P/P and FWM signals. The truncation theorem indicates that we need consider only those density matrices for which there are no more than three operators of any one type (electron or hole). One obtains the form of the relevant density matrices from equations (5.4) and (5.5) by constructing the equations of motion for the two-point operators corresponding to populations and polarizations. The equations of motion for the n-particle operators all take the form

$$i\hbar \frac{\partial \hat{A}}{\partial t} + \hbar \Omega_A \hat{A} = Q_A. \tag{5.9}$$

Here the Ω_A give the resonances of the operator (e.g. excitonic or biexcitonic energies) while the Q_A make up the sources – including the driving field, Pauli blocking, and Coulomb interactions. The equation of motion for the polarization, written $p_2^1 = \langle h_1 e_2 \rangle$ in the electron–hole representation, is found to be

$$i\hbar \frac{\partial p_2^1}{\partial t} + \hbar \Omega_p p_2^1 = \mu E - E[\langle n_{e,12} \rangle + \langle n_h^{12} \rangle] + \int d^3 r_j (V_{1j} - V_{2j})(S_{jj2}^1 - T_4^{jj3}) \tag{5.10}$$

Here we define the following two four-point operators;

$$S_{124}^3 = \langle e_1^+ e_2 h_3 e_4 \rangle$$
$$T_4^{123} = \langle h_1^+ h_2 h_3 e_4 \rangle. \tag{5.11}$$

The S operator describes an electron screened transition – that is, the correlated destruction of an electron–hole pair with the scattering of another electron. By the same logic, T describes a hole-screened transition. From the equations of motion for the electron and hole densities, n_e and n_h, we obtain three more four-point functions.

$$N_{14}^{23} = \langle e_1^+ h_2^+ h_3 e_4 \rangle$$
$$K_{1234} = \langle e_1^+ e_2^+ e_3 e_4 \rangle \tag{5.12}$$
$$L^{1234} = \langle h_1^+ h_2^+ h_3 h_4 \rangle$$

N represents correlated excitonic occupations, while K and L represent conduction and valence band density–density correlations. Note, however, that K and L are *not* included in the $\chi^{(3)}$ truncation, as both have n_e or $n_h > 3$. Finally, in the equations of motion for the four-point operators, N, S, and T, we obtain two more density matrices which contribute at the $\chi^{(3)}$ level. The first such operator is

$$B_{24}^{13} = \langle h_1 e_2 h_3 e_4 \rangle, \tag{5.13}$$

which represents the correlated destruction of two e–h pairs – or the coherent two-photon (biexciton) emission. The second is a six-point term, Z, which represents exciton-two-exciton transitions, and which is written

$$Z^{235}_{146} = \langle e_1^+ h_2^+ h_3 e_4 h_5 e_6 \rangle. \tag{5.14}$$

The coherent limit

There are five high order Coulomb operators (B, N, S, T, and Z) whose equations of motion must be solved simultaneously in order to obtain the exact $\chi^{(3)}$ result. However, Axt and Stahl (Axt, [37]) showed that under perfectly coherent circumstances, the entire dynamics at the $\chi^{(3)}$ level can be calculated in terms of biexcitonic (B) and excitonic (p) transition amplitudes, while all other operators can be factorized into products of p and B, and thus appear only as *dependent* variables. The exact factorized solutions for N and Z coincide to lowest order in the exciting fields with the results of the HF/RPA, that is,

$$N^{23}_{14} \approx p_1^{2*} p_4^3 + O(E^4)$$
$$Z^{234}_{146} \approx p_1^{1*} B^{34}_{56} + O(E^5). \tag{5.15}$$

The RPA decoupling is incorrect even to lowest order in the field for the matrices S, T, and B.

In the coherent limit one obtains particularly simple equations of motion – just two coupled equations for B and P. Though the restrictions of the perfectly coherent limit tend not to apply to real systems, some factorizations are more robust than others. In particular, the factorizations of S and T tend to hold true even for small deviations from the coherent limit, while the factorizations of N and Z break down. Nevertheless, a large fraction of the experiments in which four-particle correlations are evident agree well with the predictions of the $\chi^{(3)}$ coherent limit.

The DCT scheme can certainly be applied at the $\chi^{(5)}$ level and beyond (Victor, [38]). However, the twenty-two dynamical variables required in the general $\chi^{(5)}$ case render the problem quite numerically intractable. As is the case for $\chi^{(3)}$, factorization at the $\chi^{(5)}$ level in the coherent limit reduces the number of independent dynamical variables dramatically. In this limit only p, B, and W remain, where W describes the triexcitonic emission, $W^{135}_{246} \equiv \langle h_1 e_2 h_3 e_4 h_5 e_6 \rangle$.

Interpreting and solving the equations of the DCT

In studying the roles of n-particle correlations, it is often useful to separate explicitly the contributions of HF/RPA terms from those which are due to high-order Coulomb contributions. This is most easily done by constructing operators representing the differences between full n-particle terms and their HF factorized approximations. We will designate these "high-order correlation" terms with a bar. Thus, the correlated part of the two-photon emission is given by

$$\overline{B}_{24}^{13} = \langle h_1 e_2 h_3 e_4 \rangle - \langle h_1 e_2 \rangle \langle h_3 e_4 \rangle + \langle h_1 e_4 \rangle \langle h_3 e_2 \rangle. \tag{5.16}$$

It should be noted that $\overline{B}$ is a rapidly oscillating, phase-sensitive quantity, and contains contributions from both the bound biexcitonic states and from the unbound two-exciton continuum. $\overline{B}$ is the only four-particle term needed to describe the $\chi^{(3)}$ system in the absolute coherent limit (where N, Z, S, and T can be factorized). In this case the coupled equations of motion for p and $\overline{B}$ are as follows:

$$[-i\hbar(\partial_t + \gamma_p) + \hbar\Omega_{p_2^1}] p_2^1 = \mu E - \sum_{j,l} (\mu E p_l^{j*} p_2^j - \mu E p_l^{j*} p_l^1)$$

$$+ \sum_{j,l} V(12|jl) p_l^{j*} (p_l^j p_2^1 - p_2^j p_l^1 + \overline{B}_{l2}^{j1}) \tag{5.17}$$

and

$$[-i\hbar(\partial_t + \gamma_B) + \hbar\Omega_{B_{24}^{13}}] \overline{B}_{24}^{13} = V(12|34) p_2^1 p_3^4 - V(14|32) p_4^1 p_2^3 \tag{5.18}$$

with

$$\hbar\Omega_{p_2^1} \equiv \hbar\omega_g - \frac{\hbar^2}{2m_h}\Delta_1 - \frac{\hbar^2}{2m_e}\Delta_2 - V_{12} \text{ and } \hbar\Omega_{B_{24}^{13}} \equiv \hbar\Omega_{p_2^1} + \hbar\Omega_{p_4^3} - V(12|34). \tag{5.19}$$

Here the γ represent damping constants, while ω_g is the optical gap, V_{12} is the Coulomb potential and $V(12 | 34) = V_{14} + V_{23} - V_{13} - V_{24}$.

Equations (5.17) and (5.18) show quite clearly the role of the high-order Coulomb correlations in the $\chi^{(3)}$ coherent limit. The pair transitions, p, are driven by four types of sources: (1) a linear source, ($\propto \mu E$) that starts the dynamics when the semiconductor is initially excited, (2) sources that arise due to the Pauli exclusion principle ($\propto \mu E p^* p$), (3) Hartree–Fock Coulomb nonlinearities ($\propto Vp^*pp$) representing the mean field part of the dynamics, and (4) genuine many-body correlations ($\propto Vp^* \overline{B}$). Note that discrete (bound biexciton) as well as continuum two-exciton states appear in the equation for Ω_B, and their relative weight is determined by equation (5.19).

For systems not in the coherent limit, it is necessary to include in the equations of motion both the four-particle term N, which describes pair occupations, as well as the six-particle term Z, which describes exciton-to-exciton transitions. The contributions to N beyond the mean field limit are given by

$$\overline{N}_{14}^{23} = N_{14}^{23} - p_1^{2*} p_4^3 = \langle e_1^+ h_2^+ h_3 e_4 \rangle - \langle e_1 h_2 \rangle^* \langle h_3 e_4 \rangle. \tag{5.20}$$

$\overline{N}$ describes two physically different aspects of the dynamics – (1) coherences between different pair states related to intersubband and intraband transitions and (2) incoherent occupations of pair states. Writing the expression for $\overline{N}$ in the form

$$\overline{N} = \langle (p - \langle p \rangle)^+ (p - \langle p \rangle) \rangle \tag{5.21}$$

suggests the interpretation of $\overline{N}$ as the fluctuation in the exciton transition amplitude, p. The physical meaning of this dualism is exactly the same as in the well-known case of the optical field, where fluctuations of the field amplitude are by definition identified as the incoherent part of the intensity.

Finally, the correlated part of the six-point density matrix, $\overline{Z}$, is written

$$\overline{Z}_{146}^{235} = \langle p_1^{2+} p_4^3 p_6^5 \rangle - p_1^{2*}(\overline{B}_{46}^{35} + p_4^3 p_6^5 - p_6^3 p_4^5) - \overline{N}_{14}^{23} p_6^5$$
$$- \overline{N}_{16}^{25} p_4^3 + \overline{N}_{14}^{25} p_6^3 + \overline{N}_{16}^{23} p_4^5 \tag{5.22}$$

$\overline{Z}$ accounts for the correlated part of transitions from incoherent exciton densities to two-pair states. $\overline{Z}$ belongs to the incoherent part of the dynamics, because it can only have finite values in the presence of incoherent densities. Nevertheless, as a transition density, $\overline{Z}$ has some properties of coherent variables, e.g., it oscillates at optical frequencies and is therefore phase sensitive. In atomic physics it is well known that contributions corresponding to $\overline{Z}$ yield excited state absorption. Dephasing induced resonances are also related to transitions of the $\overline{Z}$ type (Dick, [39]; Abram, [40]).

The solutions to the coupled equations of motion in the DCT are obtained numerically by first formally inverting the equations of motion for $\overline{B}, \overline{N}$ and $\overline{Z}$, leaving only a single equation of motion for p. This results in a memory kernel representation based on Green's functions (Axt, [41]). For example, in the coherent limit the equation for $\overline{B}$ may be solved using the Green's function G_B of Ω_B as

$$\overline{B}_{24}^{13}(t) = \int_{-\infty}^{t} dt' \sum_{\substack{57 \\ 68}} G_B(1324, 5768; t - t')$$
$$\times [V(56|78) p_6^5(t') p_8^7(t') - V(58|76) p_8^5(t') p_6^7(t')]. \tag{5.23}$$

Equivalent memory kernels may be generated for the other four-particle terms. In any case, the price to be paid for a formulation where only the two-point function, p, enters explicitly is that the equations become non-local in time. As a result of this non-locality the four-particle interactions are often referred to as "memory effects" in the literature.

The effective polarization model

Although the DCT is complete and internally consistent, full numerical solutions based on this model can be both time consuming and complex. For the χ^3 case in the coherent limit, one can derive an "average polarization" or "effective polarization" model (EPM), which takes the form of a nonlinear Schrodinger equation. The EPM provides significant insight into how processes not included in the SBE influence FWM and P/P measurements, in particular demonstrating the polarization dependencies which are so critical in NLO experiments. The first step for both full numerical solutions of the DCT and for the EPM is to expand both the excitonic and biexcitonic polarizations in the basis of exciton eigenfunctions, ϕ_α, that is:

$$P = \sum_k \mu_k p_k = \sum_\alpha p_\alpha \phi_\alpha(k), \tag{5.24}$$

Substituting this expression into the equations of motion (5.17) and (5.18) for p and $\overline{B}$ leads to a nonlinear Schrodinger equation which contains sums over states in all of the Coulomb source terms. However, if we assume that the system is excited by an ultrashort pulse with significant bandwidth, we can replace the true polarizations p_k and $\overline{B}_k$ by "averaged" (over states) quantities $\mathscr{P}$ and $\mathscr{B}$, so that all sums over states are converted into averages as well. In the coherent limit the populations are related to polarizations via $n_i(k) \sim p_k^2$, and the EPM equations of motion read:

$$\left(i\frac{\partial}{\partial t} - i\gamma - \Omega_x\right)\mathscr{P}^\pm(t) = -\mu E\left[1 - \frac{|\mathscr{P}^\pm(t)|}{\mathscr{P}_s^2}\right] + V\mathscr{P}^\pm(t)|\mathscr{P}^\pm(t)|^2$$

$$+ V_{XX\text{exch}}\mathscr{P}^\pm(t)|\mathscr{P}^\pm(t)|^2$$

$$- V_{XX\text{screen}}\mathscr{P}^\pm(t)[|\mathscr{P}^\pm(t)|^2 + |\mathscr{P}^\mp(t)|^2]$$

$$+ V_{XX2}\mathscr{B}(t)\mathscr{P}^\pm(t)^* \tag{5.25}$$

and

$$\left(i\frac{\partial}{\partial t} - i\Gamma - \Omega_{x2}\right)\mathscr{B}(t) = \mathscr{P}^+(t)\mathscr{P}^-(t). \tag{5.26}$$

Here + and − represent the separate spin 1/2 manifolds excited by σ+ and σ− polarized light. The results of the EPM can be interpreted in a very straightforward fashion. The first line reproduces the results of the SBE approximation (this is what would be obtained from equation (5.6) if an averaging over states were performed). The next two lines contain the contributions of the exciton–exciton screening, exciton–exciton exchange, exciton–biexciton interaction. It is important to note that the EPM equations contain the polarization selection rules. Neither the HF terms nor the excitonic exchange terms couple the + and – spin manifolds. Thus, in the HF approximation one would predict *no* coherent contributions to the pump–probe signal for counter-circularly polarized pump and probe. The only contributions to counter-polarized P/P signals originate in the excitonic screening and the exciton–biexciton interactions.

The V for each term in equation (5.25) can be calculated explicitly if necessary (Schafer, [42]). More often the results of the EPM are used for direct comparisons with experiment, in which case the V are treated as free parameters which may be fit to the data to discern the relative contributions of various four-particle terms to the signal. It should be noted that the EPM equations pertain only to the ideal coherent limit, and thus do not contain dephasing rates. The excitonic dephasing rate, γ, and the biexcitonic dephasing rate, Γ, must be obtained from other models or extracted from experimental data.

Phonons

For samples at higher temperatures, it is clear that any description of the dynamics must include exciton–phonon couplings as well as Coulomb interactions. The introduction of phonons brings a number of important changes. For example, in a perfectly coherent system with only Coulomb interactions, there is no mechanism by which occupation densities may have longer lifetimes than transition densities. Phonons provide the relevant microscopic basis for such an evolution of transition densities into occupation densities, which then become independent dynamical quantities with independent lifetimes.

It is possible to take a DCT approach to the phonon assisted dynamics, via the addition of a phonon Hamiltonian $\hat{H}_{phonon} = \sum_q \hbar\omega_q (\hat{a}_q^+ \hat{a}_q + 1/2)$ to equations (5.1) and (5.2), followed by an expansion in terms of the number of phonon operators (Axt, [43]). Density matrices with different numbers of phonon operators are dynamically coupled, however, so that after the electronic branch of the hierarchy has been truncated one is still left with an infinite hierarchy of phonon-assisted densities. Thus far, this dilemma has only been solved in particular special cases, for example, by factorizing all doubly assisted quantities in the following manner:

$$\langle \hat{a}_q^+ \hat{a}_{q'} \hat{X} \rangle \rightarrow \delta_{q,q'} n_q \langle \hat{X} \rangle \tag{5.27}$$

Here X is any electronic operator, and n_q gives the thermal occupation of the phonon mode. Note that using the thermal occupation for the phonon distribution in this manner neglects hot phonon effects, treating the phonon system as a simple bath. A comparison of the relative contributions of phonon and Coulomb couplings for pump–probe signals in the spectral region of the biexciton has been completed using this scheme in a one-dimensional model. This work demonstrated that phonon coupling is considerably less important than Coulomb coupling for all conditions studied (Axt, [43]). Thus, it is generally sufficient to calculate many-body effects at the excitonic resonance without the phonon contribution, and this is certainly appropriate for low-temperature experiments, where $T < 50$ K.

5.4 Experimental studies of high-order Coulomb correlations

With the development of more precise experimental techniques and more sophisticated theory, it became increasingly clear that the role of high-order Coulomb correlations in semiconductors was both significant and experimentally accessible. The surprising experimental appearance of the bound biexciton in GaAs was nearly coincident with the development of the DCT theory in 1992–1994, beginning a surge of effort to understand the role of many-particle correlations in semiconductor NLO response. It is significant to note that essentially every paper demonstrating progress in this field is the result of close collaboration between theoretical and experimental teams.

The fully coherent regime

Initial work on high-order Coulomb correlations focussed on the fully coherent regime, in which the only independent dynamical variables are the excitonic and biexcitonic polarizations (p and $\overline{B}$). Early work by Schafer *et al.* evaluated the polarization dependence of FWM emission near the excitonic resonance in the fully coherent regime (Schafer, [42]). Comparison of experimental results in GaAs with this theory demonstrated that the diffraction of internal, induced fields completely dominates the FWM signal. In particular, the spectral profile of FWM emission was nearly perfectly predicted by the EPM model, *neglecting* the PSF nonlinearity. (The spectral signature of Coulomb interactions in this case is a FWM profile at zero delay with the asymptotic form $(\omega - \omega_0)^4$.) Once again, noninteracting few-level models, which assume the dominance of the external fields, were demonstrated to be completely inappropriate for describing the relevant physics of FWM. Schafer *et al.* spectrally resolved the weak biexcitonic resonance using cross-linear polarized excitation in FWM. For cross-linear excitation the excitonic screening contribution to FWM vanishes, and biexcitonic contributions become dominant. By careful study of the dependence of the FWM signal on detuning from the excitonic resonance, Schafer *et al.* concluded that the biexciton has a considerably shorter dephasing time than does the exciton, although the precise details of the dephasing time were not determined.

Spectrally resolved FWM studies performed in ZnSe quantum wells also revealed the importance of the four-particle contributions (Haase, [44]). These experiments, however, could be compared very precisely with the results of the *microscopic* DCT (in which the correlations are computed directly from the known properties of the sample, rather than introduced as free parameters as in the EPM). ZnSe has a number of very advantageous properties for such studies. First, the hh-exciton is 25.6 meV below the band edge, thus hh-excitons can be excited by ultrashort pulses without the introduction of any free carriers or lh-excitons, which significantly complicate the dynamics in lower gap semiconductors (such as GaAs and InGaAs). Furthermore, the biexciton resonance in ZnSe is spectrally well resolved from the hh-exciton (biexciton binding energies range from 4.5 meV in quasi-bulk structures to 7.3 meV in a 2.8 nm well) allowing detailed comparisons of excitonic versus biexcitonic emission. Finally, ZnSe has a very high optical nonlinearity, which allows high quality measurements to be performed in heterostructures containing only a single quantum well – thus eliminating both the distortions which can result from propagation through optically thick media and the inhomogeneous broadening present in multi-QW samples. A comparison of experimental FWM data at low excitation densities with the results of the microscopic DCT theory is shown in Fig. 5.7. The dramatic dependence of the FWM emission on excitation

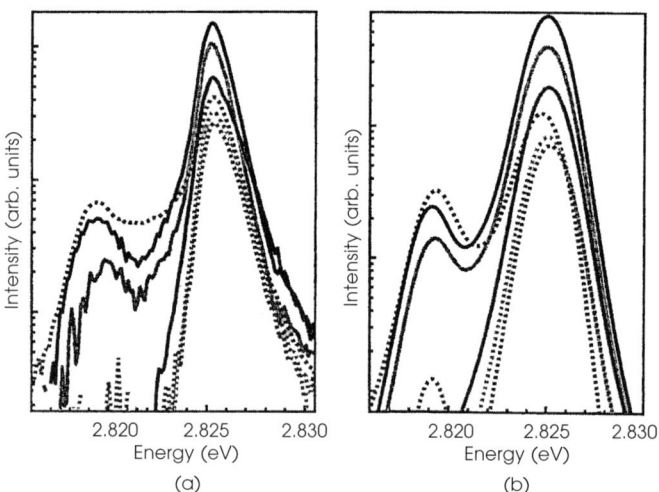

Fig. 5.7 Experimental FWM spectra, (a) measured in a ZnSe 4.8 nm single quantum well at a time delay of $\Delta t = 0$, for a pump power of 100 μW. Data are compared with the results of the microscopic DCT theory, (b) at the coherent $\chi^{(3)}$ level. From top to bottom, the curves are for the following excitation polarizations $(k_1 k_2)$: xx, $\sigma + x$, $\sigma + \sigma-$, xy $y\sigma-$, $x\sigma-$. Note that the data are presented on a logarithmic scale, and show remarkable agreement with the theory over more than three orders of magnitude! Adapted from [44].

polarization is evident, as is the superb agreement of theory with the experimental data for every polarization configuration.

For higher excitation densities Haase *et al.* observed clear violation of the $\chi^{(3)}$ selection rules and the onset of $\chi^{(5)}$–level processes (Haase, [8,44]). For example, at the $\chi^{(3)}$ level the polarization configuration $k_1 = x$, $k_2 = \sigma+$ (known henceforth as $x\sigma+$) cannot produce a bound biexcitonic signal, as this would require two photons of opposite circular polarization from k_2. On the other hand, a fifth-order contribution in which three photons come from beam 1 (final wave vector: $2k_2 - k_1 + k_1 - k_1$) does allow a bound biexciton. The onset of the $\chi^{(5)}$ processes can be clearly evaluated by examining the magnitude of signals of this type. More recent studies of the intensity dependence of signals obtained in FWM from ZnSe have demonstrated that substantial contributions above third order are generated even at the lowest excitation levels available in current experiment (Haase, [8]).

It is important to note that the DCT theory with truncation at $\chi^{(3)}$ reproduces very precisely the experimental FWM spectra even in the regime where $\chi^{(5)}$ processes become important. This is possible because the DCT uses $\chi^{(3)}$ to select the relevant many-particle correlations, but then solves the coupled equations of motion to *infinite* order in the driving field. These results demonstrate that the correlations present in the $\chi^{(3)}$ truncation in the coherent limit (p and $\overline{B}$) are sufficient to describe the NLO response of semiconductors under a significant range of excitation conditions.

In GaAs the excitons are so much less bound than in ZnSe that precise comparisons with microscopic theory prove challenging. However, exciton-exciton correlations can be enhanced by the application of a large magnetic field (Kner, [45]). With the application of such a field, the lowest energy excitons remain Lorentzian, while additional resonances appear at the origins of the Landau levels. The magnetic field does not change the nature of the Lorentzian excitons, but it strongly affects their internal structure, contracting the wave functions by a factor $\propto |B|^{1/2}$ perpendicular to the field, and by a factor $\ln |B|$ parallel to it. This contraction of the excitons yields a dramatic enhancement of the exciton–exciton correlation, demonstrated by the rise of the negative time FWM signal with increasing B shown in Fig. 5.8. It should be noted that these experiments are performed with *co-circularly polarized* excitation, thus the Coulomb-correlation induced signal is due to exciton–exciton (XX) interactions in the *two-exciton continuum*, rather than to *bound biexcitonic states*. The data are compared to a version of the EPM in which the polarization is expanded on the magnetoexciton eigenfunctions. The XX correlations beyond the SBE dominate the negative time signal, as shown in Fig. 5.9. The inclusion of screening in the model causes the negative time signal to vanish smoothly with increasing density, in complete agreement with experiments in which the density of free carriers is increased.

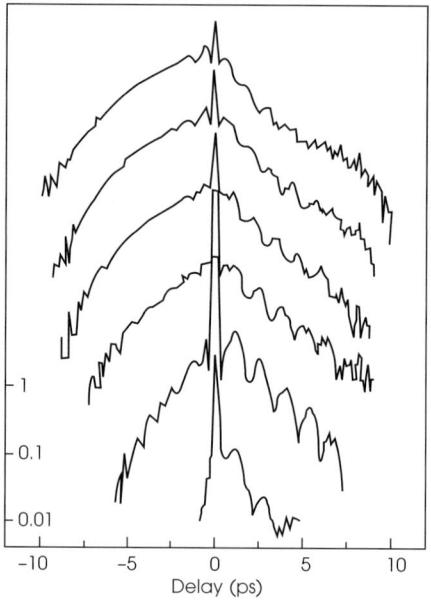

Fig. 5.8 FWM in a 0.25 μm GaAs layer at low temperature, as a function of applied magnetic field. From bottom to top $B = 0, 2, 4, 6, 8$, and 10 T. Note the rise in the negative time signal with increasing B field, demonstrating the enhancement of four-particle Coulomb correlations. Excitation is co-circularly polarized, and the carrier density is approximately 10^{15} cm^{-3}. Adapted from reference [45].

Four-particle correlations enhanced by B field have also been used to study the sensitivity of exciton–exciton correlations to temperature. Experiments show the 4PC signal to be strikingly temperature dependent, changing by more than a factor of three when the temperature is raised from 2 to 44K, as shown in Fig. 5.10 (Kner, [46]). This behavior is reproduced by the EPM model when a phenomenological parameter, γ_B, is introduced to describe the dephasing of the exciton–exciton correlation function. Increasing temperature in the experiment gives changes in emission nearly identical to those produced by increasing γ_B in the EPM. It is surprising that such a small change in T should produce such dramatic changes in 4PC, as the occupation number of LO phonons varies only very slightly over this temperature range, and theoretical work shows that phonon-induced transitions are relatively unimportant at these temperatures (Axt,[43]). A similar extreme sensitivity of the dephasing of four-particle correlations to the density of free-eh pairs was obtained by Kner et al. when the spectral overlap of the exciting pulses with the e–h continuum was increased.

Detailed studies of exciton and bound biexciton dephasing have been performed

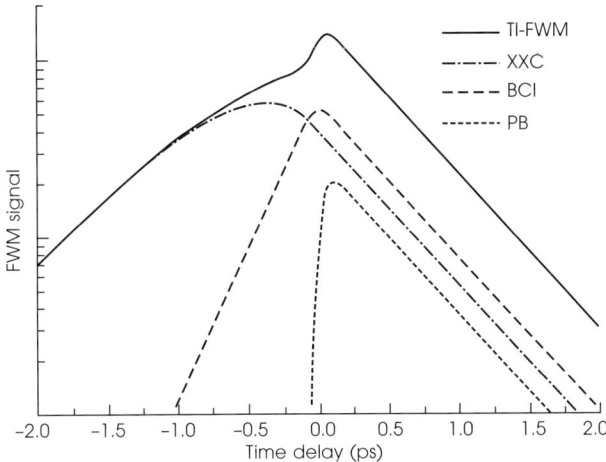

Fig. 5.9 EPM calculation of FWM emission for a single excitonic resonance, assuming co-circularly polarized excitation by delta function pulses in bulk GaAs. Note that the negative time signal is completely dominated by four-particle exciton–exciton correlations (XXC), has a small contribution from the Hartree–Fock Coulomb term (BCI) and has no contribution from Pauli blocking (BC) sources. Adapted from reference [46].

using FWM in a single GaAs quantum well with very narrow exciton linewidth (hh-exciton linewidth = 0.075 meV at 20K) (Langbein, [47]). This work indicates that exciton and biexciton dephasing are very similar over the temperature range 2–80K, and also that at higher temperatures, exciton and biexciton scattering events are highly correlated. The contrast between these results and those of Kner *et al.* may be due to sample quality effects, which have been shown to be critical in the magnitude of biexcitonic contributions to FWM (Schafer, [42]). Alternatively, the disagreement may originate in differences in the structure of four-particle correlations in the 3D GaAs system at high magnetic field vs. those in the 2D GaAs quantum well studied by Langbein *et al.*

The experiments of Kner *et al.* demonstrate the importance of exciton–exciton correlations for FWM polarizations in which the bound biexciton is forbidden. For polarizations in which the bound biexciton is created, it is often thought that a few-level model including both exciton and biexciton levels will sufficiently describe the NLO response. This has been shown to be far from correct in recent work by Axt *el al.* [48], where the coherent $\chi^{(3)}$ DCT was used to predict the spectra of FWM signals. For every choice of excitation condition, the signals at negative delay were accurately reproduced by the theory *only* if the influence of the correlations in the two-exciton continuum were included. In particular, the correlated two-exciton continuum contributions strongly compensate the mean field Coulomb contributions, as had been predicted in several

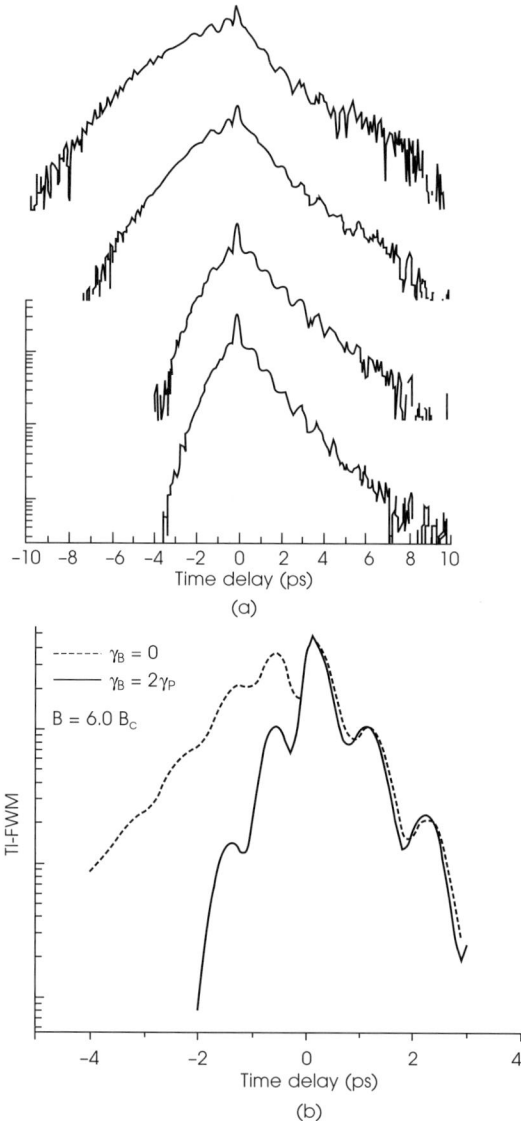

Fig. 5.10 (a) FWM data in bulk GaAs at high magnetic field (B = 10T) as a function of temperature. From top to bottom temperatures are 2K, 16K, 29K, and 44K. (b) Calculation of the FWM signal in the EPM: the solid line is for four-particle correlation dephasing rate, γ_b twice the excitonic dephasing, γ_p, while the dashed line is for $\gamma_b = 0$. Note that increasing the dephasing rate of the four-particle correlations destroys the negative time signal in the same manner as increasing temperature in the experimental data. Adapted from reference [46].

earlier theoretical studies (Ostereich, [49]; Axt, [41]). Models neglecting the influence of the correlated two-pair contribution consistently overestimate the ratio of exciton to biexciton emission by about one order of magnitude, while also missing significant spectral features originating in competition between the mean-field Coulomb nonlinearity and the two-pair transitions.

Although FWM has been the primary tool for the study of exciton–exciton correlations in the coherent regime, these correlations also play a significant role in P/P spectroscopy. When the intense pump pulse precedes the probe (positive time delay), the P/P signal is dominated by incoherent population effects, for which the coherent $\chi^{(3)}$ theory is clearly inappropriate. However, when the probe precedes the pump a P/P signal is obtained which derives from the nonlinear interaction of the pump field with the coherent macroscopic polarization created by the probe. In this regime coherent signals due to four-particle correlations are to be expected, and these can be modeled by DCT or EPM approaches. In fact, the coherent P/P signal predicted at the SBE level for counter-circularly polarized excitation is *zero*, thus measurements of the coherent P/P signal allow background free observations of the high-order Coulomb terms.

One prominent aspect of negative time P/P signals is the coherent oscillatory structures observed in the spectral and temporal domains. These result from four-photon processes, in which pump photons are scattered into the probe direction, k_1 ($= k_1 + k_2 - k_2$; with k_1 and k_2 being the respective wavevectors of the probe and pump beam). The superposition of the first-order transmitted probe and the delayed coaxial third-order scattered pump signals leads to oscillations in the spectral domain. In a non-interacting model the spectral period of these oscillations is predicted to be $\Delta E = h/\Delta t$. However, the presence of Coulomb correlations significantly alters the coherent oscillations, as observed in resonant excitation on ZnSe (Neukirch, [50]). Figure 5.11 shows a comparison of the P/P signal with the EPM theory for counter-circularly polarized excitation. At negative delay the oscillatory structure below the excitonic resonance (E_x) shows two deviations from the SBE predictions: it has significantly higher contrast than the oscillations above E_x, and it has a spectral period which is always smaller than $h/\Delta t$. Both of these features are completely explained by the presence of 4PC interactions. For counter-circular polarization the EPM model gives only two driving terms, as shown in equation (5.28), which describes the evolution of the third-order polarization in the probe direction.

$$\frac{\partial}{\partial t} P^3_{+t} - (i(\Omega_x - \omega)_o - \gamma_x) P^3_{+t} = iV_s P^1_{+t} |P^1_{+p}|^2 + iV_B B P^{1*}_{-p} \qquad (5.28)$$

with

$$\frac{\partial}{\partial t} B - (i(\Omega_{xx} - 2\omega_o) - \Gamma_{xx}) B = i P^1_{-p} P^1_{+t} \qquad (5.29)$$

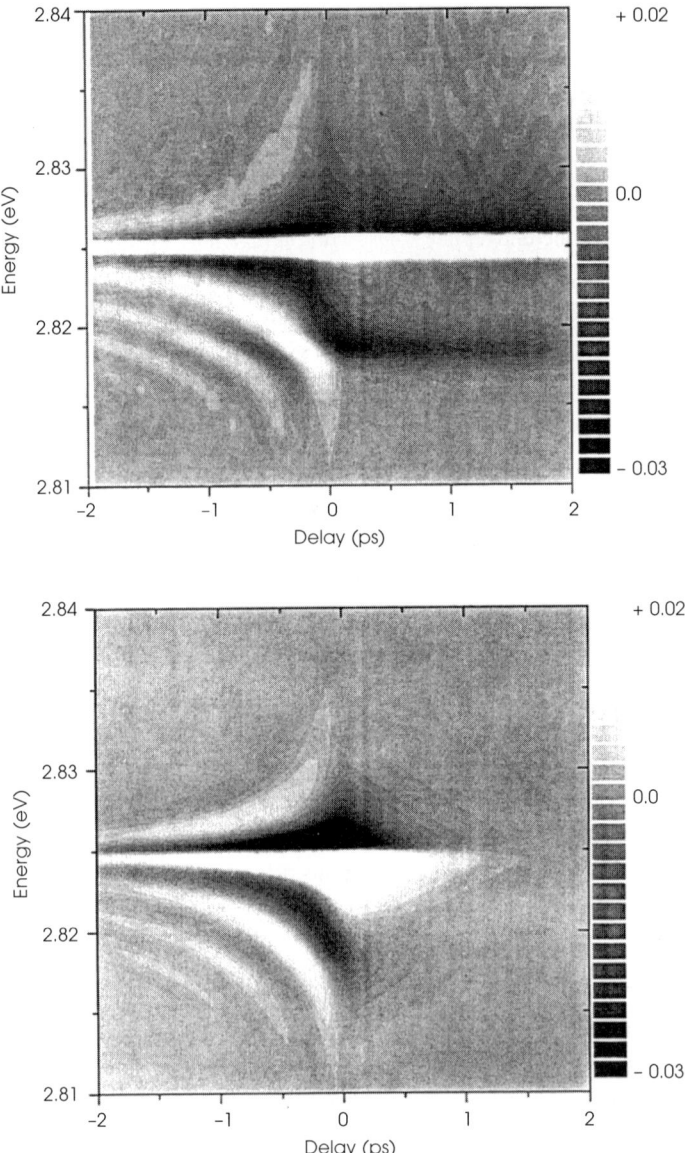

Fig. 5.11 Differential transmission data (upper panel) and EPM theory (lower panel) for P/P spectroscopy in a ZnSe 5 nm single quantum well. The excitonic resonance is at 2.825 eV, the biexciton at 2.818 eV. Note the excellent agreement between theory and experiment in the coherent ($\Delta t < 0$) regime. Adapted from reference [50].

Here the indices + and − label the polarization states, and t and p indicate test (probe) and pump, respectively. ω_o is the laser frequency, while Ω_x and Ω_{xx} are the exciton and biexciton resonances. The first driving term is proportional to the excitonic screening, V_s, while the second originates in the exciton–biexciton interaction, V_B, and contributes only for counter-circular polarization. The calculation shown in Fig. 5.11 contains *only* the exciton–biexciton interaction contribution, and clearly demonstrates that this *particular* 4PC correlation dominates the P/P response at negative delay. Again, the power of NLO spectroscopy to isolate the particular contributions of high-order Coulomb correlations is demonstrated by this result.

A similar study of polarization-resolved P/P response in a ZnSe microcavity has shown the dominance of the exciton–biexciton interaction in that instance as well (Neukirch, [51]). In a microcavity, the EPM model must be complemented with a resonance equation for the intracavity electric field. This approach takes into account the coupling of cavity electric field modes with excitons to yield cavity polaritons, which have mixed light/exciton character. The dominance of the biexcitonic contribution to the P/P spectra, as well as the appearance of a new resonance in counter-circularly polarized excitation in the microcavity, provide unambiguous evidence of polariton–biexciton transitions. It should be noted that the observation of a spectrally resolved polariton–biexciton transition requires a material with biexciton binding energy larger than all system inherent damping constants, and thus would be impossible to achieve in GaAs. Nevertheless, studies of P/P signals in GaAs microcavities *have* shown indirect evidence of exciton–biexciton interactions, such as the nonlinear *increase* of normal mode splitting observed with counter-circularly polarized excitation (Fan, [52]).

The coherent limit also pertains in P/P for the case of excitation in the gap below E_x. In this case the excitation does not produce an incoherent population of electron–hole pairs, but rather generates only virtual transitions and coherent polarizations, whose phase is determined by that of the laser. The presence of these virtual excitations causes a shift of E_x known as the optical Stark effect (OSE). Careful study of the polarization dependence of the OSE in high quality InGaAs quantum wells demonstrated a dramatic effect of high-order Coulomb correlations, as shown in Fig. 5.12. While for counter-circularly polarized excitation the OSE is a *redshift*, for all other polarization configurations the exciton experiences a *blueshift* (Sieh, [53]). Comparison of these data with a full two-dimensional tight-binding calculation based on the coherent $\chi^{(3)}$ DCT shows that the Hartree–Fock terms (Pauli blocking and first-order Coulomb interaction) give a blue shift for any polarization configuration, while the high-order Coulomb correlations always give a redshift. For counter circular polarization the HF terms are zero, and thus the correlation-induced redshift becomes evident. DCT calculations show that even in the absence of a bound biexciton this correlation induced redshift persists. Importantly, the redshift disappears in the calculations when the second Born and Markov approximations are included. Thus, the redshift is an inherently non-Markovian signature, and has at its origin correlation-induced Coulomb memory effects.

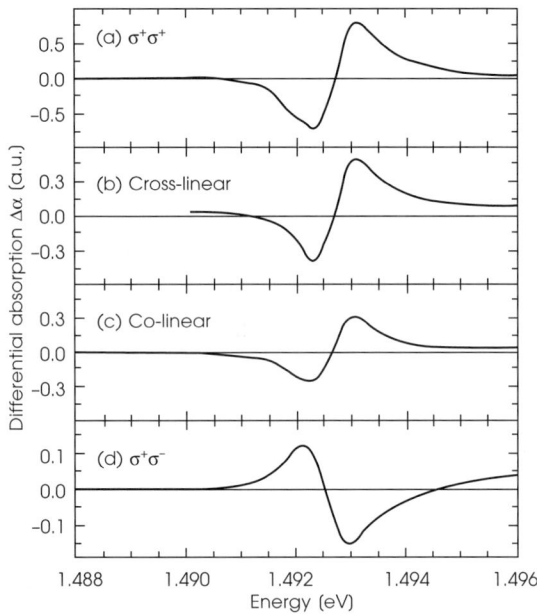

Fig. 5.12 Differential absorption data from a heterostructure of 8.5 nm $In_{0.4}Ga_{0.96}As$ quantum wells, as a function of excitation polarization. Data are shown for $\Delta t = 0$, for excitation 4.5 meV below the $1s$ hh-exciton resonance. Adapted from reference [53].

Optical experiments have been the most frequent tools for studies of high-order Coulomb correlations, however other coherent electromagnetic techniques can also probe these interactions. The emission of THz radiation from optically excited semiconductors is one such testing ground, which has demonstrated the importance of Coulomb correlations beyond the SBE even for responses of only *second*-order in the driving field. Theoretical and experimental studies were performed of the intraband current in a semiconductor superlattice with narrow minibands under static electric field bias (Axt, [54]; Haring Bolivar, [55]). This experiment is expected to be particularly sensitive to Coulomb interactions, as the shape of the Wannier–Stark ladder is strongly influenced by excitonic interactions, giving a set of anticrossings of the excitonic state with the Stark ladder states as the static field is increased. The SBE approach predicts that the THz emission well after excitation with a short pulse is essentially independent of Coulomb interactions – giving a THz signal which oscillates at the frequency corresponding to the single-particle energy separation. The DCT, by contrast, predicts excitonic effects that dominate for long times, yielding THz emission frequencies which correspond to the excitonic interband transitions. Full experimental tests of the emitted THz frequencies as a function of applied field demonstrate that the emission is indeed governed by excitonic contributions, in agreement with the DCT theory.

Contributions from incoherent densities

Although it is well known that scattering among carriers creating incoherent populations takes place whenever a semiconductor is excited above a gap, departures from the coherent $\chi^{(3)}$ limit did not become apparent in studies of high-order Coulomb correlations for several years. There are two issues which must be examined: the role of incoherent HOC terms such as $\overline{N}$ and $\overline{Z}$ in measurements dominated by coherent effects (wave mixing, pump–probe oscillations, optical Stark effect) and the contribution of HOCs to signals most sensitive to incoherent populations (e.g. resonantly excited P/P at positive delay). In the first case, it is reasonable to ask whether departures from the coherent limit of the DCT can be detected at all – do incoherent populations contribute to coherent signals? The answer to this question is, certainly, yes. In the 4PC theory the polarization, p, is driven by terms of the form $\overline{N}p$ or $\overline{N}E$, which oscillate at optical frequencies and are phase sensitive due to the p or E factors. The first clear-cut experiments where such incoherent contributions to the coherent signal were observed was a study of the transient polarization state of FWM signals in a ZnSe single quantum well (Haase, [56,57]). In this experiment, the FWM emission is projected on different polarization states (x, y, σ+, σ–, and linear polarization at 45 degrees with respect to x) by means of a sequence of two Pockels cells and a polarizer. The projected component is then either time-resolved, via upconversion with the residual 800 nm beam of the exciting laser, or spectrally resolved, using a spectrometer and CCD. Such measurements fully determine the FWM emission to within an overall phase factor. Figure 5.13 shows the comparison of the measured polarization component Re(E_y) with calculations based on the full DCT theory. The polarization of the incident beams is $k_1 = x$, $k_2 = $ σ+. It is immediately clear that the coherent limit theory, including only p and $\overline{B}$, fails to describe the experimental data, while the inclusion of the four-point correlation $\overline{N}$, which describes incoherent densities, yields excellent agreement with the experiment.

The contribution of $\overline{N}$ to the emission is density dependent. As expected, the coherent limit agrees best with the data at low excitation densities, giving nearly perfect agreement for areal carrier densities of order 10^8/cm^2. With only one order of magnitude

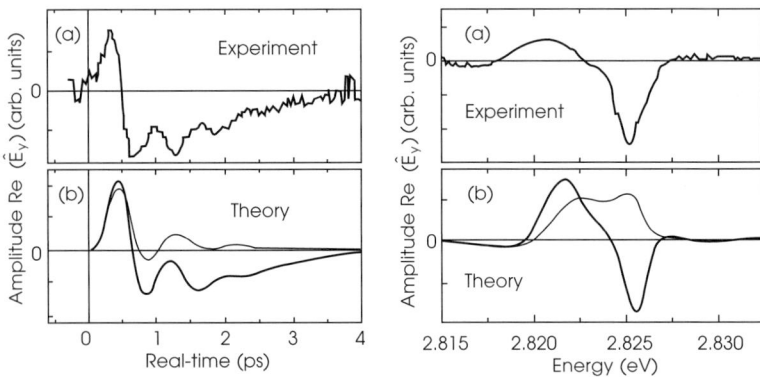

Fig. 5.13 Temporally (left panel) and spectrally (right panel) resolved FWM emission from a ZnSe 4.8 nm single quantum well at 7K, with $x\sigma+$ excitation polarization. The real part of the component of the emission polarized along the y-direction is shown here. Data are compared with theory based on the microscopic DCT. The thin lines show the results of the theory in the coherent limit, while thick lines show the results including the incoherent contribution, $\overline{N}$. Adapted from reference [56].

increase in density the coherent limit starts to break down, and the contributions of $\overline{N}$ emerge. These results demonstrate that the ellipticity of the emitted FWM signal is a very sensitive probe of high-order Coulomb correlations. This measurement technique yields distinct signatures for each of the three 4PC which contribute at $\chi^{(3)}$: the biexciton amplitude, the exciton–exciton scattering continuum, and the incoherent densities/amplitude fluctuations. These terms contribute in a phase-sensitive manner, and may interfere, as in the interference of $\overline{N}$ with the correlated part of the exciton – exciton scattering continuum observed by Bartels et al. [57].

For P/P measurements at positive delay, it is evident that the signal must include the effects of incoherent populations. Bartels et al. [58]) performed a study of differential transmission signals in semiconductors based on the DCT. Their data was modeled using p, $\overline{B}$ and $\overline{N}$ as source terms, and included both heavy- and light-hole excitation. They found that the phase of the positive time oscillations induced by heavy-hole–light-hole beats was highly sensitive to high-order correlations. More surprisingly, Bartels et al. demonstrated a dramatic influence of Coulomb interaction sources on the long-lived positive time P/P signal. This signal is usually interpreted in the HF model as due to Pauli blocking – simply the occupation of states by an incoherent population. Bartels et al. showed that the signal is in fact completely *dominated* by the Coulomb sources, with Pauli blocking giving significant contributions only exactly at the HH exciton resonance. More recent work on P/P signals in GaAs has demonstrated the importance of four-particle Coulomb correlations in coupling the HH and LH excitons – evidenced by bleaching at both exciton resonances after resonant pumping of either one (Meier, [59]). This work also shows evidence of a mixed LH–HH biexciton.

Contributions beyond the four-particle level

As discussed in section 5.3, the full $\chi^{(3)}$ DCT theory includes one six-particle correlation, $\overline{Z}$, which describes transitions from incoherent exciton densities to correlated two-pair states. FWM and P/P measurements have thus far been very well described by the $\chi^{(3)}$ theory including only p, $\overline{B}$, and $\overline{N}$. However, it is expected that signals requiring more photons in their lowest order would be more sensitive to higher-order correlations.

Recently, six-wave mixing experiments performed in the ZnSe quantum well system with spectral and polarization resolution have shown the first evidence of six-particle contributions. Six-wave mixing is a six-photon process of at least fifth order in the electric field, and is observed in the momentum conserving $3k_2 - 2k_1$ direction. (See Fig. 5.1.) (Bolton, [60]) In contrast to FWM experiments performed on the same sample at the same density, the measured SWM emission departs significantly from the results of the $\chi^{(3)}$ DCT including only p, $\overline{B}$, and $\overline{N}$. Figure 5.14 shows that the four-particle theory underestimates the strength of the biexcitonic contribution by almost two orders of magnitude, and introduces temporal beats in the excitonic emission which are absent in the experimental data. The excitation configuration for these data is x, y, y. The inclusion of the *six*-particle term, $\overline{Z}$, in the DCT theory completes the $\chi^{(3)}$ level without approximation, and vastly improves the agreement between SWM experiment and theory. This work demonstrates that SWM is a sensitive probe for Coulomb correlations beyond the 4PC level, which typically are invisible in FWM emission. (It has been verified that the inclusion of $\overline{Z}$ gives negligible changes to the predicted FWM emission.) It is particularly interesting to note that $\overline{Z}$ suppresses the temporal beats introduced to the signal by $\overline{N}$ for the x, y, y polarization configuration. Similar effects are observed in other polarization configurations, where there is frequently strong competition between the contributions of $\overline{N}$ and $\overline{Z}$ sources. This competition is somewhat analogous to the competition between Coulomb mean-field and two-exciton continuum sources observed in FWM.

It is possible to use the DCT theory to examine the relative contributions of the bound-biexciton and unbound two-exciton contributions to the SWM signal. Keeping only one contribution or the other in the $\overline{Z}$ part of the dynamics gives very poor agreement with the data. In fact, the incoherent discrete transitions contribute not only at the biexcitonic energy (which is expected) but also at the exciton energy. A delicate

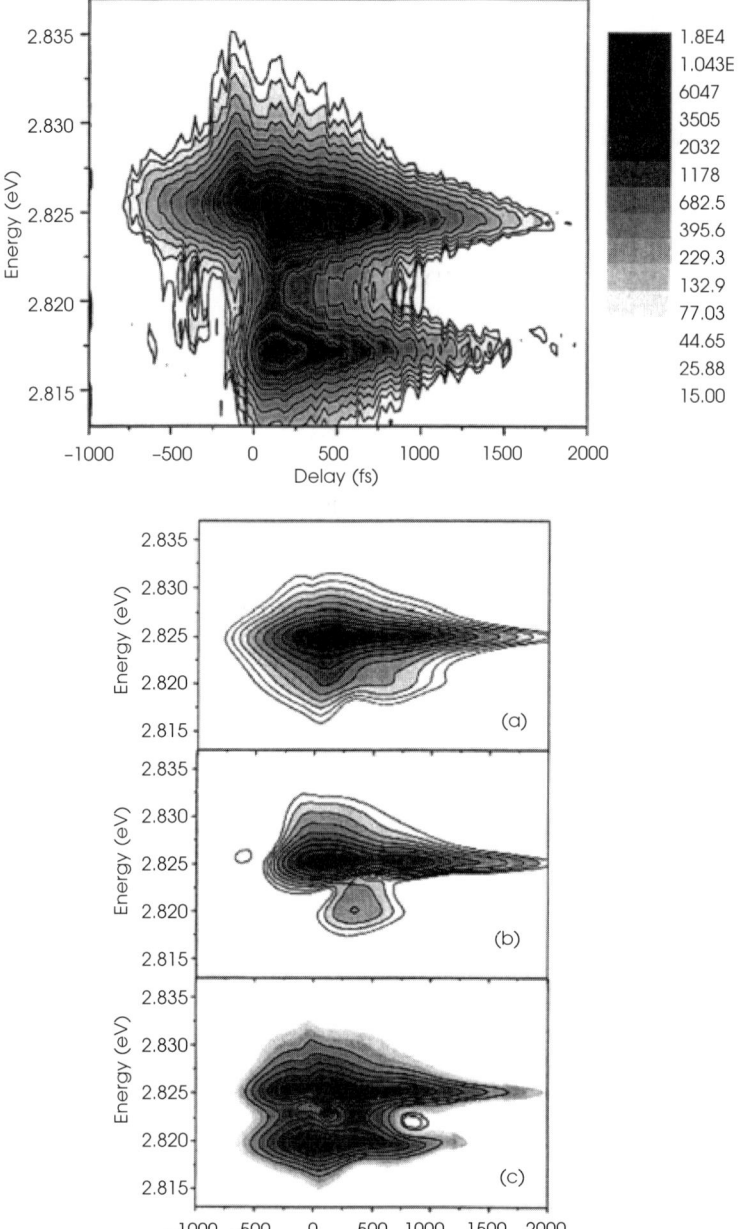

Fig. 5.14 Six-wave mixing (SWM) data and theory for a ZnSe single quantum well, measured at 10K with xy excitation polarization. The component of the emission polarized along y is shown here. The excitonic resonance is at 2.825 eV, and the biexciton at 2.818. Data are compared with the results of the DCT theory (lower panel) at three different levels: (a) coherent limit, (b) including $\bar{N}$, (c) including $\bar{N}$ and the six-particle term, $\bar{Z}$. Adapted from reference [60].

balance between the discrete biexciton and two-pair continuum contributions of $\overline{Z}$ is critical to SWM response (Axt, [61].

Contributions beyond the $\chi^{(3)}$ truncation

As has been shown in many of the studies discussed above, the use of the $\chi^{(3)}$ truncation to select the relevant n-particle density matrices often gives excellent agreement with experiment, even when higher-order terms in the driving field are clearly relevant. (Indeed, in SWM the leading order in the field is $\chi^{(5)}$, yet the full $\chi^{(3)}$ DCT scheme gives good agreement with experiment.) Recently, however, careful studies of P/P spectra in InGaAs quantum wells as a function of polarization, excitation density, and detuning have indicated the necessity of using a full $\chi^{(5)}$ truncation for high density excitation (Meier, [62]). In the coherent limit the $\chi^{(5)}$ truncation contains three independent variables: p, $\overline{B}$, and W (the triexcitonic transitions). Calculations demonstrate that the contributions of triexcitonic transitions are negligible, but that the full fifth-order equations of motion for p and $\overline{B}$ must be included in order to obtain agreement with the data. These equations of motion have the general form

$$\frac{\partial p}{\partial t} = i\begin{bmatrix} \omega_p p + \mu E - \mu E b(p^*p + p^*p^*pp + Bpp + Bp^*p^* + B^*B) \\ +V(p^*B + p^*p^*p^*B + B^*pB) \end{bmatrix},$$

$$\frac{\partial B}{\partial t} = i[\omega_B B + \mu E(p + p^*B)]$$

where ω_p and ω_b are the energies of a single and a two-exciton state, respectively. b gives the strength of the Pauli blocking nonlinearity, while V describes the Coulomb interaction. For comparison, note that within the $\chi^{(3)}$ limit, the NLO response is due to the Pauli blocking ($\mu E b p^*p$) and Coulomb interaction induced ($V p^*B$) contributions only, which involve single- and two-exciton excitation to lowest orders.

5.5   Future directions

The work discussed here has gone a long way towards elucidating the role of high-order Coulomb correlations in semiconductor NLO response. The DCT formalism has proven excellent for studies of highly correlated states at low and intermediate excitation densities in the vicinity of the absorption edge. As long as coherence dominates, the DCT allows for quantitative analysis of experiments, and permits identification of clear signatures of each Coulomb correlation. Recent work by Meier et al. [63] has taken the DCT well into the incoherent limit, considering the influence of a completely incoherent population of electron–hole pairs on the P/P spectrum of the exciton. In this work the population was assumed to be in thermal equilibrium, and distinct signatures of high-order correlations, including $\overline{Z}$, were predicted for the P/P spectra.

Despite the success of the DCT in describing highly correlated states, however, it only addresses half of the problem of the dynamics of optically excited semiconductors. The processes of scattering, energy relaxation, and dephasing are not included in the DCT, because scattering cannot be described via a perturbative expansion in the driving fields. Thus, although the DCT can describe initial coherent dynamics, and now is beginning to be used to describe the influence of the final, equilibrium populations, it cannot even begin to describe the evolution from initially coherent states through incoherently relaxing distribution functions to an ultimate quasi-thermal population.

Simultaneous with the recent burst of activity on HOCs, however, there has been dramatic progress in understanding scattering processes in semiconductors. These processes have been illuminated by ultrafast measurements made on time scales short compared to the natural scattering times. In such regimes the Markovian approximation fails, and a real-time non-equilibrium Green's function (NGF) approach must be used to understand the memory structure of the relaxation process (see, for example, Chemla, [64] and references therein). This approach has been tremendously successful in modeling ultrafast measurements of nonequilibrium populations, including phonon-scattering and carrier–carrier scattering based relaxation process (Wehner, [65]; Bar-Ad, [66]). In nearly all cases, however, the NGF approach is restricted to a second-order Born approximation, and does not include Coulomb correlation effects. Thus very little is known about the effects of high-order Coulomb correlations on scattering and relaxation, and the NGF formalism is valid only for processes high in the electron–hole continuum or at very high densities, where Coulomb correlations are negligible.

The work to join these two regimes and thus to bring light to the fundamental scattering processes at the band edge is just beginning. A first step was taken by Schafer *et al.* [67], who have obtained a formalism which includes both the DCT and the NGF as limiting cases. In order to achieve this synthesis, it was necessary to approximate the full two-time Green's functions with single-time density matrices (an approximation which is exact only in the coherent limit). Despite the fact that this formalism does not give a fully consistent picture at intermediate densities, it agrees very well with an initial experiment on dephasing of exciton populations with significant four-particle correlations in high magnetic fields. It seems likely that work to bring together the understanding of HOCs with that of incoherent scattering will be extremely fruitful and will continue for many years.

Finally, we should keep in mind that the many-body problem is ubiquitous in condensed matter systems, and shows itself in nearly every calculation. In the future it is likely that the understanding of Coulomb correlations we have gained in semiconductors, a particularly clean "model" system, will be extended to other materials with exciting new results. Work using ultrafast spectroscopic techniques in materials with highly correlated ground states (superconductors, magnetic systems, quantum Hall liquids) is just beginning.

References

1. Haug, H., and Koch, S.W. (1993). Quantum Theory of the Optical and Electronic Properties of Semiconductors. World Scientific, Singapore.
2. Shen, R.Y. (1984). The Principles of Nonlinear Optics. Wiley-Interscience, New York.
3. Elliott, R.J. (1957). Intensity of optical absorption by excitons. *Phys. Rev.* **108**, 1384–1389.
4. Tanguy, C. (1995). Optical dispersion by Wannier excitons. *Phys. Rev. Lett.* **75**, 4090–4093.
5. Lindberg, M., Koch, S.W. (1988). Effective Bloch equations for semiconductors. *Physical Review B* **38**, 3342–3350.
6. Lindberg, M., Binder, R., Koch, S.W. (1992). Theory of the semiconductor photon echo. *Phys. Rev. A* **45**, 1865–1875.
7. Wegener, M., Chemla, D.S., Schmitt-Rink, S., Schafer, W. (1990). Line shape of time-resolved four-wave mixing. *Phys. Rev. A* **42**, 5675–5683.
8. Haase, B., Neukirch, U., Meinertz, J., Gutowski, J., Axt, V.M., Bartels, G., Stahl, A., Nurnberger, J., Faschinger, W. (2000). Intensity dependence of signals obtained in four-wave-mixing geometry: influence of higher-order contributions. *Journal of Crystal Growth* **214–215**, 852–855.
9. Yajima, T., Taira, Y. (1979). Spatial optical coupling of picosecond light pulses and transverse relaxation effect in resonant media. *J. Phys. Soc. Japan* **47**, 1620–1626.
10. Leo, K., Wegener, M., Shah, J., Chemla, D.S., Gobel, E.O., Damen, T.C., Schmitt-Rink, S., Schafer, W. (1990). Effects of coherent polarization interactions on time-resolved degenerate four-wave mixing. *Phys. Rev. Lett.* **65**, 1340–1343.
11. Kim, D.S., Shah, J., Damen, T.C., Schafer, W., Janhke, F., Schmitt-Rink, S., Kohler, K. (1992). Unusually slow temporal evolution of femtosecond four-wave-mixing signals in intrinsic GaAs: Direct evidence for the dominance of interaction effects. *Phys. Rev. Lett.* **69**, 2725–2728.
12. Weiss, S., Mycek, M.-A., Bigot, J.-Y., Schmitt-Rink, S., Chemla, D.S. (1992). Collective effects in excitonic free induction decay: Do semiconductors and atoms emit coherent light in different ways? *Phys. Rev. Lett.* **69**, 2685–2689.
13. Chemla, D.S., Bigot, J.-Y. (1995). Ultrafast phase and amplitude dynamics of coherent transients in semiconductor quantum wells. *Chem. Phys.* **210**, 135–154.
14. Lindberg, M., Hu, Y.Z., Binder, R., Koch, S.W. (1994). $\chi^{(3)}$ formalism in optically excited semiconductors and its applications in four-wave-mixing spectroscopy. *Phys. Rev. B* **50**, 18060–18072.

15. Haug, H., and Koch, S.W. (1984). Electron theory of the optical properties of laser excited semiconductors. *Prog. Quant. Electron.* **9**, 3–100.
16. Wang, H., Ferrio, K., Steel, D.G., Hu, Y.Z., Binder, R., Koch, S.W. (1993). Transient nonlinear optical response from excitation induced dephasing in GaAs. *Phys. Rev. Lett.* **71**, 1261–1264.
17. Wang, H., Ferrio, K.B., Steel, D.G., Berman, P.R., Hu, Y.Z., Binder, R., Koch, S.W. (1994). Transient four-wave-mixing line shapes: Effects of excitation-induced dephasing. *Phys. Rev. A* **49**, R1551–R1554.
18. Hu, Y.Z., Binder, R., Koch, S.W., Cundiff, S.T., Wang, H., Steel, D.G. (1994). Excitation and polarization effects in semiconductor four-wave-mixing spectroscopy. *Phys. Rev. B* **49**, 14382–14386.
19. Rappen, T., Peter, U.G., Wegener, M. (1994), Polarization dependence of dephasing processes: A probe for many-body effects. *Phys. Rev. B* **49**, 10774–10777.
20. Mysyrowicz, A., Grun, J.B., Levy, R., Bivas, A., Nikitine, S. (1968). Excitonic molecule in CuCl. *Physics Letters A* **26**, 615.
21. Hanamura, E., and Haug, H. (1977). Condensation effects of excitons. *Phys. Rep.* **33**, 209–284.
22. Maruani, A., Chemla, D.S. (1981). Active nonlinear spectroscopy of biexcitons in semiconductors: Propagation effects and Fano Interferences. *Phys. Rev. B* **23**, 841–860.
23. Chemla, D.S. and Maruani, A. (1982). Nonlinear optical effects associated with excitonic molecules in large gap semiconductors. *Rep. Prog. Quant. Electr.* **8**, 1–77.
24. Bar-Ad, S., and Bar-Joseph, I. (1991). Absorption quantum beats of magneto-excitons in GaAs heterostructures. *Phys. Rev. Lett.* **66**, 2491–2494.
25. Lovering, D.J., Phillips, R.T., Denton, G.J., Smith, G.W. (1992). Resonant generation of biexcitons in a GaAs quantum well. *Phys. Rev. Lett.* **68**, 1880–1883.
26. Barad, S., and Bar-Joseph, I. (1992). Exciton spin dynamics in GaAs heterostructures. *Phys. Rev. Lett.* **68**, 349–352.
27. Mayer, E.J., Smith, G.O., Heuckeroth, V., Kuhl, J., Bott, K., Schulze, A., Meier, T., Benhardt, D., Koch, S.W., Thomas, P., Hey, R., Plook, K. (1994). Evidence of biexcitonic contribution to four-wave-mixing in GaAs quantum wells. *Phys. Rev. B* **50**, 14730–14733.
28. Denton, G.J., Phillips, R.T., Smith, G.W. (1995). Biexcitonic contribution to polarization dependent degenerate 4-wave-mixing in GaAs quantum-wells. *Appl. Phys. Lett.* **67**, 238–240.
29. Spano, F.C., and Mukamel, S. (1989). Nonlinear susceptibilities of molecular aggregates: Enhancement of $\chi(3)$ by size. *Phys. Rev. A* **40**, 5783–5801.
30. Spano, F.C., and Mukamel, S. (1991). Cooperative nonlinear optical response of molecular aggregates: Crossover to bulk behavior. *Phys. Rev. Lett.* **66**, 1197–1200.
31. Dubovsky, O. and Mukamel, S. (1991). Exciton coherence-size and phonon-mediated optical nonlinearities in restricted geometries. *J. Chem. Phys.* **95**, 7828–7845.
32. Leegwater, J.A. and Mukamel, S. (1992). Exciton-scattering mechanism for enhanced nonlinear response of molecular nanostructures. *Phys. Rev. A* **46**, 452–464.
33. Mukamel, S. (1994). Manybody Effects in Nonlinear Susceptibilities: Beyond the local-field approximation. *In* Molecular Nonlinear Optics (J. Zyss, Ed.). Academic Press, New York.
34. Maille, M.Z., Sham, L.J. (1994). Interacting electron theory of coherent nonlinear response. *Phys. Rev. Lett.* **73**, 3310–3313.

35. Ostreich, Th., Schonhammer, K., Sham, L.J. (1995). Exciton–exciton correlation in the nonlinear optical regime. *Phys. Rev. Lett.* **74**, 4698–4701.
36. Axt, V.M., Stahl, A. (1994). A dynamics-controlled truncation scheme for the hierarchy of density matrices in semiconductor optics. *Zeitschrift fur Physik B* **93**, 195–204.
37. Axt, V.M., Stahl, A. (1994). The role of the biexciton in a dynamic density matrix theory of the semiconductor band edge. *Zeitschrift fur Physik B* **93**, 205–211.
38. Victor, K., Axt, V.M., Stahl, A. (1995). Hierarchy of density matrices in coherent semiconductor optics. *Phys. Rev. B* **51**, 14164–14175.
39. Dick, B., and Hochstrasser, R.M. (1983). Resonant non-linear spectroscopy in strong fields. *Chem. Phys.* **75**, 133–135.
40. Abram, I., Maruani, A., Schmitt-Rink, S. (1984). The non-linear susceptibility of the biexciton 2-photon resonance. *J. Phys. C* **17**, 5163–5170.
41. Axt, V.M., Victor, K., Kuhn, T. (1998). The exciton–exciton continuum and its contribution to four-wave mixing signals. *Physica Status Solidi (b)* **206**, 189–196.
42. Schafer, W., Kim, D.S., Shah, J., Damen, T.C., Cunningham, J.E., Goosen, K.W., Pfeiffer, L.N., Kohler, K. (1996). Femtosecond coherent fields induced by many-particle correlations in transient four-wave-mixing 1996. *Phys. Rev. B* **53**, 16429–16443.
43. Axt, V.M., Victor, K., Stahl, A. (1996). Influence of a phonon bath on the hierarchy of electronic densities in an optically excited semiconductor. *Phys. Rev. B* **53**, 7244–7258.
44. Haase, B., Neukirch, U., Gutowski, J., Bartels, G., Stahl, A., Nurnberger, J., Faschinger, W. (1998). Coherent dynamics of excitons and biexcitons: Polarization and intensity dependence. *Physica Status Solidi (b)* **206**, 363–368.
45. Kner, P., Bar-Ad, S., Marquezini, M.V., Chemla, D.S. (1997). Magnetically enhanced exciton–exciton correlations in semiconductors. *Phys. Rev. Lett.* **78**, 1319–1322.
46. Kner, P., Bar-Ad, S., Marquezini, M.V., Chemla, D.S., Lovenich, R., Schafer, W. (1999). Effect of magneto-exciton correlations on the coherent emission of semiconductors. *Phys. Rev. B* **60**, 4731–4748.
47. Langbein, W., Hvam, J.M. (2000). Dephasing in the quasi-two-dimensional exciton–biexciton system. *Phys. Rev. B* **61**, 1692–1695.
48. Axt, V.M., Haase, B., Neukirch, U. (2001). Influence of two-pair continuum correlations following resonant excitation of excitons. *Phys. Rev. Lett.* **86**, 4620–4623.
49. Ostreich, Th., Schonhammer, K., Sham, L.J. (1998). Theory of exciton-exciton correlation in nonlinear optical response. *Phys. Rev. B* **58**, 12920–12936.
50. Neukirch, U., Bolton, S.R., Sham, L.J., Chemla, D.S. (2000). Electronic four-particle correlations in semiconductors: Renormalization of coherent pump–probe oscillations. *Phys. Rev. B* **61**, R7835–R7837.
51. Neukirch, U., Bolton, S.R., Fromer, N., Sham, L.J., Chemla, D.S. (2000). Polariton–biexciton transitions in a semiconductor microcavity. *Phys. Rev. Lett.* **84**, 2215–2218.
52. Fan, X., *et al.*, *Phys. Rev. B.* **57**, R9451 1998.
53. Sieh, C., Meier, T., Jahnke, F., Norr, A., Koch, S.W., Brick, P., Hubner, M., Ell, C., Prineas, J., Khitrova, G., Gibbs, H.M. (1999). Coulomb memory signatures in the excitonic optical Stark effect. *Phys. Rev. Lett.* **82**, 3112–3115.
54. Axt, V.M., Bartels, G., Stahl, A. (1996). Intraband dynamics at the semiconductor band edge: Shortcomings of the Bloch equation method. *Phys. Rev. Lett.* **76**, 2543–2546.

55. Haring Bolivar, P., Wolter, F., Muller, A., Roskos, H.G., Kurz, H., Kohler, K. (1997). Excitonic emission of THz radiation: Experimental evidence of the shortcomings of the Bloch equation method. *Phys. Rev. Lett.* **78**, 2232–2235.
56. Haase, B., Neukirch, U., Gutowski, J., Bartels, G., Stahl, A., Axt, V.M., Nurnberger, J., Faschinger, W. (1999). Manifestation of exciton-amplitude fluctuations in the transient polarization state of four-wave-mixing signals. *Phys. Rev. B* **59**, R7805–R7808.
57. Bartels, G., Stahl, A., Axt, V.M., Haase, B., Neukirch, U., Gutowski, J. (1998). Identification of higher-order electronic coherences in semiconductors by their signatures in four-wave-mixing signals. *Phys. Rev. Lett.* **81**, 5880–5883.
58. Bartels, G., Cho, G.C., Dekorsky, T., Kurz, H., Stahl, A., Kohler, K. (1997). Coherent signature of differential transmission signals in semiconductors: Theory and experiment. *Phys. Rev. B* **55**, 16404–16413.
59. Meier, T., Koch, S.W., Phillips, M., Wang, H. (2000). Strong coupling of heavy- and light-hole excitons induced by many-body correlations. *Phys. Rev. B* **62**, 12605–12608.
60. Bolton, S.R., Neukirch, U., Sham, L.J., Chemla, D.S., Axt, V.M. (2000). Demonstration of sixth-order Coulomb correlations in a semiconductor single quantum well. *Phys. Rev. Lett.* **85**, 2002–2005.
61. Axt, V.M., Bolton, S.R., Neukirch, U., Sham, L.J., Chemla, D.S. (2000). Evidence of six-particle Coulomb correlations in six-wave-mixing signals from a semiconductor quantum well. *Phys. Rev. B* **63**, 115303.
62. Meier, T., Koch, S.W., Brick, P., Ell, C., Khitrova, G., Gibbs, H.M. (2000). Signatures of correlations in intensity dependent excitonic absorption changes. *Phys. Rev. B* **62**, 4218–4221.
63. Meier, T., Koch, S.W. (1999). Excitons versus unbound electron–hole pairs and their influence on exciton bleaching: A model study. *Phys. Rev. B* **59**, 13202–13208.
64. Chemla, D.S. (1999). Ultrafast transient nonlinear optical processes in semiconductors. In *Nonlinear Optics in Semiconductors* (E. Garmire, Ed.) Academic Press, New York.
65. Wehner, M.U., Ulm, M.U., Chemla, D.S., Wegener, M. (1998). Coherent control of electron-LO-phonon scattering in bulk GaAs. *Phys. Rev. Lett.* **80**, 1992–1995.
66. Bar-Ad, S., Kner, P., Marquezini, M.V., El Sayed, K., Chemla, D.S. (1996). Carrier dynamics in the quantum kinetic regime. *Phys. Rev. Lett.* **77**, 3177–3180.
67. Schafer, W., Lovenich, R., Fromer, N., Chemla, D.S. (2001). From coherently excited highly correlated states to incoherent relaxation processes in semiconductors. *Phys. Rev. Lett.* **86**, 344–347.

Quantum Coherence, Correlation and Decoherence in
Semiconductor Nanostructures
T. Takagahara (Ed.)
Copyright © 2003 Elsevier Science (USA). All rights reserved.

Chapter 6
Electronic and nuclear spin in the optical spectra of semiconductor quantum dots

D. Gammon, Al.L. Efros, J.G. Tischler, A.S. Bracker

Naval Research Laboratory, Washington, DC 20375, USA

V.L. Korenev and I.A. Merkulov

A.F. Ioffe Institute, St Petersburg, Russia

Abstract

In this chapter we discuss the physics of spin in semiconductor quantum dots as measured through optical spectroscopy. We treat photoluminescence spectroscopy of single quantum dots in detail. Optical selection rules connect the electronic spin to the exciton polarization properties. Exchange and Zeeman interactions lead to fine structure of the localized exciton. We consider these interactions in detail for the cases of the exciton and the trion (the singly charged exciton). We also discuss the hyperfine interaction between the electronic and nuclear spins. Under certain experimental conditions the nuclear spin can become optically pumped into a high polarization state, which significantly affects the optical spectrum through the Overhauser effect. Fluctuations of nuclear spin polarization lead to relaxation of the electronic spin, which is measurable in the optical spectrum through the Hanle effect.

6.1 Introduction to spin in the optical spectrum

Recent interest in the physics of spin in semiconductor nanostructures is driven largely by materials research developments in the growth of quantum dots (QDs) and a concurrent flurry of activity in high spatial and spectral resolution spectroscopy of individual QDs. As examples, the extremely sharp spectral lines measured in single dot spectroscopy have led to direct observations of fine and hyperfine splittings and related phenomena in optical spectra. The application of coherent and quantum optical techniques have resulted in the demonstration of quantum superposition states composed of the pseudo-spin states of excitons in individual QDs. These materials science breakthroughs have opened up new technological frontiers, such as the possibility of constructing a quantum information processor out of solid state materials. These new materials and techniques have generated a surge of new opportunities in a subfield that began in the late 1960s with studies of shallow donors in bulk semiconductors.

There is a need for a broad introduction to the topic of spin in quantum dots for those researchers and students entering the field. In this chapter, we introduce the central physical concepts and results. Rather than provide an exhaustive review, we highlight important concepts in an introductory manner with experimental or theoretical results to illustrate our points. We focus primarily on the use of continuous wave (cw) photoluminescence spectroscopy as a tool for probing spin energy levels and dynamics. The research examples we have chosen come mostly from the natural QDs formed by well width fluctuations. However, our intention is to focus on the fundamental physics that is common to all quantum dots.

Quantum dots are particles of semiconductor crystals, often embedded in a barrier material, with dimensions that are small enough (10–100 nm) such that quantum confinement plays a central role in their physics. Their discrete energy spectra and relatively wide energy level spacings minimize many of the relaxation processes that dominate in bulk or two-dimensional samples, resulting in homogeneous linewidths that are generally narrower in quantum dots. Optical oscillator strengths are also relatively large, and in combination with narrowband lasers tuned to the discrete optical transitions of quantum dots, both linear and nonlinear spectroscopies are highly effective.

In the last decade, researchers around the world have studied several broad varieties of quantum dots. Self-assembled dots grown on a solid semiconductor surface with gas-phase epitaxial techniques are the variety of quantum dots most widely studied by optical means [1]. Part of the technological impetus for research in growth and spectroscopy of self-assembled dots is to develop highly efficient near-infrared laser diodes. Other types of quantum well-based dots that depend for their confinement on strain or surface vacuum-level barriers have also been investigated [2]. Colloidal nanocrystals, whose fabrication depends heavily on advances in inorganic synthesis, have been the subject of wide ranging spectroscopic and structural investigations [3]. Several laboratories, including our own, have achieved wide success in research on the spectroscopy of a type of quantum dot defined by changes in the well width in a GaAs quantum well, arising in part from monolayer-high interface islands [4]. In this chapter, we will consider primarily this type of natural dot.

Spectroscopy provides a powerful probe of spin and its interactions in semiconductors. Common optical spectroscopies include photoluminescence (PL), absorption, and nonlinear techniques. Magnetic resonance techniques include electron spin resonance (ESR) and nuclear magnetic resonance (NMR). In quantum dots with a strong optical response (III-V and II-VI semiconductors with a direct band gap), PL has been extremely useful because of its sensitivity and experimental simplicity. Unlike ESR, PL can be used to study undoped samples. In some cases, optical spectroscopy can be combined with magnetic resonance to provide an extremely powerful hybrid technique. PL and its variations are the most widely used techniques in semiconductor spectroscopy, and their application to the study of spin states of quantum dots will be the emphasis of this chapter.

Spin leads to several important features and phenomena in the optical spectra. One of the most obvious effects of spin in the optical spectra is the existence of spin degeneracies or near degeneracies (fine structure) in the optical transitions. This fine structure is often studied as a function of magnetic fields.

A relatively recent and powerful development is the use of microscopic techniques to measure the optical spectra of individual QDs. Inhomogeneous broadening in quantum dot ensembles is much larger than the homogeneous linewidths of individual dots and therefore masks many interesting features. The spectral linewidth of PL from a single QD can be as narrow as a few tens of μeV [5] and could approach a μeV in self-assembled QDs [6,7], which is less than many of the splittings caused by spin interactions. Therefore, this technique, in combination with high resolution laser spectroscopy, presents the opportunity to measure directly the fine structure of the spectral lines arising from spin and gain a more complete understanding of the nature and magnitude of spin phenomena in QDs. An introduction to single QD spectroscopy will be presented in section 6.2.

The origins of the fine structure splittings arise from interactions that affect the different spin components differently. These interactions consist of the spin–orbit interactions, exchange interaction between the electrons and/or holes, the Zeeman interactions between spin and the applied magnetic field, and the hyperfine interaction between the spins of the electron and those of the nuclei. One of the most powerful features of single QD spectroscopy is that these splittings can be measured directly, thus allowing new opportunities to explore the nature, the magnitudes, and the impact of these spin interactions. In section 6.3 we will discuss the fine structure arising from the exchange and the Zeeman interactions.

It should be realized that fine structure in the spectrum could arise from splitting in the initial and/or final states of the transition. For example a trion singlet transition has a two-fold degeneracy in both the excited state and the ground state. In contrast, in the exciton transition the ground state is just the exciton vacuum, and the fine structure of the exciton transition is the same as the fine structure of the exciton quantum state itself. In section 6.4 we will consider the fine structure of the trion and its PL spectrum.

Another powerful feature of optical spectroscopy is the intimate connection between the polarization of light and the polarization of angular momentum in matter. This connection is central to many of the topics discussed in this chapter. Although optical fields do not couple directly to spin, they provide an indirect probe because of coupling between the spin and orbital degrees of freedom. In fact, the PL polarization has historically provided the primary experimental window into the physics of the spin in semiconductors [8]. Optical selection rules connect the electron spin to the optical polarization in the spectra. Because of this connection, the polarization of the emitted light contains information about the population and nature of spin states, and their mixture through the interactions. One advantage of polarization spectroscopy is that it is not necessary to resolve the individual substates. In PL polarization spectroscopy the polarization is measured as a function of magnetic field, of detection frequency, excitation frequency, and so on.

This relationship between spin and polarization also works the other way – the selection rules can be used to selectively excite certain spin states. This leads to the opportunity to *optically pump* the spin system into a highly non-thermal spin state, which enables the study of spin physics even if the fine structure splitting is much less than k_BT, or even if it is zero. For example, although the exciton is only partially spin polarized when in thermal equilibrium, it can be optically pumped to nearly 100% polarization. At a magnetic field of 1 T the Zeeman splitting of an exciton in a GaAs QD is about 100 μeV, which is a fraction of k_BT at normal helium temperature (at $T = 4K$, $k_BT = 300$ μeV). To optically orient the exciton it is necessary for the recombination time to be much faster than the spin-flip processes that would tend to thermalize the exciton substates. Optical pumping, in combination with polarization spectroscopy of the emission, leads to the extremely powerful technique known generally as *optical orientation* [8]. This field was developed in atomic systems [9], and applied to bulk semiconductors starting in the late 1960s [10]. An excellent review is given in [11].

Because the exciton can be put into a nonthermal spin polarization state with a steady-state probability that depends on the spin relaxation rate, this rate can be measured through optical orientation methods. One common approach for doing this is through the Hanle depolarization effect [11]. This involves the measurement of the rate that the PL polarization is lost as a transverse magnetic field is increased under optical pumping conditions. The physics of spin relaxation and the Hanle effect will be discussed in sections 6.3 and 6.6.

The electronic spin also interacts with the underlying nuclear spin within the QD through the hyperfine interaction. The interaction of an electron with a single nuclear spin is relatively weak. However, as mentioned above, under optical pumping conditions the electron spin polarization becomes highly nonthermal, and the Overhauser effect leads also to a highly polarized nuclear spin system. As much as 70% of the nuclear spin in a QD has been pumped into orientation [12]. With such a large number of nuclear spins oriented within the electron wave function, the average hyperfine interaction becomes much larger, and the resulting hyperfine splitting can be the dominant fine structure splitting [12,13]. The physics of the Overhauser effect in QDs will be discussed in section 6.5.

The average of the hyperfine field can become important under certain conditions, but even when the average field is zero, fluctuations in the hyperfine field are important. In fact, fluctuations in the hyperfine field can become the dominant spin relaxation mechanism in QDs because the spin relaxation processes connected with spin–orbit interactions are suppressed there. Essentially, the orbital degrees of freedom of the electron and hole become frozen by the complete quantum confinement and do not react as strongly to fluctuations of the environmental fields (e.g. phonons and charge fluctuations) that usually interact with the orbital part of the wavefunction. This is understood in quantum mechanical terms through the development of gaps in the energy spectrum of the electronic system. Unlike a higher dimensional electron, which has a continuum of states into which it can scatter, a fully confined electron must be excited across a gap. Secondly, the spin–orbit matrix elements are reduced; for example, because of mismatches between the length scale of the electron wavefunction and the acoustic phonons [14,15]. As a result of the reduction of spin relaxation mediated by spin–orbit interactions, fluctuations in the hyperfine fields play a more important role. The dominant effect in the measured spectra is typically the inhomogeneity in hyperfine fields from dot to dot, or over time in the same QD. The physics of spin relaxation will be discussed in section 6.6.

To summarize, we will first give a brief overview of the structure and spectra of QDs and the technique of single QD spectroscopy in section 6.2. Then we will discuss in turn: the fine structure of excitons in section 6.3, of trions in section 6.4, the hyperfine interaction of excitons and electrons with nuclear spins in section 6.5, and the relaxation of electron spin in section 6.6.

6.2   Photoluminescence spectroscopy of quantum dots

Natural (interface fluctuation) QDs

In a quantum dot, carriers are localized in all three spatial dimensions. There are a number of different QD systems being studied [1], but much of the physics and experimental methodology is general. In most QDs, confinement results from a bandedge offset between the dot material and the surrounding matrix, which is usually a semiconductor of higher bandgap. In optically excited QDs, bound electron–hole pairs (excitons) are produced. In QDs with weak lateral confinement, such as the natural QDs to be discussed throughout this chapter, the exciton binding energy is larger than the confinement energy, and it is the center-of-mass wavefunction of the exciton that is localized. Localization changes the energy spectrum of the exciton from a continuum into a set of discrete levels (Fig. 6.1). As a result, there is a qualitative change in the relaxation dynamics of the exciton [16,17,18]. It should be noted that there are strong similarities in the properties of electrons or excitons bound to shallow donors and those localized in QD potentials. This analogy is important because there has been extensive work on shallow impurities in bulk semiconductors [19], and many of the basic concepts and techniques in the field of quantum dots were developed during these earlier studies. In this section, we discuss the morphological properties and coarse energy level structure of natural quantum dots.

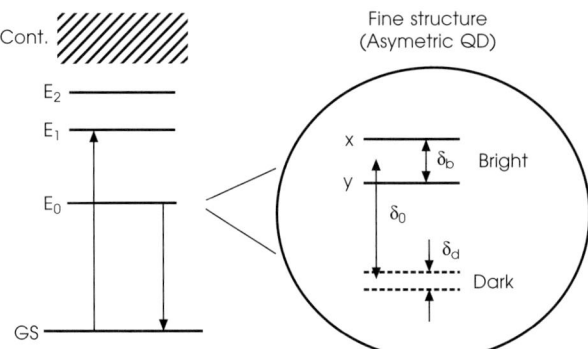

Fig. 6.1 Schematic diagram of energy levels of QD with fine structure showing exchange splittings in anisotropic disk-like QD.

Excitons in a natural quantum dot are confined in three dimensions by an imperfect GaAs quantum well (Fig. 6.2(a)). Vertical confinement (in the epitaxial growth direction) is provided by the quantum well barriers ($Al_xGa_{1-x}As$), while lateral confinement in the quantum well plane results from natural variations in the effective thickness of the GaAs quantum well. This confinement results from the level mismatch between quantum well subbands in regions of the well with different thickness (Fig. 6.2(b)). Natural QDs are also often named 'interface fluctuation quantum dots', or simply 'interface dots' [4].

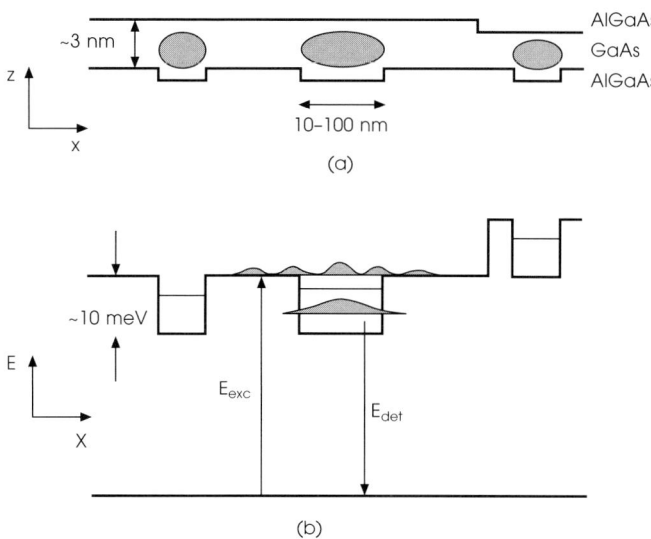

Fig. 6.2 Schematic diagram of (a) 3 nm quantum well with large monolayer-high islands at the interfaces that lead to confinement of the exciton and (b) The corresponding lateral confinement potentials associated with the interface islands.

During growth of a quantum well by molecular beam epitaxy (MBE), large monolayer-high islands develop at the well–barrier interfaces during interrupts of a minute or more under an arsenic flux. Much of this roughness persists at an interface as subsequent layers are grown. During growth interrupts, the islands can grow to lateral sizes larger than the Bohr diameter of the exciton (~20 nm in GaAs) and an order of magnitude larger than the well width, leading to a disk-like shape. Figure 6.3 shows the top surface of a GaAs quantum well imaged with scanning tunneling microscopy (STM). As seen in the figure, the structures tend to be elongated along the $[\bar{1}10]$ crystal axis.

The potential barrier formed by the monolayer-high steps is an order of magnitude less than that for the $Al_xGa_{1-x}As$ potential barriers in the vertical direction. Therefore the QD, though providing strong confinement in the vertical direction, is weak in the lateral directions. As a useful approximation, the energy spectrum of the QD can be separated into energies associated with the vertical and lateral dimensions. The strong confinement along the z-axis governs many of the properties of the exciton, such as the g-factor and exchange Coulomb energies. The low energy excited states are determined primarily by the lateral size and shape of the QD and have energy splittings on the order of a few meV. Because the light-hole exciton is shifted up by tens of meV, the spectra and other properties of these low energy QD states are derived primarily from the lowest energy heavy hole subband of the quantum well, and light-hole mixing is weak.

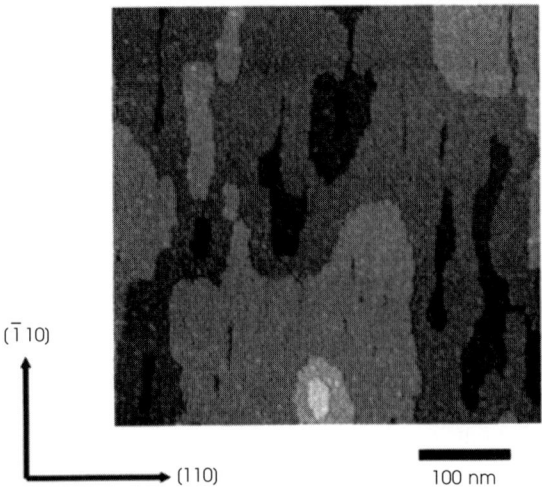

Fig. 6.3 STM image of GaAs surface showing the large monolayer-high islands. The islands tend to be elongated along the [$\bar{1}$10] axis. From [41].

It should be mentioned that there are several complications to the simple picture of lateral confinement discussed above. In addition to the large, monolayer-high islands at the interfaces, there also exists atomic-scale roughness [20,21,22], probably due to exchange of Ga and Al atoms as the interface is being formed [23,24]. The exciton energy is affected by this small-scale roughness, which will also lead to localization if there are fluctuations in the density of this small-scale roughness on the scale of the exciton Bohr diameter or larger. The same holds true of the barriers composed of $Al_xGa_{1-x}As$ rather than pure AlAs [25]. Correlation of the interface steps on the top and bottom interfaces of a well will also affect the optical properties [26]. Excitons localized by impurities will contribute sharp lines to near-field spectra. However, this will be a small contribution in undoped samples (<2 μm^{-2}) because of the small background density of impurities in this system (<10^{15} cm^3).

Early spectroscopic studies revealed strong inhomogeneous broadening and identified its origin as interface roughness [27]. Exciton localization, which has been studied since the early 1980s [27,28], can change the dynamics of the exciton considerably [16,17]. It was discovered that localization leads to a strong decrease in the homogeneous linewidth, or equivalently, an increase in the dephasing time of the exciton, which occurs because the exciton cannot interact as strongly with its environment. A mobility edge was discovered, corresponding to an energy at which the excitons become mobile and the homogeneous linewidth increased dramatically [16,17].

Microscopic and optical near-field spectroscopy have led to a more detailed understanding of these phenomena in the localized limit, and it has become clear that in certain cases, excitons localized by interface roughness can be modeled through QD potentials (Fig. 6.2). Investigation of this system is rewarding, not only because it provides a useful model system for the study of QD physics, but also because the new perspective obtained through recent single exciton spectroscopies enhances the study of real quantum wells started two decades ago [27,28].

Photoluminescence spectroscopy of single QDs

In linear optical spectroscopy, photon absorption excites an electron across the band gap of a semiconductor, from the valence band into the conduction band. The Coulomb attraction between the electron and the hole leads to the formation of an exciton, which is the elementary optical excitation of a semiconductor. Like the hydrogen atom in atomic and molecular physics, the exciton is the simplest and most-studied two particle complex in the field of semiconductor spectroscopy. The analogy between them is close – a light negative particle and a heavier positive particle orbit on another. Multiple exciton complexes (biexcitons, triexcitons, etc.) also form readily as a consequence of using high intensity lasers. The biexciton is the semiconductor counterpart of the hydrogen molecule. In the last five years, singly charged excitons (trions) confined in heterostructures have also become a topic of intense interest. In the molecular analogy, negative trions (two electrons and a hole) and positive trions (two holes and an electron) may be compared to the atomic hydrogen negative ion (H^-) and the molecular hydrogen positive ion (H_2^+), respectively.

Laser light with photon energy above the bandgap of the barrier material containing the quantum dots is directed at the sample. (Resonant excitation will be discussed below.) The electron–hole pairs have excess kinetic energy that is lost through emission of phonons. Spin polarization arising from circularly polarized excitation is also lost. Some fraction of the carriers drift into the QD confinement potential, where they form excitons or other bound complexes. The discrete energy levels of the quantum dots are spaced by several meV, and relaxation into the lowest dot levels requires emission of acoustic phonons. Photoluminescence is then emitted from the lowest dot levels. In an experiment with continuous lasers, there is a steady cascade of carriers relaxing through the various energy levels of the sample. Because photon emission is a comparatively slow process, it is possible for two or more excitons to collect within a single dot. At higher laser powers, one often observes a red-shifted biexciton peak or higher energy multi-exciton peaks produced by recombination from higher orbital states of the QD.

In a typical experiment the sample is contained within a liquid helium cryostat, and the excitation laser is brought toward the sample at slightly off normal incidence. The incoming beam may be sent through a polarization-modifying optic and a focusing lens. The photoluminescence is collected and collimated with a large f-number lens, after which optical elements may be inserted in order to determine the polarization of the PL. The PL is refocused onto the slit of a grating spectrometer. If necessary, the beam diameter and refocusing angle can be adjusted in order to match the f-number of the spectrometer grating. A photomultiplier or avalanche photodiode is used for single channel detection, while a CCD array detector may be used to collect the whole dispersed photoluminescence spectrum simultaneously. Single channel detection allows phase sensitive detection which is important in polarization spectroscopy. Multi-channel detection is highly advantageous for the weak PL signals in single QD spectroscopy.

Before discussing the energy levels and PL spectra of individual quantum dots, we briefly consider the coarse features of ensemble spectra for these dots, which are derived from GaAs quantum wells. Typical samples are grown by MBE with two minute growth interrupts under arsenic at each of the quantum well interfaces. A common structure in our work consists of five $GaAs/Al_{0.3}Ga_{0.7}As$ wells of different width, ranging from 3 nm to 14 nm. The luminescence peaks in the spectrum arise primarily from the lowest heavy-hole exciton states of each well. The spectrum for each well consists of two (and sometimes more) broad peaks with an energy splitting corresponding to a difference in well width of approximately one monolayer. The inhomogeneous linewidth of each peak results from the distribution in width and shape of the particular regions in which the excitons are confined (Fig. 6.2), which in turn governs the lateral confinement strength. The detailed features of these ensemble spectra have been extensively studied [29].

Many details of quantum dot energy levels are hidden when studying the spectra of ensembles, due to the large inhomogeneous distribution of dot size, shape, and other morphological features. Using high spatial resolution spectroscopy [30,31,32,33,34,35] to study few or even a single quantum dots, it becomes possible to uncover the detailed features that were previously hidden. In order to study individual quantum dots, it is necessary to reduce the size of the region in which the PL is excited or from which it is detected. One approach is to use a low temperature microscope [30,32] or even a scanning near-field optical microscope [33,35] to focus to micron or submicron laser spots. A less technically-demanding approach is to obtain spatial selectivity by effectively shrinking the size of the sample rather than the size of the light source. Optical lithography can be used to fabricate submicron mesas containing small numbers of quantum dots [36,37]. The etched regions contain no dots and if the mesas are well-separated, it is straightforward with a typical laser spot size to excite only a single mesa. Another approach has been to fabricate submicron apertures in a thin aluminum film, through which a laser can excite a small collection of QDs [34]. The effect of using successively smaller aperture sizes is shown in Fig. 6.4. As the apertures become smaller, the broad inhomogeneous peak profile breaks up into a collection of much sharper lines. Each of these lines arises from a single exciton localized within a quantum dot potential defined by the interface structure. From this experiment, we see that the exciton energies are clustered within the broad ensemble monolayer peaks split by the difference in quantum well width, and that within each of these groupings, the excitons have an additional distribution in energy due to variations in their confinement energy via their lateral size and other perturbations.

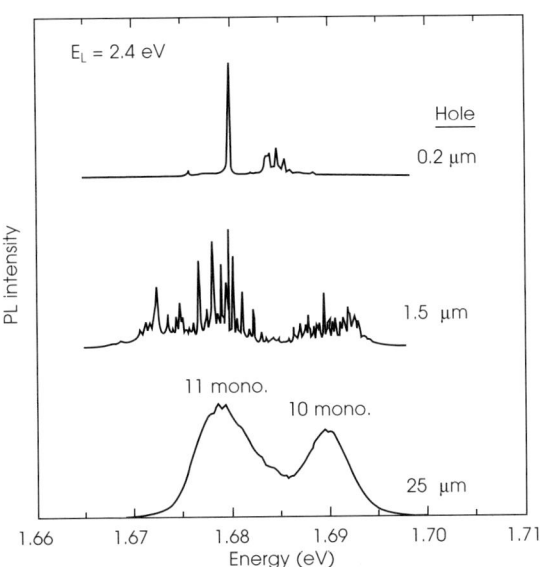

Fig. 6.4 Photoluminescence of 3 nm thick GaAs/Al$_{0.3}$Ga$_{0.7}$As quantum well grown with 2 min. growth interrupts at the interfaces. The PL was excited and detected through an aperture with diameter listed. Monolayer high islands lead to the splitting of the ensemble spectrum (bottom spectrum) into lines associated with either 10 or 11 monolayer well width. With decreasing aperture diameter these inhomogeneously broadened lines break up into a decreasing number of single exciton lines arising from single QD-like potentials. From [41].

PL excitation spectroscopy of single QDs

Beyond the use of high spatial resolution, it is possible to further reduce the number of QDs that contribute to a PL spectrum through selective excitation of specific dots with a tunable laser. The laser is scanned in frequency until it excites a local excited state of a QD, which relaxes rapidly through phonon emission to the ground state and produces a single PL line at lower energy. A photoluminescence excitation (PLE) spectrum is observed by monitoring a single PL line at fixed energy while scanning the excitation laser [31,34]. PLE gives the excited state spectrum of a quantum dot and can be performed on each of the sharp lines in a PL spectrum. An example of such a spectrum for 10-monolayer GaAs/Al$_{0.3}$Ga$_{0.7}$As single quantum well is shown in Fig. 6.5. This spectrum shows discrete absorption resonances and the onset of a continuum about 11 meV above the PL line [5].

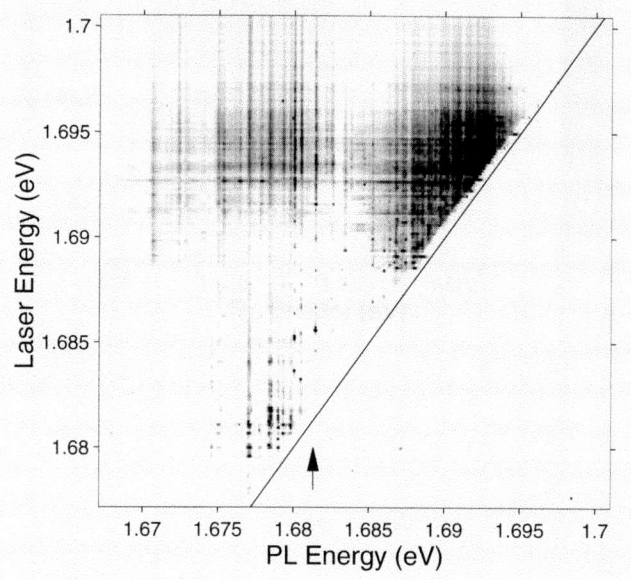

Fig. 6.5 PL intensities as a function of both spectrometer and laser frequency. A horizontal slice gives a PL spectrum for a given laser frequency and a vertical slice gives a PL excitation spectrum at a given detection frequency. The arrow points to the QD PL excitation spectrum displayed in Fig. 6.7. Adapted from [41].

The excitation spectrum provides an important fingerprint with considerable information about the QD potential. From PLE spectra, the lateral sizes of the QDs can be estimated. First, we note that the *vertical* size and depth of the potential are those of the quantum well, in this case 11 monolayers and about 300 meV, respectively. The potential depth in the *lateral* dimensions is about 11 meV and is given approximately by the energy between an exciton in an 10 and an 11 monolayer quantum well (Fig. 6.2). By fitting the lateral size and shape to calculated values, the entire QD spectrum can be fitted in principle. However, it should be noted that this is a complicated calculation that has not been done with a realistic potential. It is possible to make a rough estimate of the widest lateral dimension using a one-dimensional particle in a box calculation using the energy measured between the exciton ground state and the first excited state, assuming that the potential is rectangular and that motion in the two lateral dimensions is separable. This calculation leads to a size of 40 nm for the QD whose spectrum is plotted in Fig. 6.5 and a size of about 100 nm for the average energy splitting (~1 meV) in the PLE spectra for this sample. Because of the small lateral potential depth, low temperatures are necessary to observe zero-dimensional behavior (11 meV corresponds to 130K).

6.3   Exciton fine-structure (spin and sublevels)

There is an internal ("spin") degree of freedom associated with the Bloch function that exists in addition to those degrees of freedom associated with the envelope functions, and within each exciton spectral line there exists fine structure (sublevels). In a disk-like QD the projection of the heavy-hole exciton's spin on the strong quantization axis (z-axis) is the sum of the electron ($S_z = \pm 1/2$) and heavy hole ($J_z = \pm 3/2$) spin projections (Fig. 6.6(a)) [38,39,40]. This leads to four exciton sublevels: a degenerate doublet with spin projection along the z-axis (± 1) which is optically active (bright) and another doublet (± 2) that is optically inactive (dark) (Fig. 6.6(b)). However, even at zero field this degeneracy is often lifted by the exchange interaction into two closely spaced doublets [41] (Fig. 6.6(c)). Furthermore, the energy splitting and mixing of the spin states can be controlled by external magnetic fields through the Zeeman interaction. All these states play a key role in the fundamental physics of semiconductors, such as exciton dynamics, spin relaxation [42] and the Overhauser effect [13]. Therefore, it is important to have a good qualitative and quantitative understanding of these energy splittings. Previous work studied the fine structure of excitons in wide GaAs quantum wells [43,44,45] and several quantum dot systems [46,47].

Exchange interaction

The electron–hole exchange interaction is enhanced by the spatial confinement of electrons and holes into the same volume by the barriers of the QW structure. That is why the dark/bright exciton splitting (δ_0) is larger than in bulk [43,45]. In the disk-like natural QDs, the magnitude of the exchange splitting is dominated by the vertical confinement. However, lateral anisotropy leads also to a small splitting of the bright exciton doublet

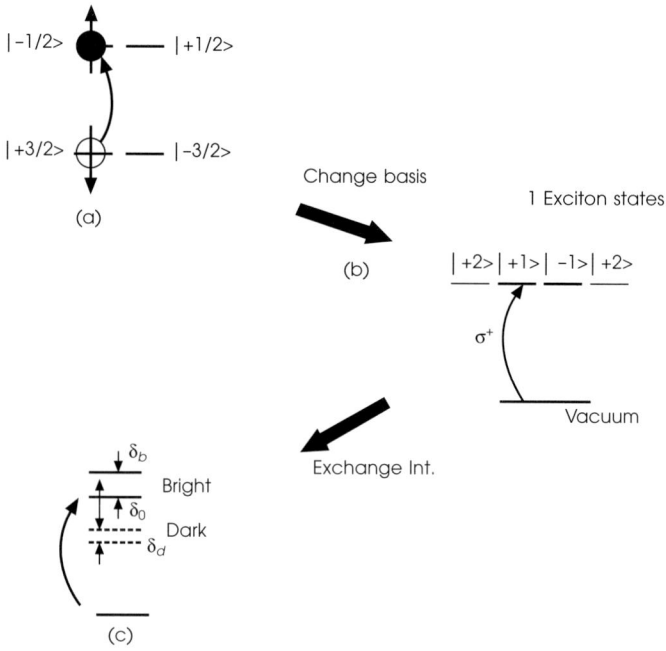

Fig. 6.6 Schematic diagram of (a) the lowest energy QD electron and hole states, (b) the same states in the exciton basis, and (c) the exciton energy levels including the exchange interaction.

in a QD due to the long-range part of electron–hole exchange interaction [48,49,50]. This splitting would be absent in islands having cylindrical shape. Moreover, the polarization of the emission changes from circular to linear in anisotropic QDs. Van Kesteren et al. [44], using optically detected magnetic resonance spectroscopy, first observed such behavior in Type II quantum wells. This behavior was observed for localized states in normal Type I GaAs quantum wells (similar to those studied here) by Blackwood et al. [43], using ensemble magneto-PL. The magnitude of the splitting observed by Blackwood et al. is about the same as that measured directly for the exciton ground states in single QDs, and likely has the same origin.

To explain these observations, we need to consider the spin states of the exciton at zero field. In bulk GaAs the Hamiltonian describing exchange is given by [44]

$$\hat{H}_{\text{exchange}}^{\text{bulk}} = - \sum_{i=x,y,z} (a\hat{J}_{h,i}\hat{S}_{e,i} + b\hat{J}_{h,i}^3\hat{S}_{e,i}), \qquad (6.1)$$

where $\hat{J}_{hj}$ and $\hat{S}_{ej}$ are the projections of the hole spin operator $J = 3/2$ [51] and the electron spin operator $S = 1/2$ correspondingly on the crystal lattice cubic axis. In a quantum well (or natural QD), confinement shifts the light-hole exciton to energies much higher than the exchange energies, and the light- and heavy-hole excitons are largely decoupled. In this case the exchange interaction in the heavy-hole exciton can be written with the help of projection, $\hat{S}_{hj}$ of an effective spin 1/2 operator, the Hamiltonian can be rewritten as [44]

$$\hat{H}_{exchange} = - \sum_{i=x,y,z} (a_i \hat{J}_{h,i} \hat{S}_{e,i} + b_i \hat{J}_{h,i}^3 \hat{S}_{e,i}) \approx \sum_{i=x,y,z} c_i \hat{S}_{e,i} \hat{S}_{h,i}. \tag{6.2}$$

In this representation the heavy-hole wave functions $|J_z = \pm 3/2\rangle$ transform to the pseudospin $|S_{h,z} = \mp 1/2\rangle$ wave functions as done in [40]. This transformation changes the sign of the sum in the second equality. It also keeps the total angular momentum projection of the bright exciton state onto the z-axis as $|M| = 1$.

Historically, two different approaches have been developed in the literature for the transformation from the heavy-hole wave function to the pseudo-spin representation. Besides the Ivchenko–Pikus representation that we used above, there is another in which the $|J_z = \pm 3/2\rangle$ heavy-hole wave function transforms to the pseudo-spin $|S_{h,z} = \pm 1/2\rangle$ wave function as introduced by Van Kesteren [44]. To switch from one representation to the other, one should replace $\hat{S}_{h,\alpha}$ by $-\hat{S}_{h,\alpha}$ in the Hamiltonian. Here we use the Ivchenko–Pikus representation.

It is convenient to parameterize the coefficients in terms of the energy splittings of the exciton states in an anisotropic QD [40]:

$$\hat{H}_{exchange} = \frac{\delta_0}{2} \hat{\sigma}_z^e \hat{\sigma}_z^h + \frac{\delta_b}{4} (\hat{\sigma}_x^e \hat{\sigma}_x^h - \hat{\sigma}_y^e \hat{\sigma}_y^h) + \frac{\delta_d}{4} (\hat{\sigma}_x^e \hat{\sigma}_x^h + \hat{\sigma}_y^e \hat{\sigma}_y^h), \tag{6.3}$$

where we have products of the Pauli matrices ($\hat{\sigma}_\alpha^{(e,h)} = 2\hat{S}_{(e,h),\alpha}$) acting on electron and heavy-hole spin variables, and $\delta_{0,b,d}$ are the exchange interaction constants in anisotropic QDs (Fig. 6.6).

Equation (6.3) can be written explicitly with the exciton states ($|+1\rangle$, $|-1\rangle$, $|+2\rangle$ and $|-2\rangle$) as a basis, giving [46]:

$$\hat{H}_{Exchange} = \frac{1}{2} \begin{pmatrix} \delta_0 & \delta_b & 0 & 0 \\ \delta_b & \delta_0 & 0 & 0 \\ 0 & 0 & -\delta_0 & \delta_d \\ 0 & 0 & \delta_d & -\delta_0 \end{pmatrix}. \tag{6.4}$$

Diagonalization of Eq. (6.4) gives the zero field energy levels and states shown in Fig. 6.6(c). The degeneracy of these levels is split due to the exchange interaction into two closely spaced doublets. As can be seen from Eq. (6.4), δ_0 is the energy splitting between the dark ($|\pm 2\rangle$) and bright ($|\pm 1\rangle$) sublevels, and $\delta_b(\delta_d)$ is the energy splitting between the bright (dark) sublevels themselves. For the weak disk-like QDs considered here, the magnitude of δ_0 depends most strongly on the well width. For a 2.8 nm quantum well this splitting is about 150 μeV [43].

The weak lateral confinement plays a minor role except for the case in which the shape of the lateral confinement is not axially symmetric. In this case, $\delta_b \neq 0$, and the remaining degeneracy is lifted. More explicitly, in a quantum well the Bloch functions of the optically active states have symmetries like

$$|+1\rangle = (|X+iY\rangle\uparrow)(|1s\rangle\downarrow)$$
$$|-1\rangle = (|X-iY\rangle\downarrow)(|1s\rangle\uparrow),$$
(6.5)

where the first and second parentheses contain Bloch functions of the heavy-hole and electron, respectively. Directions of the two perpendicular axes (X and Y) in the QW plane can be selected arbitrarily. These Bloch functions contain the spatial and spin functions and determine the circularly polarized selection rules.

Within the elongated QD the bright states are mixed together by the long-range part of the exchange interaction, which is sensitive to shape, and become

$$|X\rangle = (|+1\rangle + |-1\rangle)/\sqrt{2}, \quad |Y\rangle = -i(|+1\rangle - |-1\rangle)/\sqrt{2}.$$
(6.6)

where X and Y are no longer arbitrary directions. For example, in ellipsoidal QDs they coincide with the major and minor axis of an ellipsoid. In rectangular QDs they coincide with the side directions. In our sample they are [110] and [$\bar{1}$10] directions, and the exciton states are dipole active along these axes.

To summarize, this mixing leads to three observable effects. First, there is energy splitting of the two bright states. Second, the mixing of the two bright states leads to linear polarization. Third, the spin states of the electron are mixed and any projection of the electron and hole spins is no longer a good quantum number. As a result only coherent superpositions of these eigenstates have a magnetic moment (e.g. $|X\rangle + i|Y\rangle$, etc), which oscillate with frequencies $\delta_{0,b,d}/\hbar$. An external longitudinal magnetic field restores the circular polarization and the spin of the electron when the field is sufficiently strong so that the Zeeman interaction is much larger than the exchange interaction.

Long-range exchange interaction

The formalism of the last section provides a parameterization of the exchange splittings. We now discuss the magnitude and origin of the bright state splitting (δ_b). It turns out that calculations of the magnitude of the long range exchange interaction can account for the experimental results. The magnitude of splitting between the bright and the dark states (δ_0) will be discussed in the next section.

When individual QDs are resolved in single QD spectroscopy and the PL is analyzed with linear polarizers, it is found that the sharp lines often are strongly polarized doublets with splittings on the order of tens of μeV, as expected from the discussion above (Fig. 6.7), and with small differences in the intensities [41].

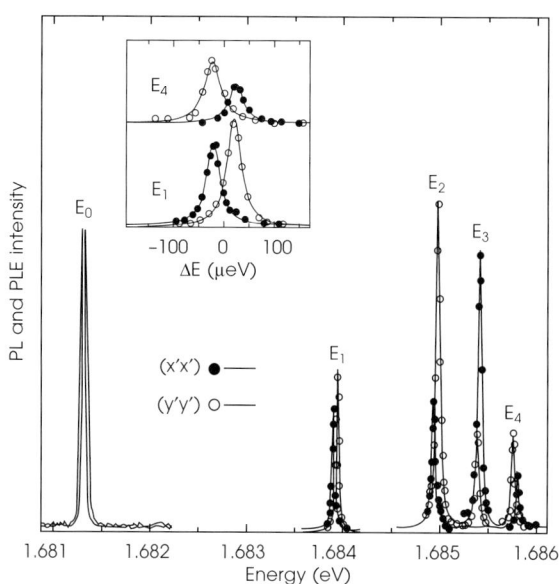

Fig. 6.7 PL and PLE spectrum of the QD indicated by the arrow in Fig. 6.5. E_0 is the ground state PL while $E_1 - E_4$ are the first four excited states as measured by excitation spectroscopy. Polarization configurations are for excitation and detection both along x' or y' ([110] crystal axes). The inset shows an expanded view of two of the PLE resonances. From [41].

More information is gained from the polarization dependence of the *PL excitation spectra*. Two linearly polarized PL excitation spectra of the QD of Fig. 6.7, in which the polarization of the laser and the detected PL are parallel and along the ⟨110⟩ crystallographic orientations, are shown in Fig. 6.7. It is seen that each of the excitation resonances is also a linearly polarized doublet. The sign and magnitude of the splittings vary from line to line (inset to Fig. 6.7).

The spectra shown in Fig. 6.7 were taken with polarizations along the ⟨110⟩ axes. If the polarization of exciting and emitted light are rotated to lie along the ⟨100⟩ axes, this complete polarization is no longer observed. With polarizations along the ⟨100⟩ each component of the doublet is observed in both parallel polarization configurations, and intensities in the crossed configurations are comparable to the parallel. From this it was concluded that the polarization of the QD is linearly polarized with principal axes along the ⟨110⟩. This result is not surprising in view of the elongation of the typical interface islands along the [$\bar{1}$10] axis as seen in Fig. 6.3.

There are differences in the PL intensities for the two polarizations of the exciton ground states (≤20%) and much larger differences for some of the excited states (Fig. 6.7). These results likely arise from mixing of the light and heavy holes due to the QD potential [52,53,50].

The fine structure splittings seen in Fig. 6.7 arise from the exchange interaction between the electron and hole as mentioned above. The long-range part of this interaction is sensitive to the shape of the wavefunction. Explicit calculations of the magnitude of the energy splitting due to the long-range exchange interaction were carried out using a simple model for the exciton in an elongated rectangular quantum dot, which though small for narrow quantum wells does lead to a significant polarization for some excited states [48,49,50]. In this model the exciton is localized laterally in a rectangular potential with dimensions L_x and L_y. Results for the fine structure splitting [δ_b] as a function of L_y, keeping L_x fixed, are reproduced in Fig. 6.8. The calculated results for both the magnitudes and signs of the fine structure splittings are in relatively good agreement with the experimental data [48].

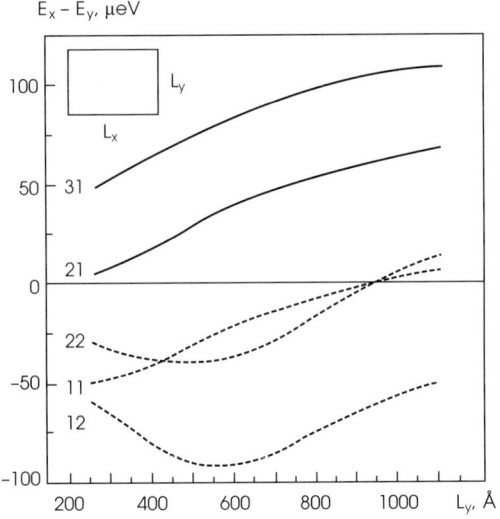

Fig. 6.8 Calculated fine structure splitting arising from the long range part of the exchange interaction as a function of the elongation of a rectangular QD potential. The z- and x-dimensions are kept fixed at $L_z = 2.8$ nm and $L_x = 95$ nm, respectively. Results are shown for the lowest energy exciton (11), corresponding to E_0, and excited states (nm), n,m >1. From [48].

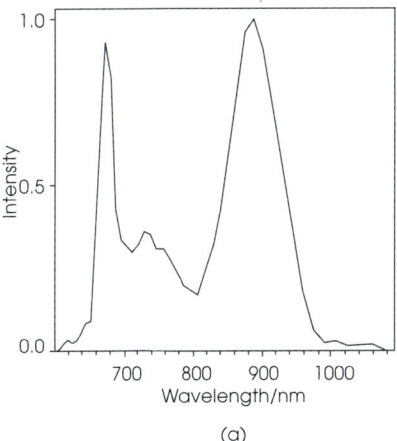

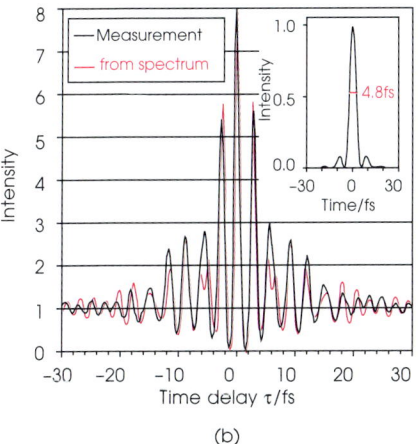

Plate 1 Experiment: (a) measured laser spectrum, (b) measured interferometric autocorrelation. The red curve in (b) is the autocorrelation computed from the spectrum, (a), under the assumption of a constant spectral phase (no chirp). The inset in (b) depicts a 4.8 fs full width at half maximum real time intensity profile computed under the same assumptions.

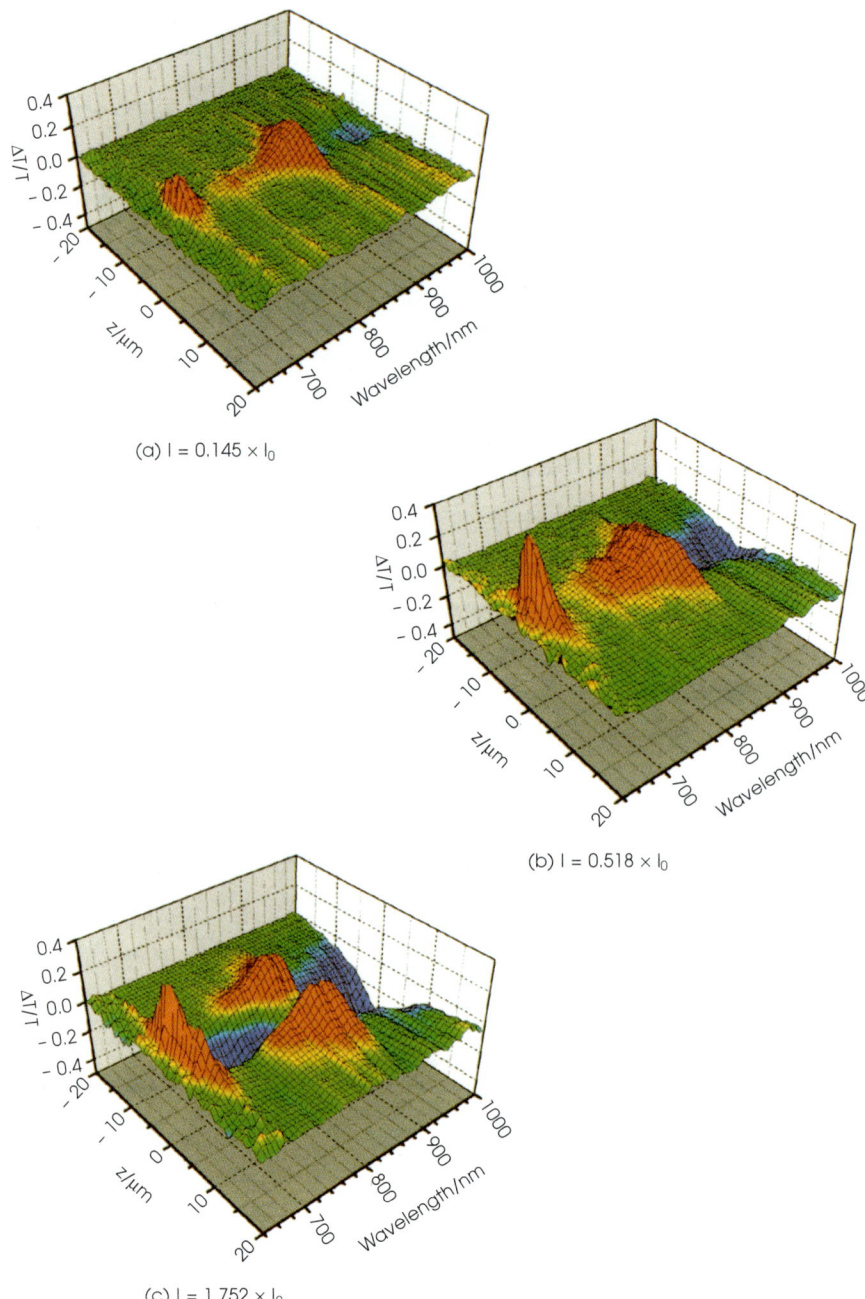

Plate 5 Differential transmission $\Delta T/T$ as a function of the sample coordinate z for three different incident intensities I (referring to $z = 0$). (a) $I = 0.145 \times I_0$, (b) $I = 0.518 \times I_0$, (c) $I = 1.752 \times I_0$.

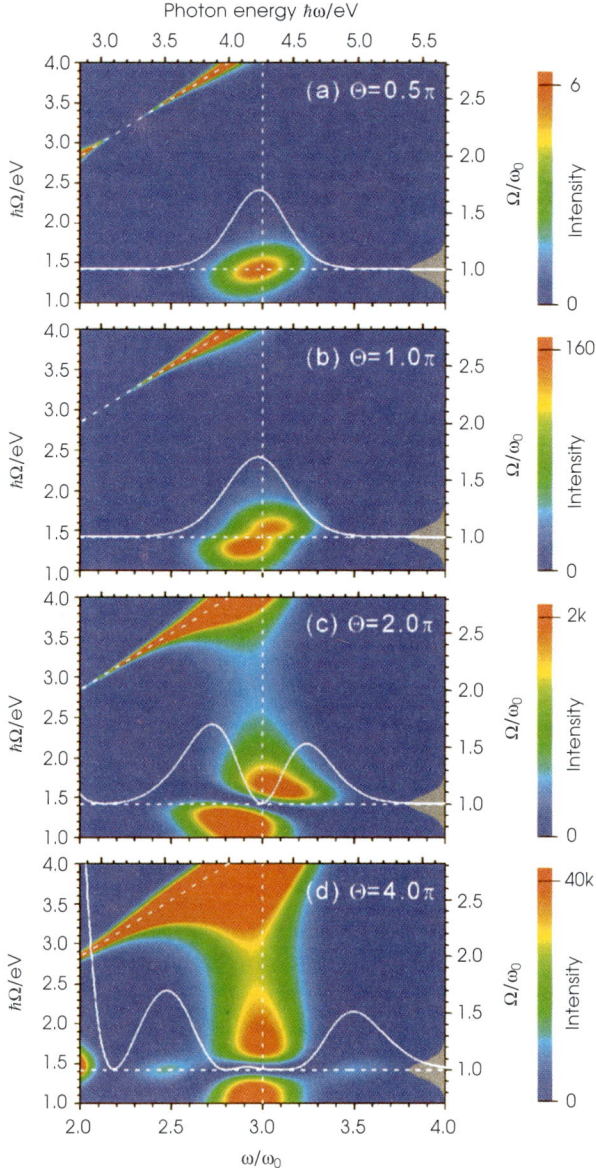

Plate 6 Theory: False color plot of the intensity as a function of photon frequency ω and transition frequency Ω. The center frequency ω_0 of the optical pulses (see grey areas on the RHS) is centered at the band gap frequency, i.e. we have $\hbar\omega_0 = E_g$. The spectrum for a transition right at the band gap, i.e. $\hbar\Omega = E_g$, is highlighted by the white curve. The diagonal dashed line corresponds to $\Omega = \omega$. Excitation with sech^2-shaped 5 fs pulses. The envelope pulse area Θ is indicated and increases from (a) to (d).

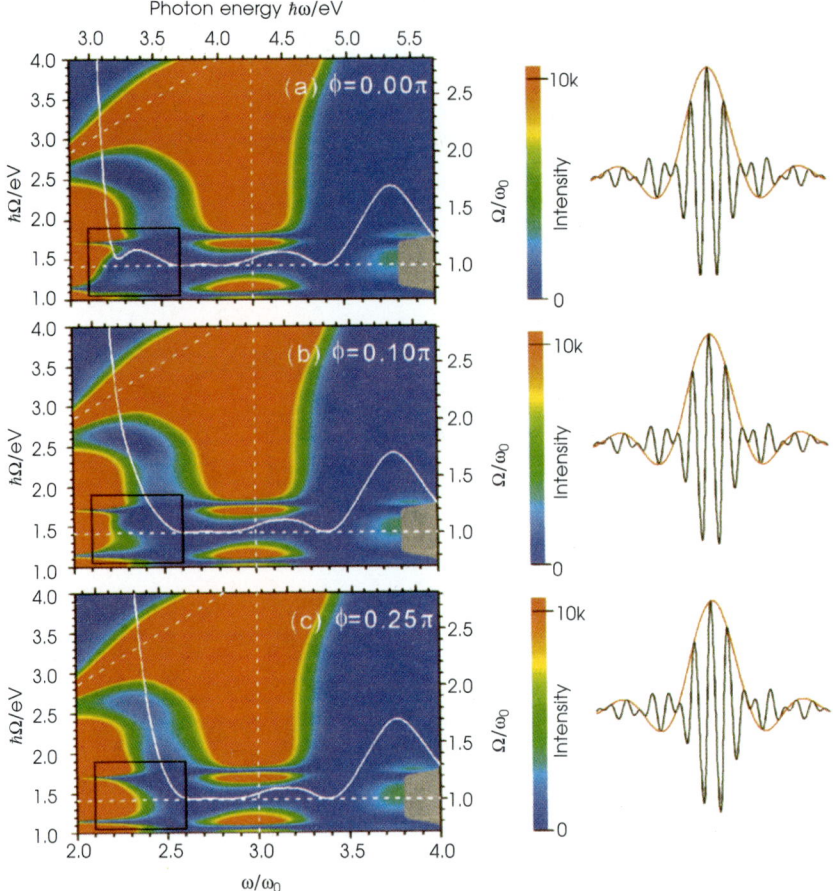

Plate 7 Theory: As Plate 6, however for sinc^2-shaped t_{pulse} = 5.6 fs pulses with envelope pulse area Θ = 4.0π. The optical phase ϕ, i.e. the phase between carrier envelope and carrier-wave, is parameter. Note that significant changes occur within the black rectangles. This might be observable in future experiments. (a) ϕ = 0.00π, (b) ϕ = 0.10π, and (c) ϕ = 0.25π. The corresponding electric fields versus time are depicted on the RHS. The red curves are the field envelopes.

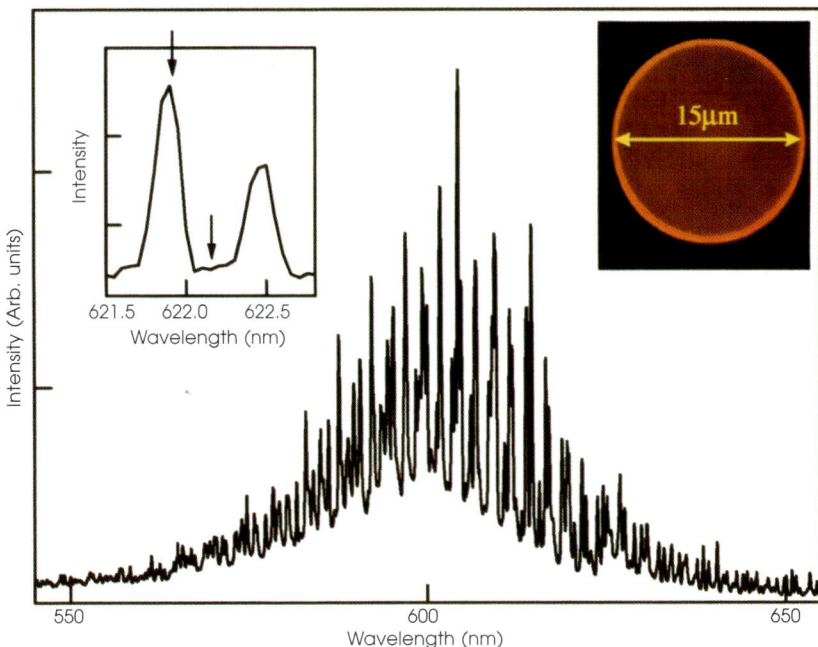

Plate 8 Photoluminescence spectrum from CdSe/ZnS core/shell nanocrystals with $R = 2.7$ nm embedded in the interior surface of a polystyrene sphere. The right inset shows an optical image of a doped polystyrene sphere. The left inset shows an expanded plot of the spectrum near 622 nm. The arrows in the inset indicate the spectral positions used for the time-resolved PL in Fig. 8.7. (From [32] with permission.)

Plate 9 An optical image of a fused silica microsphere with CdSe/ZnS core/shell nanocrystals deposited on the surface of the microsphere. The radius of the microsphere is near 50 μm.

A more complete theoretical study of this effect that included also the mixing of the heavy and light hole was published recently [50]. By including this mixing it was possible to account for the observed differences in intensity in the different spectral lines. Agreement of this theory with experiment was found.

An intuitive picture of the long range exchange interaction can be obtained from the following. The magnitude of the exchange term is given by

$$\delta_b \sim \frac{e^2}{\varepsilon} \iint dr\, dr'\, \frac{\psi_e^*(r)\psi_h(r)\psi_e(r')\psi_h^*(r')}{|r-r'|}, \tag{6.7}$$

which can be separated into a short range and long range part [50,54]. The long range part corresponds physically to the Coulomb interaction between different unit cells in the QD. In lowest order in a multipole expansion, it reduces to the interaction between transition dipole moments [50]:

$$\delta_b^{LR} = \frac{e^2}{\varepsilon} \sum_{r_e \neq r_e'} f_{t\sigma}^*(r_e, r_h) f_{t\sigma}^*(r_e', r_h') \vec{u}_{ct,v\sigma} \frac{[1 - 3\vec{n} \cdot {}^t\vec{n}]}{|r_e - r_e'|^3} \vec{u}_{v\sigma',ct'} \tag{6.8}$$

where $f_{\tau,\sigma}(r_e, r_e)$ are the envelope functions, $\vec{n}$ is the unit vector between dipole positions, and $\vec{u}_{ct,v\sigma}$ is the transition dipole moment between electron and hole Bloch functions ($\phi_{c\tau R(r)}$),

$$\vec{u}_{c\tau,v\sigma} = \int d^3r\, \phi_{c\tau R}^*(r)(\vec{r} - \vec{R})\phi_{v\sigma R}(r). \tag{6.9}$$

Equation (6.8) for the long range part of the exchange lends itself to a useful classical analogy. This equation is the interaction energy between two polarization densities arising from the transition dipoles. Takagahara [50] has plotted the polarization densities for several of the exciton transitions for a QD modeled on the one discussed in Fig. 6.7.

Taking this classical model a little further, one can describe the long range part of the exchange interaction in terms of the interaction of the polarization density associated with the transition dipole moment with a *depolarization field*, a concept often used in the physics of dielectrics [55]. Eq. (6.8) can be written as [54]

$$\delta_b^{LR} \sim \int dr\, E_{\text{dep}}(r) P(r), \tag{6.10}$$

in which $E_{\text{dep}}(r)$ is the depolarization field that accompanies the transition dipole density of the exciton. Equation (6.10) is the electrostatic equation for the depolarization energy associated with a dielectric medium. This formalism provides an intuitive method for visualizing the long-range exchange interaction in terms of a depolarization field associated with a dipolar field, in analogy with the picture often used for a dielectric medium in a capacitor or the optical phonon in a polar medium [54]. The additional energy of the depolarization field can be visualized as an interaction between surface charges set up by the depolarization field (see Fig. 6.9). This electrostatic picture is also helpful in understanding the large exchange shift (of order 1 meV) in the energy of the z-polarized light-hole exciton compared to the x- or y-polarized light-hole exciton in a 2-D quantum well [56]; the depolarization of a dipole density normal to a thin dielectric layer is large and constant, whereas that for polarization in the plane is small. Similar pictures are used in other areas of physics also. For example, depolarization fields are helpful in intuitively understanding the increase in energy of the longitudinal optical phonon with respect to the transverse phonon in polar materials. In fact, it is possible to intuitively understand the anisotropy in LO-TO splitting in quantum wells in terms of their depolarization field [57]. It is also useful in the case of intersubband transitions of the quasi-2D electron gas in quantum wells [58].

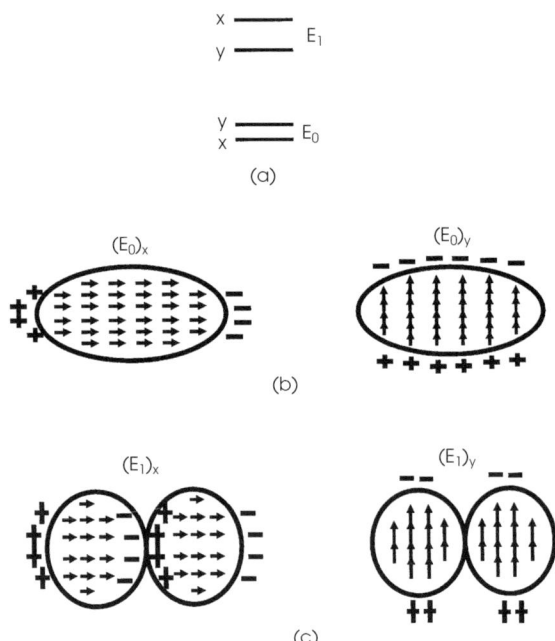

Fig. 6.9 (a) Schematic energy level diagram showing the order of the fine structure splitting of the first two optically allowed exciton states. Diagrams of the dielectric response associated with a polarized exciton for both the (b) lowest exciton state and the (c) first excited exciton state. This picture leads to an appreciation of the relative magnitudes of the energy associated with the depolarization field and explains the flip in sign of the fine structure splitting for the two states.

This picture allows one to understand intuitively why the energy of the exciton depends on the shape of the quantum dot. The exciton polarized along the long axis of the dot will have a smaller depolarization field and therefore a smaller energy than that polarized along the short axis. In a similar manner the larger splittings of the excited states and the reversal in sign can be understood (Fig. 6.9).

The diagonal part of the exchange interaction in Eq. (6.4), the term that gives the splitting between the bright and dark states (δ_0), is dominated by the short range part of the exchange interaction. This corresponds to the part of the exchange Coulomb interaction within a single unit cell and summed over all unit cells with a weight determined by the exciton envelope function [50,43]:

$$\delta_0 \propto \int d^3r f_{\tau\sigma}^*(r,r) f_{\tau'\sigma'}(r,r). \tag{6.11}$$

In the next section we show how this exchange parameter can be measured directly in tilted magnetic fields.

Zeeman interaction

Let us consider the dynamics of a spin's motion in a magnetic field, B. The magnetic moment, $\mathbf{M}$, of the spin, $\mathbf{S}$, interacts with the magnetic field. This interaction is described by the Hamiltonian,

$$\hat{H}_{\text{Zeeman}} = -(\hat{\mathbf{M}}\mathbf{B}) = \mu_B g (\hat{\mathbf{S}}\mathbf{B}), \tag{6.12}$$

where

$$\mu_B = \frac{|e|\hbar}{2m_e c} \approx 5.795 \cdot 10^{-5} \frac{eV}{T} \approx 60 \mu eV/T$$

is the Bohr magnetron, and g is the particle's g-factor. For free electrons the electron g-factor $g \approx 2$. Often $\hat{\mathbf{M}} = -\mu_B g \hat{\mathbf{S}}$, but in the general case of low symmetry structures the g-factor may be an anisotropic tensor of the second rank with elements whose magnitudes depend on the composition and size of the structure.

The interaction of the magnetic moment with the magnetic field leads to spin precession around the direction $g\mathbf{B}$ with a frequency, $\Omega = \mu_B g B/\hbar$:

$$\frac{d\hat{\mathbf{S}}}{dt} = [\mathbf{\Omega} \times \hat{\mathbf{S}}]. \tag{6.13}$$

The spin precession of particles is also affected by the magnetic fields of other particles, and in general the field is the vector sum of the external field and the local field. For the electron spin, these local fields are the exchange fields of holes, other electrons and paramagnetic impurities, as well as the hyperfine field of nuclei. The average value of these fields leads also to the spin precession of the carriers, and their fluctuating parts result in spin polarization relaxation. Spin–orbit interactions also affect the spin relaxation because the elastic or inelastic scattering of the particles changes their spin orientation. Here we consider the effect of the external field and the average of the exchange field. Fluctuations in the local fields will be treated in sections 6.5 and 6.6.

In the presence of both exchange and an external magnetic field, the total spin Hamiltonian is $\hat{H} = \hat{H}_{\text{exchange}} + \hat{H}_{\text{Zeeman}}$. In the bulk the Zeeman Hamiltonian is given by

$$\hat{H}_{\text{Zeeman}}^{\text{bulk}} = \mu_B \sum_{i=x,y,z} [g_i^e \hat{S}_{e,i} - 2(\kappa \hat{J}_{h,i} + q \hat{J}_{h,i}^3)] B_i. \tag{6.14}$$

where g^e is the electron g-factor and κ and q are the Luttinger parameters that describe the Zeeman effect in the degenerate valence band of zinc-blende semiconductors. B_i is the projection of the magnetic field along the crystal axes. Confinement shifts the light-holes to energies large compared to the Zeeman energies, and the spin hole operator projection, $\hat{J}_{h,j}$ ($J = 3/2$) can be replaced by a pseudo-spin heavy-hole operator projections $\hat{S}_{h,j}$ ($S = 1/2$), as discussed above on the exchange interaction. As a result, the Zeeman Hamiltonian in the Ivchenko–Pikus representation is written as:

$$\hat{H}_{\text{Zeeman}} = \mu_B \sum_{i=x,y,z} (g_i^e \hat{S}_{e,i} + g_i^{hh} \hat{S}_{h,i}) B_i = \frac{\mu_B \vec{B}}{2} \cdot (\overline{\overline{g}}^e \cdot \hat{\sigma}^e + \overline{\overline{g}}^h \cdot \hat{\sigma}^h). \tag{6.15}$$

The g-factor tensor ($\overline{\overline{g}}$) has only diagonal terms in the reference coordinate system given by the symmetry axes of the QD, in our case the crystal axes [001], [110], and [$\bar{1}$10]:

$$\overline{\overline{g}}^e = \begin{pmatrix} g_x^e & 0 & 0 \\ 0 & g_y^e & 0 \\ 0 & 0 & g_z^e \end{pmatrix}; \quad \overline{\overline{g}}^h = \begin{pmatrix} g_x^{hh} & 0 & 0 \\ 0 & g_y^{hh} & 0 \\ 0 & 0 & g_z^{hh} \end{pmatrix}, \tag{6.16}$$

where $g_x^e = g_y^e$, and g_x^{hh}, g_y^{hh} are usually negligibly small [44]. The direct measurements of the perpendicular component of g-factor show that it is ten times smaller than its longitudinal components [60].

Using the same basis as in Eq. (6.4), ($|+1\rangle$, $|-1\rangle$, $|+2\rangle$ and $|-2\rangle$), this can be written explicitly as

Electronic and nuclear spin in the optical spectra 229

$$H_{Zeeman} = \frac{\mu_B \cdot B}{2} \begin{pmatrix} (g_z^e + g_z^{hh}) \cdot \cos\theta & 0 & g_x^e \cdot \sin\theta & g_x^{hh} \cdot \sin\theta \\ 0 & -(g_z^e + g_z^{hh}) \cdot \cos\theta & g_x^{hh} \cdot \sin\theta & g_x^e \cdot \sin\theta \\ g_x^e \cdot \sin\theta & g_x^{hh} \cdot \sin\theta & (g_z^e - g_z^{hh}) \cdot \cos\theta & 0 \\ g_x^{hh} \cdot \sin\theta & g_x^e \cdot \sin\theta & 0 & -(g_z^e - g_z^{hh}) \cdot \cos\theta \end{pmatrix}$$

(6.17)

where we take $\vec{B}$ in the x-z plane, and θ is the angle between $\vec{B}$ and the z-axis. g_z^e (g_z^{hh}) is the g-factor of the electron (heavy hole) in the z-direction and g_x^e (g_x^{hh}) is the g-factor of the electron (heavy hole) in the x-direction.

In the Faraday ($\theta = 0$) and Voigt ($\theta = 90°$) configurations the total Hamiltonian can easily be diagonalized. It can be seen that in the Faraday configuration the bright and dark states mix only among themselves while in the Voigt geometry all four states mix together. This means that in the Faraday configuration only the two bright states are observed (Fig. 6.10(a)), while in the Voigt configuration dark and bright mixtures can be observed as the magnetic field is increased (Fig. 6.10(b)). The reason for not observing splittings in the latter geometry for the high energy peaks (bright-related states) or low

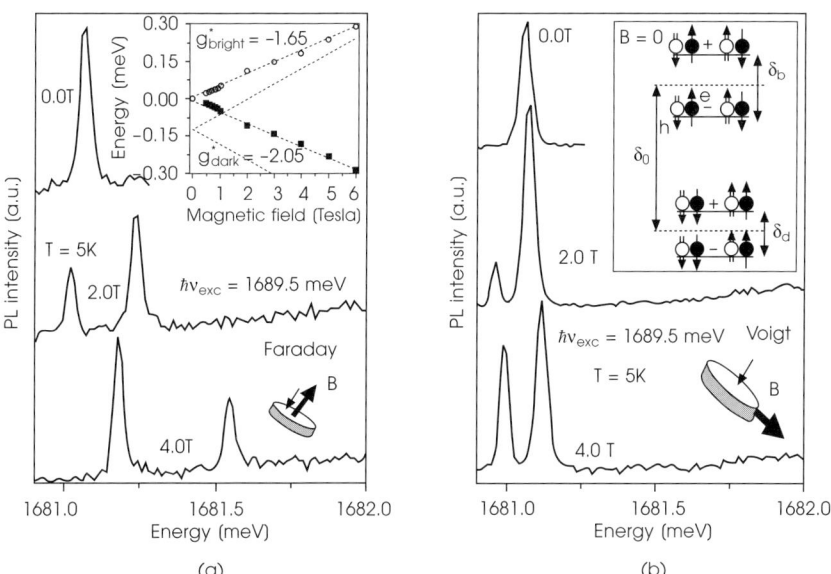

Fig. 6.10 PL spectra of a single QD as a function of magnetic field in the (a) Faraday geometry and the (b) Voigt geometry.

energy peaks (dark-related states) is mainly due to the decrease in the heavy hole g-factor (i.e. $g_x^{hh} \ll g_z^{hh}$).

Furthermore, by working at other polar angles (θ) it is possible to observe all four transitions (Fig. 6.11) and to obtain all the above parameters in Eqs. (6.4) and (6.17). For example we can fix the magnitude of the magnetic field and measure all four states as a function of θ. This is shown in Fig. 6.11 for a fixed magnetic field of 4.0T, where

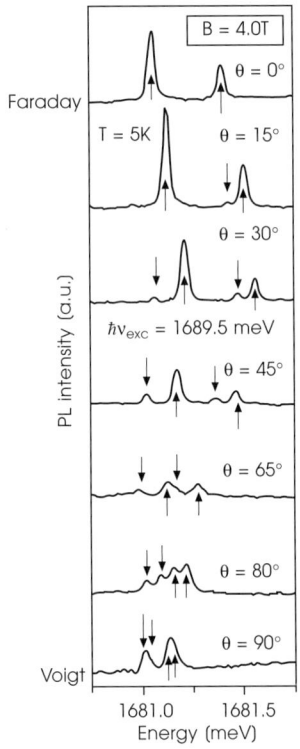

Fig. 6.11 PL spectra of a single QD in a tilted magnetic field as a function of the polar angle (θ) at B = 4 T. In the Faraday geometry (θ = 0) only the two bright states are detected (up arrows), but as the angle is increased the dark-related transitions (down arrows) are turned on.

the bright-related transitions are indicated with upward arrows while the dark-related transitions are indicated by downward arrows. Figure 6.12 is a summary plot of the energies obtained after removing the diamagnetic shift. Open symbols are the bright-related states data while solid symbols are the dark-related energies. The lines show the theoretical fittings obtained by numerical diagonalization of the Hamiltonians, $\hat{H}_{exchange} + \hat{H}_{Zeeman}$ in Eqs (6.92) and (6.17), correspondingly. The parameters utilized

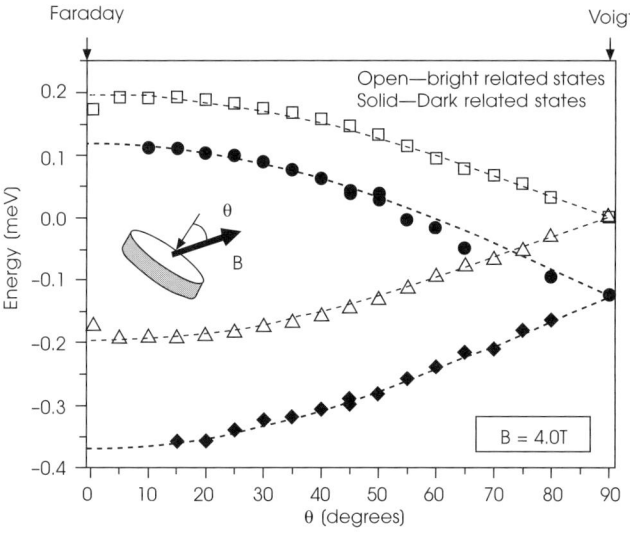

Fig. 6.12 Peak energies as a function of the angle of the magnetic field relative to the z-axis. Taken from the data represented in Fig. 6.11.

in this fitting, for a 3.4 nm quantum well, are $\delta_o = (125 \pm 9)\mu eV$, $\delta_b = (0 \pm 25)\mu eV$, and $\delta_d = (0 \pm 12.5)\mu eV$ for the exchange energies and $g_z^e = (0.20 \pm 0.05)$, $g_x^e = (0.2 \pm 0.1)$, $g_z^{hh} = (-1.85 \pm 0.05)$ and $g_z^{hh} = (0.0 \pm 0.01)$, for the g-factors. In Fig. 6.13 we depict the magnetic field (B) and polar angle (θ) dependence of the exciton states as well as the ground state (vacuum state, $|0\rangle$). As can be seen, the angle dependence shown in Fig. 6.12 is mainly due to the highly anisotropic heavy-hole g-factor. The heavy hole g-factor value in this system varies from -1.85 in the Faraday geometry to approximately zero in the Voigt geometry (but not exactly; see [60]).

After studying several dots we conclude that the g-factor variation from dot to dot is about ±15% and that their values are in good agreement with previous results for quantum wells of the same size [59,60], which indicates that the lateral confinement of the interface fluctuations do not influence the g-factors as strongly as the vertical confinement. Similarly, δ_o does not vary significantly from dot to dot with fixed well width and is in good agreement with previous measurements done in quantum wells [43]. This can be understood by considering that the value of δ_o is given by the overlap of the electron and hole wave function [50] and this overlap is basically ruled by the strong confinement in the quantum well direction (lateral confinement is much weaker). Furthermore, as can be seen in Fig. 6.14, the agreement of the values obtained in this work for δ_o as a function of well width with previous calculations [43] is good, indicating once more the negligible effect of the lateral confinement on δ_0. It is also worth mentioning

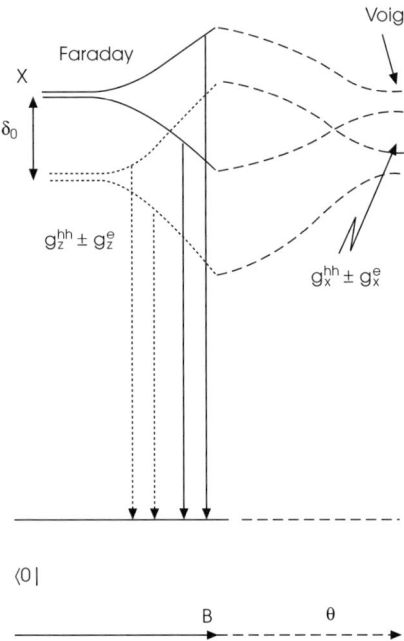

Fig. 6.13 Schematics of the exciton states as a function of magnetic field strength in the Faraday configuration *and* as a function of the polar angle at a constant magnetic field. Notice the decrease of the exciton *g*-factor in the Voigt geometry mainly due to the decrease of the heavy hole *g*-factor.

that, in contrast to ensemble measurements [43], single dot spectroscopy allows a direct measurement of the fine structure splittings, which provide a more accurate and objective determination of the exchange energies. Because δ_b depends on the in-plane elongation of the dot, its value varies significantly from one dot to another. Typical values for δ_b are in the tens of µeV [41]. No finite values for δ_d were observed in this study. In the case of self assembled dots [46] the *g*-factors and exchange energies (δ_o, δ_b and δ_d) are affected by confinement in all three directions since the lateral confinement is comparable to the vertical one.

Pseudo-spin model

In the special cases of zero magnetic field or with field along the *z*-axis, the bright and dark states can be separated into two independent two-level systems, and each can be described simply with the use of a pseudo-spin 1/2 model [61]. The essence of this model is that any two-level system can be modeled as a quasiparticle having pseudo-

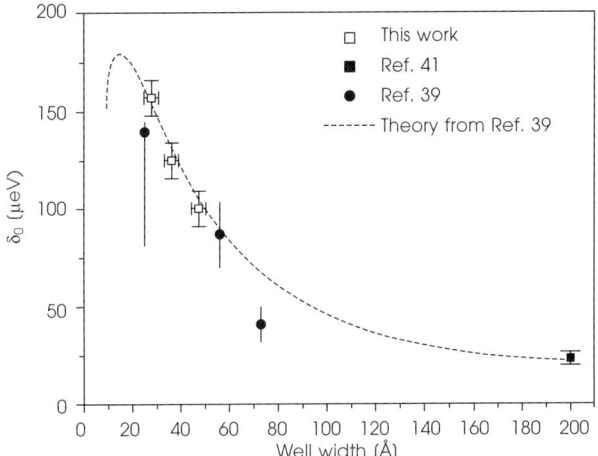

Fig. 6.14 Comparison between experiments and theory of the exchange energy δ_0 as a function of quantum well width.

spin 1/2 and can be represented as a vector in physical space [62]. We outline this useful model.

It is natural to consider the bright states with angular momentum projection +1 and −1 as the states having up $|z, +1/2\rangle$ and down $|z, -1/2\rangle$ projections of the pseudo-spin onto the z-axis in the pseudo-spin space (x, y, z). The z-component, S_z, of the mean (ensemble averaged) pseudo-spin characterizes the population difference between +1 and −1 exciton states and determines the degree of circular polarization of the PL;

$$P_c = \frac{I_{\sigma^+} - I_{\sigma^-}}{I_{\sigma^+} + I_{\sigma^-}} = 2 S_z.$$

The exciton states, $|Y\rangle = -i(|+1\rangle - |-1\rangle)/\sqrt{2}$ and $|X\rangle = (|+1\rangle + |-1\rangle)/\sqrt{2}$ (dipole active along the [110] and [1$\bar{1}$0] axes), are described by the x-component of pseudo-spin, $|x, -1/2\rangle$ or $|x, +1/2\rangle$, respectively. For example, the state $|x, +1/2\rangle = (|z, +1/2\rangle + |z, -1/2\rangle)/\sqrt{2}$ corresponds to $(|+1\rangle + |-1\rangle)/\sqrt{2} = |X\rangle$. The x-projection of the mean pseudo-spin determines the degree of linear polarization with respect to the [110]/[1$\bar{1}$0] axes;

$$P_l = \frac{I_{110} - I_{1\bar{1}0}}{I_{110} - I_{1\bar{1}0}} = -2S_x.$$

The states, $|X'\rangle = (|X\rangle + |Y\rangle)/\sqrt{2}$ and $|Y'\rangle = (-|X\rangle + |Y\rangle)/\sqrt{2}$, polarized along the [100] and [010] directions, correspond to y-components of pseudo-spin +1/2 and −1/2. Hence the y-projection of mean pseudo-spin gives the PL polarization with respect to the [100]/[010] axes:

$$P_{l'} = \frac{I_{100} - I_{010}}{I_{100} + I_{010}} = 2S_y.$$

Polarization of excitons along the [001]-axis (usually giving rise to circularly polarized light) is known as orientation, while those polarized in the [001] plane (usually giving rise to linearly polarized light) is known as alignment.

The Hamiltonian describing this two-level system of bright states is the sum of the exchange and Zeeman terms

$$\hat{H} = \frac{\hbar}{2}(\omega_b \hat{\sigma}_x + \Omega_{ext}\hat{\sigma}_z) = \frac{\hbar}{2}\vec{\Omega}\cdot\hat{\vec{\sigma}}, \quad (6.18)$$

where $\omega_b = \delta_b/\hbar$, and we have included a magnetic field along the z-axis (Faraday geometry) with Larmor frequency, $\Omega_{ext} = \mu_B(g_z^e + g_z^{hh})B/\hbar$. The polarization dynamics of bright excitons in the presence of both magnetic field and anisotropic exchange can be described classically as the precession of the pseudo-spin vector in an effective magnetic field, $\vec{\Omega} = (\omega_b, 0, \Omega_b)$ – that is, the vector sum of the external field and an effective exchange field – with the familiar equation of motion, Eq. (6.13). Thus pseudo-spin, and therefore the optical polarization, can be physically understood in a highly intuitive way in terms of the evolution of this vector in space. Figure 6.15 is a schematic representation of the process.

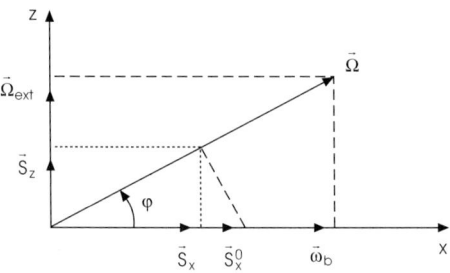

Fig. 6.15 Schematic illustration of pseudo-spin model. $\vec{S}_x^0$ is the initial pseudospin vector under [110] linearly polarized excitation; S_x and S_z are the components of the average pseudo-spin [61].

Consider, for example, excitation by light linearly polarized along the $[\bar{1}10]$ axis. It creates excitons in the $|X\rangle = (|+1\rangle + |-1\rangle)/\sqrt{2}$ state. Using the pseudo-spin model we can say that the quasiparticles with initial pseudospin state $|x, +1/2\rangle$ are excited. Thus the mean pseudo-spin $\vec{S}$ is created initially along the x-axis ($\vec{S} = \vec{S}_x^0$). It rotates around the total vector $\vec{\Omega} = (\omega_b, 0, \Omega_{ext})$ with Larmor frequency $\Omega = \sqrt{\Omega_{ext}^2 + \omega_b^2}$. If the exciton lifetime is much longer than the inverse of this Larmor precession frequency, the average value of the spin component perpendicular to $\vec{\Omega}$ vanishes. Only the spin projection of $\vec{S}$ onto $\vec{\Omega}$ is conserved, $\mathbf{S} = (\mathbf{S}^{(0)}\boldsymbol{\Omega})\boldsymbol{\Omega}/\Omega^2$.

The steady-state pseudo-spin $\vec{S}$ determines the polarization of PL by the relations, $P_c = 2S_z$, $P_l = -2S_x$, $P_{l'} = 2S_y$, which are the same as those for excitation. One can see from Fig. 6.15 that the S_x component decreases when external field ($\Omega_{ext} \neq 0$) is applied, as does P_l (alignment decreases). Simultaneously the z-component of pseudo-spin appears in magnetic field, and P_c appears in the PL emission (circular polarization of excitons). We can say that the alignment is converted into orientation. In a similar way one can consider the excitation by circularly polarized light (orientation) creating initially the quasiparticles with pseudo-spin along z-axis | z, +1/2⟩. The steady-state orientation of excitons is absent ($S_z = 0$) in zero external field due to fast precession (with frequency ω_b) in the effective field of anisotropic exchange interaction. The external field restores the orientation of excitons. Simultaneously the S_x component appears. This leads to the appearance of linear polarization of exciton PL. Thus, in this case we have the conversion from orientation into alignment.

With this model one can get simple relations between the polarization of the exciton PL (P_l, P_l', P_c) and the polarization of exciting light for the case of arbitrary polarization of incident light (P_l^0, $P_{l'}^0$, P_c^0) (see Fig. 6.15 for a graphical derivation). Here we show these relationships for circular and linear excitation (P_c^0 and P_l^0, respectively). S_y (and therefore, $P_{l'}$) are orthogonal to the effective magnetic field and averages to zero.

$$P_l = P_l^0 \frac{\omega_b^2}{\omega_b^2 + \Omega_{ext}^2}, \quad P_c = -P_l^0 \frac{\omega_b \Omega_{ext}}{\omega_b^2 + \Omega_{ext}^2} \tag{6.19}$$

$$P_c = P_c^0 \frac{\Omega_{ext}^2}{\omega_b^2 + \Omega_{ext}^2}, \quad P_{l'} = -P_c^0 \frac{\omega_b \Omega_{ext}}{\omega_b^2 + \Omega_{ext}^2} \tag{6.20}$$

This simple model has been used to describe experimental results [61] (see also [63,64]). The model allows one to find the value of the exchange splitting of the bright states, characterizing the fine structure of bright excitons. Typical values are of the order of a few μeV for type-II quantum wells and tens of μeV for type-I quantum wells. In the more tightly confined self-assembled QDs, values on the order of 100 μeV have been reported [64,65].

A magnetic field in the Faraday geometry restores the optical orientation (circular polarization) of exciton states and simultaneously suppresses the optical alignment (linear polarization) of excitons. As a result the degree of circular polarization of the PL under circularly polarized excitation (orientation of excitons) is also restored. Similarly, the magnetic field decreases the linear polarization of PL under linearly polarized excitation (alignment of excitons). In the intermediate field region the bright states are not purely circular or linear, but some superposition. In this case the new effect of conversion of orientation to alignment and vice versa is possible, as can be seen from Figs 6.16(a)–(d) [61,63,64]. Namely, the excitation by circular polarized light induces the linearly polarized PL (orientation-to-alignment conversion), whereas the linearly polarized excitation leads to circularly polarized PL (alignment-to-orientation conversion).

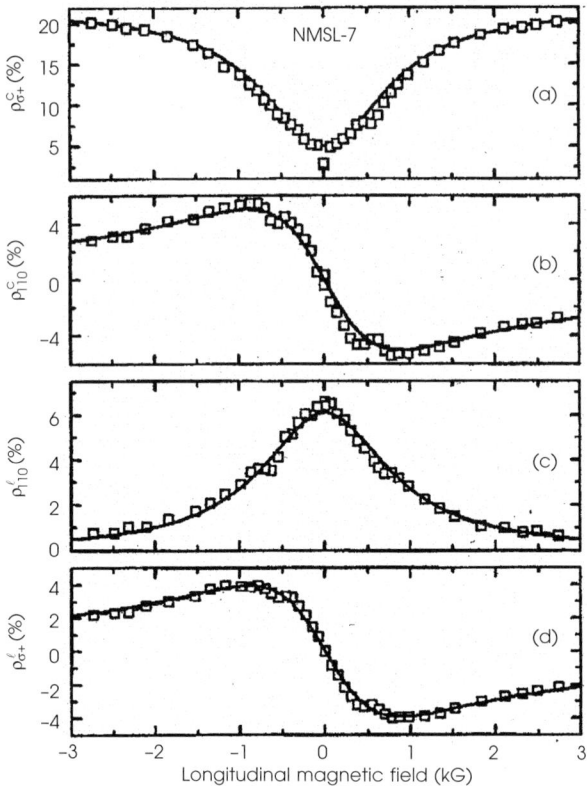

Fig. 6.16 Effect of magnetic field in Faraday geometry on optical orientation and alignment of excitons localized in type-II GaAs/AlAs superlattices [61]. (a) optical orientation: circular polarizer and circular analyzer, (b) orientation-to-alignment conversion: circular polarizer and linear analyzer, (c) optical alignment: linear polarizer and linear analyzer, (d) alignment-to-orientation conversion: linear polarizer and circular analyzer.

We considered the case of bright excitons above, however, the same can be done for the dark states. The electron-hole exchange interaction also mixes the +2 and −2 dark states. Therefore, spin orientation of dark excitons is suppressed in zero magnetic field as it was in the case of bright excitons, and a magnetic field in Faraday geometry will restore the polarization of dark excitons. However, this effect is difficult to observe because dark states do not contribute directly to PL due to selection rules. Nonetheless, the fine structure of dark states has been found in self-organized InP/InGaP QDs containing additional electrons as discussed in section 6.4 [65].

Relaxation

Relaxation can be introduced into the pseudo-spin model providing an intuitive way to picture finite relaxation (or linewidth) versus precession. The evolution of the pseudo-spin vector, S, is described by the equation of motion in the total effective magnetic field $\vec{\Omega} = (\omega_b, 0, \Omega_{ext})$, with relaxation included through phenomenological damping terms,

$$\frac{d\mathbf{S}}{dt} = [\mathbf{\Omega} \times \mathbf{S}] - \frac{\mathbf{S}}{\tau_s} + \frac{\mathbf{S}_0 - \mathbf{S}}{\tau}. \tag{6.21}$$

The relaxation terms are governed by the lifetime of the (nonequilibrium) exciton (τ), and the spin relaxation between the two levels (τ_s). In the case of the bright exciton doublet the spin relaxation time is the effective spin relaxation time which includes scattering via the dark states as intermediate states [42]. $\mathbf{S}_0$ is the spin polarization generated by excitation. This well known Bloch equation is often used to describe the suppression of optical orientation by an external magnetic field (Hanle effect).

The average spin for arbitrary directions of initial polarization and external magnetic field is found from Eq. (6.21) under the steady-state condition, $\frac{d\mathbf{S}}{dt} = 0$:

$$\mathbf{S} = \frac{\mathbf{S}_i + (\mathbf{S}_i \mathbf{\Omega})\mathbf{\Omega} T_s^2 + [\mathbf{\Omega} \times \mathbf{S}_i] T_s}{1 + (\Omega T_s)^2}, \tag{6.22}$$

where the total spin relaxation time is $T_s^{-1} = \tau^{-1} + \tau_s^{-1}$. The steady-state polarization of the pseudo-spin, in the absence of precession, depends on the ratio of the exciton lifetime to the spin relaxation, $\mathbf{S}_i = \mathbf{S}_0/(1 + \tau/\tau_s)$. For example, if the exciton lifetime is long, spin flip processes will lead to low average pseudo-spin polarization. Equation (6.22) connects the average and initial value of exciton pseudo-spin, or in other words, the polarization of the photoluminescence with initial polarization of carriers created by the exciting light. The two-level description remains appropriate if mixing with the dark states remains negligible, i.e. in tilted fields we must have, $\Omega_\perp^{ext} \ll \delta_0/\hbar$. Otherwise, the dynamics of the coupled four-level system must be considered. Also, if the external field becomes large then the spin relaxation rates for the pseudo-spin components parallel and perpendicular to the magnetic field may differ, in which case the above vector equation (6.21) must be replaced by equations of motion for the individual vector components.

Polarization including finite relaxation

It is obvious that fine-structure splitting in a homogeneously broadened spectral line is resolved only when the fine structure splitting is larger than the homogeneous linewidth. The homogeneous linewidth is determined by the total relaxation rate of the pseudo-spin ($\gamma = \hbar T_s^{-1} = \hbar(\tau^{-1} + \tau_s^{-1})$ where τ is the lifetime and τ_s is the spin relaxation time), if there is no pure dephasing. Therefore, thinking in terms of the pseudo-spin vector, well resolved fine structure splitting occurs when the precession frequency of the vector is larger than the relaxation rate of the pseudo-spin. We will show how to calculate the *polarization* of the pseudo-spin in this case of only partially resolved fine-structure splitting.

An exciton with initial polarization ($\vec{S}_i$) processes around the effective field, $\vec{\Omega}$ = (ω_b, 0, 0). Here we make the external field equal to zero for clarity. If $\omega_b \gg T_s^{-1}$ then we can neglect relaxation, and we obtain the picture described previously. In this case, $\vec{\Omega}$ is aligned along the x-axis at zero field, and the average of the pseudo-vector over many precession periods is aligned along the x-axis. Therefore, the polarization is linear. The emission spectrum would consist of a linearly polarized doublet with splitting ($\Delta E = \hbar\omega_b$).

However, if $\omega_b \leq T_s^{-1}$, then the exciton pseudo-spin does not complete a precession period before it relaxes or recombines, and the emitted light is no longer purely linear, but elliptical. Of course, we must average over many optical lifetime cycles in the measurement of the spectrum. This corresponds to averaging over the distribution in exciton lifetimes, $W(t) = \tau^{-1}\exp(-t/\tau)$. Thus, taking the example of circularly polarized generation, the emission polarization is a mixture of polarizations, with an average circular polarization given by

$$P_c = 2S_z = 2\int_0^\infty S_z^i \cos(\omega_b t) W(t) \, dt = \frac{2 S_z^i}{1 + (\omega_b T_s)^2} = \frac{2 S_z^0 (1 + \tau/\tau_s)^{-1}}{1 + (\omega_b T_s)^2}.$$

The average linear component is found in a similar fashion. The same result can be obtained directly from Eq. (6.22).

Therefore, the average emitted polarization is mixed and dependent on the initial polarization when the linewidth is comparable to the anisotropic exchange splitting. The classical picture is that of a spin generated along z, beginning to precess, but recombining before completing a period and before losing its polarization; and then being regenerated along the z-direction. After averaging over many cycles the average spin is determined by the phase angle, $\omega_b \tau$. In the limit when the exchange splitting is much less than the linewidth, the emission has the same polarization as the excitation.

In an inhomogeneous system of quantum dots in which the anisotropic exchange splitting varies and is greater or less than the linewidth, each line will have a polarization dependence that depends on how well the lines are resolved; i.e. on the relative magnitudes of ω_b and the homogeneous linewidth, γ. To find the polarization of the ensemble of QDs, the result for an individual QD is averaged over the distribution in exchange energies. In addition, if the average lifetime varies appreciably from dot to dot, this also is averaged over. This may explain the result that, although an ensemble measurement can give little linear polarization, some individual exciton lines can be strongly linearly polarized with a doublet structure [41].

We note that the average over lifetime is an example of the general theoretical concept of the *correlation time* (τ_c). Often it is useful to consider the microscopic evolution of the spin under a specific interaction (perhaps modeled as a precession in an effective magnetic field as is done here). However, the finite lifetime, or more generally, the random fluctuation of the environment of the QD allows only a limited time for this coherent evolution to proceed. It could be that the microscopic origin of this dephasing process is complicated or not important. In this case random fluctuations often are introduced phenomenologically in a way similar to that above. The exciton lifetime, then, is replaced with a correlation time, which may or may not be microscopically derived. Examples of this will be given in sections 6.5 and 6.6.

Hanle effect

It is interesting to compare the polarization behavior of the exciton with the well-known Hanle effect of electron spin in semiconductors in a transverse magnetic field. As we have just discussed, the polarization dependence varies as the line splitting becomes larger than the homogeneous linewidth. Thus, a useful *cw* method to measure the relaxation time (and/or the precession frequency – that is, the anisotropic exchange interaction or *g*-factor) is to measure the polarization as the splitting is changed – typically by changing a magnetic field. This effectively compares the spin relaxation rate (T_s^{-1}) with the frequency of spin precession in a magnetic field (Ω). For an electron (e.g., in bulk GaAs [11]) the usual Hanle effect is obtained by turning on a transverse magnetic field. In this case the initial polarization is circular (along the *z*-axis) and perpendicular to the total magnetic field (along the *x*-axis). Equation (6.22) reduces to

$$S_z = \frac{1}{1+(\Omega T_s)^2} \frac{S_0}{1+\tau/\tau_s}; \quad \text{for } \mathbf{S}_0 \parallel \hat{z}; \; \mathbf{\Omega} \parallel \hat{x}. \tag{6.23}$$

The dependence of the polarization on magnetic field is Lorentzian with halfwidth given by ΩT_s. The two lifetimes (τ and τ_s) can be found from the measured halfwidth and the zero-field polarization. Unfortunately, the role of the transverse magnetic field is played by the anisotropic exchange interaction, which is fixed in the case of the exciton. Moreover, an external transverse field would lead to mixing of the bright and dark states and a breakdown in the 2-level model.

In the next section we will consider a QD doped with a single electron. In this case the electron's lifetime can be much longer because the spin is in the ground state instead of the exciton state, and the total spin relaxation rate can be dominated by pure spin relaxation. In any case, if the measurement is made over an inhomogeneous ensemble of spins, Eq. (6.23) must be averaged over the distribution in frequencies or lifetimes. This becomes especially relevant for the ground state spin of trions or electrons bound to dilute concentration of donors where the spin lifetimes can become very long. As will be discussed in section 6.6, in that case the Hanle lineshape is typically inhomogeneously broadened by the distribution in the internal hyperfine fields.

6.4 Trions (singly charged excitons)

We have discussed excitons and their pseudo-spin states in a neutral QD. We now consider the optical properties of a QD doped or otherwise charged with a single electron. This case is interesting, in part, because here it is the ground state of the QD that has a spin 1/2; therefore, its spin lifetime is not dominated by the exciton's recombination time and can become much longer. Furthermore, the QD can be excited optically into a charged exciton state (trion) and the electron's spin state probed and controlled optically, for example, through optical pumping. We will show that one important difference between a neutral and charged exciton is that the charged exciton lacks an exchange interaction in its singlet state, which is the lowest energy trion state. This is especially important in view of the fact that much of the fine structure dynamics of a QD exciton is dominated by its large exchange interaction. After a brief overview of the optical study of this system we consider the physics of the optical fine structure and polarization. A charged QD is in many ways analogous to a neutral donor-bound exciton, and we will take advantage of this close analogy to extend our treatment to the study of electrons bound to shallow donors, primarily in section 6.6, when we consider relaxation in detail.

Trions in natural QDs

Recent spectroscopic work on trions in quantum wells and quantum dots is the result of a natural progression from the study of two-dimensional electron gases in quantum wells, carried to the limit of a single electron. At low density, individual electrons in a quantum well are sufficiently well-separated that collective effects are unimportant and spectroscopic features reflect properties of isolated electrons and their optically-excited trions. An excitation or emission spectrum for a quantum well with low electron density shows characteristic peaks for both trions and excitons. The main task for early investigators studying these systems was to identify characteristic signatures that allowed the two species to be distinguished. Although the negatively charged trion in semiconductors was originally predicted in 1958 by Lampert [66], a proper identification of the X^- was not achieved until the early 1990s in quantum well structures doped with electrons [67,68,69]. The excess electrons were introduced by remote doping, which consists of introducing donors in the barrier (e.g., silicon donors in AlGaAs barriers). Since then, extensive work has been carried out on the two dimensional (2D)X^- in wide quantum wells and more recently on the zero dimensional X^- in QDs [70,71,72,73,74,75,76].

An obvious although not unique signature for the presence of quantum well trions in an absorption or photoluminescence spectrum is a doublet arising from the simultaneous presence of excitons and trions. This structure emerges out of the broad peak of a two-dimensional electron gas as electrons are removed from the quantum well, a change that may be observed through photo-depletion of charge in the quantum well [77,78], through depletion by an electrical bias [68], or by a decrease in modulation doping densities [79].

Recently, investigations of trions in quantum wells have been pushed into the regime of narrow wells, where monolayer variations in well thickness lead to confinement of trions in all three dimensions. Figure 6.17 shows typical photoluminescence spectra for four modulation-doped GaAs quantum wells with different widths. The spectra were obtained under experimental conditions (excitation laser wavelength and power density) chosen so that each well simultaneously displays features due both to neutral excitons and trions. For the narrower wells, one also observes the monolayer splittings discussed in section 6.2. In Fig. 6.17, these splittings are indicated separately above the spectrum for each quantum well.

The trion peak is redshifted from that of the exciton by the trion binding energy, which corresponds to the reduction in the exciton's energy due to the presence of an electron. The trion binding energy is a strong function of the well confinement strength, increasing by a factor of 3 in going from a bulk GaAs sample to a 2.8 nm quantum well. This dependence is shown in the inset in Fig. 6.17. The reason for the increase in binding energy is the stronger Coulomb interactions imposed by the closer proximity of the constituent charges, with the electron–hole attraction dominating the repulsion between like-charged carriers.

The power density and wavelength of the excitation laser affect the relative number

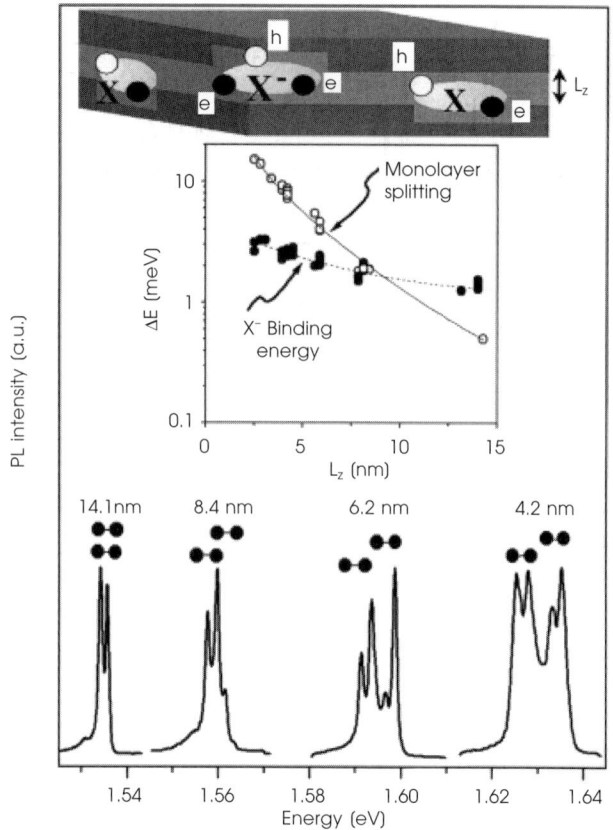

Fig. 6.17 Ensemble photoluminescence spectra of modulation doped quantum wells for several quantum well widths. Both trions and excitons are observed. Monolayer splittings are indicated. Inset: well width dependence of the binding energy of the trion and the monolayer splittings.

of electrons and excitons and the absolute number of excess electrons in a quantum well. For excitations below the barrier bandgap, high laser powers tend to increase the number of electron–hole pairs relative to excess electrons, enhancing the exciton spectral features. When electron hole pairs are excited above the barrier bandgap, they often separate due to local electric fields, neutralizing ionized donors or acceptors as well as the charges in the wells, with the result that trion spectral features are diminished [77].

Additional signatures, including optical polarization, temperature dependence, magnetic field shakeup processes and laser excitation dependence, make a convincing argument for the presence of trion ensembles in quantum wells. At low temperature in a magnetic field, the ground state electrons are thermally polarized, and optical selection rules permit transitions involving only a specific optical polarization. A zero-field polarization effect for positive trions and excitons has also been reported [78], which allows them to be distinguished from negative trions. The temperature dependence of peak intensities is a straightforward signature for the weakly-bound trion. Typical binding energies for trions in quantum wells are 1.1–3.5 meV, and so they ionize at temperatures much lower than for excitons. Nevertheless, the most important discovery for a clear distinction between positively and negatively charged trions has been the observation of shakeup transitions [80,81], which also allows a proper distinction with respect to excitons bound to donors.

Fine structure in single trion spectroscopy

Three-dimensionally confined trions may be studied by resolving photoluminescence from narrow GaAs quantum wells through apertures in a metal shadow mask. As with excitons, monolayer fluctuations in well thickness may confine trions to produce the discrete spectral lines expected in quantum dots. In the smallest apertures, sharp lines from individual zero-dimensional excitons and trions are apparent, and one can no longer distinguish the broad inhomogeneous trion and exciton features of the ensemble spectra. As in the case of ensemble measurements, identification of trions becomes an issue, and in the same way, magnetic fields and temperature dependencies provide an important identification tool. Shown in Fig. 6.18(a) are PL spectra obtained in the Faraday geometry (i.e. magnetic field perpendicular to the quantum well plane: along the z-axis) at several field strengths. As can be seen from this figure, in the absence of a magnetic field these PL spectra are characterized by two predominant transitions that correspond to the recombination of an exciton (X) and a trion (X_s^-) from the lower monolayer. In this magnetic field geometry all lines split into doublets with the same magnetic field dependence (diamagnetic shifts, g-factors, etc). When the temperature is raised (Fig. 6.18(b)), the trion-related lines decrease, similar to the result with ensemble measurements. By 30K all lines related to the trion are gone, which is a signature for the weakly-bound trion.

Although the temperature dependence provides an identification of the trion, the magnetic field dependence in different geometries is the one that provides the clearest signatures for a trion. In order to understand these differences, it is helpful to study the energy level diagram for trions and excitons (Fig. 6.19). As discussed earlier, the optically-excited exciton exhibits fine structure at zero field, with two closely spaced pairs of states (bright $| M = \pm 1 \rangle$ and dark $| M = \pm 2 \rangle$) split by the exchange energy δ_0, where M is the momentum projection on the quantization axis. The peak separations observed in the spectrum are due solely to the excited state structure, since there is only a single

244 D. Gammon et al.

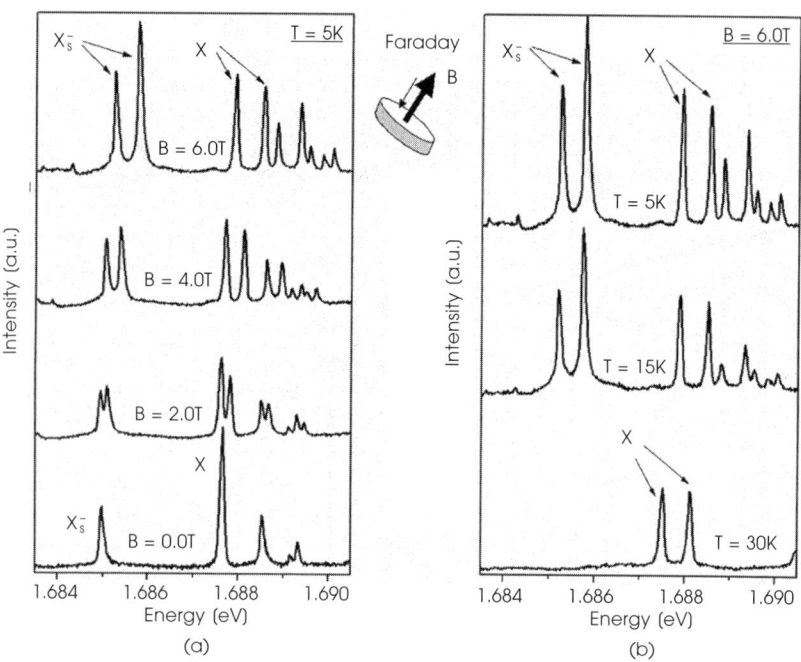

Fig. 6.18 Single dot magneto-photoluminescence of trion singlet (X_s^-) and exciton (X) transitions in the Faraday geometry (a) at several magnetic fields and 5K (b) at 6.0 T as a function of temperature.

vacuum ground state level. Unlike the exciton, the lowest optically-excited trion state (the "singlet") consists of just two spin states, corresponding to the two orientations of the unpaired hole ($|X_s^-\rangle = |M = \pm 3/2\rangle$). Because the two electrons are paired, there are no further exchange splittings. The ground state associated with the trion is a single unpaired conduction electron and also exhibits a spin doublet structure ($|e\rangle = |M = \pm 1/2\rangle$). The structure in the optical spectrum therefore reflects the level spacings in both the ground (electron) and excited (trion single) states.

Both the exciton and the trion thus have four possible recombination channels, shown as down arrows in Fig. 6.19. Considering the selection rule ($\Delta M = \pm 1$), we expect two allowed transitions (solid lines) and two forbidden transitions (dotted lines) for each in the Faraday geometry or at zero field. Faraday geometry and zero-field PL spectra, each containing peaks from a trion and an exciton, are shown in Figs 6.18(a) and 18(b). In these spectra, it is difficult to distinguish the exciton from the trion, because the lower component of the exciton fine structure doublet is not observed. On the other hand, when the magnetic field has a component parallel to the quantum well plane, the eigenstates evolve into mixtures of the basis states $|M\rangle$, allowing all four

Electronic and nuclear spin in the optical spectra   245

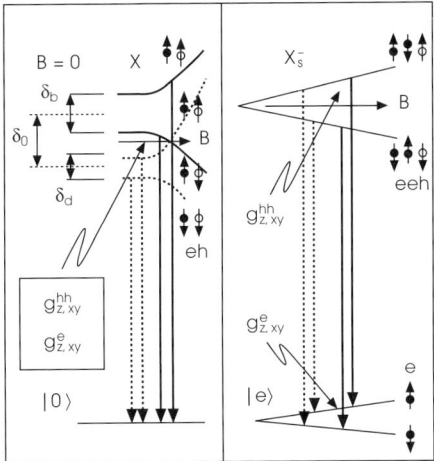

Fig. 6.19 Schematic diagram of the excited and ground states involved in the exciton (*X*) and trion singlet (X_s^-) transitions studied as a function of magnetic field (*B*) strength and orientation. The solid arrows indicate bright-related transitions while the dotted arrows indicate dark-related transitions. These schematics illustrate the case when $g^h < 0$, $g^e > 0$ and $g^h > g^e$.

transitions to be observed for both species. A Voigt geometry spectrum is shown in Fig. 6.20. Here, the lower fine structure component for the exciton is seen clearly. The trion peak appears just slightly broadened, due to a small unresolved Zeeman splitting ($g_x^e \sim 0.2$ and $g_x^{hh} \sim 0$).

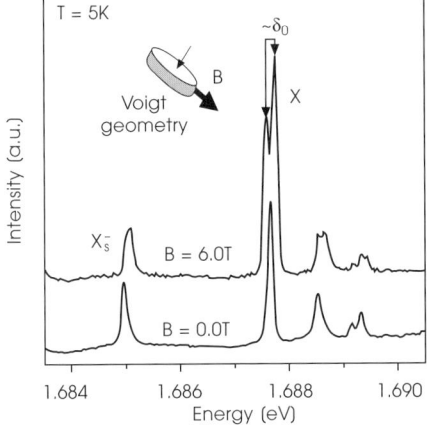

Fig. 6.20 Single dot magneto-photoluminescence in the Voigt geometry.

The full dependence of the PL peak positions on the magnetic field polar angle (θ) makes the distinction between excitons and trions quite clear (Fig. 6.21). In Fig. 6.21 we indicated bright-related transitions by open symbols and dark-related transitions by solid symbols. The similarity of the two species' spectra in the Faraday geometry ($\theta = 0°$) is seen clearly, as is the contrast between the exciton fine structure splitting and the small trion Zeeman splitting in the Voigt geometry ($\theta = 90°$) [Note that the effective trion g-factor (Fig. 6.21) goes to zero around 85°]. For intermediate angles, where both bright and nominal dark transitions are observed, there is a clear qualitative difference in behavior, the origin of which is evident from the energy level diagrams. For the exciton, the dark Zeeman components have a larger effective g-factor that is offset to lower energies (by the exchange energy δ_0) from the bright states. In contrast, the trion dark transitions have the highest and lowest energy (see Fig. 6.21), because they correspond to transitions between Zeeman components of the trion and the ground state electron, that add together in the same direction (i.e. both "spin up" or both "spin down").

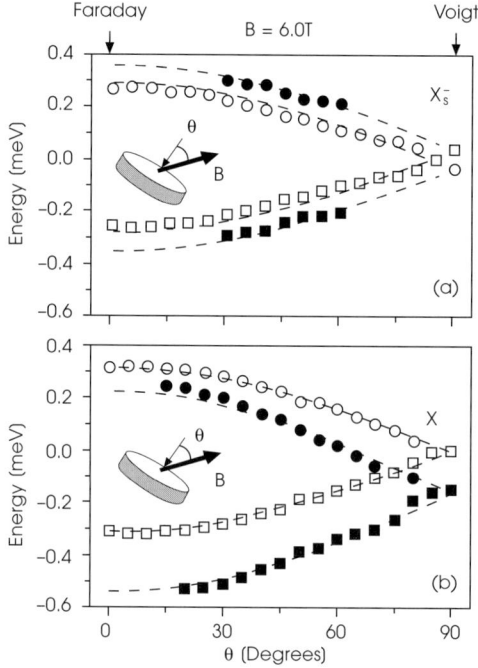

Fig. 6.21 Summary plot of the dependence on polar angle (θ) of the dark (solid symbols) and bright-related (open symbols) transitions for the (a) trion and (b) exciton at constant field of 6.0 T.

The mathematics describing the dependence of the fine structure of the X as a function of the polar angle (θ) was already discussed in section 6.3 [Eqs (6.4) and (6.15)]. In the case of the trion we need a Hamiltonian describing the mixing of the excited ($| X_{\bar{s}}^- \rangle$) and another one for the mixing of the ground state ($| e \rangle$).

In the case of the $X_{\bar{s}}^-$, the Hamiltonian can be written as

$$\hat{H}_{X_{\bar{s}}^-} = \frac{\mu_B \vec{B}}{2} \cdot \bar{\bar{g}}^h \cdot \hat{\sigma}^h = \frac{\mu_B \cdot B}{2} \begin{pmatrix} g_z^{hh} \cdot \cos\theta & g_x^{hh} \cdot \sin\theta \\ g_x^{hh} \cdot \sin\theta & -g_z^{hh} \cdot \cos\theta \end{pmatrix}, \quad (6.24)$$

We also assume that the in-plane elongation of the quantum dot does not affect significantly the in-plane g-factors, which is a good assumption for this system. Similarly, the final state Hamiltonian is:

$$\hat{H}_e = \frac{\mu_B \vec{B}}{2} \cdot \bar{\bar{g}}^e \cdot \hat{\sigma}^e = \frac{\mu_B \cdot B}{2} \begin{pmatrix} g_z^e \cdot \cos\theta & g_x^e \cdot \sin\theta \\ g_x^e \cdot \sin\theta & -g_z^e \cdot \cos\theta \end{pmatrix} \quad (6.25)$$

After diagonalizing both matrices and taking all possible differences between the different eigenvalues, we obtain the expression for the $X_{\bar{s}}^-$ bright-related and dark-related transition energies ($E_{X_{\bar{s}}^-}^b, E_{X_{\bar{s}}^-}^d$):

$$E_{X_{\bar{s}}^-}^b = \pm \frac{\mu_B \cdot B}{2} (\sqrt{(g_z^{hh} \cdot \cos\theta)^2 + (g_x^{hh} \cdot \sin\theta)^2} - \sqrt{(g_z^e \cdot \cos\theta)^2 + (g_x^e \cdot \sin\theta)^2})$$

(6.26)

$$E_{X_{\bar{s}}^-}^d = \pm \frac{\mu_B \cdot B}{2} (\sqrt{(g_z^{hh} \cdot \cos\theta)^2 + (g_x^{hh} \cdot \sin\theta)^2} + \sqrt{(g_z^e \cdot \cos\theta)^2 + (g_x^e \cdot \sin\theta)^2})$$

(6.27)

The dashed lines shown in Figs 6.21(a) and 6.21(b) are the fits to the data using expressions (21) and (22) in the case of the $X_{\bar{s}}^-$, and using Eqs (6.4) and (6.17) for the X (section 6.3). The values utilized in the fitting of the $X_{\bar{s}}^-$ data were $g_z^e = (0.20 \pm 0.05)$, $g_x^e = (0.2 \pm 0.1)$, $g_z^{hh} = (-1.85 \pm 0.05)$ and $g_x^{hh} = (0.0 \pm 0.1)$. In the case of the X the parameters obtained were $\delta_o = (157 \pm 9)\mu eV$, $g_z^e = (0.20 \pm 0.05)$, $g_x^e = (0.2 \pm 0.1)$, $g_z^{hh} = (-2.00 \pm 0.05)$ and $g_x^{hh} = (0.0 \pm 0.1)$.

As shown in Fig. 6.22, and as in the case of the exciton (see Fig. 6.13), the polar angle (θ) dependence of the trion transitions is mainly given by the highly anisotropic heavy hole g-factor, which affects the excited state. The ground state is not affected within the accuracy of our experiment.

The rich single dot magneto-photoluminescence spectra obtained for both the X^- and X provide detailed information on their internal structure, and the lack of exchange splitting provides a signature for clear identification of charged and uncharged quantum dots.

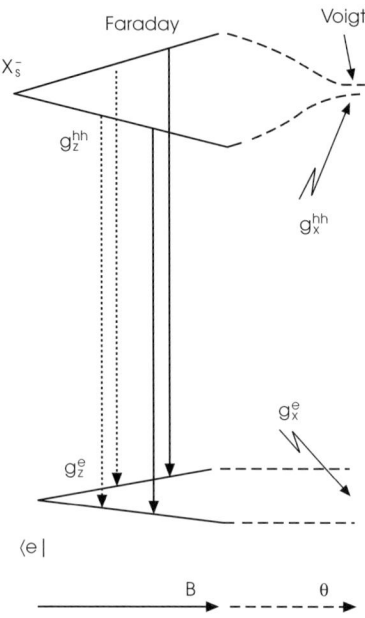

Fig. 6.22 Schematics of the trion states as a function of magnetic field strength in the Faraday configuration *and* as a function of the polar angle at a constant magnetic field. Notice the decrease in the heavy hole *g*-factor in the Voigt geometry, while the electron *g*-factor remains constant.

Optical orientation of negatively charged excitons

As mentioned before, the spin properties of negatively charged excitons in nanostructures are reminiscent of excitons bound to neutral donors D^0X in bulk GaAs-type semiconductors. The ground state of both have two electrons in a singlet state and a single hole spin that determines the polarization of the complex. Compared to bulk GaAs, the optical orientation of trions in nanostructures can be traced over a large temperature range due to the increased binding energy and the suppressed spin relaxation of holes due to heavy hole – light hole subband splitting. We shall consider now the optical polarization of three-dimensionally confined trions (and/or D^0X complexes) in nanostructures. In this case the hole spin relaxation is suppressed and the trion can be polarized.

There is a striking difference between the optical polarization of trions and that of excitons (considered in IIID). The exciton spin projections are determined by the spin projections of both the electron and hole making up the exciton. However, the spin projection of the trion singlet is determined only by the hole, because the electron spins are antiparallel. Moreover, if the trion is excited through the optical generation of an exciton followed by localization into a QD charged with a single electron, the hole spin projection of the trion provides information about the spin projection of the electron and the exciton (both bright and dark) *before* the formation of the trion [65]. The kinetics of the formation depends strongly on the spatial confinement within the dot. If the particles are confined within a QD with radius less than the exciton Bohr radius it is a three-particle problem from the start. We consider here the opposite case, which is the one realized experimentally when the size of lateral confinement is much larger than the characteristic size of single electron and exciton Bohr radius (~10 nm). These dots can be imagined as large areas (islands) of quantum wells. This happens for naturally grown GaAs QDs (Fig. 6.2(a)) and for self-organized InP/InGaP QDs [65]. In this case the excitons and single electrons behave like independent quasiparticles before the trion formation. The hole spin memorizes their spin polarizations at the moment of binding, and therefore contains the prehistory of the exciton and single electron spins. The PL polarization is determined in this case by the bright and dark excitons, with the dark states polarized opposite to the bright ones.

A simple example illustrates this point (see Fig. 6.23). The σ^+-light creates the free bright excitons $|+1\rangle = |+3/2, -1/2\rangle$. If the hole spin in the exciton relaxes faster than the electron spin, then there are both $|+1\rangle = |+3/2, -1/2\rangle$ bright excitons and $|-2\rangle = |-3/2, +1/2\rangle$ dark excitons. Binding of a $|+1\rangle = |+3/2, -1/2\rangle$ exciton with a $|+1/2\rangle$ electron creates the singlet trion with total momentum +3/2 (the electron spins are antiparallel). The recombination of such a complex leads to the σ^+ PL. In contrast, the dark exciton forms the −3/2 trion that emits a σ^- photon, which is opposite to the

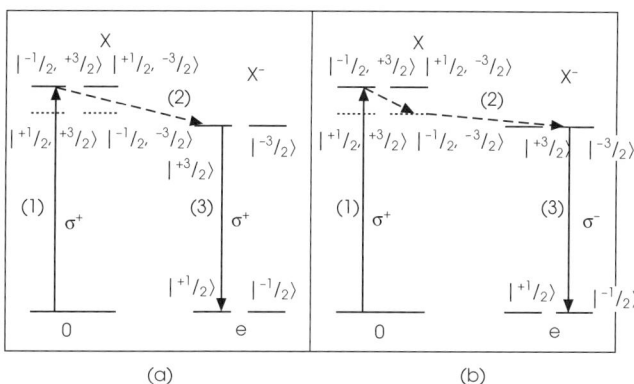

Fig. 6.23 Illustration of optical pumping of the trion in the absence of exchange mixing. (a) In step 1, circularly polarized light creates a spin polarized free exciton, which is then bound into trion state in step 2, and then emits light with same polarization in step 3. (b) Alternatively, after excitation, the hole of the exciton flips its spin and transfers to a dark state before being bound into the trion state. The PL emission has the opposite polarization in this case.

polarization of the incident light. Thus, the contributions of optically active and forbidden excitons have opposite signs.

If the trion formation time is long enough (much longer than $\hbar/\delta_b$, $\hbar/\delta_d$), the excitons lose their orientation in zero magnetic field as a result of the anisotropic exchange interaction, which mixes the +1 and −1 bright exciton states (and the +2 and −2 dark exciton states) (section 6.3). Therefore, at zero field the trion polarization depends mainly on the polarization of the single electron at formation. The magnetic field in Faraday geometry restores the orientation of excitons in a way similar to that described in section 6.3. However, the trion polarization changes with magnetic field in a way different from that of the exciton (see Fig. 6.24 and compare it with Fig. 6.16(a)). The magnetic field restores first the optical orientation of dark excitons (usually $\delta_d <$ δ_b), and the degree of circular polarization ρ_c becomes more negative (B ~ 100 G for Fig. 6.24). The orientation of optically active excitons is restored at a higher magnetic field value (B ~ 2 T on Fig. 6.24), which causes the degree ρ_c to increase. Thus, the dependence, $\rho_c(B)$, is nonmonotonic, with the orientation of bright and dark excitons being restored at different magnetic field values. This enables one to separate the contributions of single electrons, bright excitons and dark excitons to the trion PL.

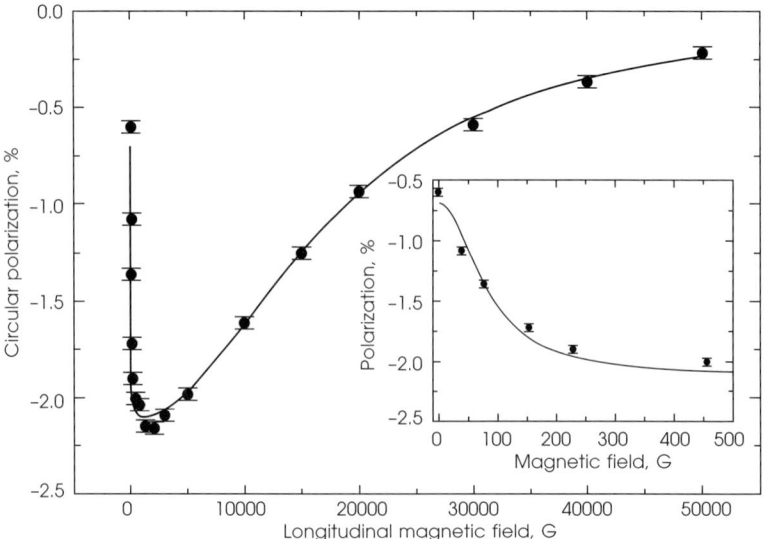

Fig. 6.24 Circular polarization of bright and dark (inset) trion PL as a function of the longitudinal magnetic field in the Faraday polarization. From [65].

We would like to calculate now the degree ρ_c of circular polarization of trion PL when the trion is formed in the singlet state from bright excitons with orientation

$$P_c^b = \frac{N_{+1} - N_{-1}}{N_{+1} + N_{-1}},$$

dark excitons with polarization

$$P_c^d = \frac{N_{+2} - N_{-2}}{N_{+2} + N_{-2}}$$

and single electrons having polarization

$$P_s = \frac{N_\uparrow - N_\downarrow}{N_\uparrow + N_\downarrow},$$

where $N_{\pm 1}$ ($N_{\pm 2}$) are the population of bright (dark) excitons in the $|\pm 1\rangle$ ($|\pm 2\rangle$) state, and $N_{\uparrow(\downarrow)}$ is the population of single electrons with spin up (down), respectively. The population of the $|\pm 3/2\rangle$ trions, $N_{+3/2}$ and $N_{-3/2}$, are proportional to $N_{+1}N_\uparrow + N_{+2}N_\downarrow$ and $N_{-2}N_\uparrow + N_{-1}N_\downarrow$, respectively. Taking into account the fact that the $|\pm 3/2\rangle$ trions emit $\delta^\pm$-light, one can find the PL polarization degree

$$\rho_c = P_T = \left(N_b \frac{P_c^b + P_s}{1 + P_c^b P_s} + N_d \frac{P_c^d - P_s}{1 - P_c^d P_s} \right) \Big/ (N_b + N_d) \qquad (6.28)$$

where the polarization of trions is

$$P_T = \frac{N_{+3/2} - N_{-3/2}}{N_{+3/2} + N_{-3/2}},$$

and $N_b = N_1 + N_{-1}$ and $N_d = N_{-2} + N_2$ are the total populations of the bright and dark excitons, respectively. Under σ^+-excitation, the hole state $|+3/2\rangle$ and electron state $|-1/2\rangle$ are created preferentially. One may then expect that $P_c^b > 0$, $P_s < 0$. Usually the hole spin relaxes faster than the electron one; therefore $N_{-2} > N_{+2}$, and the dark excitons are polarized negatively ($P_c^d < 0$), in agreement with the qualitative discussion given above.

The magnetic field dependence of $\rho_c(B)$ is determined by the magnetic field dependence of the bright $P_c^b(B)$ and dark $P_c^d(B)$ excitons as a result of anisotropic exchange mixing of the $|+1\rangle$ and $|-1\rangle$ bright states and the $|+2\rangle$ and $|-2\rangle$ dark ones. If the excitons take a long time (much longer than $\hbar/\delta_b$, $\hbar/\delta_d$) before forming a trion, the polarizations, $P_c^b(B)$ and $P_c^d(B)$, are given by the first equation in Eq. (6.20):

$$P_c^b = P_b^0 \frac{\Omega_b^2}{\omega_b^2 + \Omega_b^2}, \quad P_c^d = P_d^0 \frac{\Omega_d^2}{\omega_d^2 + \Omega_d^2} \qquad (6.29)$$

where P_b^0, P_d^0 are the bright and dark exciton polarization in the absence of anisotropic splitting. The Zeeman splitting of dark states is $\hbar\Omega_d = \mu_B (g_z^e - g_z^h) B$. We fit the data in Fig. 6.24 according to Eqs (6.28) and (6.29) [65]. Using the electron $g_z^e = 1.6$ and hole $g_z^h = -2.9$ [82] g-factors, we estimate $\delta_b \sim 100$ μeV, $\delta_d \sim 2$ μeV.

We have considered the effect of optical orientation of trions formed by optically oriented electrons and excitons. Unlike the case of excitons considered in section 6.3, the effects of optical alignment with linearly polarized light are absent for trions because the electron-hole spin correlation is destroyed under trion singlet formation.

6.5 Hyperfine interaction

When a magnetic field is applied in the Faraday geometry (along the z-axis) the spectral lines show both Zeeman-like splittings that increase linearly with field (at sufficiently high fields) and diamagnetic shifts that increase quadratically with field. However, the behavior is more complicated under closer inspection, especially at low fields. This richness arises from additional effects associated with exchange and hyperfine interactions, and from competition between the interactions. Previously, we considered exchange and Zeeman interaction – we now include the hyperfine interaction between the electron and nuclear spins.

Hyperfine effects can become important under excitation with circularly polarized and nearly resonant excitation. The circularly polarized light creates spin polarized electrons because of spin-orbit interaction and the resulting selection rules. The excitation must be quasi-resonant so that the electron polarization remains during the time it takes the exciton to relax into the luminescing QD state and recombine. The nuclear spin polarization, I_N, is created through the dynamic part of the contact hyperfine interaction and is very large compared to that expected at thermal equilibrium. The spin-polarized electron has a relatively small probability of flipping a nuclear spin during each optical cycle, however, because the nuclear spin lifetimes are so much longer than the electron lifetime, the nuclear spin system can be pumped into a high state of polarization. The hole spin is not affected because its Bloch wave function is p-like with nodes at the positions of the nuclei, and so the holes do not interact directly with the nuclear spin.

A large average nuclear polarization, $\langle I_N \rangle$, exerts an effective magnetic field back on the electron through the average part of the contact hyperfine interaction. This effect can be understood by considering the static spin-dependent terms in the exciton Hamiltonian, where we discuss only the Faraday geometry with $\vec{B}_{ext} = B_{ext}\vec{z}$.

$$\langle \hat{H}_{spin}^{exciton} \rangle = \langle \hat{H}_{exchange} \rangle + \langle \hat{H}_{Zeeman} \rangle + \langle \hat{H}_{hyperfine} \rangle \tag{6.30}$$

The first term is due to the exchange interaction. The second term is the Zeeman term and depends on the g-factors for the electron and hole. The third term is the hyperfine term, averaged over the nuclear spin, and describes the effect that the average nuclear spin polarization has on the electron spin.

$$\langle \hat{H}_{\text{hyperfine}} \rangle_N = \frac{1}{2} A \langle I_z \rangle \hat{\sigma}_z^e = \frac{\mu_B}{2} g_z^e \hat{\sigma}_z^e B_N, \quad (6.31)$$

where $\langle I_z \rangle$ is the average spin of the polarized nuclei, and the average hyperfine constant A describes the electron spin interaction with all nuclei in the unit cell ($A \approx 90$ μeV in GaAs [19]). The hyperfine term has been rewritten in terms of an effective internal nuclear magnetic field, $B_N = A\langle I_N \rangle/g_z^e \mu_B$, that gives a Zeeman-like spin splitting (Overhauser shift) but does not contribute to the diamagnetic field. Note that a small g-factor leads to a large average hyperfine field. If the nuclear spins are disordered, $B_N = 0$.

Diagonalization of the average spin Hamiltonian, $\langle \hat{H}_{\text{spin}}^{\text{exciton}} \rangle_N$, gives the fine structure of the exciton level in the Faraday geometry:

$$E_{\pm}^{(1)} = \frac{1}{2}(\delta_0 \pm \sqrt{h_1^2 + \delta_b^2}), \quad E_{\pm}^{(2)} = \frac{1}{2}(-\delta_0 \pm \sqrt{h_2^2 + \delta_d^2}), \quad (6.32)$$

where $h_n = \mu_B g_n B_{\text{ext}} + (-1)^n A \langle I_z \rangle$, $g_n = g_z^{hh} - (-1)^n g_z^e$, and $n = 1$ or 2. The splitting of both the bright ($n = 1$) and dark ($n = 2$) exciton doublet states, $\Delta E^{(1),(2)} = E_+^{(1),(2)} - E_-^{(1),(2)}$, is determined by the sum ($h_{1,2}$) of an external magnetic field and the effective magnetic field of the nuclei, as well as the anisotropic exchange interaction, $\delta_{b,d}$. To fully describe the electronic spectra we must find the average nuclear spin polarization, $\langle I_z \rangle$, which is determined by the balance between dynamical nuclear polarization and depolarization. These processes are governed by fluctuations of the electron and nuclear polarization from their average values and will be discussed later.

The PL spectrum of a single QD at several fields for σ^+ polarized excitation is shown in Fig. 6.25. With narrow spectral lines it is possible to measure the energy splittings even at small fields. The energies of the two lines of the doublet as a function of applied magnetic field are plotted in Fig. 6.26. There is an overall energy shift that can be traced to the diamagnetic shift, but also there is energy splitting which has a complicated dependence on field. And there is lack of symmetry in the energies with a change in the sign of the applied field. Moreover, if the excitation polarization is changed to σ^-, the spectra change dramatically. However, the peak energies are symmetric under the simultaneous inversion of both the excitation polarization and magnetic field $E(\sigma^+, B_{\text{ext}}) = E(\sigma^-, -B_{\text{ext}})$.

The origin of this behavior is more easily recognized by plotting separately the average energy of the doublet and its splitting. The average energy of the doublet shows a quadratic dependence on external field that is independent of polarization of the

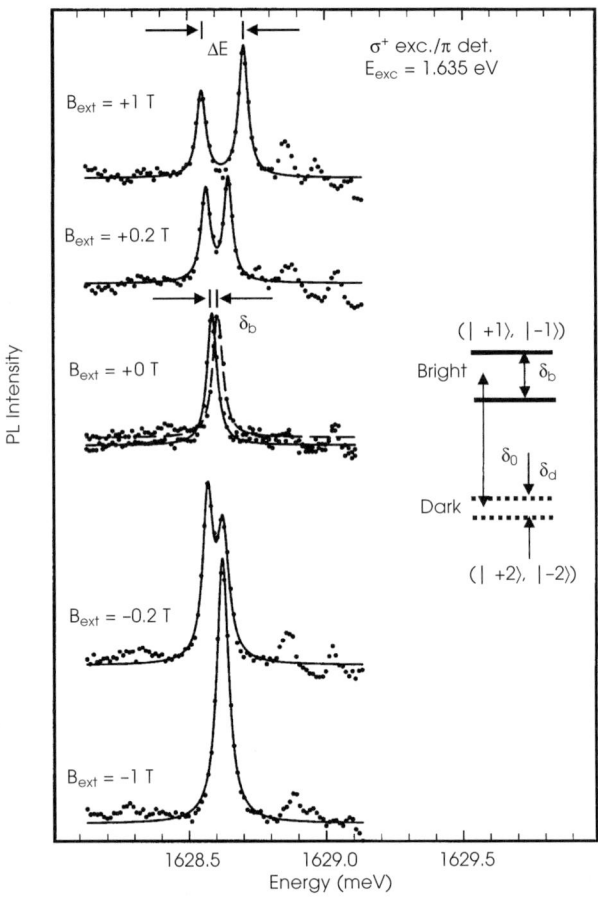

Fig. 6.25 Photoluminescence spectra of exciton as a function of external magnetic field. At nonzero fields, circular excitation (σ^+) and linear detection (π) was used. At $B_{ext} = 0$ both excitation and detection was with vertical (solid) or horizontal (dashed) linear polarization. Inset: energy level diagram of the QD fine structure at zero field. From [13].

exciting light (Fig. 6.26: inset). However, the energy splitting depends strongly on the exciting polarization as shown in Fig. 6.27. The splitting shows roughly a linear dependence on field as expected for the Zeeman effect, but with a constant offset from the origin with magnitude $B_n = \pm 1.2 T$ along the horizontal axis ($\Delta E = \pm 90$ μeV along the vertical); where the sign depends on the polarization of the exciting light. Moreover there is a strong nonlinear dependence on field around $B_{ext} = 0$ (Fig. 6.27: upper inset).

The Overhauser effect leads to the offset in magnetic field observed in the measured spin splitting. A large optically pumped nuclear polarization (65%) gives rise to a large

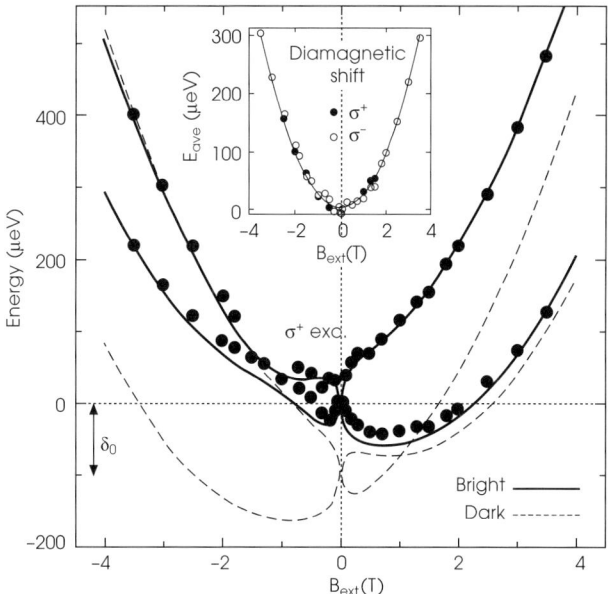

Fig. 6.26 Peak energies of the spectral lines for σ^+ excitation as a function of external magnetic field. The lines are calculated values of the bright (solid) and dark (dashed) energies. Inset: the average of the two components of the doublet, showing a quadratic dependence on B_{ext} (diamagnetic shift). Values for σ^+ and σ^- excitation show no difference, as expected. From [13].

internal magnetic field of $B_N = 1.2T$. The sign of the nuclear polarization, and therefore the sign of the field offset, depends on the sign of the circular polarization of the exciting light. The diamagnetic shift is independent of polarization because the internal field affects only the splitting.

Note that the Overhauser effect has been found also in self-assembled InP/InGaP quantum dots [83,84], using the optical orientation technique. The characteristic nuclear fields acting on bright and dark excitons are 1 kG and 100 G, respectively. The different field values result from different g-factors of bright $(g_h - g_e)$ and dark $(g_h + g_e)$ states (see Eq. (6.25)).

This initial discussion provides a good understanding at high magnetic fields but does not explain the nonlinear dependence at low fields. For this it is necessary to consider the dynamics of the nuclear spin system and the hyperfine interaction in more detail, which we do now. It turns out that at low fields nuclear dipole-dipole interactions as mediated by the electrons destroy the nuclear polarization.

Hyperfine interaction: static and dynamic

All the major mechanisms of electron spin relaxation active in the bulk and in QWs are suppressed in quantum dots, as we will discuss in the next section. As a result the hyperfine interaction of localized electrons with nuclear spin plays the major role in the electron spin relaxation in QDs. For free electrons the interaction with nuclei leads to an electron spin relaxation time on the order of 3×10^{-5} s even at helium temperatures. Localization enhances the overlap of the electron on individual nuclear spins and the contact hyperfine interaction of the electron spin with macroscopic numbers of nuclear spins has a significant effect both on the exciton fine structure and the dynamics of electron and exciton spin polarization.

The contact hyperfine interaction between the electron and the nuclear spins is given by

$$\hat{H}_{hf} = \frac{v_0}{2} \sum_j A^j |\psi(\mathbf{R}_j)|^2 (\hat{\mathbf{I}}^j \hat{\boldsymbol{\sigma}}^e), \tag{6.33}$$

where $v_0 = a_0^3$ is the volume of the unit cell, $\mathbf{I}^j$ and $\mathbf{R}_j$ are the spin and coordinate of the jth nucleus ($I = 3/2$), $|\psi(\mathbf{r})|^2$ is the electron density at the jth nuclear site and the sum goes over all nuclei. A^j is the constant of the hyperfine interaction:

$$A^j = \frac{16\pi \mu_B \mu_I^j}{3 I^j v_0} |u(\mathbf{R}_j)|^2 \tag{6.34}$$

$|u(\mathbf{R}_j)|^2$ is the square of Bloch function, and μ_I^j is a magnetic moment of a nucleus with number j. In zinc-blende semiconductors the contact interaction with nuclei is important for electrons because the Bloch amplitude has s-symmetry and $|u(\mathbf{R}_j)|^2 \gg 1$. The hole Bloch functions at the Γ-point of the Brillouin zone has p-symmetry, $|u(\mathbf{R}_j)|^2 = 0$ and the contact hyperfine interaction is not important for holes. The spin-spin interactions between the electron and nuclei are conveniently written as

$$(\hat{\mathbf{I}}^j \hat{\boldsymbol{\sigma}}^e) = (\langle \mathbf{I} \rangle \hat{\boldsymbol{\sigma}}^e) + (\hat{\mathbf{I}}^j \langle \boldsymbol{\sigma}^e \rangle) + ((\hat{\mathbf{I}}^j - \langle \mathbf{I} \rangle)(\hat{\boldsymbol{\sigma}}^e - \langle \boldsymbol{\sigma}^e \rangle)) - (\langle \mathbf{I} \rangle \langle \boldsymbol{\sigma}^e \rangle), \tag{6.35}$$

where the first two terms describe the electron and nuclear spin motion in the average magnetic fields of the nuclei and electron respectively, and the third one describes the interactions between fluctuations of the electron and nuclear spin polarization. If the average nuclear and electronic spin $\langle \mathbf{I} \rangle$ and $\langle \boldsymbol{\sigma}^e \rangle$ align along the magnetic field (z-direction, $\langle \mathbf{I} \rangle = \langle I_z \rangle$ and $\langle \boldsymbol{\sigma}^e \rangle = \langle \sigma_z^e \rangle$), the electron is affected by an effective magnetic field, B_N, arising from the nuclear spins, that was introduced in Eq. (6.31),

$$B_N = \frac{A \langle I_N \rangle}{g_z^e \mu_B}; \tag{6.36}$$

where the hyperfine constant A was introduced in Eq. (6.31) ($A = \sum_j A^j$, where the sum goes over all nuclei in the unit cell), and likewise, the nuclei are affected by the mean hyperfine field of the electron:

$$B_e^j = \frac{IA^j}{2\mu_I} v_0 |\psi(\mathbf{R}_j)|^2 \langle \hat{\sigma}_z^e \rangle. \tag{6.37}$$

As we showed above, under conditions of polarized nuclei, their effective magnetic field B_N leads to the additional splitting of the exciton sublevels known as the Overhauser shift.

The third term in Eq. (6.35), which describes the interactions between fluctuations, can be presented in the form

$$\left((\hat{I}_z^j - \langle I_z \rangle)(\hat{\sigma}_z^e - (\sigma_z^e)) + \frac{\hat{I}_+^j \hat{\sigma}_-^e + \hat{I}_-^j \hat{\sigma}_+^e}{2} \right), \tag{6.38}$$

where $\hat{I}_\pm^j = \hat{I}_x^j \pm i\hat{I}_y^j$, and $\hat{I}_x^j, \hat{I}_y^j, \hat{I}_z^{(j)}$ are the projections of the nuclear spin. The fluctuations of the hyperfine field arising from the nuclear spins lead to the broadening of electron spin sublevels. In zero magnetic field and in the absence of nuclear polarization, the hyperfine magnetic field acting on an electron averaged over time or over an ensemble of dots is zero. However, fluctuations of this field from dot to dot or from one time to another in the same QD do not vanish. The magnitude of this fluctuation can be estimated by straightforward calculation of the second order correlation function of the Hamiltonian that describes the hyperfine interaction (Eq. 6.33). Neglecting correlations between different nuclear spins and considering their directions as random we obtain the expression for the electron level broadening δ_s:

$$\delta_s^2 = v_0 \int |\psi(\mathbf{r})|^4 d^3r \sum_j (A^j)^2 I^j (I^j + 1), \tag{6.39}$$

where the sum goes over all nuclei in the unit cell. δ_s is the inhomogeneous linewidth arising from fluctuations of the hyperfine energy from dot to dot or in the same dot at different times. We can also view $\hbar/\delta_s$ as the time for the spin precession of a localized electron in a typical fluctuation of the nuclear hyperfine field, and thus it is the characteristic time for the ensemble electron spin relaxation.

If the time of electron interaction with the nuclei, τ_c, is significantly shorter than $\hbar/\delta_s$, the rate of electron spin relaxation is reduced below $(\delta_s/\hbar)^2 \tau_c$. As a result, $\hbar/\delta_s$ is the shortest time in the theory of electron spin dephasing by nuclei in an ensemble of localized electrons. This time is proportional to the square root of the QD volume, and for the typical GaAs QD is of order 1–10 ns (in GaAs $\sum_j (A^j)^2 I^j (I^j + 1) = 3.8 \cdot 10^{-3}$ (meV)^2).

Dynamical polarization of nuclei: Overhauser effect

The interaction between the z-components of the electron and nuclear spin fluctuations $(\hat{I}_z^j - \langle I_z \rangle)(\hat{\sigma}_z^e - \langle \sigma_z^e \rangle)$ does not change the average values of the spin z-component. However, the term

$$\hat{I}_x \hat{\sigma}_x^e + \hat{I}_y \hat{\sigma}_y^e = \frac{\hat{I}_+^j \hat{\sigma}_-^e + \hat{I}_-^j \hat{\sigma}_+^e}{2}$$

generates flip–flop transitions and leads to the polarization flow between the nuclear and electron spin systems. This is the process for nuclear dynamical polarization, which is responsible for the creation of the average magnetic field arising from the nuclei B_N.

The rate of flip–flop transitions in an external magnetic field depends strongly on the electron's interactions with its surrounding media. The huge difference in Zeeman energy of nuclei and electrons usually requires some type of assisting process to conserve the total energy of the system, such as an emission/absorption of phonons (see, for example, [85]), photons or inelastic scattering with charge carriers. The rate of dynamic nuclear polarization can be described by the following equation:

$$\left.\frac{\partial \langle I \rangle}{\partial t}\right|_{\text{flip-flop}} = -W(B)[\langle I \rangle - Q(\langle S \rangle - S_T)] \quad (6.40)$$

where $W(B)$ is the transition rate of a flip–flop process,

$$S_T = -\frac{1}{2}\tanh\left(\frac{\mu g_e B}{2kT}\right) \quad (6.41)$$

is the equilibrium value of the electron spin at temperature T, and

$$Q = \frac{\langle I_x^2 \rangle + \langle I_y^2 \rangle}{\langle S_x^2 \rangle + \langle S_y^2 \rangle} = 2(\langle I_x^2 \rangle + \langle I_y^2 \rangle). \quad (6.42)$$

In GaAs, where the nuclear spins have $I = 3/2$, $Q \approx 5$. In the case when the electrons are completely depolarized ($\langle S \rangle = 0$), and from Eq. (6.33) the nuclear polarization in steady state is given by

$$\langle I \rangle = -Q S_T = \frac{Q}{2}\tanh\left(\frac{\mu g_e B}{2kT}\right). \quad (6.43)$$

One can see that the steady-state value of dynamic nuclear polarization is determined by the large splitting of the electron spin sublevels instead of the small splitting of the nuclear spin sublevels in the same magnetic field (the Overhauser effect). In optical orientation conditions, $\langle S \rangle \gg S_T$, and one can usually neglect the equilibrium electron polarization in Eq. (6.40): then $\langle I \rangle = Q \langle S \rangle$ (see also [86,87]).

The energy of the electron nuclear system is not conserved during the flip–flop processes. These transitions are possible because the electron and nuclear interactions are a function of time. In this case the probability depends on the duration time of this interaction (τ_c) and the precession frequency of the electron in the magnetic field (Ω_e), and can be written as (see the Appendix),

$$W(B) \approx \frac{\omega_N^2 \tau_c}{1 + (\Omega_e \tau_c)^2}, \qquad (6.44)$$

where

$$\omega_N = \frac{2\mu_I \langle B_e \rangle}{\hbar I}$$

is the average precession frequency of a nuclear spin in the hyperfine field of the localized electron (Eq. (6.37)), and $\Omega_e = \mu_B g_e B/\hbar$ is the electron spin precession frequency. Equation (6.44) shows that the rate of nuclear dynamical polarization decreases drastically when the Zeeman splitting of the electron spin sublevel is larger than its broadening (($\Omega_e \tau_c) > 1$). The role of the large magnetic field can be played by the exchange electron-hole interaction (δ_0), which splits the bright and dark exciton levels. The experimental appearance of the suppression of the dynamical polarization of the nuclear spin due to this effect will be discussed next.

Nuclear dipole–dipole interactions

The kinetic equation (40) does not take the nuclear spin interactions into account. The nuclear spins interact with each other as magnetic dipoles with the Hamiltonian,

$$H_{d-d} = \frac{\mu_I^2}{I^2 r_{1,2}^3} \left((\hat{\vec{I}}_1 \cdot \hat{\vec{I}}_2) - 3 \frac{(\hat{\vec{I}}_1 \vec{r}_{1,2})(\hat{\vec{I}}_2 \vec{r}_{1,2})}{r_{1,2}^2} \right), \qquad (6.45)$$

where $r_{1,2}^3$ is the vector describing the relative positions of nuclei. This Hamiltonian does not conserve the total spin of nuclei; it is transferred to the crystal lattice. As a result, the coupling of neighboring nuclear spins through the dipole–dipole interaction leads to nuclear spin depolarization. In a magnetic field larger than the dipole field, $B_{ext} > B_L \sim 0.15$ mT, the energy of the nuclear dipole–dipole interaction is not enough to drive the transition between the two nuclear spin sublevels split by the Zeeman energy ($\mu_I B_L/I \ll \mu_I B_{ext}/I$, where μ_I is the magnetic moment of the nucleus), and this mechanism should be negligible. However, fluctuations of the z-component of the electron polarization can provide the necessary energy, leading to depolarization of the total nuclear spin, even in a relatively strong magnetic field. Calculations show that the nuclear dipole–dipole interaction weakly mixes the wave functions of different nuclear spin projection states, of order B_L/B_{ext} [13]. Transitions between these mixed states are induced by the hyperfine magnetic field of the electrons, B_e^j, acting on the nucleus, j, during the interaction time, τ_c, leading to the following rate of nuclear depolarization:

$$\left.\frac{\partial \langle I \rangle}{\partial t}\right|_{rel} = -W(0)\langle I \rangle \frac{B_L^2}{B_{ext}^2} \xi, \qquad (6.46)$$

where ξ is a numerical coefficient of order 1, and $W(0) = \omega_N^2 \tau_c$ from Eq. (6.44). As a result, the full equation describing the rate of dynamic nuclear polarization should be written as

$$\frac{d \langle I \rangle}{dt} \approx -W(B)\left[\langle I \rangle\left(1 + \tilde{\xi}\frac{B_L^2}{B_{ext}^2}\right) - \langle S \rangle Q\right], \qquad (6.47)$$

where

$$\tilde{\xi} = \xi\frac{W(0)}{W(B)} = \xi[1 + (\Omega_e \tau_c)^2]. \qquad (6.48)$$

In calculating the dynamical polarization of nuclei by localized electrons, for example, an electron bound to a donor in bulk GaAs, one can always neglect the second term in Eq. (6.48). In a small magnetic field ($B \leq B_L$) the correction of $\tilde{\xi}$ is nonessential because $(\Omega_e \tau_c)^2 \sim (\mu_B g_e B_L \tau_c / \hbar)^2 \ll 1$. In a high magnetic field ($B \gg B_L$) the contribution of the local fields to Eq. (6.47) is negligible. In this case the steady-state value of the nuclear spin is given by the following equation:

$$\langle I \rangle = Q\frac{\langle S \rangle B_{ext}^2}{B_{ext}^2 + \xi B_L^2}. \qquad (6.49)$$

One can see that nuclear polarization reaches its maximum value in magnetic fields $B_{ext} > B_L$.

The situation changes dramatically if the nuclear polarization is created by localized excitons in disk-like QDs; in particular, the region where the nuclear dipole–dipole interaction suppresses the dynamical nuclear polarization is greatly extended. This is because, in this case, Ω_e in Eqs (6.44) and (6.48) is determined by the exchange field of the hole ($\delta_0/\hbar$) instead of an external magnetic field. As a result, ξ in Eq. (6.49) is replaced by $\tilde{\xi} = \xi[1 + (\delta_0 \tau_c/\hbar)^2] \approx \xi(\delta_0 \tau_0/\hbar)^2 \approx 10^5 \xi$, and the nuclear spin polarization is suppressed up to magnetic fields significantly larger than the characteristic magnetic field of the nuclear dipole–dipole interactions (B_L). The width of the dip seen in the Overhauser shift of natural QDs (see upper inset to Fig. 6.27) is 300 times wider than

Electronic and nuclear spin in the optical spectra   261

in bulk GaAs [13]. In Ref. [13] the correlation time was taken as $\sqrt{\tau_b \tau_d}$, and from a fit to the data was found to be ~3.5 ns.

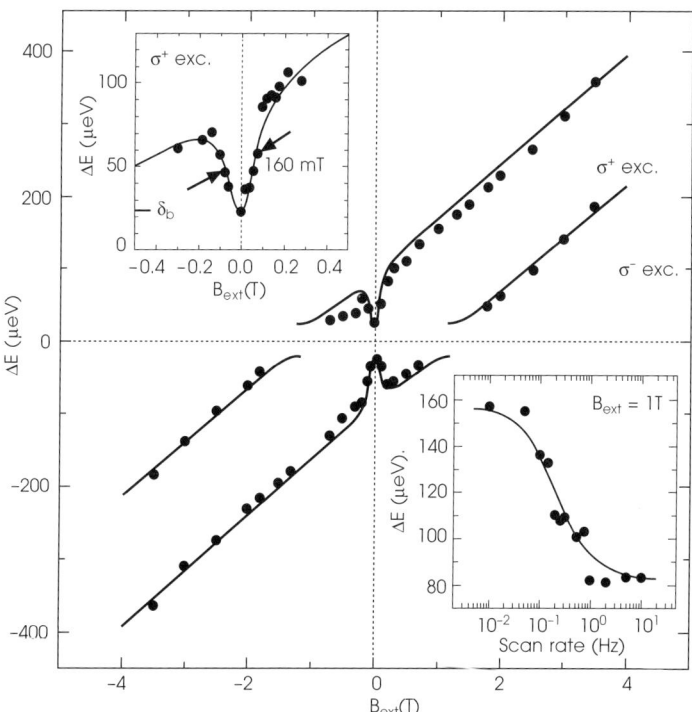

Fig. 6.27 Energy splitting for both σ^+ and σ^- excitation as a function of external magnetic field. Curves are calculated values. Upper inset: Higher resolution data for the region around $B_{ext} = 0$, showing the dip in the energy splitting. Lower inset: Energy splitting at $B_{ext} = 1$ T as a function of the sweep rate of a transverse *rf* field through the nuclear spin resonances. At a rate of (3 s)$^{-1}$ the optical alignment rate equals this external heating rate. From [13].

Optical nuclear magnetic resonance

The nuclear polarization, and therefore the Overhauser shift, can be erased by applying a transverse magnetic field of approximately 1 G with a radio frequency (RF) that is resonant with the nuclear magnetic resonances (Fig. 6.28) [88]. By sweeping the RF field through the As and the Ga resonances while the PL spectrum is obtained, the Overhauser shift is almost completely erased, and the Zeeman splitting reverts to what it would be in the absence of the Overhauser effect.

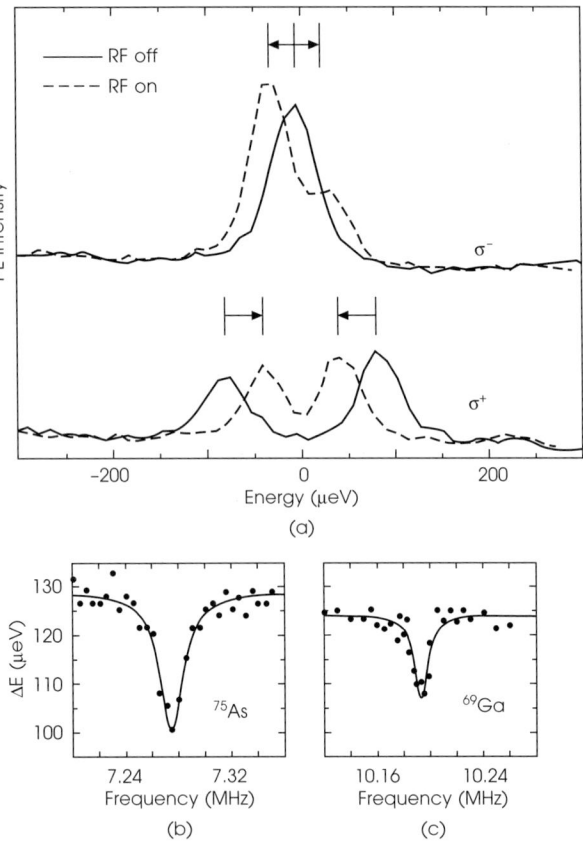

Fig. 6.28 Energy splitting in a single QD PL spectral line in a longitudinal external magnetic field of $B_{ext} = 1$T excited with either σ+ or σ− excitation polarized light with (dashed lines) and without (solid lines) a transverse RF magnetic field continuously scanning through all the Ga and As nuclear magnetic resonances. (b) and (c) NMR spectra from a QD, showing both the ^{75}As and ^{69}Ga resonances, respectively, at $B_{ext} = 1$T. From [88].

The capability to monitor the nuclear polarization through the Overhauser shift and to depolarize it with the RF field provides the opportunity to do nuclear magnetic resonance (NMR) spectroscopy on individual QDs, as shown in Fig. 6.28. The Overhauser shifts measured as the RF field is stepped through the As and the Ga resonant frequencies gives NMR spectra associated with a single QD. This can be done for several QDs simultaneously because we can measure the PL lines from several QDs simultaneously in the same optical spectrum. In some cases there are shifts in the NMR frequencies that may arise from Knight shifts [89]. Knight shifts are shifts in the nuclear spin resonance due to an effective magnetic field associated with the electronic spin through the hyperfine interaction. Furthermore, there are large differences in linewidths from dot to dot that are not understood.

Because of the local nature of the hyperfine interaction, the NMR spectroscopy is sensing only those nuclei within the wave function of the QD exciton, corresponding to approximately 10^4 nuclei [88]. The capability to measure the NMR spectra associated with individual QDs provides a very local probe of strain and composition. Many fascinating questions remain about the dynamics of the exciton and nuclear spin that perhaps can be addressed with the new perspectives obtained from such experiments.

6.6 Spin relaxation

A phenomenological description of spin relaxation was given in section 6.3. Here we consider the origin of spin relaxation for a *localized electron*. In contrast to previous sections we do not treat excitons but focus entirely on electrons. An electron spin in a remotely doped QD (or neutral donor) can be probed through the trion (or neutral donor-bound exciton) PL transition.

The interaction of the magnetic moment with a magnetic field leads to spin precession around the vector, $g\mathbf{B}$, with a frequency, $\Omega = \mu_B g B/\hbar$. Other interactions can be treated as local, fluctuating magnetic fields. The average value of these fields leads also to spin precession of the carriers, and their fluctuating part results in spin relaxation.

Spin-orbit interactions lead to spin relaxation because the elastic or inelastic scattering of the particles changes their spin orientation. For the electron spin, other interactions arise from the exchange fields of holes, other electrons and paramagnetic impurities, and from the hyperfine field of the nuclei. We will show that for an electron localized in a QD (or donor), the interaction with the hyperfine field of the nuclei limits the electronic spin lifetime. Because the precession periods of the nuclei are so much longer than the electron recombination and spin relaxation times, the nuclei are effectively frozen during the electron's lifetime. However, fluctuations in the nuclear polarization, either from QD to QD or from one optical cycle to the next in the same QD, lead to an inhomogeneous broadening of the lifetime that dominates the measured spin lifetime. At the end of the section we show how this time is measured via the Hanle effect.

Spin relaxation: spin–orbit interactions

Spin–orbit interactions generate the dominant spin relaxation mechanisms of free carriers in bulk semiconductors and 2D QW heterostructures. This interaction is described by various terms in an effective mass Hamiltonian that depend on the symmetry of the crystal lattice and on the heterostructure potential. It strongly affects the motion of free electrons and holes. For example, the energy dispersion of holes in zinc-blende semiconductors depends on the hole spin projection on the wave vector **k**, as described by the Luttinger Hamiltonian, as a result of the strong spin–orbit interaction in the valence band. This strong spin–orbit coupling leads to very fast spin relaxation of the hole that accompanies its momentum relaxation.

The spin–orbit interaction also lifts the spin degeneracy of the free electron states in the conduction band of bulk zincblend semiconductors, which do not have a center of inversion symmetry. This splitting is proportional to the third power of the electron wavevector projections, k_α ($\alpha = x, y, z$), along the cubic crystal axes [100], [010], and [001] and depends strongly on the vector direction. The spin–orbit interaction leads to the electron spin precession (Eq. (6.13)) in an effective magnetic field [90]

$$\mathbf{B}_{Dr} \propto (k_x(k_y^2 - k_z^2)\mathbf{e_x} + k_y(k_z^2 - k_x^2)\mathbf{e_y} + k_z(k_x^2 - k_y^2)\mathbf{e_z}),$$

where $\mathbf{e}_\alpha$ are the unit vectors along the cubic crystal axes.

In asymmetrical quantum wells with different hetero-interfaces, or with an external or built-in electric field normal to the QW plane, the spin-orbit interaction leads to an effective magnetic field (the Rashba field) that acts on the spin of two-dimensional electrons. This field is proportional to $\mathbf{B}_{Ras} \propto [\mathbf{n} \times \mathbf{k}]$ [91], where **n** is the polar vector normal to the QW plane and **k** is the two-dimensional wave vector of the electron.

There are two relaxation mechanisms of the spin polarization of *free electrons* connected with the spin–orbit interactions. The first one, the Elliott–Yafet (EY) mechanism, is connected to the spin-flip transitions that result from momentum changing scattering processes. The second one, the D'ykonov–Perel (DP) mechanism, is connected with the precession of the electron spin in the effective magnetic field connected with the spin–orbit interaction between the two scattering events. The DP mechanism of the electron spin relaxation is dominant in semiconductors with high mobility.

For a *localized electron* (or hole) the direction and the value and direction of its spin averaged over a unit cell depends on the position of the cell in the localization volume as a result of the spin–orbit interaction. The distribution of the electronic spin density and orientation varies from state to state. However, these dependences do not change with time. As a result, the spin relaxation of the localized states arising from spin–orbit interaction occurs only via transitions between the different quantum size levels; for example, phonon assisted or free electron gas assisted transitions. For electrons these spin relaxation processes have been analyzed by Khaetzkii and Nazarov [92], and for excitons by Takagahara [50]. The exciton spin relaxation time in natural GaAs quantum dots for inter-level one-phonon transitions has been estimated to be 100 ps at a temperature of $T = 30$K. The transitions between the degenerate spin sublevels of the same localized state are allowed only through excited quantum size levels via the two-phonon-assisted real or virtual transitions. The time of these processes has been estimated in Ref. [93] to be on the order of 1–9 ns at $T = 30$K. The number of phonons and the rate of phonon-assisted recombination therefore depends strongly on temperature: the rate of one-phonon and two-phonon processes decreases with temperature as T and T^2, respectively (see also Ref. [94]). The exciton relaxation in QDs is mainly connected with the relaxation of holes. For localized electrons this time is significantly longer and exceeds microseconds; the spin–orbit interaction in the conduction band is of order one hundred times weaker than in the valence band [92]. As a result, the hyperfine interaction with the spin of the nuclei becomes a dominant mechanism for spin relaxation of localized electrons.

Spin relaxation: hyperfine interaction

The interaction of localized electrons with nuclei was studied early on for electrons localized at donors in bulk semiconductors (see, for example, [8,95]). There, the electron interacts with a large number of nuclei and feels the hyperfine magnetic field of the nuclei, $\mathbf{B}_N$, (Eq. 6.36) located in the region where the electron is localized; this, of course, is also true for electrons localized in QDs. The interaction time of the electron and nuclear spin for donor-localized electrons is usually limited by the time of shallow donor ionization, or by tunneling between donors. This can be considered as the correlation time for the electron spin motion, τ_c (see the Appendix). For quantum computation and spin storage it is important to have a large value of τ_c, and therefore, to eliminate tunneling and ionization, and to maximize this time, we will consider only the electrons localized in QDs or on shallow donors at sufficiently low temperature.

Electron spin relaxation via its interaction with the spins of the nuclei in QDs and donors is facilitated by the disparity of the characteristic time scales of the three processes that determine the relaxation: (1) the period of electron precession in the frozen fluctuation of the hyperfine field of the nuclei, (2) the period of the nuclear spin precession in the hyperfine field of the electron, and (3) the nuclear spin relaxation time in the dipole–dipole field of its nuclear neighbors. Estimates of these time scales can be made for the case of GaAs, whose hyperfine constants are well known. For QDs containing 10^5 nuclei they are: $\sim$ 1ns, $\sim$ 1µs and $\sim$100µs, respectively. Below we describe the electron spin relaxation as a precession in the quasi-stationary frozen fluctuation of the hyperfine field of the nuclear spins without and with external magnetic field. In a strong external magnetic field the precession of the nuclear spins does not affect the electron spin motion, and the model of the frozen fluctuation of hyperfine fields of nuclei is limited only by nuclear dipole–dipole interactions. These interactions do not conserve the total nuclear spin, and the third time scale provides a natural limit to the coherence of the electron-nuclear spin system.

The effective nuclear hyperfine field, $\mathbf{B}_N$, acting on a localized electron spin is the sum of contributions from a large number of nuclei:

$$\mathbf{B}_N = \frac{v_0}{\mu_B g_e} \langle \sum_j A^j |\psi(\mathbf{R}_j)|^2 \hat{\mathbf{I}}^j \rangle_N, \qquad (6.50)$$

where $\langle...\rangle_N$ denotes the quantum mechanical average (trace) over the nuclear wave function, and g_e is the isotropic electron g-factor. Note that the nuclear field introduced in Eq. (6.36) is the statistical average of Eq. (6.50) over the distribution function that reflects the probability of different nuclear spin configurations. Usually the nuclear spin temperature is rather high so that the nuclear spins are uncorrelated. The magnitude and direction of this hyperfine field are randomly distributed, and described by a Gaussian probability density distribution function:

$$W(\mathbf{B}_N) = \frac{1}{\pi^{3/2} \Delta_B^3} \exp\left[-\frac{(\mathbf{B}_N)^2}{\Delta_B^2}\right], \qquad (6.51)$$

where Δ_B is the dispersion of the nuclear hyperfine field distribution, which is determined by the following equation:

$$\Delta_B^2 = \frac{2}{3} \frac{\sum_j I^j (I^j + 1)(A^j)^2}{(\mu_B g_e)^2} \frac{v_0}{V_L}, \qquad (6.52)$$

where the sum in this equation goes over only those nuclei in a unit cell, and $V_L = \left(\int d^3 r \psi^4(r)\right)^{-1}$ is the typical localization volume of the electron. In GaAs, $I = 3/2$, for all nuclei, $\sum_j (A^j)^2 \approx 1.2 \cdot 10^{-3}$ meV^2 and $\Delta_B = 54G$.

The equation of the spin's motion in a fixed magnetic field, $\mathbf{B}$, is given by:

$$\mathbf{S}(t) = (\mathbf{S}_0 \cdot \mathbf{n})\mathbf{n} + \{\mathbf{S}_0 - (\mathbf{S}_0 \cdot \mathbf{n})\mathbf{n}\} \cos \omega t + [\{\mathbf{S}_0 - (\mathbf{S}_0 \cdot \mathbf{n})\mathbf{n}\} \times \mathbf{n}] \sin \omega t, \qquad (6.53)$$

where $\mathbf{S}_0$ is the initial spin, $\mathbf{n} = \mathbf{B}/B$ is a unit vector in the direction of the magnetic field, and $\omega = \mu_B g_e B/\hbar$ is the Larmor frequency of the electron precession in this field. The equation also describes the coherent electron spin precession in a single QD due to the magnetic field, $\mathbf{B}_N$ of the frozen fluctuation of the nuclei ($\mathbf{n} = \mathbf{B}_N/B_N$, and $\omega = \mu_B g_e B_n/\hbar$). Averaging Eq. (6.53) over the magnetic field distribution of Eq. (6.51), we obtain the time dependence of the ensemble averaged electron spin polarization in the absence of an external field [93]:

$$\langle \mathbf{S}(t) \rangle = \frac{\mathbf{S}_0}{3} \left\{ 1 + 2\left[1 - 2\left(\frac{t}{T_d}\right)^2\right] \exp\left[-\left(\frac{t}{T_d}\right)^2\right] \right\} \quad (6.54)$$

averaged over a large number of measurements. Here

$$T_d = \frac{\hbar}{\mu_B g_e \Delta_B} = \sqrt{\frac{3}{2}} \frac{\hbar}{\delta_s} \quad (6.55)$$

is the ensemble dephasing time which arises from the random electron precession frequencies in the randomly distributed frozen fluctuation of the nuclear hyperfine field in the dots. This time is on the order of 1 ns for GaAs quantum dots with 10^5 nuclei. The spin dephasing time is proportional to $\sqrt{V_L}$. One can see that the average electron polarization relaxes to 10% of its original value after a time equal to the dephasing time and then increases to a steady-state value of 33% of its initial polarization.

A strong external magnetic field, B, $(B \gg B_N)$ significantly changes the process of electron spin relaxation. In this large field the Zeeman splitting of the electron spin levels is larger than their inhomogeneous broadening in the hyperfine nuclear magnetic field. The total magnetic field acting on the electron is now effectively directed along the external magnetic field. The nuclear hyperfine fields only perturb the precession frequency of the electron spin about the external magnetic field direction.

Consider, now, the effect of a strong external magnetic field on the electron spin polarization. The motion of the spin in the total magnetic field is again described by Eq. (6.53) where, now, $\mathbf{n} = (\mathbf{B} + \mathbf{B}_N)/|\mathbf{B} + \mathbf{B}_N|$. Averaging Eq. (6.53) over the ensemble, using the distribution of nuclear magnetic fields in Eq. (6.51), we obtain:

$$\langle \mathbf{S}(t) \rangle = R_\parallel(t)(\mathbf{S}_0 \cdot \mathbf{b})\mathbf{b} + R_\perp^0(t)\{\mathbf{S}_0 - (\mathbf{S}_0 \cdot \mathbf{b})\mathbf{b}\} + R_\perp^1(t)[\{\mathbf{S}_0 - (\mathbf{S}_0 \cdot \mathbf{b})\mathbf{b}\} \times \mathbf{b}] \quad (6.56)$$

where $\mathbf{b} = \mathbf{B}/B$ is a unit vector along the external magnetic field, and R_α^k are the time dependent coefficients, with explicit dependence on the parameter, $\beta = B/\Delta_B$, which one can find in Ref. [93]. Equation (6.56) simplifies considerably in strong magnetic fields. In the limit $\beta \gg 1$ Eq. (6.56) can be written:

$$\langle \mathbf{S}(t) \rangle \approx \left\{ 1 - \frac{1 - \cos(\omega_B t)}{\beta^2} \exp\left[-\left(\frac{t}{2T_d}\right)^2\right] \right\} (\mathbf{S}_0 \cdot \mathbf{b})\mathbf{b}$$

$$+ \left\{ \left[\cos(\omega_B t) + \frac{1 - \cos(\omega_B t)}{2\beta^2}\right][\mathbf{S}_0 - (\mathbf{S}_0 \cdot \mathbf{b})\mathbf{b}] \right. \\ \left. + \sin(\omega_B t)[(\mathbf{S}_0 - (\mathbf{S}_0 \cdot \mathbf{b})\mathbf{b}) \times \mathbf{b}] \right\} \exp\left[-\left(\frac{t}{2T_d}\right)^2\right] \quad (6.57)$$

One can see that in strong magnetic fields, $B \gg \Delta_B$ the component of spin along **B** is conserved, while its two transverse components precesses with a frequency $\omega_B = \mu_B g_e B/\hbar$ and decay as a result of the inhomogeneous broadening of the levels in the random magnetic field of nuclei, respectively. The dephasing arises from the dispersion of the nuclear field along the external magnetic field, which leads to an inhomogeneous dispersion of the electron precession frequency.

In a strong external magnetic field ($B \gg \Delta_B$) the average electron spin is directed along this strong field, independent of the nuclear hyperfine fields ($B_N \ll B$). Although the nuclei precess with different frequencies in the inhomogeneous electron field, the electron is affected only by the component of the nuclear field along the external field. As a result the nuclear magnetic field acting on the electron spin is frozen for times much longer than T_N. Thus a frozen fluctuation model of the nuclear hyperfine field is valid when describing the dephasing dynamics of the electron spin polarization in an ensemble of quantum dots in strong magnetic fields. As we mentioned above this consideration is limited by the condition of sufficiently low temperature and by the time scale of the nuclear dipole–dipole interaction.

In each QD, the motion of the electron spin in the "frozen" hyperfine field of the nuclei is coherent. The dephasing is a result of inhomogeneous broadening of the electron spin levels in the ensemble of quantum dots. This makes it possible to recover the transverse electron spin polarization using the spin echo technique, which also can be used for quantum computation [96].

Hanle effect for localized electrons

We previously considered the Hanle effect for the neutral exciton pseudo-spin in section 6.3. For the exciton case the circular polarization of the PL is simply related to the z-component of the pseudo-spin. One can also measure the Hanle effect in a doped n-type system and thereby probe the spin relaxation of electrons through the recombination of the electrons with a photo-excited hole. If the hole spin relaxes quickly to an unpolarized state, the Hanle effect in the polarization of the PL is determined by the dynamical behavior of the electron spin as governed by the Bloch equation (6.21). One can also probe the spin dynamics of an electron in an isolated neutral donor or in a doped QD through the polarization of the PL, but not necessarily so directly. The polarization of the PL is proportional to the spin polarization of the trion (or neutral donor-bound exciton) which is related to the polarization of the electron through the dynamics of the trion (donor-bound exciton) formation or through exchange interactions (see section 6.4). Nevertheless, the spin dynamics of the localized electron can be described by the

Bloch equation (6.21), and in some cases, directly determines the PL polarization. The initial electron spin polarization for the ground state of a neutral donor or doped QD results from optical pumping of the ground state spin. Experimentally, this requires the hole spin flip rate in the optically excited state to be faster than the electron spin flip rate as discussed in section 6.4.

Spin relaxation can be viewed as the result of the action of random local magnetic fields on the electron spin [97] and depends both on the amplitude of the random field and on its fluctuation rate (τ_c). In the limit of short correlation time (the so-called motional narrowing case) the precession period of the electron spin in the local field ($\omega_f^{-1} = \hbar/g\mu_B B_N$) is much longer then the characteristic time of the random field fluctuation (τ_c). The dynamic averaging of local random fields takes place and the spin relaxation slows down: $\tau_s^{-1} \approx \omega_f^2 \tau_c$. In this case, the transverse magnetic field that depolarizes the electron spin in Voigt geometry (Hanle effect) satisfies the condition, $\Omega_{ext} \geq T_s^{-1} = \tau_s^{-1} + \tau^{-1}$, and the magnetic depolarization curve (Hanle curve) has a Lorenzian shape (see Eq. (6.55)). This situation is typical for spin relaxation of free electrons (see above) as well as for the donor-bound electrons with relatively larger donor concentration ($N_d \approx 10^{16}$ cm^{-3}) in bulk GaAs at low temperature [8]. In the latter case the hyperfine interaction with lattice nuclei is responsible for relaxation; the random nuclear fields in the vicinity of the donor sites are averaged as a result of fast jumps of electrons from donor to donor. The fast jumps (short τ_c) strongly suppress the spin relaxation of the donor bound electrons, leading to an extremely long spin relaxation time [98]. This case may be important in the field of spintronics where it is necessary to obtain giant spin diffusion and drift lengths.

The limit of long correlation times ($\tau_c > \omega_f^{-1}$) is realized for electrons localized in QDs (or on shallow donors) with low concentrations at sufficiently low temperatures such that the jumping of spins between the localized states becomes much slower. In this case the role of the correlation time is played by the electron lifetime, and relaxation of an ensemble of electron spins can be determined by fluctuations in the local precession frequency (ω_f) as it was described in Eq. (6.55).

Hanle studies of self-assembled QDs in the steady-state regime and using the quantum beat spectroscopy have been reported in Ref. [65,99] and Ref. [100], respectively. Unfortunately, no work was done on specific spin relaxation mechanisms in the papers cited above. One of the problems is related to the high anisotropy of self-assembled QDs, leading to anisotropy of all the main parameters that determine the Hanle effect (g-factor anisotropy and/or anisotropy of spin lifetimes). Also, the main parameters can vary from dot to dot, which induces additional inhomogeneous broadening.

The simplest case that is free of the above complications and has been realized experimentally [101] is the spin relaxation of donor bound electrons in bulk n-GaAs with $N_d \sim 10^{14}$ cm^{-3}. The shallow donor impurity in GaAs is spherically symmetric with good accuracy, and its binding energy has rather small inhomogeneous broadening (if compensation is not too high). This is especially true for low-doped GaAs samples where the distance between isolated donors greatly exceeds the Bohr radius of electron. In this case the electrons are trapped on donors at low temperatures and spend a relatively long time (long correlation time limit $\tau_c > \omega_f^{-1}$) before escaping as a result of ionization, tunneling, hopping, or recombination with holes. The electron spin system breaks up into an ensemble of individual donor-bound electrons, and the electronic spin relaxation is the result of dephasing of the ensemble spin by the randomly distributed hyperfine fields as we discussed in above. In this case the longitudinal external field (the Faraday geometry) eliminates spin relaxation, and the transverse field (the Voigt geometry) depolarizes electrons when it overwhelms the local field, i.e. $\Omega \geq \omega_f$.

In Fig. 6.29, polarization data taken in the Faraday (upper curve) and Voigt (bottom curve) geometry in n-GaAs are shown [101]. One can see that the longitudinal field increases the electron polarization by a factor of 2.5 whereas the transverse magnetic field decreases the electron polarization (Hanle effect) down to zero. The characteristic transverse and longitudinal magnetic field values that decrease and increase, respectively, the electron polarization are similar in Fig. 6.29 and equal to $\approx 54G$ which corresponds to an inhomogeneous dephasing rate of 5 ns. This number is in good agreement with the theoretical value of T_d (see Eq. (6.55)) calculated for a shallow donor in GaAs.

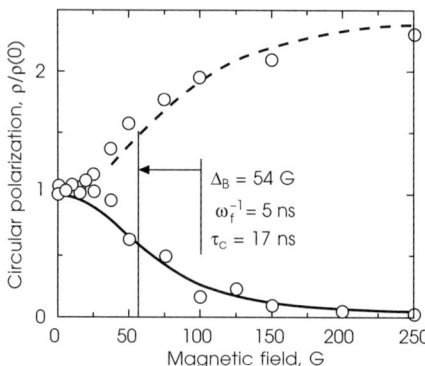

Fig. 6.29 The magnetic field dependence (open circles) of the circular polarization in Voigt (lower points) and Faraday (upper points) geometries. W = 40 mW/cm^2, T = 4.2K. Polarization degree is normalized to the zero field values, $\rho(0) = 2\%$. The dashed and solid lines are the theoretical dependences of the electron spin calculated from Eq. (6.57) with the *only* fitting parameter, $\tau_c = 17$ ns.

To describe the average electron spin polarization in the low concentration limit we consider the evolution of the electron spin, $\vec{S}_n$, localized at the nth donor. It is described by the Bloch equation (6.21) with the precession frequency, $\vec{\Omega}_n = \vec{\Omega}_{ext} + \vec{\omega}_{fn}$. During the correlation time (τ_c) the electron spin dynamics are determined by the external and the local nuclear magnetic fields only. However, each electron sees a different static nuclear hyperfine field, described by a Gaussian distribution (Eq. (6.51)) with a dispersion, $\Delta = 54G$, given by the root-mean-square of the random nuclear field. Averaging the (Lorentzian) solution to the Bloch Eq. (6.21) over the nuclear magnetic field distribution (see section 6.6) results in the theoretical dependence of circular polarization on the longitudinal and transverse magnetic field that are presented in Fig. 6.29 by dashed and solid lines, respectively. The *only* fitting parameter for the curves in Fig. 6.29 is $\tau_c = 17$ ns, which satisfies the static condition, $\tau_c > \omega_f^{-1}$. This value is determined by the processes of donor electron ionization, recombination, exchange scattering by free electrons, etc. The good agreement between experiment and theory confirms that the interaction with nuclei is the main mechanism of the electron spin relaxation for the localized carriers.

6.7 Conclusions

In this chapter we have discussed the most important features in the optical spectra of quantum dots that arise from spin. We have discussed how the recent introduction of single quantum dot spectroscopy and the ability to probe individual excitons has made it possible to measure the fine-structure of single excitons. We have discussed how this fine structure and the polarization depends on the exchange interaction, on the Zeeman interaction, and on the hyperfine interaction. We also considered how spin relaxation occurs and one way that it can be measured through the Hanle effect.

Of course there are many experimental techniques that we did not discuss that also are interesting and important. The most obvious are time-resolved techniques. One popular method in time resolved spectroscopy is that of *quantum beats*. In this approach the energy splitting of two fine-structure states can be measured as a beating frequency, either in the signal intensity as a function of the delay time between two pulses, or in the time-development of the PL after a laser excitation pulse. As an example we refer to the experiment of Bonadeo *et al.* [102], in which the fine structure splitting of the two bright states of a single quantum dot was measured as a beating frequency in the spectrum. As expected, the beating frequency and the rate at which the beats decay (dephasing rate) are in good agreement with the fine structure splitting and the linewidths as measured in frequency-domain PL excitation spectroscopy. This experiment is discussed in the following chapter.

There has been, and continues to be, a great deal of work on bulk semiconductors and more recently on quantum wells. It is interesting to consider what the new opportunities might be in QDs. One example, recently discussed by Korenev [87], is the phase transition of the nuclear spin of a QD into a self-polarized state. Nuclear self-polarization was predicted long ago in bulk semiconductors, but has not yet been detected, possibly because of the difficulty in obtaining the predicted conditions (very low temperatures under optical pumping conditions). The required conditions are much more relaxed in a QD. The essence of this theory is that the nuclear polarization depends strongly on the splitting of the exciton spin levels. In turn, the splitting is sensitive to the nuclear polarization by the average hyperfine nuclear field. Under certain conditions the positive feedback becomes strong enough to produce self-polarization. Moreover, the nonlinear behavior can bring about bistability in the system that should manifest itself in hysteretic behavior of nuclear polarization versus magnetic field. A self-polarized QD would look like a tiny nuclear magnet that could be used to control electron spin or possibly decoherence in future quantum mechanical devices.

Another exciting direction is toward the design and demonstration of QD structures that have special functionality. One example is the possibility of designing a quantum bit for exploring the concepts in quantum information science. There has been considerable progress on the coherent control of excitons in single quantum dots which themselves can act as qubits (see Chapter 7). The problem with excitons is that they are excited states of the quantum dot, and so their coherence time is ultimately limited by recombination (radiative lifetime). One possible way around this problem is to design a system in which a spin degeneracy in the ground state can act as the qu-bit. This can be done by putting a single electron in a quantum dot and using its two spin states as the qu-bit. The qu-bit could be optically controlled by using the trion as an intermediate state in what would then be a three level system. In such a system it would be possible to optically control and probe the ground state spin coherence of individual electrons. The spin coherence would have a relatively long decoherence time, because it is in the ground state and because of the relative insensitivity of spin to environmental noise.

These examples of current research ideas illustrate the exciting opportunities that exist in the field. We anticipate that there will continue to be much activity as QD materials and experimental methodologies are further developed.

Electronic and nuclear spin in the optical spectra   273

Acknowledgments

We thank Roslan Dzhioev for helpful discussions. This work was supported by ONR, DARPA/SpinS, NSA/ARDA/ARL CRDF, and RFBI.

Appendix   Relaxation of the nuclear spin due to the fluctuating electronic spin

We will derive Eq. (6.44) for the nuclear relaxation rate. Let us consider the relaxation rate of the nuclear polarization, $W(B)$, in a random, rapidly alternating hyperfine field of a localized electron. For simplicity we assume at the beginning that there is not an external magnetic field acting on the electron spin, and that the random hyperfine field of the electron is perpendicular (for example, along the y-axis) to the nuclear spin (for example, along the z-axis) (see Fig. 6.30). After the time, t, the z-projection of the nuclear spin decreases by the value $I(0)(\omega_{N1}t)^2/2 = I(0)\vartheta^2/2$. At the same time, the nuclear spin gains an x-projection equal to $I(0)\vartheta$. However, the average value of this projection is equal to zero because the positive and negative values of ω_N occur with equal probability. Averaging over random directions of the hyperfine electron field leads to the additional factor $2/3$ in the time dependence of z-projection of the nuclear spin because the z-projection of the hyperfine field does not lead to its relaxation, and the contribution of its X and Y projection are equal to each other, so that

$$\langle \omega_N^2 \rangle = \langle \omega_{Nx}^2 \rangle + \langle \omega_{Ny}^2 \rangle + \langle \omega_{Nz}^2 \rangle = 3\langle \omega_{N1}^2 \rangle.$$

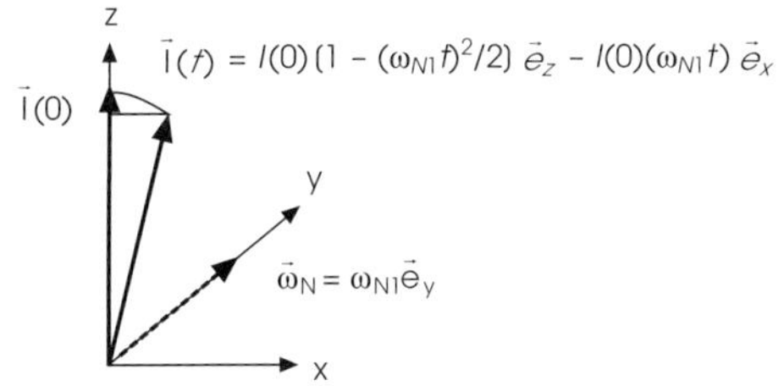

Fig. 6.30 Schematic of precession of the nuclear spin in the hyperfine field of the electron spin.

Now let us consider the result of a hyperfine field rotation caused by the rotation of the electronic spin in an external magnetic field along the z-axis: $\vec{\omega}_{N1}(t) = \omega_{N1}(\vec{e}_x \sin \Omega t + \vec{e}_y \cos \Omega t)$. One can write with the same accuracy:

$$\vec{I}(t) = [(I(0) - (I_x^2(t) + I_y^2(t))/(2I(0)))\vec{e}_z + I_x(t)\vec{e}_x + I_y(t)\vec{e}_y]$$

where

$$I_x(t) = I(0)\omega_{N1}\int_0^t \sin\Omega_e \tilde{t}\, d\tilde{t} = -I(0)\omega_{N1}\frac{\cos\Omega_e t - 1}{\Omega_e}.$$

$$I_y(t) = I(0)\omega_{N1}\int_0^t \sin\Omega_e \tilde{t}\, d\tilde{t} = I(0)\omega_{N1}\frac{\sin\Omega_e t}{\Omega_e}$$

Then

$$I_x^2(t) + I_y^2(t) = I(0)^2\omega_{N1}^2\frac{2(1-\cos\Omega_e t)}{\Omega_e^2}$$

One can get the average loss of the nuclear polarization during the time of coherent precession in the random field of the electron after averaging this expression over the exponential distribution ($\tau_c^{-1}\exp(-t/\tau_c)$) of the acting time of the random field:

$$\langle\Delta I_z\rangle = -\frac{I(0)\omega_{N1}^2}{\Omega_e^2\tau_c}\int_0^\infty (1-\cos\Omega_e t)\exp\left(-\frac{t}{\tau_c}\right)dt$$

$$= -\frac{I(0)\omega_{N1}^2}{\Omega_e^2}\left[1 - \frac{1}{2(1+i\Omega_e\tau_c)} - \frac{1}{2(1-i\Omega_e\tau_c)}\right]$$

$$= -\frac{I(0)\omega_{N1}^2}{\Omega_e^2}\left[1 - \frac{1}{1+(\Omega_e\tau_c)^2}\right]$$

$$= -\frac{I(0)\omega_{N1}^2\tau_c^2}{1+(\Omega_e\tau_c)^2},$$

which leads to the resulting expression in the case of filling factor 1:

$$\frac{dI_z}{dt} = \frac{\langle\Delta I_z\rangle}{\tau_c} = -\frac{I(0)\omega_{N1}^2\tau_c}{1+(\Omega_e\tau_c)^2}.$$

Therefore the relaxation rate (W), which is defined by $\frac{dI_z}{dt} = -W\langle\Delta I_z\rangle$, is

$$W(B) = \frac{\omega_{N1}^2\tau_c}{1+(\Omega_e\tau_c)^2} = \frac{2}{3}\frac{\omega_N^2\tau_c}{1+(\Omega_e\tau_c)^2}$$

This concludes the derivation of Eq. (6.44).

References

1. See, for example, Special Issue on Semiconductor Quantum Dots. (1998). *MRS Bul.* **23**.
2. Lipsanen, H., Sopanen, M., and Ahopelto, J. (1995). Luminescence from excited states in strain-induced $In_xGa_{1-x}As$ quantum dots. *Phys. Rev. B* **51**, 13868–13871.
3. Special Issue, *J. Luminescence* Spectroscopy of isolated and assembled semiconductor nanocrystals. v. 70, 1–484, eds. L.E. Brus, Al. L. Efros and T. Itoh (1996).
4. Gammon, D. (1998). High resolution spectroscopy of quantum dots in quantum wells. *MRS Bul.* **23**, 44–48.
5. Gammon, D., Snow, E.S., Shanabrook, B.V., Katzer, D.S., and Park, D. (1996). Homogeneous linewidths in the optical spectrum of a single quantum dot. *Science* **273**, 87–90.
6. Borri, P., Langbein, W., Schneider, S., Woggon, U., Sellin, R.L., Ouyang, D., and Bimberg, D. (2001). Ultralong dephasing time in InGaAs quantum dots. *Phys. Rev. Lett.* **87**, 157401–157404.
7. Birkedal, D., Leosson, K., and Hvam, J.M. (2001). Long lived coherence in self-assembled quantum dots. *Phys. Rev. Lett.* **87**, 227401–227404.
8. Meier, B., and Zakharchenya, B.P. (eds) *Optical Orientation*, (North-Holland, Amsterdam, 1984).
9. Kastler, A (1967). Optical methods for studying Hertzian Resonances. *Science* **158**, 214.
10. Ref. 8, Chap. 1.
11. Ref. 8, Chap. 2.
12. Brown, S.W., Kennedy, T.A., Gammon, D., and Snow, E.S. (1996). Spectrally resolved Overhauser shifts in single $GaAs/Al_xGa_{1-x}As$ Quantum dots. *Phys. Rev. B* **54**, R17339–R17342.
13. Gammon, D., Efros, Al. L., Kennedy, T.A., Rosen, M., Katzer, D.S., Park, D., Brown, S.W., Korenev, V.L., and Merkulov, I.A. (2001). Electron and nuclear spin interactions in the optical spectra of single quantum dots. *Phys. Rev. Lett.* **86**, 5179–5179.
14. Bockelmann, U., and Bastard, G. (1990). Phonon scattering and energy relaxation in two, one, and zero-dimensional electron gases. *Phys. Rev. B* **42**, 8947–8951.
15. Benisty, H., Sotomayor-Torres, C., and Weisbuch, C. (1991). Intrinsic mechanism for the poor luminescence properties of quantum-box systems. *Phys. Rev. B* **44**, 10945–10948.
16. Hegarty, J., and Sturge, M.D. (1985) Studies of exciton localization in quantum-well structures by nonlinear-optical techniques. *J. Opt. Soc. Am. B* **2**, 1143.
17. Takagahara, T. (1989) Excitonic relaxation processes in quantum well structures. *J. Lumin.* **44**, 347–366.
18. Zimmermann, R., and Runge, E. (1997) Excitons in narrow quantum wells: Disorder localization and luminescence kinetics. *Phys. stat. sol. (a)* **164**, 511–516; Runge, E. and Zimmermann, R. (1998). Spatially resolved spectra, effective mobility edge, and level repulsion in narrow quantum wells. *Phys. stat. sol. (b)* **206**, 167.
19. Paget, D., Lampel, G., B. Sapoval, B., and Safarov, V.I. (1977). Low field electron-nuclear spin coupling in gallium arsenide under optical pumping conditions. *Phys. Rev. B* **15**, 5780–5796.
20. Ourmazd, A., Taylor, D.W., Cunningham, J., and Tu, C.W. (1989). Chemical mapping of semiconductor interfaces at near atomic resolution. *Phys Rev. Lett.* **62**, 933–936.

21. Warwick, C.A., Jan, W.Y., Ourmazd, A., and Harris, T.D. (1990). Does luminescence show semiconductor interfaces to be atomically smooth? *Appl. Phys. Lett.* **56**, 2666–2668.
22. Gammon, D., Shanabrook, B.V., and Katzer, D.S. (1991). Excitons, phonons, and interfaces in GaAs/AlAs quantum-well structures. *Phys. Rev. Lett.* **67**, 1547.
23. Moison, J.M., Guille, C., Houzay, F., Barthe, F., and Van Rompay, M. (1989). Surface segregation of third-column atoms in group III-V arsenide compounds: Ternary alloys and heterostructures. *Phys. Rev. B* **40**, 6149–6162.
24. Katzer, D.S., Shanabrook, B.V., and Gammon, D. (1994). Raman-scattering study of the intermixing of alas monolayers in gaas grown by molecular-beam epitaxy. *J. Vac. Sci. Technol.* **12**(2), 1056–1058.
25. Ramsteiner, M., Hey, R., Kann, R., Jahn, U., Gorbunova, I., and Ploog, K.H. (1997). Influence of composition fluctuations in Al(Ga)As barriers on the exciton localization in thin GaAs quantum wells. *Phys. Rev. B* **55**, 5239.
26. Belousov, M.V., Yu, A., Chernyshov, I.V., Ignatev, I.E., Kozin, A.V., Kavokin, H.M., Gibbs, H., Khitrova, G. (1998). Statistical model explaining the fine structure and interface preference of localized excitons in type-II GaAs/AlAs superlattices. *J. Nonlinear Optical Physics and Materials* **7**, 13–35.
27. Weisbuch, C., Miller, R.C., Dingle, R., Gossard, A.C., and Wiegman, W. (1981). Intrinsic radiative recombination from quantum states in GaAs-GaAlAs multi-quantum well structures. *Solid State Comm.* **37**, 219–222.
28. Hegarty, J., Sturge, M.D., Weisbuch, C., Gossard, A.C., and Wiegmann, W. (1982). Resonant Rayleigh scattering from an inhomogeneously broadened transition: a new probe of the homogeneous linewidth. *Phys. Rev. Lett.* **49**, 930; Hegarty, J., Goldner, L., and Sturge, M.D. (1984). Localized and delocalized two-dimensional excitons in GaAs-AlGaAs multiple-quantum-well structures. *Phys. Rev. B*, **30**, 7346.
29. Gammon, D., Shanabrook, B.V. and Katzer, D.S. (1990). Interfaces in gaas alas quantum-well structures. *Appl. Phys. Lett.* **57**, 2710–2712.
30. Brunner, K., Bockelmann, U., Abstreiter, G., Walther, M., Bohm, G., Trankle, G., and Weimann, G. (1992). Photoluminescence from a single GaAs/AlGaAs quantum dot. *Phys. Rev. Lett.* **69**, 3216–3219.
31. Brunner, K., Abstreiter, G., Böhm, G., Tränkle, G., and Weimann, G. (1994). *Appl. Phys. Lett.* **64**, 3320; Brunner *et al.* (1994). Sharp-line photoluminescence and two-photon absorption of zero-dimensional biexcitons in a GaAs/AlGaAs structure. *Phys. Rev. Lett.* **73**, 1138.
32. Zrenner, A., Butov, L.V., Hagn, M., Abstreiter, G., Böhm, G., and Weimann, G. (1994). Quantum dots formed by interface fluctuations in AlAs/GaAs coupled quantum well structures. *Phys. Rev. Lett.* **72**, 3382–3385.
33. Hess, H.F., Betzig, E., Harris, T.D., Pfeiffer, L.N., and West, K.W. (1994). Near-field spectroscopy of the quantum constituents of a luminescent system. *Science* **264**, 1740–1745.
34. Gammon, D., Snow, E.S., and Katzer, D.S. (1995). Excited-state spectroscopy of excitons in single quantum dots. *Appl. Phys. Lett.* **67**, 2391–2393.
35. Guest, J.R., Stievater, T.H., Gang Chen, Tabak, E.A., Orr, B.G., Steel, D.G., Gammon, D., and Katzer, D.S. (2001). *Science* **293**, 2224.

36. Marzin, J.-Y., Gerard, J.-M., Izrael, A., Barrier, D., and Bastard, G. (1994). Photoluminescence of single InAs quantum dots obtained by self-organized growth on GaAs. *Phys. Rev. Lett.* **73**, 716–719.
37. Bayer, M. *et al.* (1999). Electron and hole g factors and exchange interaction from studies of the exciton fine structure in In0.60Ga0.40As quantum dots. *Phys. Rev. Lett.* **82**, 1748–1751.
38. Bastard, G. (1988). *Wave Mechanics Applied to Semiconductor Heterostructures* (Les Editions de Physiques, Les Ulis, France).
39. Weisbuch, C., and Vinter, B. (1991). *Quantum Semiconductor Structures: Fundamentals and Applications* (Academic Press, San Diego, CA).
40. Ivchenko, E.L., and Pikus, G. (1995). *Superlattices and Other Heterojunctions: Symmetry and Other Optical Properties*, Springer Ser. Solid State Sci., Vol 110 (Springer, Berlin, Heidelberg).
41. Gammon, D., Snow, E.S., Shanabrook, B.V., Katzer, D.S., and Park, D. (1996). Fine structure splitting in the optical spectra of single GaAs quantum dots. *Phys. Rev. Lett.* **76**, 3005–3008.
42. Maialle, M.Z., Andrade de Silva, E.A., and Sham, L.J. (1993). Exciton spin dynamics in quantum wells. *Phys. Rev. B* **47**, 15776–15778.
43. Blackwood, E., Snelling, M.J., Harley, R.T., Andrews, S.R., and Foxon, C.T.B. (1994). Exchange interaction of excitons in GaAs heterostructures. *Phys. Rev. B* **50**, 14246–14254.
44. Van Kesteren, H.W., Cosman, E.C., van der Poel, W.A.J.A., and Foxon, C.T. (1990). Fine structure of excitons in type-II GaAs/AlAs quantum wells. *Phys. Rev. B* **41**, 5283–5292.
45. Glasberg, S., Shtrikman, H., Bar-Joseph, I., and Klipstein, P.C. (1990). Exciton exchange splitting in wide GaAs quantum wells. *Phys. Rev. B* **60**, R16295–R16298.
46. Bayer, M., Stern, O., Kuther, A., and Forchel, A. (2000). Spectroscopic study of dark excitons in $In_xGa_{1-x}As$ self-assembled quantum dots by a magnetic-field-induced symmetry breaking. *Phys. Rev. B* **61**, 7273–7276.
47. Puls, J., Rabe, M., Wünsche, H.-J., and Henneberger, F. (1999). Magneto-optical study of the exciton fine structure in self-assembled CdSe quantum dots. *Phys. Rev. B* **60**, R16303–R16306; Besombes *et al.* (2000). *Phys. Rev. Lett.* **85**, 425.
48. Goupolov, S.V., Ivchenko, E.L., and Kavokin, A.V. (1998). Fine structure of localized exciton levels in quantum wells. *J. Exp. and Theor. Phys.* **86**, 388–394; (1997) *JETP Lett.* **65**, 804.
49. Ivchenko, E.L. (1997). Fine structure of excitonic levels in semiconductor nanostructures. *Phys. Stat. Sol. (a)* **164**, 487–492.
50. Takagahara, T. (2000). Theory of exciton doublet structures and polarization relaxation in single quantum dots. *Phys. Rev. B* **62**, 16840–16855.
51. Luttinger, J.M. (1956). Quantum theory of cyclotron resonance in semiconductors: general theory. *Phys. Rev.* **102**, 1030–1041.
52. Tanaka, T., Singh, J., Arakawa, Y., and Bhattacharya, P. (1993). Near band edge polarization dependence as a probe of structural symmetry in GaAs/AlGaAs quantum dot structures. *Appl. Phys. Lett.* **62**, 756–758.
53. Willatzen, M., Tanaka, T., Arakaw, Y., and Singh, J. (1994). *IEEE J. Quantum Elect.* **30**, 640.

54. Cho, K. (1999). Mechanisms for LT splitting of polarization waves: a link between electron–hole exchange interaction and depolarization shift. *J. Phys. Soc. Jpn.* **68**, 683–692.
55. Kittel, C. (1976). *Introduction to Solid State Physics* (Wiley, New York), p. 399.
56. Andreani, L.C. (1995). Optical transitions, excitons and polaritons in bulk and low-dimensional semiconductor structures in *Confined Electrons and Photons*, E. Burstein and C. Weisbuch, Eds. (Plenum Press, New York), p. 90.
57. Zucker, J.E., Pinczuk, A., Chemla, D.S., Gossard, A., and Wiegmann, W. (1984). Optical vibrational modes and electron–phonon interaction in GaAs quantum wells. *Phys. Rev. Lett.* **53**, 1280–1283.
58. Gammon, D. (2000). Raman scattering in semiconductor heterostructures in *Raman Scattering in Materials Science*, eds. W.H Weber and R. Merlin (Springer-Verlag, Berlin), p. 109.
59. Snelling, M.J., Blackwood, E., McDonagh, C.J., Harley, R.T., and Foxon, C.T.B. (1992). Exciton, heavy-hole, and electron g factors in type-I GaAs/Al$_x$Ga$_{1-x}$As quantum wells. *Phys. Rev. B* **45**, 3922–3925.
60. Marie, X., Amand, T., Le Jeune, P., Paillard, M., Renucci, P., Golub, L.E., Dymnikov, V.D., and Ivchenko, E.L. (1999). Hole spin quantum beats in quantum-well structures. *Phys. Rev. B* **60**, 5811–5817.
61. Dzhioev, R.I., Gibbs, H.M., Ivchenko, E.L., Khitrova, G., Korenev, V.L., Tkachuk, M.N., and Zakharchenya, B.P. (1997). Determination of interface preference by observation of linear-to-circular polarization conversion under optical orientation of excitons in type-II GaAs/AlAs superlattices. *Phys. Rev. B* **56**, 13405–13413.
62. Feynman, R.P., Vernon, F.L., and Helwarth, R.W. (1957). Geometrical representation of the Schrodinger equation for solving maser problems. *J. Appl. Phys.* **28**, 49–52.
63. Dzhioev, R.I., Zakharchenya, B.P., Ivchenko, E.L., Korenev, V.L., Kusrayev, Yu.G., Ledencov, N.N., Ustinov, V.M., and Zhukov, A.E. (1997). Fine structure of excitonic levels in quantum dots. *JETP Lett.* **65**, 804–809.
64. Paillard, M., Marie, X., Renucci, P., Amand, T., Jbeli, A., and Gerard, J.M. (2001). Spin relaxation quenching in semiconductor quantum dots. *Phys. Rev. Lett.* **86**, 1634–1637.
65. Dzhioev, R.I., Zakharchenya, B.P., Korenev, V.L., Pak, P.E., Vinokurov, D.A., Kovalenkov, O.V., and Tarasov, I.S. (1998). Optical orientation of donor-bound excitons in nanosized InP/InGaP islands. *Physics of the Solid State* **40**, 1587–1593.
66. Lampert, M.A. (1958). Mobile and immobile effective-mass-particle complexes in nonmetallic solids. *Phys. Rev, Lett.* **1**, 450–453.
67. Kheng, K. *et al.* (1993). Observation of negatively charged excitons X- in semiconductor quantum wells. *Phys. Rev. Lett.* **71**, 1752–1755.
68. Finkelstein, G. *et al.* (1995). Optical spectroscopy of a two-dimensional electron gas near the metal-insulator transition. *Phys. Rev. Lett.* **74**, 976–979.
69. Shields, A.J. *et al.* (1995). Quenching of excitonic optical-transitions by excess electrons in Gaas quantum-wells. *Phys. Rev. B* **51**, 18049–18052.
70. Karlsson, K.F. *et al.* (2001). Temperature influence on optical charging of self-assembled InAs/GaAs semiconductor quantum dots. *Appl. Phys. Lett.* **78**, 2952.
71. Warburton, R.J. *et al.* (2001). Optical emission from a charge-tuneable quantum ring. *Nature* **405**, 926–929.

72. Findeis, F., Baier, M., Zrenner, A., Bichler, M., Abstreiter, G., Hohenester U., and Molinari, E. (2001). Optical excitations of a self-assembled artificial ion. *Phys. Rev. B* **63**, 121309(R).
73. Haft, D. *et al.* (2001). Luminescence quenching in InAs quantum dots. *Appl. Phys. Lett.* **78**, 2946–2948.
74. Hartmann, A. *et al.* (2000). Few-particle effects in semiconductor quantum dots: observation of multicharged excitons. *Phys. Rev. Lett.* **84**, 5648–5651.
75. Finley, J.J. *et al.* (2001). Charged and neutral exciton complexes in individual self-assembled In(Ga)As quantum dots. *Phys. Rev. B* **63**, 073307.
76. Regelman, D.V. *et al.* (2001). Optical spectroscopy of single quantum dots at tunable positive, neutral, and negative charge states. *Phys. Rev. B* **64**, 165301.
77. Tischler, J.G., Weinstein, B.A., and McCombe, B.D. (1999). *Phys. Stat. Solidi (b)* **215**, 263.
78. Huard, V., Cox, R.T., Saminadayar, K., Arnoult A., and Tatarenko, S. (2000). *Phys. Rev. Lett.* **84**, 187; Volkov, O.V., Kukushkin, I.V., Kulakovskii, D.V., von Klitzing, K., and Eberl, K. (2000). Bistable charge states in a photoexcited quasi-two-dimensional electron–hole system. *JETP Letters* **71**, 322–326.
79. Astakhov, G.V., Kochereshko, V.P., Yakovlev, D.R., Ossau, W., Nurnberger, J., Faschinger, W., and Landwehr, G. (2000). Oscillator strength of trion states in ZnSe-based quantum wells. *Phys. Rev. B* **62**, 10345–10352.
80. Finkelstein, G., Shtrikman, H., and Bar-Joseph, I. (1997). Mechanism of shakeup processes in the photoluminescence of a two-dimensional electron gas at high magnetic fields. *Phys. Rev. B* **56**, 10326.
81. Glasberg, S., Shtrikman, H., and Bar-Joseph, I. (2001). Photoluminescence of low-density two dimensional hole gas in a GaAs quantum well: observation of valence-band Landau levels. *Phys. Rev. B* **63**, 201308(R).
82. Sirenko, A.A., Ruf, T., Kurtenback, A.K., and Eberl, K. (1996). Spin flip Raman scattering in InP/InGaP quantum dots. *23rd Int. Conf. Phys. Semicond*. Berlin. Vol. 2, p. 1385.
83. Dzhioev, R.I., Zakharchenya, B.P., Korenev, V.L., Pak, P.E., Tkachuk, M.N., Vinokurov, D.A., and Tarasov, I.S. (1998). Dynamic polarization of nuclei in a self-organized ensemble of quantum-size n-InP/InGaP islands. *JETP Lett.* **68**, 745–749.
84. Dzhioev, R.I., Zakharchenya, B.P., Korenev, V.L., and Lazarev, M.V. (1999). Interaction between the exciton and nuclear spin systems in a self-organized ensemble of InP/InGaP size-quantized islands. *Physics of Solid State* **41**, 2041.
85. Erlingsson, S.I., Nazarov, Yu. V., and Fal'ko, V.I. (2001). Nucleus-mediated spin-flip transitions in GaAs quantum dots. *Phys. Rev. B* **64**, 195306.
86. Dyakonov, M.I., and Perel, V.I. (1972). *Zh. Eksp. Teor. Fiz. Pis'ma* **16**, 563.
87. Korenev, V.L. (1999). Dynamic self-polarization of nuclei in low-dimensional systems. *JETP Let.* **70**, 129–134.
88. Gammon, D., Brown, S.W., Snow, E.S., Kennedy, T.A., Katzer, D.S., and Park, D. (1997). Nuclear spectroscopy in single quantum dots: nanoscopic Raman scattering and nuclear magnetic resonance. *Science* **277**, 88.
89. Brown, S.W., Kennedy T.A., and Gammon, D. (1998). Optical NMR from single quantum dots. *Solid State NMR* **11**, 49–58.
90. Dresselhaus, G. (1955). Spin–orbit coupling effects in zinc blende structures. *Phys. Rev.* **100**, 580–586.

91. Bychkov, Yu. A., and Rashba, E.I. (1984). Oscillatory effects and the magnetic susceptibility of carriers in inversion layers. *J. Phys. C* **17**, 6039–6045.
92. Khaetskii, A.V., and Nazarov, Y.V. (2001). Spin relaxation in semiconductor quantum dots. *Phys. Rev. B* **61**, 12639–12642.
93. Merkulov, I.A., Efros, Al. L. and Rosen, M. (2001). Electron spin relaxation by nuclei in semiconductor quantum dots. *Phys. Rev. B.* **65**, 205309.
94. Abragam, A., and Bleaney, B. (1970). *Electron Paramagnetic Resonance of Transition Ions*. (Clarendon Press, Oxford).
95. Lampel, G. (1968). Nuclear dynamic polarization by optical electronic saturation and optical pumping in semiconductors. *Phys. Rev. Lett.* **20**, 491–493.
96. Gershenfeld, N., and Chuang, I.L. (1997). Bulk spin-resonance quantum computation. *Science* **275**, 350–356.
97. Dyakonov, M.I., and Perel, V.I. (1972). Dynamics of self-polarization in solids. *JETP Letters* **16**, 398–401.
98. Dzhioev, R.I. *et al.* (1997). Spin diffusion of optically oriented electrons and photon entrainment in n-gallium arsenide. *Phys. Solid State* **39**, 1765–1768; Kikkawa, J.M., and Awshalom, D.D. (1998). Resonant spin amplification in n-type GaAs. *Phys. Rev. Lett.* **80**, 4313–4316; R.I. Dzhioev *et al.* (2001). Long electron spin memory times in gallium arsenide. *JETP Lett.* **74**, 182–185.
99. Epstein, R.J., Fuchs, D.T., Shoenfield, W.V., Petroff, P.M., and Awschalom, D.D. (2001). Hanle effect measurements of spin lifetimes in InAs self-assembled quantum dots. *Appl. Phys. Lett.* **78**, 733–735.
100. Kalevich, V.K., Tkachuk, M.N., Le Jeune, P., Marie, X., and Amand, T. (1999). Electron spin beats in InGaAs/GaAs quantum dots. *Phys. Solid State* **41**, 789–792.
101. Dzhioev, R.I., Korenev, V.L., Merkulov, I.A., Zakharchenya, B.P., Gammon, D., Efros, Al.L., and Katzer, D.S. (2002). Manipulation of the spin memory in n-GaAs. *Phys. Rev. Lett.* **88**, 256801.
102. Bonadeo, N.H., Erland, J., Gammon, D., Park, D., Katzer, D.S., and Steel, D.G. (1998). Coherent optical control of the quantum state of a single quantum dot. *Science* **282**, 1473–1476.

Quantum Coherence, Correlation and Decoherence in
Semiconductor Nanostructures
T. Takagahara (Ed.)
Copyright © 2003 Elsevier Science (USA). All rights reserved.

Chapter 7
Coherent optical spectroscopy and manipulation of single quantum dots

Gang Chen,* T.H. Stievater,[†] J.R. Guest, D.G. Steel

Harrison M. Randall Laboratory of Physics, University of Michigan, Ann Arbor, MI 48109-1120, USA

D. Gammon

Naval Research Laboratory, Washington DC 20375, USA

Pochung Chen, C. Piermarocchi, L.J. Sham

Department of Physics, University of California San Diego, La Jolla, CA 92093, USA

Abstract

Semiconductor quantum dots (QDs) represent the simplest nanostructures where the 3D quantum confinement imposed on the electron and hole motion leads to strongly modified electronic and optical properties. The most remarkable feature that makes QDs so desirable for future generations of optical, electronic and quantum logic devices is their fully quantized atomic-like energy structure (although they normally contain 10^3–10^6 atoms). Compared to atoms and ions that can only be captured for a limited period of time via sophisticated cooling and trapping techniques, QDs are structurally robust and are backed by a strong industrial base for semiconductor processing. An understanding of the fundamental physical origin of the electronic and optical properties associated with quantum confined systems is a prerequisite for more advanced research activities aimed at devices. Studies of simple QD structures are also a stepping stone to designing and understanding more sophisticated QD-based nanostructures, such as QD molecules and coupled QD arrays. This chapter reviews and discusses physics of QDs based on various optical spectroscopy experiments performed at the single QD level. The discussions will be focused on a particular model gallium arsenide (GaAS) QD system in which the experimental efforts of the authors reside. Of specific interest is the optical response from single QD excitons and correlated exciton–exciton molecules (biexcitons) as a result of coherent

---

*Current address: Bell Labs, Lucent Technologies, Murray Hill, NJ 07974, USA.
[†]Current address: Naval Research Laboratory, Washington DC 20375, USA.

optical excitation. Such coherent preparation and manipulation of QD states are among the core elements for devices utilizing quantum phase, such as quantum computers based on optically driven QDs [1–4]. Experiments show that the single QD exciton and biexciton states can be coherently manipulated in a similar way to atoms. Two single excitons confined to the same dot that are distinguishable can be optically entangled. The discrete QD energy structure due to quantum confinement substantially reduces the elastic scattering of excitons/biexcitons and therefore the coherences within a single exciton or between the constituent excitons of a biexciton are maintained during the exciton/biexciton lifetime. Discussions on the direct spectroscopic signatures of these relatively long-lived coherences under fully resonant optical excitation are the central part of sections 7.2 to 7.6. Section 7.7 will describe theoretically an implementation of a real quantum algorithm based on optically driven single QD excitons and biexcitons.

7.1 Introduction

This section starts with an overview of semiconductor QDs. Excitons and biexcitons, of importance to the optical response of single QDs, are briefly introduced. A simple model used to successfully explain many experiments in various QD systems is discussed. Optical spectroscopy techniques of relevance are described at the end of this section.

Semiconductor QDs

A semiconductor QD is a simple nanostructure that normally contains 10^3–10^6 atoms and thus the Bloch functions do not deviate very much from the bulk case. However, the envelope of the electron/hole wavefunction is strongly modified due to the finite size of the structure, giving rise to a discrete density of states.

To avoid surface states which trap electrons/holes and degrade the electrical and optical properties, semiconductor QDs must be produced so that they are passivated by a surrounding medium [5]. By engineering the passivating medium, the boundary condition of the dots can be altered, providing further control of the quantum confinement and the energy structure.

The two dominant technologies for the production of quantum dots are chemical synthesis and epitaxy. Chemically synthesized QDs are also known as nanocrystals [5]. Typically, II-VI materials, such as CdSe, CdS, CoO and ZnS, form the QD core and are passivated to form a core-shell structure [6–8]. Some techniques for engineering the size and shape of nanocrystals have been developed [9–12]. Nanocrystals are also incorporated into other structures, such as a variety of polymers as well as thin films of bulk semiconductors [13], and packed into QD lattices [14]. The epitaxial growth of

QDs is dominated by molecular beam epitaxy (MBE). MBE grown QDs are mostly from III-V group materials, such as GaAs, InGaAs, InAs and InP, and occasionally from other groups, such as PbSe and CdSe. The MBE technique allows for control within one or two monolayers in the growth direction. Nearly perfect heterojunctions produced this way give good passivation and therefore most MBE grown nanostructures are free from surface states. These heterojunctions provide quantum confinement along the growth direction. In the plane of the epilayer, quantum confinement is produced either by naturally formed interface fluctuations [15–18], self-organization [19–22] or via patterning and lithography [23–28]. There are also attempts to produce coupled QD structures and QD superlattices using the MBE technique [29–36].

The optical response from semiconductor QDs is directly mediated by the effect of quantum confinement on the motion of the electrons and holes. It therefore contains rich physics that can be extracted using various optical spectroscopy techniques. An overview of QD optical spectroscopy research will be given below.

The unique optical properties of QDs have led to QD-based semi-classical light sources such as lasers [37–43] and LEDs [44,45], and non-classical light sources, such as single photon trains [46–51] and entangled photon pairs [52]. QDs, especially nanocrystals, were used in biological labeling [53,54]. Lundstrom *et al.* [55] showed that excitons in QDs can be engineered to store information. Using quantum dots as spin memory in the Coulomb blockade regime was proposed by Recher *et al.* [56]. Photovoltaic cells were produced from nanocrystals [57,58]. In a proposal by Aguado *et al.* [59], double quantum dots can be used as quantum noise detectors in mesoscopic conductors. Using the hyperfine interaction between the electron spin and lattice nuclear spin, optical NMR was achieved with high spatial resolution (within a single nanodot) [60,61] and can be used as a sensitive probe of defect nuclei. Based on their atomic-like coherent nonlinear response (see sections 7.2 to 7.6), many authors have identified optically driven QDs as potential candidates of fundamental information carriers (quantum-bits, or qubits) in quantum computing and quantum information processing [1,3,4].

For more complete reviews on the fabrication and application of QDs, see References [5,22,62–64].

Excitons and biexcitons

In bulk semiconductors, due to the Coulomb interaction between optically excited electrons and holes, the excitonic effects dominate the optical response below the bandedge. As has been shown by many authors [65–67], under the effective mass approximation (EMA), an exciton can be regarded as a well-defined single quasi-particle containing an electron and a hole. Its envelope function can be separated into the center-of-mass (COM) motion (plane wave in bulk semiconductors) and the relative motion between the electron and the hole. The equation that determines the electron-hole relative motion is called the Wannier equation which has a solution that resembles the electron–positron relative motion in a positronium atom. The Bohr diameter (a_0) of the exciton, determined by material parameters such as the effective mass of the electron and hole, must be significantly larger than the lattice constant for the EMA to hold. This is indeed the case for most materials. For example, GaAs has an a_0 of about 250 Å, compared to the lattice constant of 5.6 Å. This type of exciton is known as the Wannier exciton. In the case that a_0 is comparable to or smaller than the lattice constant, the Frenkel limit is reached and the problem is treated differently [68].

In semiconductor QDs, an interacting electron–hole pair is subject to 3D quantum confinement. Complete localization of the exciton takes place when the Bohr diameter is comparable to the QD size. The problem is, in general, nontrivial to solve. Three regimes have been identified to simplify this problem using perturbation theory [62]. In the weak confinement regime, the quantum confinement is treated as a perturbation on the electron–hole Coulomb interaction that gives rise to excitons. In the strong confinement regime, the electron and hole motions are solved separately under the QD confinement and the Coulomb interaction then acts as a perturbation. The intermediate regime, however, requires non-perturbative methods. Despite these complications, in each of the three cases, the lowest single electron–hole pair levels are discrete and are referred to as QD exciton states. Each state represents a well defined single particle. For more details on the general treatment of excitons in QDs, see References [62–64].

The above analysis assumes that only the Coulomb attraction within each particular electron–hole pair state needs to be considered. These discrete states are the eigen-states of the system where only one bound electron–hole pair is created. In the case that two excitons are excited, the Coulomb interaction between the two electron–hole pairs needs to be included. Therefore the energy of the two-exciton state does not equal the sum of the two single exciton states. The difference reflects the interaction energy between the two excitons. In the perturbation theory, it appears as a higher-order correction and is therefore referred to as a higher-order Coulomb correlation [69]. Its sign and strength is determined by factors such as the electron/hole effective mass ratio, the quantum confinemeent and the spin of the electrons and holes.

The problem of the stable binding of the exciton–exciton molecule (with negative binding energy) has a rich history that will be briefly reviewed in section 7.5. Such a bound molecule is referred to as a biexciton. In the presence of strong confinement, the spatial separation between the electron-hole pairs is greatly reduced, leading to enhanced binding energy. The exciton and biexciton binding energies become comparable. In this case, the problem must be considered as a four-body (two electrons and two holes) problem, as opposed to a two-body (two excitons) problem. We will focus on the lowest such four-body state in a quantum dot.

Modeling single QDs

Due to the discrete nature of the energy structure, the modeling of QDs is relatively simple compared to higher-dimensional systems despite many variables, such as the shape and the size of the QDs. This section takes a well-studied GaAs QD system as an example. Other types of dots can be modeled similarly.

Figure 7.1 shows a schematic of this GaAs QD structure formed by large scale interface disorder in a GaAs layer MBE grown between $Al_{0.3}Ga_{0.7}As$ barriers [18]. The growth conditions and an example of topographical images of the interfaces can be found in Reference [70]. The response from two quantum well regions that differ in thickness by one monolayer is well separated in energy due to the difference of confinement in the growth (z) direction. Experimental studies are focused on the wider monolayer (islands) which occupies a few percent of the total area [70].

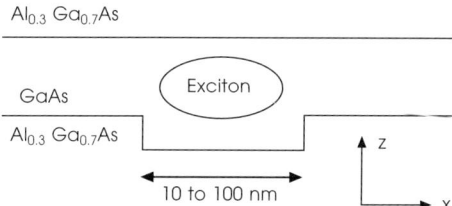

Fig. 7.1 The schematic for a QD naturally formed by interface disorder. The monolayer fluctuations form islands that lead to the localization of excitons. The size of the islands is on the order of tens of nanometers. The growth direction is represented by z.

While the GaAs layer is typically tens of angstroms thick, the lateral size of the islands due to the monolayer fluctuations is about tens of nanometers [70]. Therefore, the quantum confinement in the lateral direction is not as strong as in the case of, for example, self-assembled QDs. However, evidence for complete localization of excitons and biexcitons has been found using different approaches, including the quantification of the confinement energy [18], the direct microscopy image of excitons via the emission [15,71] and resonant coherent nonlinear response [72] of excitons (see sections 7.2 and 7.5). Evidence for localized complex states containing more than two excitons are not observed, possibly due to the limited confinement (see, for example, References [73, 74] for studies of such states in self-assembled QDs).

The level diagrams that include the crystal ground state (exciton vacuum), the two lowest orthogonal heavy-hole bright exciton states and the bound biexciton state of relevance to optical studies are shown in Fig. 7.2.

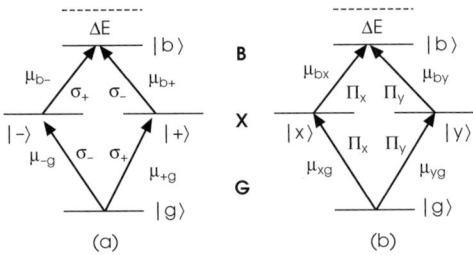

Fig. 7.2 Model for an elongated single GaAs QD. Symbols G, X and B denote the ground, the exciton and the biexciton states respectively. ΔE is the biexciton binding energy. The optical selection rules for various transitions with (a) and without (b) an external magnetic field applied in the Faraday configuration are indicated. Π_y (Π_x) is linear polarization perpendicular (parallel) to the QD elongation in the plane of the GaAs layer. Without the magnetic field, the two exciton states are excited using linearly polarized light and are labeled as $|x\rangle$ and $|y\rangle$. The magnetic field diminishes the mixing within the heavy hole states. The transitions become circularly plarized. The exciton states are represented by $|+\rangle$ and $|-\rangle$. The transition frequency and dipole moment of various transitions are denoted by ω_{ij} and μ_{ij} respectively, where i and j are the final and initial states of the dipole transition.

In the ideal situation, the optical transitions for QDs of zinc-blende semiconductors such as GaAs are circularly polarized, as indicated by the solid arrows in Fig. 7.2(a). The right-hand circularly polarized (σ_+) transition leading to the $|+\rangle$ exciton state is between the spin-up state of the s-like conduction bandedge ($m_j = +\frac{1}{2}$) and the heavy-hole bandedge state with $m_j = +\frac{3}{2}$. The left-hand circularly polarized (σ_-) transition leading to the $|-\rangle$ exciton state, however, is between the $m_j = -\frac{1}{2}$ conduction bandedge state and the $m_j = -\frac{3}{2}$ heavy-hole bandedge state. For the GaAs interface fluctuation QDs under study, however, the QDs are elongated along the $[\bar{1}10]$ axis due to the dynamics in the growth, leading to band mixing and modified optical selection rules due to the long range part of the exchange interaction. This problem has been considered in detail in References [67,75–77]. It is found that the two excitonic states become mixed and the optical transitions become linearly polarized [70,77,78]. This is shown in Fig. 7.2(b). An external magnetic field applied in the Faraday configuration can be used to restore the circularly polarized optical selection rules, as will be shown in section 7.2. The experimental studies of optical selection rules for exciton transitions will be discussed in sections 7.2 and 7.6. The optical selection rules involving the biexciton state are given in section 7.5. These selection rules are not unique to the GaAs interface fluctuation QDs but are generic to other QD systems with identical asymmetry [49,77–79].

Quantum coherence and quantum computing based on optically driven QDs

Quantum computing proposals in References [1–4] have considered the model in Fig. 7.2 as the simplest system able to implement two-qubit operations, taking advantage of two excitons coupled via the Coulomb interaction. Each exciton is considered as a basic unit to carry one bit of quantum information (qubit). A feasible scheme of building two-qubit quantum logic gates and implementing a simple quantum algorithm will be discussed in section 7.7. While in this chapter we focus on coupled excitons confined to a single QD, such studies serves as a starting point to understand the more general scalable systems consisting of excitons confined to coupled QDs.

Of particular relevance are various optically induced quantum coherences in a single QD system, including both the exciton dipole coherence (coherence between the exciton and ground states, or coherence between the biexciton and exciton states) and the nonradiative coherence between the two excitons, such as the coherence within the excitonic doublet or the ground-biexciton coherence. The latter type of coherence is nonradiative because it cannot be induced using one photon and must rely on two-photon processes (see Fig. 7.2) and the transition between the two states involved is dipole forbidden. Both types of coherences will be discussed in this chapter. Experiments show that they can be optically manipulated.

These quantum coherences are essential in any proposal for quantum computing, including those based on optically driven quantum dots, and therefore the ability to optically induce these coherences must be carefully explored. As one of the core requirements for quantum computing [80–82], these coherences must be manipulated in a controlled fashion in order to prepare arbitrary coherent superposition states within each qubit and between qubits. The decay of these coherences leads to an accumulation of errors in the operation. It is therefore critical to understand the decoherence dynamics of QDs.

Optically induced coherence between two states degrades over time by two mechanisms. First, the amplitude of either state could reduce to zero due to relaxation to other states in the system. The rate at which this event happens is half the energy relaxation rate Γ (Γ_i denotes the energy relaxation rate of state $|i\rangle$; Γ_{ij} denotes the energy relaxation rate from state $|i\rangle$ to state $|j\rangle$). In the second process, known as pure dephasing, the coupling of the states to other modes (e.g., phonons in solids) leads to a change of the relative phase between the two states without decay of the individual probability amplitudes. This random change in phase over time can cause a rapid decoherence at a rate of γ_{ij}^{ph} even though the amplitude of each state is relatively long-lived. These two decoherence processes are illustrated in Fig. 7.3 and detailed discussions can be found in References [83–85]. γ_{ij} is used to denote the total rate (including both process) of decoherence between states $|i\rangle$ and $|j\rangle$

$$\gamma_{ij} = \frac{\Gamma_i}{2} + \frac{\Gamma_j}{2} + \gamma_{ij}^{ph} \tag{7.1}$$

where the factor of two takes into consideration that the decoherence is determined by the state amplitude, not the probability amplitude, which is the absolute value of the state amplitude squared.

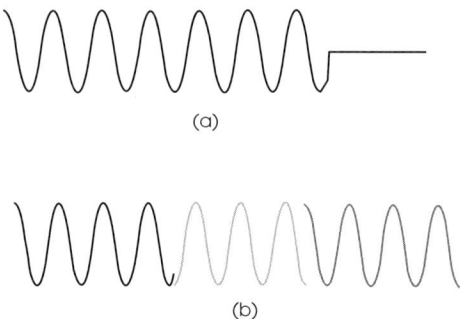

Fig. 7.3 The two decoherence mechanisms: (a) Energy relaxation and (b) pure dephasing. The curves contain the amplitude as well as phase information of a particular quantum state. In pure dephasing, the relative phase between two states experiences random changes over time.

It has been shown in higher dimensional semiconductor structures, such as bulk and quantum wells, that pure dephasing due to exciton-phonon and exciton–exciton interaction dominates the decoherence even at relatively low temperature [86,87]. For QD systems, however, the discrete density of states substantially reduces the phase-changing elastic scattering between excitons/biexcitons and phonons. In addition, the isolation of QDs dramatically decreases the probability of scattering between uncorrelated excitons, leading to relatively long-lived coherences [49,88,89]. In this chapter, experimental studies verify that these optically induced coherences are maintained during the exciton/biexciton lifetime, dramatically different from those of bulk and quantum well structures.

Single QD optical spectroscopy

The physics behind the unique optical properties of QDs can be extracted using various optical spectroscopy techniques based on the strong light-QD interaction. For many applications, it is essential that QDs be addressed individually, requiring that optical spectroscopy be performed at the single QD level.

Most QD systems available, however, have a relatively high dot density and a large amount of inhomogeneous broadening of transition energies due to the distribution of QD sizes. In traditional diffraction limited far-field optical spectroscopy, spectra of many dots merge together, preventing the extraction of information related to each individual QD. In recent years, however, several techniques yielding submicron spatial resolution have been developed. Some are based on near-field techniques using either coated fiber tips or masks with apertures [15,23,70,72,90]. Some are based on traditional/ confocal far-field microscopy in combination with low-density sample [50,74,91] or mesa structures fabricated using lithography [73,92,93]. Others used a solid immersion lens to achieve sub-wavelength resolution [51,71]. As a result, only a limited number of QDs are subject to optical excitation. Since each QD in general has its own distinct resonance frequency, individual dots can be resolved spectrally. Relying on these techniques, sharp spectral features due to single QDs have been observed in different types of systems, including nanocrystals [94,95], interdiffusion QDs [23], self-organized QDs (see, for example, References [49,73,78,90–92,96–98]), cleaved-edge QDs [26,32] as well as interface fluctuation QDs [15,16,18,99]. These single QD studies confirm that the transition energies are discrete and the greatly reduced exciton scattering rates due to localization lead to sharpened homogeneous linewidths compared to higher dimensional structures. For most of the studies on GaAs interface fluctuation QDs of relevance to this chapter, single QD resolution is achieved using an aluminum mask with a series of submicron sized apertures as detailed in [18].

Among the variety of optical spectroscopy techniques, photoluminescence (PL) is the most widely used. In PL, the nonresonantly excited electron–hole pairs either in the continuum of a dot or in the barriers of the quantum well nonradiatively decay and the energy can be captured in the confined QD exciton/biexciton states. They then radiatively recombine by emitting photons. A quantitative interpretation of the PL process is complicated due to the many unknowns in the nonradiative relaxation processes of the carriers [100]. Nevertheless, by monitoring the frequency of these emitted photons using a spectrometer, PL spectra provide information about the eigen-energies of QDs in a very straightforward way.

A simple variation of PL spectroscopy is PL excitation (PLE), in which the absorption spectrum of the optical excitation field is indirectly measured by recording the intensity of the PL at a particular energy as a function of the excitation frequency. This is very useful in studying a QD system because by monitoring the emission from the QD ground state populated due to the relaxation from the excited states driven by the optical field, PLE maps out its excited state spectrum [16,70,90,101]. Again, due to the unknowns in the nonradiative relaxation processes, it is difficult to gain a quantitative understanding of exciton decay dynamics from such experiments.

For a direct and quantitative measure of absorption of various states, resonant linear absorption spectroscopy can be used. However, observing linear absorption from a single QD requires extremely high signal-to-noise ratio. It is not until recently that such experiments are successfully performed in single dots [102,103]. These experiments will be discussed in section 7.2. The linear absorption coefficient of a QD state is directly related to the transition dipole moment and dephasing rate, the most important parameters of a single QD transition.

Nonlinear optical spectroscopy, on the other hand, explores the optical nonlinearities in semiconductor QDs and provides a measure of many additional critical parameters. In higher dimensional structures, complicated optical nonlinearities are expected, such as phase space filling and screening, most of which are due to many-body interactions (see Reference [104]). Although QD systems are considerably simpler, the atomic-like energy structure of a QD (verified using linear measurements such as PL and PLE) does not guarantee an atomic-like nonlinear optical response due to the complications of solid-state systems. The presumed atomic-like nonlinear optical response of QDs [104] must be subject to experimental verification. Investigations along this direction are therefore important, especially when optical nonlinearities involving quantum coherences of importance to devices such as QD-based quantum logic gates are concerned. In this regard, nonlinear optical spectroscopy becomes an ideal testbed for such speculations [72,88,89,105–107].

For example, it is interesting to attempt to measure in a QD system novel phenomena observed in atomic systems, such as Rabi oscillations [106,108], Mollow splitting/AC Stark shift [85,109,110] and Zeeman coherence [89], some of which are prerequisites for quantum logic operations. Nonlinear optical studies on single QDs require relatively large transition strength (oscillator strength) to give a reasonable signal-to-noise ratio and have currently only been achieved in single GaAs interface fluctuation dots.

Among many useful nonlinear optical techniques [100] employed in the investigation of semiconductor nonlinearites, we will focus on the differential transmission (DT) technique in which the third order nonlinear optical fields are homodyne-detected. The homodyne-detection improves the signal-to-noise ratio over the direct detection of the nonlinear response that occurs, for example, in either self-diffracted or phase-conjugated four-wave-mixing (FWM). Both frequency and time domain (transient) DT experiments are performed in interface fluctuation QDs. The experimental setup is shown in Fig. 7.4, using either two tunable continuous-wave (CW) lasers (for frequency domain DT) or pulsed picosecond (PS) lasers (for transient DT), labeled as $E_1(\Omega_1)$ and $E_2(\Omega_2)$. In degenerate DT experiments, the two fields are from the same laser source and are therefore degenerate in frequency. For the nondegenerate DT, however, the two beams come from two independently tunable lasers. The two CW lasers are frequency stabilized to about 4 neV and their mutual coherence bandwidth is measured to be ~20 neV, much smaller than the dephasing rates in typical QD systems (tens of μeV in energy units). Therefore, for CW measurements, the two fields are considered mutually coherent (within the decay of optically induced coherences). The two PS fields come from 76 MHz tunable mode-locked dye lasers with time-resolution of about 6 picosecond determined by the pulse width.

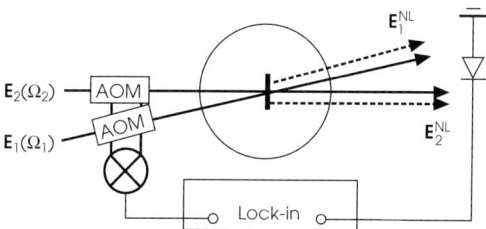

Fig. 7.4 DT experimental setup. The two beams labeled as $E_1(\Omega_1)$ and $E_2(\Omega_2)$ are either from one laser (degenerate) or two independently tunable lasers (nondegenerate). The two beams are amplitude modulated using acousto-optic modulators (AOM). The signal depending on both E_1 and E_2 is homodyne detected with transmitted E_1 or E_2 (can be experimentally chosen) at the difference frequency of the amplitude modulation using a lock-in amplifier.

In both CW and transient experiments, individual QDs are resonantly excited. To detect the third order nonlinear optical response as a result of such exitation, E_1 and E_2 are amplitude modulated and the optical response is homodyne detected with the transmitted E_i ($i = 1, 2$) at the difference frequency of the amplitude modulation using a lock-in amplifier.

Such experiments are not sensitive to the second order nonlinear optical response. In the weak field limit and keeping only the first order (linear) and third order nonlinear optical response (denoted by $E^{(1)}$ and $E^{(3)}$ respectively), the total photo current from a square-law photo-detector when detecting E_i is proportional to

$$|E_i + E^{(1)} + E^{(3)}|^2 \simeq 2\mathrm{Re}\,(E^{(3)} E_i^*) + 2\mathrm{Re}\,(E^{(3)}(E^{(1)})^*) + |E_i + E^{(1)}|^2 \quad (7.2)$$

where $E^{(3)} \ll E^{(1)} \ll E_i$ for typical QDs and thus $|E_1^{(3)}|^2$ is neglected. The second term can also be neglected compared to the other terms. In addition, since the lock-in amplifier only keeps the signal component at the modulation difference frequency, it is not sensitive to the last term in Eq. (7.2) (note that $E^{(1)} = \sum_{i=1}^{2} \chi_{E_i}^{(1)} E_i$, where $\chi_{E_i}^{(1)}$ is the linear response function due to E_i). Therefore, only the first term of Eq. (7.2) needs to be retained. In order to get the difference frequency, $E^{(3)}$ in the first term must be of the form $i\chi_{E_j,E_j^*,E_i}^{(3)} |E_j|^2 E_i$ ($i \neq j$ and note that $iE^{(3)}$ is proportional to the third order optically induced polarization $P^{(3)}$ for an optically thin sample). The signal that is homodyne-detected with E_i becomes

$$I_i^{NL} \propto \mathrm{Im}(\chi_{E_j,E_j^*,E_i}^{(3)}) \tag{7.3}$$

Here, $\chi^{(3)}$ represents the resonant third order susceptibility. Depending on the excitation (CW or PS, degenerate or nondegenerate) and detection (E_1 or E_2) scheme of a specific measurement, the $\chi^{(3)}$ responsible for the nonlinear signal may contain terms that are due to not only incoherent nonlinearities involving exciton population, such as state-filling (saturation), but also various coherent effects, such as population pulsation [88], Zeeman coherence [89,111] and two-photon coherence [112]. These experiments will be detailed later in this chapter.

The technique discussed above can also be understood as a measure of the change of the intensity of the transmitted beam (which is being homodyne-detected) induced by the presence of the other beam (differential transmission). As a convention, the phase-sensitive electronics within the lock-in amplifier are set such that an induced transmission appears positive.

A few important points about CW and transient DT are summarized below:

1. Due to the discrete nature of the QD energy structure, with a CW or transform limited PS laser field, it is possible to create only one localized exciton at a time (the bandwidth of a femtosecond pulse, however, would be too large to avoid other localized states).
2. Radiative as well as nonradiative coherences can be optically induced and then probed using CW DT and therefore their decoherence dynamics can be measured.
3. In the nondegenerate case, two laser fields can be tuned to resonantly excite two energetically distinguishable transitions, providing information regarding the coupling between states. This will be shown in sections 7.5 and 7.6.
4. Transient DT has lower spectral resolution compared to CW DT due to the larger bandwidth, but it can be used for time resolved experiments using another degree of freedom, the delay between the pulses E_1 and E_2. The choice of pulse width (6 ps) represents a compromise between two competing constraints: too much bandwidth resulting from a shorter pulse will cause excitation of multiple QD states, whereas

Coherent optical spectroscopy and manipulation of single quantum dots 293

a longer pulse will diminish the temporal resolution necessary to see decay dynamics, which is on the order of tens of picoseconds.

While most of the experiments in this chapter are in the weak field limit, an exception will be discussed in section 7.4, where the intensity of one beam is made high to induce a large probability amplitude of the exciton state of a QD (leading to Rabi oscillations). The analysis becomes very different since the higher order nonlinear optical response becomes significant and must be kept to all orders. Details will be given in section 7.4 and can also be found in Reference [85].

7.2   Single exciton optical spectroscopy

In this section, experiments that characterize single QD excitons are discussed. While the focus will be on GaAs interface fluctuation QDs, progress made in other types of QDs of relevance will also be discussed. It is found that the single exciton states in various QD systems are indeed localized and discrete. The energy structure as well as the linear and coherent nonlinear optical response of single excitons are atomic-like.

PL and PLE

The initial evidence for single localized exciton states in QD systems was found in PL spectra taken under high spatial resolution. Specifically, extremely sharp and isolated PL peaks identified as arising from single QD exciton recombination were observed in a variety of systems [15,16,18,23,26,32,49,73,78,90–92,94–99]. A typical PL spectrum taken from GaAs interface fluctuation QDs confined to a 42 Å well with a spatial resolution of 500 nanometers is shown in Fig. 7.5.

Further evidence showing that each peak in the PL spectra indeed arises from the recombination of a two-level quantum mechanical system (as opposed to a classical anharmonic oscillator) was demonstrated by Michler *et al.* [46], Becher *et al.* [50] and Zwiller [51] by analyzing these emission peaks using the technique of correlated photon counting. For a true two-level system, the emitted photons should obey nonclassical sub-Poissonian statistics and show photon antibunching, as can be determined by the second order intensity correlation function. This is due to a dead-time between successive photon emission events and has no classical counterpart. Photon antibunching was indeed observed in nanocrystals [46] and self-assembled InAs QDs [50,51], showing the nonclassical nature of various single QD systems. Experimental confirmation of photon-antibunching in single GaAs interface fluctuation QDs, however, has not been attempted largely because the shorter exciton lifetime in these systems requires the correlated photon counting apparatus to have a time resolution that is difficult to achieve (much better than 20 picoseconds).

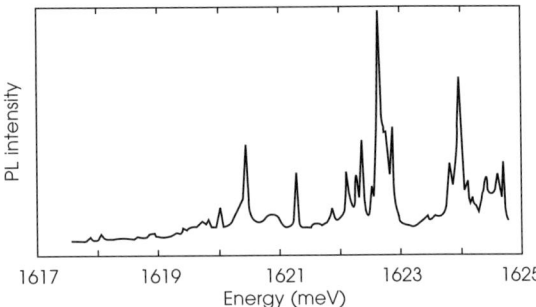

Fig. 7.5 Photoluminescence taken through a 0.5 μm aperture of a GaAs interface fluctuation QD sample. The spectrum corresponds to the wider region of a 42 Å well.

PL excitation spectrum of a QD is expected to show discrete excited states followed by a continuum, similar to that of an atomic system. This was demonstrated in all QD systems studied at the single QD level (see, for example, References [16,70,90,101]). A typical PLE spectrum is shown in Fig. 7.6 (open circle) for a GaAs interface fluctuation

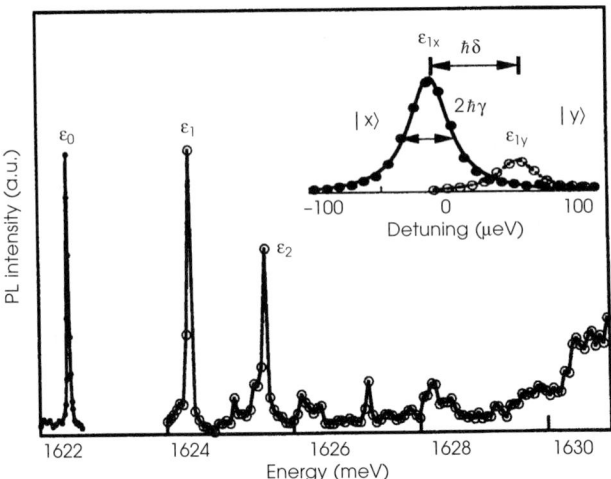

Fig. 7.6 PL and PLE spectra of a typical QD. The solid dots represent the PL spectrum of a typical QD in a 42 Å growth interrupted GaAs/Al$_{0.3}$Ga$_{0.7}$As well. The open circles represent its excitation spectrum. The polarization of the excitation beam is π_x. The inset shows the fine structure splitting of the first excited state at ε_1, with the excitation polarization indicated. The fine structure splitting in the PL state ε_0 is not resolved by the spectrometer. The figure is taken from Reference [105].

QD emitting at ε_0 (the emission spectrum is represented by solid circles). The spacing between the excited states and the PL state can be used to estimate the size of a QD with the knowledge of the band structure of the QD material and the passivating medium. These estimates are typically consistent with direct structural measurements based on techniques such as STM [18].

For an ideal dot, each peak is two-fold degenerate due to the spin degeneracy. As discussed in section 7.1, the shape of the QD is often asymmetric in many systems, such as interface fluctuation QDs, some types of self-assembled QDs [49,77–79] and nanocrystals with controllable shape [9–12], leading to a fine-structure splitting and modification of optical selection rules due to a nonzero exchange interaction and band mixing. As an example, the inset of Fig. 7.6 shows the fine structure splitting of the first excited state at ε_1. The two sub-transitions are orthogonally linearly polarized.

The nonradiative relaxation between an excited state and ground exciton state preserves exciton spin, i.e., the excitation of the excited states using Π_x (Π_y) polarized light only leads to Π_x (Π_y) polarized luminescence from the ground exciton state, as has been shown in interface fluctuation QDs [70] and self-assembled QDs [113]. This suggests that the spin relaxation of excitons is much slower than both the radiative recombination of the ground exciton state and the nonradiative relaxation between the excited exciton state and ground exciton state.

The QD can be excited resonantly into a coherent superposition of these two nearly degenerate excited states. A coherent manipulation of such a superposition using two time-delayed pulses leads to a control of the excited state population, which can then be monitored via the PL from the ground exciton states [105], such experiments will be detailed in section 7.3. In Reference [49], a coherent superposition of the two nearly degenarate ground exciton states was induced using quazi-resonant excitation of the QD into a exciton-phonon complex. The two decay paths to the crystal ground state are not distinguished and cause single photon interference in the spontaneous emission, giving rise to quantum beats in time-resolved PL. This coherent superposition within the exciton doublet also leads to unique signatures in coherent nonlinear response from single QDs, as will be detailed in section 7.6.

Linear absorption from single QD excitons

The strength of the QD-light interaction is determined by the electric dipole moment. One of the simplest ways to measure the dipole moment is to observe the amount of resonant laser light absorbed by the quantum dot. An accurate measurement can only be obtained from experiments performed on the single QD level. Linear absorption of a single QD is weak and rides on a large transmission background. Such measurements therefore require an excitation beam with extremely stable intensity, and have only been achieved in GaAs interface fluctuation QD systems. In these experiments, the excitation laser is linearly polarized such that it only excites one of the exciton doublet. Specifically, the vertically polarized exciton transition of a QD is selected using Π_y polarized light, providing a measure of the dipole moment, $\mu_{yg} = \langle y | \mu | g \rangle$ (μ_{xg} is measured using Π_x polarized light in a similar way).

In principle, a determination of the transition strength of a resonantly excited confined excitonic state using PLE can be obtained but would require knowledge of the coupling strength between the initial state and the final state (the luminescing state). In addition, all of the emitted photons would need to be counted in a time integrated experiment. Coupling strengths of this type are not typically known, and confidence that all of the emitted photons from a single QD have been measured is difficult to obtain.

Direct linear absorption monitors the transmission of a CW laser as its wavelength is scanned through a QD resonance. A typical coarse wavelength scan over the spectrum of the entire lower energy monolayer of a 6.2 nm growth interrupted quantum well obtained from a ~0.4 μm aperture is shown in Fig. 7.7. The degenerate CW DT from the same aperture is shown in Fig. 7.7 for spectral comparison. A detailed discussion of DT measurements will be given in the next section. The laser was Π_y polarized, so that only $|g\rangle \rightarrow |y\rangle$ transitions were probed. Individual QD states appear as sharp dips in the transmission spectrum of the laser shown in the middle panel of the figure; these dips are evident as resonances in the CW DT spectrum.

The spatial resolution of these measurements (~500 nm) is larger than both the center-of-mass wavefunction of the system [18,71] (~40 nm lateral extension) and the Bohr radius of the exciton ($a_0 \approx 10$ nm), allowing the quantum dots to be treated as point particles. Under this assumption, combined with resonant excitation of only the $|g\rangle \rightarrow |y\rangle$ transition, absorption from a single QD can be written as

$$\alpha(\delta) = \frac{\alpha_0}{1 + (\delta/\gamma)^2}$$

$$\alpha_0 = \frac{\Omega |\mu|^2}{cn\varepsilon_0 A_{ap} \hbar \gamma} \tag{7.4}$$

where δ is the laser detuning from the QD resonance, γ is the dephasing rate ($1/T_2$), $\Omega/(2\pi)$ is the laser frequency, ε_0 is the background dielectric constant, and A_{ap} is the area of the aperture. The total transmitted optical power, T, is then given by $T = T_0 - T_{abs} = T_0(1 - \alpha)$ where T_{abs} is the power absorbed by the QD. The above equation assumes

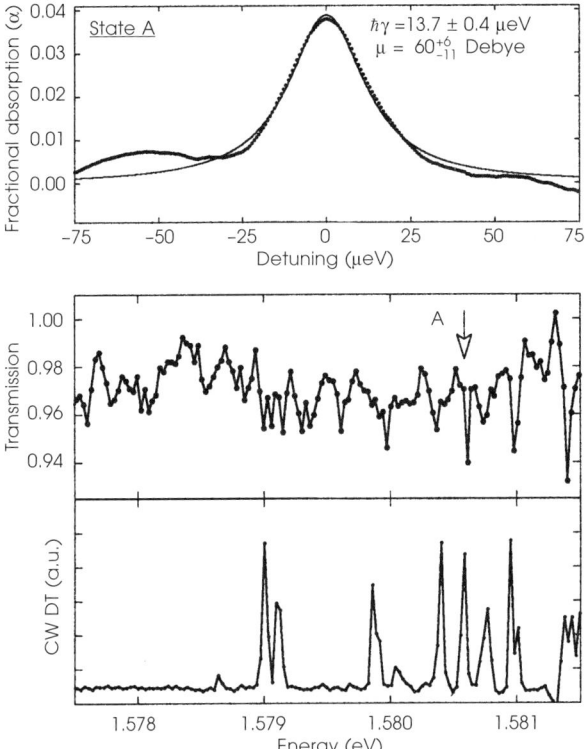

Fig. 7.7 Direct absorption from single QD states. The spectra are obtained by tuning the laser wavelength. The high resolution spectrum of a typical resonance (top) can be fit to the Lorentzian lineshape of Eq. (7.4), giving the dipole moment and linewidth shown. The bottom curve is the degenerate DT spectrum and will be discused in the next subsection. Data are taken from the 62 Å well of a growth interrupted GaAs QD sample.

the excitation stays in the linear regime and the third order nonlinear absorption is much smaller than the linear absorption.

Higher resolution spectra of individual QD absorption resonances can be fit for both the linewidth (γ) and line strength (α_0) using $T = T_0(1 - \alpha(\delta))$. As shown by Eq. (7.4), the half-width at half maximum is exactly the dephasing rate, γ, whereas α_0 is a function of both the dephasing rate and the dipole moment, allowing for an accurate measurement of the dipole moment for each resonance studied. This method ignores the contribution of reflections from the QD to the transmitted lineshape, since the spectral dependence of the reflectance from a single QD is expected to be small compared to that of the absorbance.

The measured absorption for a typical single QD state is plotted at the top of Fig.

7.7. $\alpha = 1 - T/T_0$ is the quantity plotted where T_0 is the background (off-resonace) transmission and T is the overall measured transmission. The dephasing rate is found to be $\hbar\gamma_{yg} = 13.7 \pm .4$ μeV, corresponding to a dephasing time of 48 ps. The fractional absorption for this state is $\alpha_0 = .039^{+.007}_{-.012}$. Using an aperture diameter of 385 nm and an index of refraction of 3.66, the dipole moment is found to be $\mu = 60^{+6}_{-11}$ Debye, corresponding to an oscillator strength of $f = 65^{+13}_{-23}$ (where $f = 2m_0\Omega\mu^2_{yg}/e^2\hbar$ and m_0 is the free electron mass). The notation X^{+du}_{-dl} denotes lower (dl) and upper (du) error bars on X. The errors on the linewidths are simply from the statistical error in the nonlinear fitting procedure. The origin of the error in the fractional absorption and the dipole moment is due to uncertainty in the aperture area as well as potential etaloning effects in the sample. Measurements on ~10 additional resonances revealed linewidths ranging from 12 → 29 μeV and dipole moments ranging from 50 → 100 Debye (oscillator strengths from 45 → 180).

By comparison, Reference [72] examined the 6.2 nm QW of a growth-interrupted sample using CW DT. The dephasing rates ($\hbar\gamma$) were found to be in the range 17 → 29 μeV, in excellent agreement with those measured using direct absorption and PL.

The large magnitude of the dipole moment is a manifestation of the mesoscopic enhancement of the coupling to the light field expected from semiconductor nanostructures. Theoretical predictions of the dipole moment for quantum dots formed by interface fluctuations agree well with the measured values presented here, though calculations for the exact structure of a growth-interrupted 6.2 nm QW have not been carried out. Andreani calculates a dipole moment of 67 Debye [114] for a 4 nm quantum well width and 40 nm quantum dot diameter, and Takagahara [115] calculates a dipole moment of 81 Debye for a 3 nm quantum well width and 30 nm quantum dot diameter. For comparison, atomic optical dipole moments are typically of order $e \times a_{bohr} \approx$ a few Debye. The larger dipole moment associated with these interface fluctuation quantum dots is critical to achieving the QD-cavity strong coupling regime in semiconductor microcavities [114,116].

The absorption described here allows for a straightforward calculation of the steady state occupation probability for an exciton in the QD, ρ_{yy}. Assuming that these states are not broadened by extra dephasing processes ($\gamma = \Gamma/2$), a linewidth of $\hbar\gamma = 15$μeV leads to $1/\Gamma \approx 22$ ps. With an absorbed power of $T_{abs} = T_0\alpha \approx 200$ pW, $\rho_{yy} = .017$, implying that the QDs are occupied by, on average, much less than one exciton.

CW and transient nonlinear optical response from single QD excitons

In this section, the simplest CW and transient DT experiments performed on single QDs are discussed. Nonlinear optical response at the single QD level was only measured in

GaAs interface fluctuation QDs. Copolarized laser fields E_1 and E_2 are used for resonant excitation of one of the two excitonic doublet transitions in a single QD and show that the single exciton behaves like an atom in the nonlinear regime. The exciton lifetime and dephasing rate are measured. These experiments are originally reported in References [88,107,117].

CW coherent nonlinear optical response from single excitons

By using narrow-band mutually coherent CW fields (see section 7.1 for specification of the lasers), single exciton states can be unambiguously selected. Experiments are usually done in the weak field ($\chi^{(3)}$) limit for easier interpretation of the results.

The two CW laser fields are colinearly polarized in these experiments and no external magnetic field is applied. For QDs that are asymmetric (such as the GaAs interface fluctuation QDs), only one of the linearly polarized exciton doublet transitions is excited. Figure 7.8 is a degenerate spectrum taken through a typical ~0.5 μm aperture, showing the response from many dots (42 Å well). In Fig. 7.9(c), the dotted curve are the fine degenerate DT spectra of a typical single QD exciton.

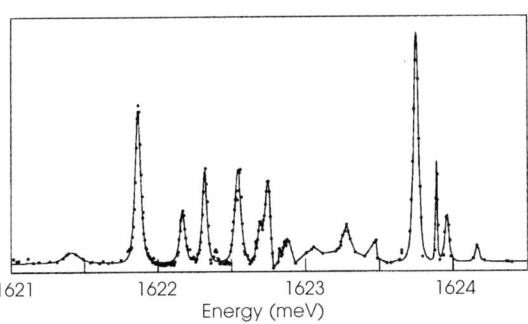

Fig. 7.8 Degenerate DT spectrum of a typical aperture at zero magnetic field. The two beams are colinearly polarized. Data are taken from the 42 Å well of a growth interrupted GaAs quantum well sample. Taken from Chen *et al.* [89].

It is shown in [85] that by homodyne detecting E_2 in a simple two level system, the third order nonlinear optical signal is due to two distinct quantum mechanical perturbation pathways. In the incoherent pathway, the first field creates an exciton population and therefore saturates the transition, causing a reduced absorption of E_2. This process is incoherent since it contributes even if the excitation fields are incoherent. Contribution from the second pathway, which leads to population pulsations as discussed by Meystre and Sargent in Reference [85], is due to the interference between the two fields leading to the excitation of an exciton grating that scatters E_1 into E_2. It follows that this process requires the two fields to be coherent.

Assuming the fields are Π_y polarized and excite the $|y\rangle$ state only, a derivation similar to Reference [88] yields a coherent nonlinear optical signal (detecting E_2)

$$I_2^{NL} = I_2^{\text{incoherent}} + I_2^{\text{coherent}} \propto \text{Im} \frac{iA}{(\gamma_{yg} - i\Delta_2)}$$
$$\times \left[\frac{2\gamma_{yg} \Gamma^{-1}}{\gamma_{yg}^2 + \Delta_1^2} + \frac{1}{\Gamma_{yg} - i(\Delta_2 - \Delta_1)} \left(\frac{1}{\gamma_{yg} + i\Delta_1} + \frac{1}{\gamma_{yg} - i\Delta_2} \right) \right] \quad (7.5)$$

where $A = |\mu_{yg}|^4 |E_1 E_2|^2$, $\Delta_1 = \Omega_1 - \omega_{yg}$, $\Delta_2 = \Omega_2 - \omega_{yg}$. ω_{yg} is the center frequency of the exciton transition. γ_{yg} and Γ_{yg} represent the dephasing and energy relaxation rate of the excitonic dipole respectively.

For the degenerate configuration, the two contributions become exactly the same with a Lorentzian squared lineshape determined by the dephasing rate, γ_{yg}. In the nondegenerate case, however, the lineshape of the incoherent terms is purely characterized by γ_{yg}, but the population pulsation term is determined by both γ_{yg} and the exciton energy relaxation rate, Γ_{yg}. Note in the single exciton case, the two decay rates are related by $\gamma_{yg} = \Gamma_{yg}/2 + \gamma_{yg}^{ph}$ (see Reference [85] and section 7.1), where γ_{yg}^{ph} is the pure dephasing of the exciton dipole coherence as discussed in section 7.1. Therefore, by examining the DT spectrum it is possible to evaluate both γ_{yg} and Γ_{yg}, and thus study the role of pure dephasing.

The nondegenerate DT lineshape is sensitively to the relative contribution from pure dephasing. Solid curves in Fig. 7.9(a) and (c) show the calculated signal for two cases as a function of scanning Ω_2 while Ω_1 is fixed at various frequencies. In the case that $\gamma_{yg}^{ph} = 10\Gamma_{yg}$, the incoherent component of the signal is much broader than the coherent one, as can be clearly seen Fig. 7.9(a). As $E_1(\Omega_1)$ is tuned across the exciton resonance, the narrow coherent component tracks it. In the second case, where $\gamma_{yg}^{ph} = 0$, the two components are comparable in linewidth, leading to an interference lineshape, as shown in (b) of Fig. 7.9.

The experiments shown in Fig. 7.9(c) suggests that pure phase changing interactions in a quantum dot are not the main sources of decoherence, due to the strong localization and isolation. The exciton dipole coherence is maintained within the exciton lifetime. The exciton dephasing rate can be extracted from the Lorentzian-square shaped degenerate spectrum to be $\gamma_{yg} \sim 19 \pm 3$ ps^{-1}, consistent with linear absorption measurements discussed in section 7.2. Measurements on other dots typically fall between 10–50 ps^{-1}. From the above data, Γ_{yg} can also be obtained although its determination needs to rely on fitting a complex nondegenerate spectrum. For this particular dot, $\Gamma_{yg} \sim 32 \pm 1$ ps^{-1}.

Coherent optical spectroscopy and manipulation of single quantum dots   301

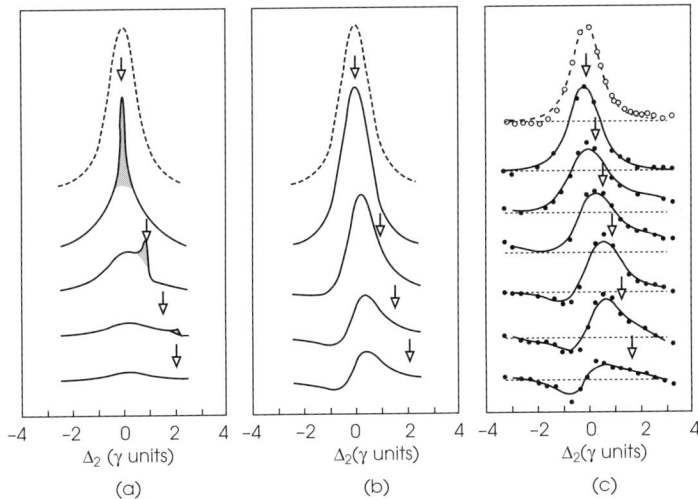

Fig. 7.9 Nondegenerate CW DT of a typical single QD exciton. (a) and (b) are calculations for two cases, $\gamma_{yg}^{ph} = 10\Gamma_{yg}$ and $\gamma_{yg}^{ph} = 0$, respectively. Dotted curves are the degenerate DT spectra. The others are nondegenerate DT spectra as a function of Ω_2. Ω_1 is fixed at the positions indicated by the arrows. Δ_2 is the detuning of E_2 from the resonance center. (c) Experimental data taken from a typical QD exciton from the 42 Å growth interrupted GaAs well. Beams are colinearly (Π_y) polarized. The nonlinear optical signal is homodyne detected with E_2. Taken from Bonadeo et al. [88].

By combining the CW DT and near-field microscopy, the nonlinear optical response can be imaged. Compared to STM images in Reference [70] which report on the structure of the interface fluctuation QDs, images based on coherent nonlinear optical response of QDs are direct evidence of strong localization of the center of mass wavefunction of excitons [72]. An example of such an image is shown in Fig. 7.10. This spectroscopy (both linear and nonlinear spectroscopy) and microscopy combination offers even more possibilities. For example, the spatial and spectral correlation of QDs in a system can be studied using these techniques. Phenomena such as level repulsion predicted by Runge and Zimmermann [118] has been observed in disordered GaAs systems [119–121].

Transient nonlinear optical response from single excitons

Differential transmission can also be obtained by exciting single QD states with pulsed lasers. A typical spectrum, obtained by scanning the center wavelength of the ~6 ps (degenerate in energy) E_1 and E_2 fields at zero delay, is shown at the top of Fig. 7.11(a). The Π_y co-polarized E_1 and E_2 fields travel through a submicron aperture, and the third order nonlinear signal is homodyne detected with E_2. Narrow lines from individual QD

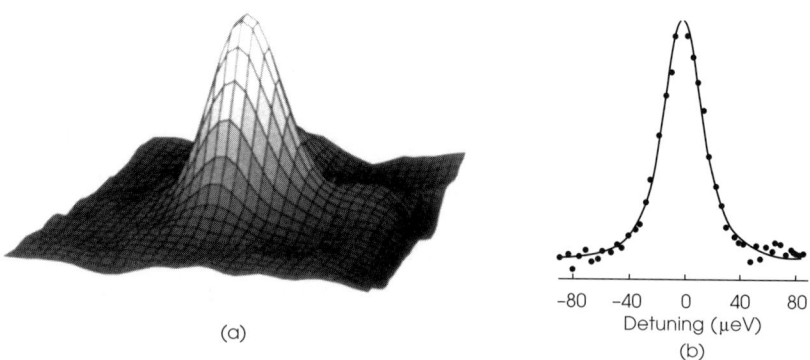

Fig. 7.10 Near-field microscopy and coherent nonlinear optical spectroscopy. (a) The image of the degenerate DT response of a typical QD obtained at low temperature using a near-field scanning optical microscope. Both laser fields are fixed at the center of the resonance. The image is 2 μm by 2 μm. (b) The degenerate coherent nonlinear optical spectrum of the same dot. E_1 and E_2 are spatially placed at the center of the image in (a). Data are taken from the 62 Å GaAs well with growth interruptions.

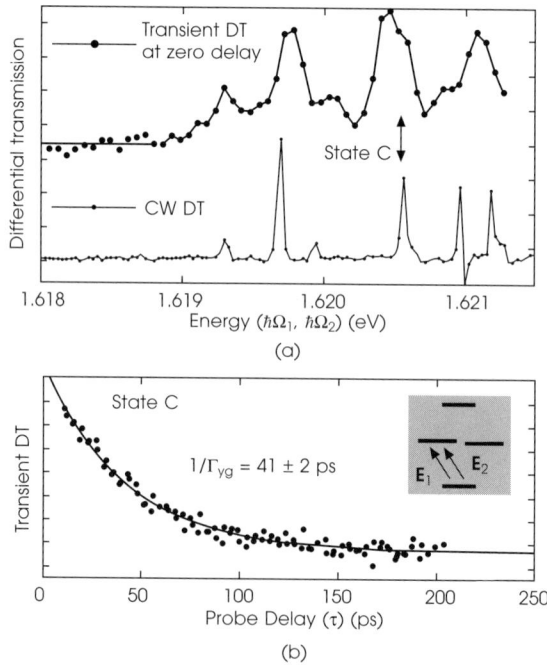

Fig. 7.11 Transient DT from single QD states. The transient DT spectrum obtained at zero probe delay is compared to the CW DT spectrum in (a). The decay of the DT from a single state as the probe delay is varied is shown in (b). All data is obtained with a ~ 0.5 μm aperture. Data were taken from the 42 Å growth interrupted GaAs quantum well.

states have a width of approximately 350 μeV, the transform limited bandwidth of a 6 ps pulse. Further confirmation that the narrow resonances observed in the PS DT of Fig. 7.11 correspond to excitons localized to single QDs comes from a comparison with degenerate CW DT, shown below the PS DT in Fig. 7.11(a). By fixing the PS laser position at one of these resonances (e.g. state C), one can therefore resonantly excite and probe a single excitonic QD state.

The decay of the DT as the delay is scanned with the laser frequency fixed at the state labeled C is shown in Fig. 7.11(b). An analysis of the DT signal from a two-level system consisting of only the ground state and the $|y\rangle$ single exciton state yields

$$DT \propto \int_{-\infty}^{\infty} \mathrm{Im}[\rho_{yg}^{(3)}(t)e^{i\Omega_2 t}]dt$$

$$\propto (\rho_{gg}^{(0)} - \rho_{yy}^{(0)})\theta(\tau)e^{-\Gamma_{yg}\tau} \qquad (7.6)$$

for the dependence of the DT on the delay. The integration is carried out over one pulse and δ-function pulses have been used. τ is the $E_1 - E_2$ delay and θ(t) is the Heavyside (step) function. Coherent artifact terms within a pulse width of zero delay are ignored. Thus, the decay of the homodyne detected DT is simply a measure of the exciton lifetime, $T_1 = 1/\Gamma_{yg}$.

The DT vs. probe delay in Fig. 7.11(b) is fit to a single exponential, giving a lifetime for state C of 41.2 ± 2 ps. Data within one pulse width (6 ps) from zero delay is omitted from the fit. The decay of the DT from other single QD states investigated have all been consistent with exponential decay with lifetimes in the range 25–50 ps.

The exciton relaxation rate can be written $\Gamma = \Gamma^{sp} + \Gamma^{nr}$, where Γ^{sp} and Γ^{nr} account for radiative (spontaneous emission) and nonradiative decay processes, respectively. Calculations in interface fluctuation QDs find radiative lifetimes of about 75 ps for a 5 nm thick QW and a lateral extent of about 40 nm [122], about 100 ps for a 3 nm thick QW and a lateral potential depth of 8 meV [123], 175 ps for a 10 nm thick QW and a lateral extent of about 50 nm [124], and 27 ps for a 3 nm thick QW and a lateral extent of about 35 nm [125]. Note that the first three of these values are radiative lifetimes, whereas the last value (by Takagahara) is an overall relaxation rate, including both radiative recombination and phonon assisted exciton migration. In Takagahara's model, phonon assisted decay is the dominant decay channel, whereas the radiative lifetime alone is calculated to be about ~200 ps. The measured lifetimes are systematically smaller than the calculated radiative lifetimes, indicating that nonradiative decay may play an important role in the decay of excitons confined by interface fluctuations in thin QWs.

From both the CW and transient DT results, it can be concluded that the pure dephasing of the exciton dipole coherence is at most comparable to the exciton energy relaxation. This is distinct from higher dimensional structures such as bulk and wide quantum wells without growth interrupts, in which the extended nature of the excitonic states makes them more susceptible to interacting with the lattice and with other excitons, leading to a quick loss of the optically induced coherence before the decay of the exciton [86].

Magneto-excitons

An external magnetic field provides an extra important experimental knob in exploring the physics behind the QD optical properties. It has been essential in studying the exciton g factor, diamagnetism and dark states.

In the Faraday configuration, both the heavy-hole and the conduction band split under nonzero magnetic field, leading to a Zeeman splitting as well as a diamagnetic shift of the exciton doublet of a QD. More importantly, the magnetic field diminishes the band mixing and restores the circularly polarized optical selection rules for excitonic transitions, as shown in Fig. 7.2. This behavior was indeed observed in many QD systems, including self-assembled QDs [78,79,93,113] and interface fluctuation QDs [77,126]. Zeeman splitting in these systems is easily resolved, providing a direct measure of exciton g factor. The diamagnetic shift of the excitons is a direct result of the perturbation to the exciton envelope function by the magnetic field and is sensitive to the size of exciton wavefunction. In References [127,128], it has been shown that the size of a QD estimated using the magnitude of its diamagnetism is consistent with that determined by other means (such as STM).

Figure 7.12(a) is the PL spectra of a typical GaAs interface fluctuation QD as a function of the magnetic field, showing a Zeeman splitting and a diamagnetic shift. The optical selection rules are determined by analyzing each emission peak, which is found to be circularly polarized. The exciton g factor and diamagnetic coefficient for this particular QD are found to be ~1.2 and 25 $\mu eV/T^2$ respectively. Details of such measurements are discussed in Reference [126]. Similar information can be obtained using degenerate DT. This is shown in Fig. 7.12(b), where co-circularly polarized E_1 and E_2 are used.

The direction of the magnetic field relative to the growth direction of a QD system can be arranged to break the symmetry and introduce band mixing [79,129]. The two optically forbidden (dark) transitions between the heavy hole and the conduction levels (angular momentum change by $2\hbar$) become optically active. Bayer *et al.* has shown in [129] in a self-assembled quantum dot system that by tilting the magnetic field away from the growth direction, the transition energies of all four possible transitions

Coherent optical spectroscopy and manipulation of single quantum dots 305

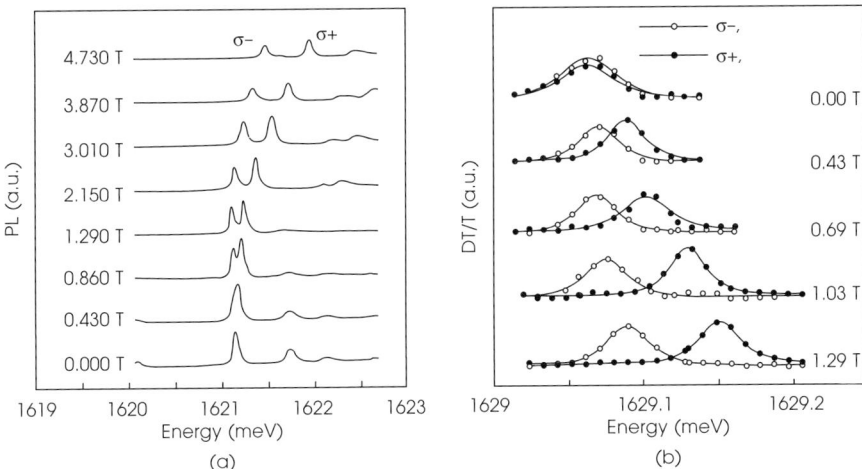

Fig. 7.12 Optical spectroscopy of magneto-excitons. (a) PL and (b) degenerate CW DT spectra of a typical QD as a function of an external magnetic field, showing Zeeman splitting and diamagnetic shift. The higher (lower) energy state is determined to follow σ_+ (σ_-) polarized selection rules. In (a), the spectrometer accepts linearly polarized light and therefore is sensitive to both states. In (b), E_1 and E_2 are co-circularly polarized (either σ_+ or σ_-). Data were taken from the 42 Å growth interrupted GaAs well. Taken from Reference [111].

as a function of the field strength can be mapped. By doing that, both the electron and hole g factors were extracted. In the Faraday configuration, only the exciton g factor, which is a combination of electron and hole g factors, can be obtained.

One of the particularly interesting phenomena under the static magnetic field is known as the Overhauser effect [130,131] and is a result of the hyperfine interaction between the electron spins and the nuclear magnetic moments of the lattice. With a circularly polarized CW excitation beam, an optically induced steady state population of spin oriented electrons creates a local magnetic field, which polarizes the lattice nuclear magnetic moments via the hyperfine interaction. The depolarization of the nuclear spin is typically very long, allowing for an effective nuclear magnetic field to build up. The total mangetic field acting on the exciton becomes a combination of both the external static magnetic field and the effective nuclear field. The effective nuclear field is zero if an unpolarized excitation beam is used. The difference of the exciton transition energy between the polarized and unpolarized excitation cases is the Overhauser shift. This effect was clearly observed in GaAs interface fluctuation QDs and was used to realize optical nuclear-magnetic-resonance (NMR) [60,61]. Since the optical excitation and therefore the nonzero net hyperfine interaction can be confined within a particular dot, the spatial resolution of the optical NMR measurements can be exceptionally high. These experiments are discussed in a separate chapter in this book (see Chapter 6).

7.3 Coherent optical control of single exciton states

In section 7.2, it was shown that the linear and coherent nonlinear optical response from single QDs resemble those of atoms in many ways. Single QD excitons under resonant excitation can be regarded as two-level quantum mechanical systems, exhibiting behaviors that have no classical analog (such as photon antibunching in emission and Rabi oscillations in excitation, see section 7.4). This implies the possibility of controlling single exciton states via constructive and destructive quantum interference using coherent and resonant optical fields, in ways analogous to those employed in atomic and molecular systems [132–139].

Early experiments in higher dimensional semiconductor materials have shown that the photocurrent [140], electron-LO-phonons scattering [141], charge oscillations [142], cyclotron emission [143], exciton population and/or orientation [144–146], coherent acoustic phonon [147], exciton-polariton [148] and the electron-plasmon [149] can be controlled using coherent optical fields. This section discusses the simplest scheme in which a coherent optical manipulation of single exciton state population is realized [105] in interface fluctuation QDs. Identical experiments have also been reported for self-assembled QDs [150–152].

A typical GaAs QD is chosen for this study, whose PL, PLE spectra are shown previously in Fig. 7.6. The fine structure splitting of the first excited states at ε_1 is ~60 μeV. The transition frequencies of these substates, $|x\rangle$ and $|y\rangle$ are ω_{xg} and ω_{yg}. This is shown in Fig. 7.13(a). Coherent optical control and wavepacket interferometry were achieved using a sequence of two phase-locked laser pulses with a controllable delay

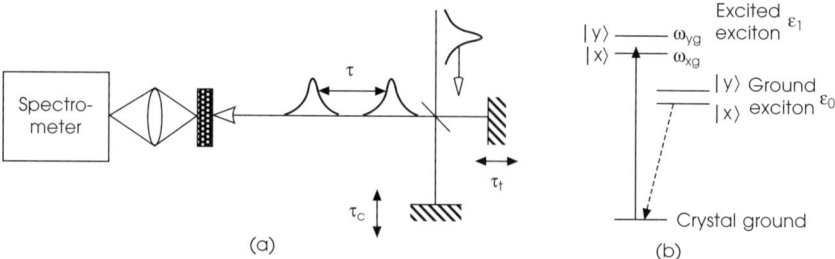

Fig. 7.13 Wavepacket interferometry setup. (a) Two optical pulses excite the excited states $|x\rangle$ and $|y\rangle$ at ε_1, which relax nonradiatively to the ground exciton states. The photons due to the recombination of the ground exciton states are detected. (b) Wavepacket interferometry setup. The two pulses with controlled delay τ are produced by sending a laser beam into this subwavelength-stable Michelson interferometer. The total delay τ is controlled by two components, a coarse delay (τ_c) and fine delay (τ_f) provided by a mechanical and piezoelectric translation stages respectively, giving $\tau = \tau_c + \tau_f$.

relative to one another [153–157] (see Fig. 7.13(b)). The laser is tuned to the $|x\rangle$ and $|y\rangle$ states at ε_1 and has a 420 μeV bandwidth (5-ps pulsewidth), broad compared to the $|x\rangle$ and $|y\rangle$ splitting but sufficiently narrow so as not to excite other higher excited states of the QD.

In the low excitation limit (only single photon processes are considered) and for a delay of τ between the two pulses, the total excited state wavefunction is a coherent superposition of the excited state wavefunctions created by both pulses with a time evolution:

$$|\Psi(t,\tau)\rangle = |\Phi^{(I)}(t)\rangle + |\Phi^{(II)}(t+\tau)\rangle \tag{7.7}$$

where $|\Phi^{(i)}(t)\rangle$ is the excited state wavefunction generated by the pulse i (i = I or II). Equation (7.7) represents the sum of two quantum mechanical paths connecting the initial and final states. Quantum interference between these two paths is observed by measuring the excited state population generated by the pulse pair as a function of the time delay between the two pulses. The population is proportional to

$$\int_{-\infty}^{\infty} \langle \Psi(t,\tau)|\Psi(t,\tau)\rangle dt$$

$$= \int_{-\infty}^{\infty} [\langle \Phi^{(I)}|\Phi^{(I)}\rangle + \langle \Phi^{(II)}|\Phi^{(II)}\rangle + 2\mathrm{Re}\langle \Phi^{(I)}(t)|\Phi^{(II)}(t+\tau)\rangle] dt \tag{7.8}$$

As τ is changed, the first two terms in this expression remain unchanged while the last term oscillates as the interference between the two quantum mechanical paths goes from constructive to destructive.

The last term in Eq. (7.8) is the cross-correlation function of the excited state wavefunction generated by the first and second pulses. A measurement of this function yields important information about the temporal evolution of the wavefunction and is the basis for wavepacket interferometry. A complete characterization (amplitude and phase) of the wavefunction generated by one pulse can be obtained if the wavefunction generated by the other pulse is known. In the case that both pulses are identical, Eq. (7.8) becomes the auto-correlation function of the excited state wavefunction and allows us to extract the decoherence time of the system as well as the dynamics of the wavefunction's temporal evolution.

In the following, three experiments are discussed with increasing degrees of complexity in the final state wavefunction. In these experiments, the excited state population including the quantum interference effect is measured by monitoring the PL from the exciton ground states at ε_0 (see Fig. 7.13(a)), to which the state $|x\rangle$ and $|y\rangle$ states at ε_1 decay via acoustic phonon emission [18]. The quantum interference can be controlled by changing the delay τ and, as will be shown below, the polarization between the pulses. The control of τ is accomplished by sending a laser pulse through a subwavelength-stable Michelson interferometer, as shown in Fig. 7.13(b).

In the simplest experiment, both pulses ($E_1(t)$ and $E_2(t)$) are co-polarized along the x-axis (see lower inset of Fig. 7.14) and thus only excite the $|x\rangle$ state. The excited state wavefunction created by both pulses is $|\Phi^{(I)}(t)\rangle = |\Phi^{(II)}(t)\rangle = e^{-i\omega_{xg}t}c_x|x\rangle$. The constructive and destructive interference between $|\Phi^{(I)}(t)\rangle$ and $|\Phi^{(II)}(t+\tau)\rangle$ leads to oscillations of $\langle\Psi(t,\tau)|\Psi(t,\tau)\rangle$ as a function of τ according to Eq. (7.8), which causes similar oscillations of the PL intensity from states at ε_0. Such oscillations around $\tau \sim$ 40 ps are shown in the upper inset of Fig. 7.14. The oscillation period is the inverse of $2\pi\omega_{xg}$. The oscillation amplitude as a function of τ is shown in Fig. 7.14 as large filled circles. The auto-correlation function of the laser pulse is represented by open circles. Strong oscillations persist after the duration of the first pulse.

The reduction in the amplitude of the oscillations for increasing pulse delay observed in Fig. 7.14 is a consequence of the decoherence processes discussed earlier.

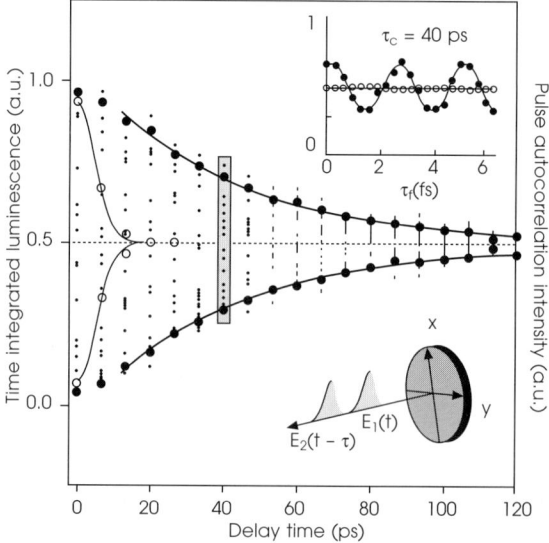

Fig. 7.14 The amplitude of the oscillations in PL as a function of τ (large filled circles) when both pulses are co-polarized along the x-axis, thus exciting just the $|x\rangle$ state (lower inset). This measures the auto-correlation function of the excited state wavefunction. The details of the oscillations due to quantum interference are represented by small dots which can only be resolved with an expanded scale. The upper inset show such an expansion around $\tau \sim 40$ ps (corresponding to the shadowed region in the main plot). The amplitude of the oscillations shows an exponential decay over time. The auto-correlation function of the pulse is also plotted for reference (open circles). Figure is taken from [105].

A calculation using the density matrix equations which includes dephasing of the optically induced quantum coherence shows that the time integrated PL intensity as a function of τ is $I(\tau) \propto \frac{1}{2}(1 + \cos(\omega_{xg}\tau)\exp(-\tau/T_2))$, where $1/T_2$ is the dephasing rate. As seen in Fig. 7.14, the envelope of the oscillation fits an exponential decay and provides an alternative way of measuring T_2. In this particular case, $T_2 \sim 40$ ps, in excellent agreement with the value obtained from the PLE linewidth ($\gamma_{xg}^{-1} = 39$ ps). The loss of coherence limits the ability to coherently manipulate the wavefunction for time scales $> T_2$.

In the second experiment, the polarization of the co-polarized pulse sequence with respect to the sample's eigen-axis (see lower inset in Fig. 7.15(a)) is rotated to generate a non-stationary (time dependent) wavefunction composed of a superposition

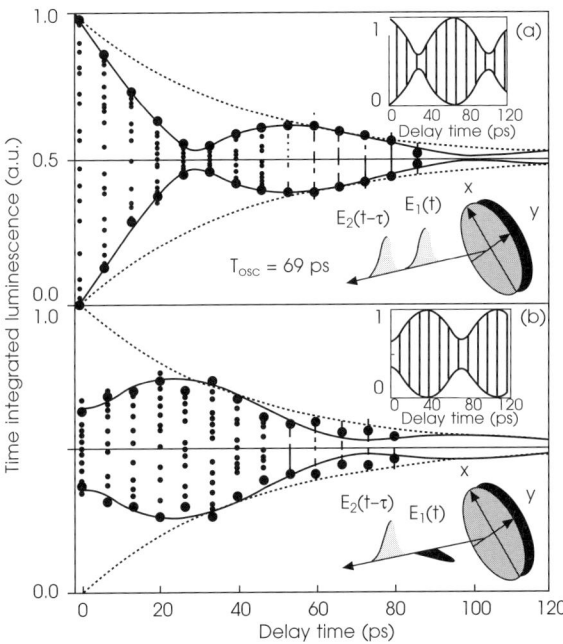

Fig. 7.15 Auto-correlation and cross-correlation function. (a) The excited state auto-correlation function of the excited state wavefunction, as in Fig. 7.14, but for both pulses co-polarized and rotated to equally excite both the $|x\rangle$ and $|y\rangle$ states. The temporal evolution shows oscillations of the wavefunction between $|x\rangle$ and $|y\rangle$. The oscillation period corresponds to the inverse of the difference frequency between the ω_{xg} and ω_{yg}. (b) The cross-correlation function between two excited state wavefunctions generated by orthogonally linearly polarized optical pulses. The relative phase of the two superposition of states produced by each pulse differs by π. The top inset in each figure shows the corresponding calculations in the absence of dephasing. The PL is detected near 45° relative to the y-axis to ensure that the emission is proportional to the total population and not just the population in one eigenstate. Figure is taken from [105].

of states $|x\rangle$ and $|y\rangle$. The polarization of the pulses is adjusted to compensate for the difference in the oscillator strength of the x- and y-transition to equally excite both states. The wavefunction generated by each pulse is now

$$|\Phi^{(I)}(t)\rangle = |\Phi^{(II)}(t)\rangle = e^{-i\omega_{xg}t}c_x|x\rangle + e^{-i\omega_{yg}t}c_y|y\rangle \qquad (7.9)$$

where the coefficients c_x and c_y are similar in magnitude and are defined to be real. The nonstationary dynamics of the excited state wavefunction are seen as a slow oscillation that modulates the rapid oscillation shown in the upper inset of Fig. 7.15(a). More specifically, the auto-correlation function of the excited state wavefunction (see Eq. (7.8)) shows how it oscillates between two orthogonal states, $|x\rangle + |y\rangle$ and $|x\rangle - |y\rangle$, as the envelope function goes from a maximum to a minimum as a function of time. The slow oscillation period, $T_{osc} = 69$ ps, as determined, by this experiment (the main plot of Fig. 7.15(a)), is in excellent agreement with the fine structure splitting. Again, the exponential decay of the envelope is due to the loss of coherence.

In the third experiment, the polarization of the first pulse is rotated by $\pi/2$ relative to the previous experiment (see the lower inset of Fig. 7.15(b)) and therefore generates a wavefunction where the relative quantum phase between $|x\rangle$ and $|y\rangle$ is shifted by π relative to that created by the first pulse. That is, the first pulse leads to $|\Phi^{(I)}(t)\rangle = c_x e^{-i\omega_{xg}t}|x\rangle + c_y e^{-i\omega_{yg}t}|y\rangle$ while the second pulse leads to $|\Phi^{(II)}(t)\rangle = c_x e^{-i\omega_{xg}t}|x\rangle + c_y e^{-i\omega_{yg}t}|y\rangle e^{i\pi}$. The quantum interferogram shown in Fig. 7.15(b) now reports on the cross-correlation function of the two wavefunctions as described by Eq. (7.8). The plot shows a minimum at $t = 0$ and a maximum at $t = T_{osc}/2$, 180^{circ} out of phase with the second experiment.

In these experiments, the relative quantum phase of the superposition of states is controlled and a simple target wavepacket is produced by varying the polarization of the excitation pulses. This demonstrates the feasibility of more complex wavefunction engineering in single quantum dots.

7.4 Rabi oscillations of single quantum dots

Results presented thus far in this chapter have shown that many different optical properties of excitons confined to single QDs can be understood in terms of the discrete-level model of Fig. 7.2. These experiments used weak optical fields, characterized by a Rabi rate for a CW field that is smaller than the QD dephasing rate, or a time-integrated Rabi rate for the fields in the transient experiments that was smaller than π. Strongly-driven QDs, on the other hand, should show evidence of Rabi oscillations if the two-level model is an appropriate description.

Rabi oscillations are one of the most fundamental strong-field phenomena observed in resonantly driven two-level systems. They are a direct result of the nonlinearities inherent in a two-level system and have no classical analog [158]. They were first described in the context of nuclear magnetic resonance [159,160], and observed in atoms soon after the development of laser spectroscopy [161,162]. Rabi oscillations are more difficult to observe in the solid state due to the fast decoherence times of solids compared to atoms and molecules. With the advent of femtosecond spectroscopy, Rabi oscillations in semiconductors were predicted [163], and experimental evidence has been reported in free-carrier transitions in bulk [164], excitonic transitions in QWs [165–168], and most recently in impurity-doped systems [169]. For QDs, Rabi oscillations are only observed in interface fluctuation QD systems so far [106].

Rabi oscillations in delocalized excitonic systems are complicated by the presence of many-body effects, especially at high excitation intensity. These effects modify the discrete-level model used to understand the optical properties of excitonic transitions [87]. However, an isolated discrete-level model is expected to be more accurate for an exciton confined to a QD due to its decreased interactions with its local environment.

The observation of Rabi oscillations from QDs would not only further the understanding of the optical properties of localized excitons, but also make possible the control of the state of excitation of an individual quantum dot. A π-pulse corresponds to one-half of a Rabi oscillation, and can be used to fully invert a two-level system. In the case of excitons in quantum dots, a π-pulse represents the creation of an exciton with unity probability. Pulse areas less than π can be used to create coherent superpositions of the crystal ground state (no exciton state) and the one exciton state. This type of coherent control of the state of a quantum dot has potential applications for QD-based optoelectronic nanoscale devices, wavefunction engineering with QDs, and quantum computing. In particular, a quantum-controlled NOT gate as described in [1,170] is based on the ability to shape the state of excitation in single QDs with π-pulses and π/2-pulses.

Rabi oscillation theory for two-level systems

A pulsed pump field (E_1) nearly resonant with a two-level system turned on at $t = 0$ will induce coherent oscillations of the populations of both the ground and excited state as a function of pulse area, Θ. It is defined as

$$\Theta(t) = \int_{-\infty}^{t} \mathscr{R}(t')dt' \qquad (7.10)$$

where the Rabi rate, $\mathscr{R}(t) = \sqrt{R^2(t) + \delta^2}$, $R(t)$ is the on-resonance Rabi frequency, and δ is the laser detuning. For the data shown in this section, $\delta = 0$, so $\mathscr{R}(t) = R(t)$. These oscillations persist for times shorter than the lifetime of the excited state, $T_1 = 1/\Gamma$, and can be expressed simply as $\sin^2(\Theta(t)/2)$ for the excited state (exciton) population. A straightforward way to detect the oscillations is to measure the absorption of a weak probe beam (E_2). A DT measurement that homodyne-detects the nonlinear optical signal with E_2 and corresponds to the probe transmission with the pump on minus the probe transmission with the pump off, will show Rabi oscillations because the DT is proportional to the exciton population.

The experiment performed measures the change in transmission of E_2 induced by E_1. A strong E_1 field can contribute to excess detector noise as well as mix with the E_2 at the detector leading to an undesirable background signal. To avoid these complications, the Π_x transition of the exciton fine structure doublet is probed while the Π_y transition is pumped. A linear polarizer in front of the detector is set to block the pump (E_1). Neglecting spin relaxation within the exciton doublet (which can be confirmed via other means), this configuration only allows for detection of signal that is based on a nonequilibrium population distribution created by the pump, followed by absorption of the weak probe (E_2). There are no terms homodyne detected with the pump or zero delay coherent artifact terms to complicate the analysis. The binding energy of the biexciton state keeps the exciton to biexciton transition well detuned from the resonant ground state to exciton transition (see section 7.5). Though the biexciton level is never directly probed in these strong-field experiments, the presence of the Coulomb correlation between the $|y\rangle$ exciton and the $|x\rangle$ exciton is responsible for the observation of cross-polarized signal [89,171]. The DT is monitored as a function of pump (E_1) power, and uses degenerate E_1 and E_2 fields, since the splitting $\omega_{yg} - \omega_{xg} \approx 20$ μeV is much less than the laser bandwidth.

The pump pulse $\mathbf{E}_1(t) = 1/2(E_1(t)e^{-i\Omega t} + \text{c.c.})\,\hat{y}$ is used to excite the single QD exciton, $\rho_{yy}(t)$, which then decays back to the crystal ground state, ρ_{gg}, at the relaxation rate, Γ_{yg}. Here, $\Omega_1 = \Omega_2 \equiv \Omega$. The weak probe pulse $\mathbf{E}_2(t) = 1/2(E_2(t)e^{-i\Omega t} + \text{c.c.})\,\hat{x}$ (delayed with respect to the pump by a time τ) upon absorption by the excitonic resonance creates an induced nonlinear optical polarization field, $\mathbf{P}^{(NL)}(t) = 1/2(P^{(NL)}(t)e^{-i\Omega t} + \text{c.c.})\,\hat{x}$, where

$$P^{(NL)}(t) \propto \int_{-\infty}^{t} E_2(t')(\rho_{gg}(t') - \rho_{xx}(t'))e^{(i\delta - \gamma_{xg})(t-t')}dt'. \tag{7.11}$$

where $\delta = \Omega - \omega_{xg}$ is the laser detuning. The quantity $(\rho_{gg}(t) - \rho_{xx}(t))$ is the nonequilibrium population induced by the pump, zeroth order in the probe. Thus, $\rho_{xx}(t)$ is zero since the pump is Π_y polarized. Time-integrated homodyne detection of the polarization field with the probe field is given by

$$DT \propto \int_{-\infty}^{\infty} P^{(NL)}(t) E_2(t) dt \qquad (7.12)$$

Equations (7.11) and (7.12) are used in this section to numerically calculate the theoretical DT signal from a two-level excitonic system, assuming that $\delta = 0$, and the pump and probe field amplitudes are real and described by hyperbolic secants. The decay rates are chosen to be $1/\Gamma_{yg} = 1/\gamma_{yg} = 1/\gamma_{xg} = 35$ ps (similar to measured rates shown in section 7.2). The DT can be measured as a function of pump pulse area, accomplished by evaluating Eq. (7.12) for a range of pump field amplitudes and a probe delay less than the lifetime but larger than the pulsewidth. Calculations of the DT for three different probe delays, 10 ps, 15 ps, and 20 ps, all clearly show that Rabi oscillations should be observed with this type of experimental configuration from an ideal two-level system.

Strong-field differential transmission: Rabi oscillations of single QD excitons

The strong-field response of a single QD exciton based on the DT configuration discussed above was investigated monitoring the DT of E_2 as a function of both the probe delay (τ) and the pump field strength (E_1).

Figure 7.16 shows the DT as a function of probe delay, for various E_1 powers, obtained by tuning the center wavelength of the laser to the QD state under investigation. Each case may be separated into a part driven by the pulse which is consistent with Rabi rotation in a two-level system, and a part after the pulse duration which is due to relaxation. The decay time at the lowest power is a measure of $1/\Gamma_{yg}$ for the $|y\rangle$ exciton state. The level of excitation induced by the E_1 is given approximately by $\sin^2(\Theta(t)/2)$. Thus, the differential probe (E_2) absorption (transmission) is minimal (maximal) for $\Theta(t) = \pi, 3\pi, 5\pi, \ldots$ For a pump (E_1) pulse with a power that corresponds to $\Theta(\infty) = \pi$, the DT signal increases until the pump pulse is completely gone, then it decays at the relaxation rate. For powers that correspond to $\pi < \Theta(\infty) < 2\pi$, the DT signal increases, then decreases as the exciton is stimulated back to the crystal ground state. The remaining probability of finding an exciton then decays to the ground state at the usual relaxation rate. At $\Theta(\infty) = 2\pi$, $\sin^2(\Theta(\infty)/2) = 0$, so the exciton that was created with unity probability by the first half of the pump (E_1) pulse is then stimulated (or driven) back to the crystal ground state by the second half of the pulse. This type of behavior is exactly what is shown by the data to the left in Fig. 7.16. The data is an indication that the pump pulse can be tuned to have pulse areas between zero and 2π.

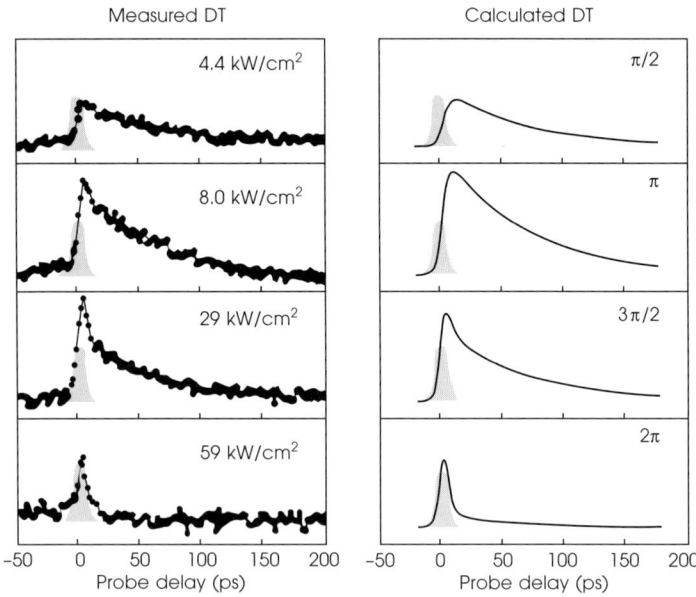

Fig. 7.16 DT vs. probe (E_2) delay for various pump (E_1) powers, obtained from a typical QD in the 42 Å well. The experimental data is shown on the left, with the peak intensity for the E_1 pulse shown in the upper right of each plot. The calculated DT is shown to the right, with the E_1 power expressed as a total pulse area in the upper right of each plot. The vertical axis is the same for all the data plots, and for all the theory plots. The shaded region shown with each plot is the pulse autocorrelation (for a constant power pump), in arbitrary vertical units. Taken from Stievater et al. [106].

The calculated DT as a function of probe delay on the right of Fig. 9.16 clearly illustrates the predicted DT as a function of probe (E_2) delay in the presence of a strong E_1 pulse. Here, the power of E_1 is indicated in each plot by the total pulse area, $\Theta(\infty)$. The calculation is based on a numerical solution of Eq. (7.11) and Eq. (7.12) from the density matrix equations with a hyperbolic secant pump pulse used as the source term. The calculation for each plot of Fig. 7.16 is obtained with a constant pulse area while the probe delay is varied. The relaxation rate for the calculation was taken from a fit to the low power data. The agreement between the measured DT and the calculated DT further establishes that the data is consistent with the onset of Rabi oscillations.

The Rabi oscillations can be seen more explicitly by examining the DT as a function of E_1 power for fixed probe delays. For probe delays such that $T < \tau < 1/\Gamma_{yg}$, the DT should show oscillations as the pump (E_1) field strength is increased, since the level of excitation is proportional to $\sin^2(\Theta(\infty)/2)$ for probe delays longer than the pulse width.

Figure 7.17(a) shows the data obtained from a single excitonic QD resonace. The DT is measured at two different probe delays, $\tau \approx 11.5$ ps, and $\tau \approx 18.5$ ps. The *x*-axis is the square root of the measured optical power, proportional to the amplitude of the electric field at the QD. Both delay values show an oscillatory behavior, with the first peak corresponding to a pump (E_1) pulse with $\Theta(\infty) \approx \pi$, the first trough corresponding to $\Theta(\infty) \approx 2\pi$ and so on. The oscillations imply that the excitation of the $|y\rangle$ QD exciton can be coherently controlled by the strength of the pump pulse, and are consistent with the DT shown as a function of delay in Fig. 7.16. The pulse can be tuned to have pulse areas between zero and 2π.

The oscillation period in Fig. 7.17(a) is proportional to μ_{yg}, the $|g\rangle \to |y\rangle$ dipole moment. Measuring this dipole moment from the data is therefore predicated upon an

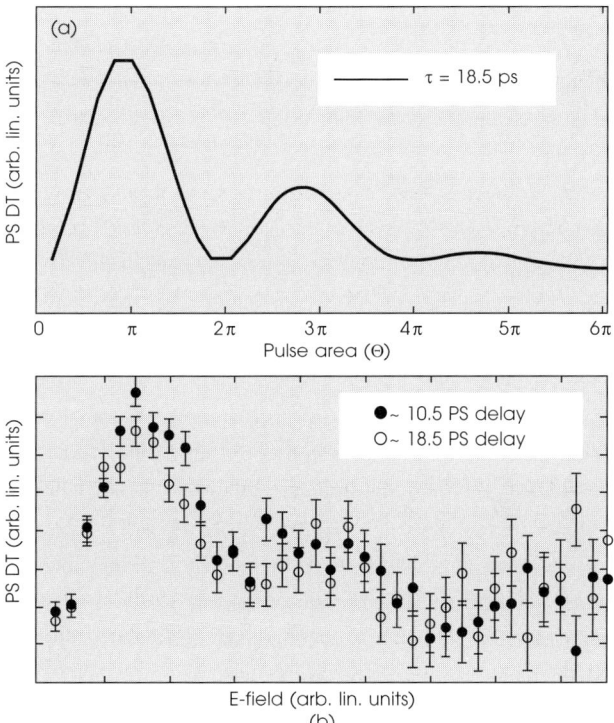

Fig. 7.17 (a) Rabi oscillations of a single QD state. Measured DT vs. pump (E_1) field amplitude for $\tau \approx 10.5$ ps and $\tau \approx 18.5$ ps. Both probe delays show behavior consistent with Rabi oscillations. Data are taken from a typical QD in the 42 Å well. (b) Calculated DT in the presence of delocalized excitons. By including a density dependent scattering rate from delocalized excitons in the QD relaxation rate, the qualitative features of the decay in (a) can be reproduced. Taken from Stievater *et al.* [106].

accurate knowledge of the pump field shape and amplitude inside the sample. Using Eq. (7.10) and the Poynting theorem for a field inside a dielectric, the average pump power incident on the cryostat (before chopping) can be related to the pulse area using a refractive index of 3.6, a beam waist of 20 µm (diameter) and $T = 3$ ps. This yields a dipole moment of about 78 Debye for this state. This value compares well with predictions of dipole moments for interface fluctuation QDs with similar sizes, as discussed in section 7.2.

Understanding the decay: coupling to delocalized excitons

Though the data in Fig. 7.17(a) show the oscillatory behavior expected for a two-level quantum system, the strength of the oscillations decreases faster as a function of pulse area than predicted by the simple two-level model. Numerical calculations based on Eqs (7.11) and (7.12) were re-evaluated for the full four-level system of Fig. 7.2, with a biexciton binding energy of 3.5 meV (see section 7.5). The results of these calculations showed that the inclusion of the biexciton state alone cannot explain the loss of signal. Also, calculations including center-of-mass excited states with reasonable state separations of ~meV did not result in a decay of the oscillations consistent with the data.

In ensemble measurements, an increase in the exciton relaxation rate with relatively high pump power has been observed [172]. An increase in relaxation rate with pump intensity could be due to a local heating effect, or could arise from exciton–exciton interactions. The former is unlikely due to the low average powers used for these measurements. Though these dots are adequately described as simple isolated systems at the low excitation intensities typically used for spectroscopy in the $\chi^{(3)}$ limit, exciton–exciton interactions may become significant at higher excitation level due to their weak lateral confinement. Indeed, PL imaging studies of narrow GaAs/AlGaAs quantum wells find broad resonance features (degenerate with sharp PL lines) that emerge with an increase in power, and show data that suggests the presence of delocalized excitons, especially in the interface fluctuation type of structures [71,173]. Also, biexponential photon echo decays attributed to a class of delocalized excitons nearly degenerate with localized excitons have been observed in similar structures [172]. Though direct spectroscopic evidence of such states is absent when the QDs are probed with high spatial resolution with weak fields, they may indirectly affect the QD resonances by creating potential scattering channels, especially when strongly excited. A QD characterized by stronger confinement may turn out to be less susceptible to these effects.

The role of those delocalized excitons has been investigated by adding an equation of motion for an incoherent population of delocalized states. The equation of motion for the delocalized exciton density is taken to be

$$\dot{n}_{del} = -\Gamma_{del} n_{del} + \frac{\alpha_{del} I(t)}{\hbar \Omega} \quad (7.13)$$

The properties of delocalized excitons relating to Coulomb scattering and the absorption coefficient were assumed to be close to the two dimensional case for the interface fluctuation structures. The following is an analysis of the effect of delocalized states on the Rabi flop experiment discussed in the previous section.

The lifetime of these delocalized states is assumed to be about 100 ps, based on measurements from a 4.5 nm QW [174]. The absorption coefficient for the delocalized excitons is taken to be $\alpha_{del} = 0.1\%$. This value is approximated from the measured value of 5×10^{-3} in a 8.0 nm QW [175], with the additional assumption that our value would be smaller because the QD resonances are in the tail of the absorption continuum of these delocalized states [18]. The peak laser intensity is given by $I(t) = E_1^{(0)} \text{sech}^2(t/T)$. For the peak intensity of 4.4 kW/cm² (approximately a $\pi/2$-pulse, see Fig. 7.16) the delocalized exciton density is only about $8 \times 10^7/\text{cm}^2$, consistent with an overall optical response that is dominated by the QD resonance.

The localized excitons can scatter with the delocalized excitons and a density dependent relaxation term results. Such nearby states could also contribute to a shift in the exciton resonance frequency. However, the density dependent scattering was found to be more effective in causing the damping in the DT and thus the effects of the shift were neglected. The density dependent relaxation rate is modeled as

$$\Gamma_{yg}(t) = \Gamma_{yg}^{(0)} + b n_{del}(t) \quad (7.14)$$

where $b = 0.75 \times 10^{-7}$ μeV cm² [174,176,177] and $\hbar \Gamma_{yg}^{(0)} = 10$ μeV, taken from the low power measurements of the relaxation rate of the QD state. For peak intensities corresponding to a 2π-pulse for the data in Figs 7.17 and 7.16, the relaxation rate would increase by about a factor of eight, according to Eq. (7.14).

Even if the total population of delocalized excitons is small, the DT can be sensitive to a change in the scattering rate of few tens of μeV for the localized exciton. Figure 7.17(b) shows the calculated DT in the presence of scattering from delocalized excitonic states, which reproduces the qualitative features of the data in Fig. 7.17(a).

For the purpose of one-bit rotation, the pulse area only needs to be between 0 and 2π. As long as the effects from delocalized states are negligible within this excitation range, there would be no concerns from a quantum logic operation standpoint. For QD structures that are more isolated, such as self-assembled QDs and nanocrystals, it is expected that the delocalized states be greatly suppressed, making it possible to achieve full Rabi oscillations with pulses greater than 2π. To date, however, Rabi oscillations have not been observed in those structures, possibly because of their relatively small dipole moments which require a π pulse to have very high field intensity.

7.5 Biexcitons in single QDs

So far, only studies on one of the exciton doublets in a dot have been discussed. Since the two orthogonally polarized ground state excitons, once excited, are both confined to the same QD with size comparable to the exciton Bohr diameter, it is predicted that they strongly interact via Coulomb coupling. Due to this interaction, the total energy of the bound two-exciton state, or the biexciton state, differs from the sum of individual transition energies by ΔE, defined as the biexciton binding energy.

The history of the concept of biexcitons goes back to 1947, when Hylleraas and Ore [178,179] predicted the existence of a stable bound positronium molecule (with negative binding energy). This was then extended by Lampert and Moskalenko [180,181] in predicting bound exciton–exciton molecules in semiconductors, similar to the positronium case except that the mass of the electrons and positrons (holes) in solids is different (under the effective mass approximation). Various calculations for the 3D case based on variational methods have shown that the ratio between the biexciton and exciton binding energies is a function of the ratio between the electron and hole effective masses [182–187], known as the Haynes' rule [188]. The exciton–exciton binding problem in quantum wells, quantum wires and quantum dots is dealt with in [189–196]. It is shown that quantum confinement increases the electron/hole overlap and therefore $|\Delta E|$ increases (ΔE becomes more negative) as long as the confinement size is not too much smaller than the exciton Bohr diameter; when the size of the biexciton envelope function becomes too small, the Coulomb repulsive core of the four particle system takes over, leading to a rapid decrease of $|\Delta E|$ towards zero and then an increase as ΔE crosses the zero from negative to positive [190].

Experimental evidence of biexcitons based primarily on PL has been found in various QD systems, including self-assembled QDs (see for example, References [74,79]) and interface fluctuation QDs [16,173]. It was found that the 3D quantum confinement leads to enhanced exciton–exciton binding. This section, however, will focus on coherent and resonant manipulation of the biexciton states. The coherent nonlinear optical spectroscopy of biexcitons at the single QD level has only been achieved in GaAs interface fluctuation QD systems [112]. Due to the discrete nature of the density of states, the coherent optical response from a single QD biexciton is expected to be similar to that of a diatomic molecule following the model discussed in Fig. 7.2.

Coherent nonlinear optical spectroscopy of single QD biexcitons is important since it directly probes the coherences between the two correlated excitons, which cannot be studied using PL-based techniques. In the language of quantum computing, these are the coherences between two qubits. One such example is the two-photon coherence (TPC) between the ground and the biexciton states introduced in section 7.1. The relatively long dipole coherence of excitons discussed in the previous sections does not guarantee a long-lived TPC, due to the exciton–exciton Coulomb interaction that is not present when only a single exciton is involved in the optical excitation. Therefore, an explicit measurement of the TPC decay dynamics is important. We show in this section that the TPC of the interface fluctuation QDs has a decay rate comparable to that of the exciton dipole coherence.

Other important parameters include the biexciton lifetime and the biexciton transition dipole moment. The role of the exciton–exciton Coulomb correlation to these parameters can be explored using nonlinear optical spectroscopy.

Excitation of single QD biexcitons using CW fields

Biexcitons can be formed using nonresonant optical excitation. The PL from the biexciton recombination can be monitored to provide evidence for the presence of the biexciton state. In PL spectroscopy, a single QD biexciton is formed by capturing two nonresonantly excited electron–hole pairs. The recombination of a biexciton is a stepwise process. In the first step, the biexciton recombines to form an exciton by emitting a photon. In the second step, the resulting exciton radiatively recombines. The two photons are of different energies; the first photon is lower in energy by $|\Delta E|$ due to the exciton–exciton binding. Since it takes two electron–hole pairs to form a biexciton and an exciton is formed in a dot primarily by directly capturing one e–h pair, the emission intensity of the biexciton-to-exciton transition depends nonlinearly (quadratically in the ideal case) on the excitation intensity whereas the emission intensity of the exciton recombination depends linearly on the excitation intensity. This has been considered as the main signature of the existence of biexciton states in various QD systems [16,73,77,78,93,173,197].

An example is shown in Fig. 7.18 for a 42 Å growth interrupted GaAs quantum well. The inset is the PL spectrum from a 0.5 μm region of the sample. The excitation field is tuned to 1632 meV (continuum of the QDs of interest). The low energy portion of the spectrum is magnified by a factor of 10 for clarity. The integrated PL intensity of three peaks at energies ε_X, ε_B and ε_{B2} as a function of the excitation intensity is plotted on a log-log scale. The peaks at ε_B and ε_{B2} show quadratic behavior and are attributed to the biexciton-to-exciton recombination. The peak at ε_X, however, shows linear dependence and is therefore from the exciton recombination. ε_B and ε_{B2} are energetically below ε_X and in the low energy tail of the overall response, suggesting that the exciton–exciton binding energy is negative. It is, however, not possible in this measurement to determine which exciton emission peak and biexciton-to-exciton emission peak are from the same dot. In other words, the binding energy cannot be precisely determined using only nonresonant PL spectroscopy.

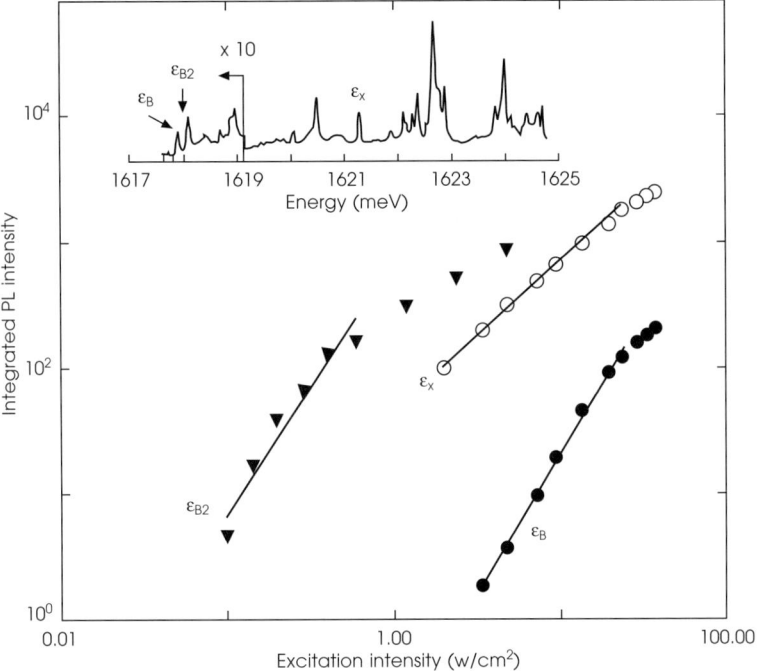

Fig. 7.18 Integrated PL intensity of peaks ε_X, ε_B and ε_{B2} as a function of the excitation intensity. Data are taken from a 42 Å growth interrupted GaAs quantum well.

The resonant excitation of a single QD biexciton can be achieved using two-photon absorption, i.e., the system can simultaneously absorb two photons from the same laser beam to form a biexciton using the exciton state as a virtual intermediate state. In this case, the absorption spectrum consists of not only the exciton resonance but also a *degenerate* two-photon absorpton resonance | $\Delta E/2$ | below the exciton resonance (half of the total biexciton energy). In the case that | ΔE | is substantially larger than the linewidth of the resonances, the excitation beam cannot be simultaneously resonant with both dipole transitions involving the ground, biexciton and the intermediate exciton states. In other words, the virtual intermediate state is far off the real exciton states. The degenerate two-photon absorption is therefore extremely weak.

The two-photon absorption cross-section can be increased dramatically by using two photons of nondegenerate energies, one near the exciton transition and the other near the exciton-to-biexciton transition. This can be achieved using the excitation scheme of nondegenerate DT. In this configuration, the excitation of the biexciton via the virtual intermediate state becomes enhanced because the virtual state now lies close to the real exciton state. In addition, an extra contribution involving a real excitation of the exciton population becomes significant due to the resonance condition.

Both two-photon absorption processes give rise to nonlineear optical signal in the CW DT experiment, where E_1 and E_2 are tuned to the ground-to-exciton and exciton-to-biexciton transitions respectively (inset of Fig. 7.19). The $|y\rangle$ exciton state is chosen by using Π_y polarized beams and the differential transmission of E_2 is monitored. Both contributions are negative, meaning that the transmission (absorption) of E_2 is decreased (increased) due to E_1. Taking the second contribution as an example, the presence of E_1 creates real exciton population, causing induced absorption of E_2, which is absent without E_1.

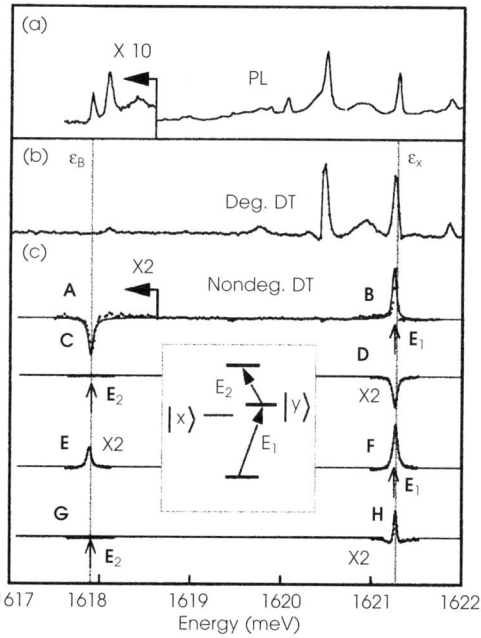

Fig. 7.19 DT spectra of a typical single QD biexciton. (a) PL spectrum taken from a 0.5 μm aperture. (b) Degenerate CW DT spectrum taken from the same aperture. (c) Nondegenerate CW DT spectra demonstrating the coupling between the resonances at ε_X and ε_B, which are identified as coming from the exciton-to-ground and biexciton-to-exciton transitions respectively. In curve A, B, E and F (C, D, G and H), E_1 (E_2) is fixed at ε_X (ε_B) and E_2 (E_1) is scanned. In curve A, B, C and D (E, F, G and H), the differential transmission of E_2 (E_1) is monitored. Dots are experimental data and solid lines are theoretical calculations. Data are taken from the 42 Å GaAs growth interrupt well. Taken from Reference [112].

This coherent nonlinear optical signal, labeled as I_2^{NL} (the subscript indicates that E_2 is monitored) can be calculated based on the equations of motion [198]. The two contributions mentioned above are a result of following two distinct perturbation paths leading to the third order polarization between the exciton and the biexciton state. The one that involves the real excitation of the exciton population via $|E_1|^2$ is known as the stepwise contribution and follows the path of

$$\rho_{gg}^{(0)} \underset{E_1^*}{\overset{E_1}{\nearrow}} \begin{matrix} \rho_{yg}^{(1)} & E_1^* \\ & \searrow \\ \rho_{gy}^{(1)} & E_1 \end{matrix} \nearrow \rho_{bb}^{(2)} \xrightarrow{E_2} \rho_{by}^{(3)}. \qquad (7.15)$$

It is incoherent because the phase of E_1 disappears in the modulus squared. The contribution involving a virtual intermediate state, however, follows the path of

$$\rho_{gg}^{(0)} \xrightarrow{E_1} \rho_{yg}^{(1)} \xrightarrow{E_2} \rho_{bg}^{(2)} \xrightarrow{E_1^*} \rho_{by}^{(3)} \qquad (7.16)$$

It is coherent and is due to the nonradiative two-photon coherence (between the ground state and biexciton state) induced jointly by E_1 and E_2. For this term to contribute, the two fields must be mutually coherent within the decay bandwidth of the two-photon coherence.

The calculation yields

$$I_2^{NL} = I_2^{\text{incoherent}} + I_2^{\text{coherent}}$$

$$\propto \text{Im} \frac{-i\alpha/\Gamma_{yg}}{(\gamma_{by} - i\Delta_2)} \left(\frac{1}{\gamma_{yg} - i\Delta_1} + c.c \right) \qquad (7.17)$$

$$+ \text{Im} \frac{-i\alpha}{(\gamma_{yg} - i\Delta_1)(\gamma_{by} - i\Delta_2)[\gamma_{bg} - i(\Delta_1 + \Delta_2)]}$$

where $\alpha = |\mu_{bx}\mu_{yg}E_2 E_1|^2$, $\Delta_1 = \Omega_1 - \omega_{yg}$ and $\Delta_2 = \Omega_2 - \omega_{by}$. μ_{ij} is the dipole moment.

It is interesting to point out that if the differential transmission of E_1 instead of E_2 is monitored, only the two-photon coherence contributes to the coherent nonlinear signal. This is because, in this case, $|E_2|^2$ cannot create a real exciton population since E_2 is greatly detuned from the ground-to-exciton transition by $|\Delta E|$. The perturbation path is

$$\rho_{gg}^{(0)} \xrightarrow{E_1} \rho_{yg}^{(1)} \xrightarrow{E_2} \rho_{bg}^{(2)} \xrightarrow{E_2^*} \rho_{yg}^{(3)} \qquad (7.18)$$

yielding a signal of

$$I_1^{NL} \propto \text{Im} \frac{i\alpha}{(\gamma_{yg} - i\Delta_1)^2 [\gamma_{bg} - i(\Delta_1 + \Delta_2)]} \qquad (7.19)$$

Therefore, by detecting E_1, the signal that is due exclusively to the coherent pathway is measured. The strength of I_1^{NL} is similar to the coherent (second) term in Eq. (7.17). The importance of the two-photon coherence can be examined by comparing I_1^{NL} and I_2^{NL}. If the two-photon coherence decay is fast ($\gamma_{bg} \gg \Gamma_{yg}$), then I_2^{NL} would be dominated by the incoherent (first) term and I_1^{NL} would be much smaller than I_2^{NL}.

Figure 7.19(b) is the degenerate DT spectrum taken from the same set of dots as in Fig. 7.18 with both beams Π_y polarized, selecting only the $|y\rangle$ states. The PL spectrum is shown again in (a) for comparison. The single exciton state $|y\rangle$ at ε_X will be focused on in the search for the exciton-to-biexciton resonance using nondegenerate CW DT. All experimental data are plotted as dots in Fig. 7.19(c). Solid lines are theoretical calculations based on Eqs (7.17), (7.19) and (7.6). As discussed in section 7.1, due to the QD elongation in the interface fluctuation QD systems, the resonant excitation of the biexciton state follows $\Pi_x - \Pi_x$ (or $\Pi_y - \Pi_y$) polarization [77,78,171]. For these nondegenerate experiments, E_1 and E_2 are both Π_y polarized. Detailed discussions of the optical selection rules of biexcitons confined to asymmetric QDs will be presented later in this section.

In the first experiment, the DT spectrum is taken as a function of Ω_2 with $\hbar\Omega_1$ fixed at ε_X. The differential transmission of E_2 is monitored. The excitonic optical nonlinearities led to peak B of Fig. 7.19(c) when E_2 is tuned to ε_X, as expected. However, due to the nondegenerate two-photon absorption discussed above, a negative resonance (induced absorption of E_2) at ε_B is observed, as indicated by A in Fig. 7.19(c). Recall that the same resonance exhibits a quadratic PL dependence on excitation density. Therefore, this resonance can be unambiguously attributed to the exciton-to-biexciton transition in the same QD that gives the excitonic recombination peak ε_X. The exciton–exciton binding energy can be extracted by taking the difference between ε_B and ε_X, yielding a ΔE of -3.360 ± 0.001 meV, about two orders of magnitude larger than the excitonic linewidth. The binding energy ΔE is in general a function of the QD size, shape, the electron-to-hole mass ratio and the dielectric constant ratio between the QD material and the barrier material. The calculation in Reference [189] shows that for an ideal 40 Å GaAs quantum well, the biexciton binding energy is about -1.5 meV. The measured $|\Delta E|$ is more than two times larger than this value, indicating a strong effect of the disorder and localization on the biexciton binding energy. This value is also larger than other experimental results on GaAs quantum wells with comparable well width. In this case, the quantum confinement reduces the spatial separation between the electron–hole pairs and enhances their mutual attraction [190].

Also by monitoring the differential transmission of E_2, but now fixing $\hbar\Omega_2$ at ε_B and scanning Ω_1, it can be verified that the exciton resonance at ε_X is the only state that couples to the state at ε_B. This is shown in curve C and D of Fig. 7.19(c). Again, curve D shows induced absorption of E_2. The absence of signal in curve C is because the exciton-to-biexciton transition cannot be directly accessed without excitation involving the ground state.

In the next set of experiments, the differential transmission of E_1 is monitored. The corresponding spectra under various conditions are shown in E, F, G and H of Fig. 7.19(c).

In E, I_1^{NL} is plotted as a function of Ω_2 near ε_B with $\hbar\Omega_1$ fixed at ε_X. $\hbar\Omega_2$ can also be tuned to ε_X, recovering the excitonic coherent nonlinear optical response, as shown in curve F. In curve G and H, however, Ω_2 is fixed at ε_B and Ω_1 is tuned around ε_X. The response in curve E and H is predominantly positive, corresponding to reduced absorption of E_1, opposite to that of curve A and D. The lineshapes are in excellent agreement with Eq. (7.19), as shown by the solid curves.

This signal in curves E and H results entirely from the two-photon coherence. Its strength is measured to be of the same order as that of curves A and D and shows that the decoherence rate of the biexciton (γ_{bg}) is comparable to the exciton relaxation rate (Γ_{yg}). A more quantitative discussion is given below. Combining the signal strength of I_1^{NL} and I_2^{NL} and the decay parameter derived above, the exciton-to-biexciton transition dipole moment can be studied. This will be discussed later in this section.

The two-photon coherence and stepwise exciton population are also the sources for the conventional degenerate two-photon absorption (as discussed earlier), where one color excitation is used and the two-photon absorption resonance appears | $\Delta E/2$ | below the exciton resonance. At this spectral position, the two-photon coherence as well as the incoherent population term is significantly reduced due to the off-resonant condition. This can be seen by setting $\Omega_1 = \Omega_2$ in Eqs (7.17) or 7.19), yielding a coherent nonlinear optical signal that is maximum midway between the two resonances [190] but is greatly reduced in magnitude (compared to the fully resonant case) by $(\Delta E/2\gamma)^2 \simeq 2500$. Similar calculations were carried out in Reference [190]. Brunner et al. [16] reported a nice experiment where the degenerate two-photon absorption resonance leading to the excitation of a localized biexciton in a disordered GaAs quantum dot is identified by detecting the subsequent PL. Such an experimental scheme, however, does not allow for the determination of the portion in the signal that is due to the two-photon coherence which makes it difficult to obtain information on the decoherence of the biexciton state. Due to the limited signal-to-noise ratio for degenerate DT measurements, the two-photon resonance between ε_B and ε_X is not observed in Fig. 7.19(b) for dots under investigation (cases where the two-photon resonance is strong enough to be observable are discussed in Reference [72]). The observation of two-photon peak for these particular dots (where the ratio between the fully resonant peak signal of curve E and the laser transmission, $\Delta T/T$, is about 10^{-4}) would require the ratio between the noise (shot noise and the noise from the transmitted field that is homodyne detected) and laser transmission be of order 10^{-8}, which is difficult to achieve. It has been shown in this section that the nondegenerate and fully resonant two-photon absorption, however, leads to a greatly enhanced signal stength. This allows for a study of the biexciton decoherence dynamics, which will be detailed in the next section.

Dephasing of biexcitons

The previous section has discussed the excitation of single QD biexcitons via resonant nondegenerate two-photon absorption and the subsequent coherent nonlinear optical response. From these data, information regarding the dephasing dynamics of the biexcitonic system can be extracted. This type of information is critical for QD-based quantum information processing, where long-lived quantum coherence is an essential element.

As discussed in section 7.2, in higher dimensional semiconductor systems and in the absence of disorder, the excitonic and biexcitonic states are characterized by extended Bloch wavefunctions, which make them susceptible to purely phase changing interactions with other excitons (excitation induced dephasing, EID [87]) as well as with the surrounding crystal, leading to relatively fast dephasing of various coherences [199], even though the excitonic and biexcitonic lifetimes are relatively long. In QD structures, 3D confinement results in strong localization of the exciton/biexciton wavefunction and therefore reduces the potential for interactions with the crystal. Since the density of states becomes discrete and the excitation is resonant, EID type of interactions do not dominate in experiments using narrow-band weak-field CW excitation, although evidence for such interactions resulting from excitation of more extended excitonic states was found in strong-field PS spectroscopy discussed in section 7.4. It is therefore expected that the broadening of the biexcitonic state is dominated by energy relaxation in the weak field limit.

From Eqs (7.17) and (7.19), it is straightforward to extract the decay parameters of the biexcitonic system by studying the lineshape of the curves in Fig. 7.19. Specifically, by setting $\Delta_1 = 0$ in Eq. (7.19), the lineshape for the experiment producing curve E of Fig. 7.19(c) can be expressed as

$$I_1^{NL} \propto \mathrm{Im} \frac{i\alpha}{\gamma_{ug}^2 [\gamma_{bg} - i\Delta_2]}, \qquad (7.20)$$

a simple Lorentzian with half width half maximum determined by the two-photon decay rate γ_{bg}. Using the experimental data, it is found that $\gamma_{bg} = 22.0 \pm 0.7$ ps^{-1}, comparable to the exciton dipole coherence decay rate of $\gamma_{yg} = (19.5 \pm 4.8$ ps$)^{-1}$ and exciton relaxation rate of $\Gamma_{yg} = (13.3 \pm 5$ ps$)^{-1}$ obtained by repeating the analysis seen in section 7.2 on the excitonic features in (b) and B and F of (c) in Fig. 7.19.

Analysis of the induced absorption lineshape in curve A of Fig. 7.19(c) based on Eq. (7.17) gives γ_{by} of 10.5 $\pm$ 1 ps^{-1}. Based on the model of Fig. 7.2 and an approximation involving weak Markovian scattering of excitons [200, 201], the biexciton–exciton dipole coherence decay rate

$$\gamma_{by} = \gamma_{bg} + \gamma_{yg} - 2R\sqrt{\gamma_{bg}\gamma_{yg}} \qquad (7.21)$$

where R denotes the correlation between the phase changing scattering events of two different excitons confined in a QD. This correlation term is to account for the fact that the phase changing scattering events for state $|b\rangle$ and state $|y\rangle$ may not be completely unrelated. Taking the experimental values for γ_{yg}, γ_{bg} and γ_{by}, we note that the correlation term is negligible. This is because pure dephasing events are insignificant. It is therefore inferred that the biexciton state is radiatively broadened, as expected.

Assuming only radiative broadening of the biexciton state, the biexciton lifetime can be inferred. However, the experiments discussed above are not suitable for a direct measurement of the biexciton lifetime, since the decay of the biexciton population does not enter until the fourth-order in the perturbation chain and these experiments measure the third order response. In the next section, transient DT methodology is used to directly measure the biexciton lifetime.

The dephasing dynamics discussed above are independent of the excitation density within the $\chi^{(3)}$ limit, verifying that excitation-induced-dephasing type of interactions are insignificant for CW experiments due to the discrete density of states which results from the quantization of the confined exciton/biexciton center of mass motion.

Direct measurement of biexciton lifetime

The biexciton lifetime is $\Gamma_b^{-1} = (\Gamma_{by} + \Gamma_{bx})^{-1}$ according to Fig. 7.2 (assuming that radiative recombination dominates). Γ_{by} or Γ_{bx} represents the rate that a biexciton decays to produce an exciton and a photon. From a physics standpoint, it is interesting to study how Γ_{by} or Γ_{bx} compares to Γ_{yg} or Γ_{xg}, since this ratio contains rich physics related to, for example, the superradiant effect and the coherent volume concept of localized biexcitons [197,202]. For excitons in higher dimensional systems, such as bulk crystals, it is often assumed intuitively that the presence of one exciton does not affect the recombination of the other (due to relatively weak binding energy) and therefore the biexciton lifetime is half the single exciton lifetime (assuming $\Gamma_{yg} = \Gamma_{xg}$). This intuition, however, does not include the supperradiant effect which greatly enhances biexciton recombination rate when the exciton–exciton separation is much smaller than the wavelength of the emitted light. Citrin *et al.* [202], however, pointed out that the radiative lifetime of localized biexcitons in quantum wells is longer than the single exciton lifetime, despite the inclusion of the supperradiance. The predicted ratios of the biexciton to the exciton lifetimes in single QDs are presently not in agreement [190, 197, 203].

Experimental studies on bulk or quantum wells show that the biexciton lifetime is half of the exciton lifetime (see References [180,204]), seemingly in agreement with the simple intuition discussed in the previous paragraph. But it was pointed out later in Reference [197] that such experiments measure a biexciton lifetime that is due to not only the radiative recombination but also the biexciton dissociation. The measured ratio of 0.5, therefore, does not represent the ratio of the radiative part of the biexciton and exciton lifetimes. Recent results from high quality QWs show that the radiative biexciton lifetime is in fact almost equal to the single exciton lifetime [201]. For QD systems, the biexciton dissociation is suppressed and the measured biexciton lifetime is assumed to be radiative. In Reference [197], Bacher *et al.* measured the ratio of the biexciton and exciton lifetimes to be slightly larger than 1 for CdSe/ZnSe QDs grown by migration enhanced epitaxy.

In light of the discrepancy discussed above, measurements of biexciton lifetime of GaAs interface fluctuation QDs become intriguing. One straightforward way to measure the biexciton lifetime, as used in Reference [197], is to time resolve the biexciton PL emission. For GaAs interface fluctuation QDs, however, the shorter lifetimes impose a stringent requirement on the time resolution of the experimental apparatus. The direct measure of the biexciton lifetime discussed below is based on transient DT which measures effects due to real population in the biexciton level, ρ_{bb}.

In this measurement, a significant exciton population (for example, $|y\rangle$ exciton) is created using a strong prepulse, $E_{pre}(\Omega_{pre})$, of order $\sim \pi$. In section 7.4, it was shown that such an effective preparation of $|y\rangle$ population is possible in GaAs interface fluctuation QD systems. After a few picoseconds, E_1 and E_2 are tuned to excite the $|y\rangle \rightarrow |b\rangle$ transition and are arranged in the copolarized DT configuration to measure the biexciton population $\rho_{bb}^{(2)}$ decay rate. The two-level system $|y\rangle \rightarrow |b\rangle$ in this case is analogous to the $|g\rangle \rightarrow |y\rangle$ two-level system (studied in section 7.2 using two-beam transient DT) except that $|g\rangle$ does not decay whereas $|y\rangle$ does.

Figure 7.20(a) shows the three beam transient DT spectrum taken under identical conditions as the spectra in Fig. 7.11(a), except for the addition of the prepulse. The Π_y

polarized E_{pre} is tuned to dot C (see Fig. 7.11) as indicated by the arrow in Fig. 7.20(a). The DT from dot C in the presence of the prepulse appears to be slightly negative (gain of E_2), indicating an efficient population transfer from the crystal ground state to the exciton state by the prepulse.

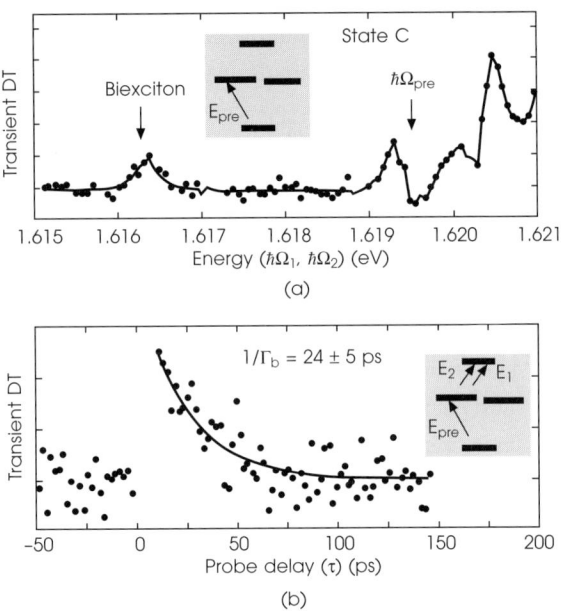

Fig. 7.20 Single QD transient degenerate DT with prepulse. (a) Same as Fig. 7.11(a) but with a strong Π_y prepulse tuned to the excitonic state C. The $|y\rangle \to |b\rangle$ biexcitonic transition can be clearly seen 3.3 meV lower in energy than the $|g\rangle \to |y\rangle$ excitonic transition. (b) The E_1 and E_2 are fixed in energy at the biexciton transition energy as the E_2 delay is scanned. The biexciton lifetime $1/\Gamma_b$ is found from fitting Eq. (7.22). Data are taken from the 42 Å growth interrupt well.

The biexciton lifetime can be measured by fitting the decay of the DT as E_2 delay is increased with both Ω_1 and Ω_2 fixed at the biexcitonic resonance ($\Omega_1 = \Omega_2 = \omega_{by}$) and the prepulse fixed at the excitonic resonance ($\Omega_{pre} = \omega_{yg}$) of dot C. The third order polarization that gives rise to the signal field that is homodyne detected with the E_2 is given by $\mathbf{P}^{(3)}(t) \propto (\vec{\mu}_{by} \rho_{yb}^{(3)}(t) + c.c.)$. Again assuming δ-function pulses and keeping only those terms for the DT of E_2 more than a pulse width from zero delay, the time-integrated homodyne detected signal is then

$$DT \propto (\rho_{yy}^{(0)} - \rho_{bb}^{(0)})\theta(\tau)$$

$$\times \left[e^{-\Gamma_b \tau}\left(1 + \frac{\Gamma_b}{2(\Gamma_b - \Gamma_{yg})}\right) + e^{-\Gamma_{yg}\tau}\left(1 - \frac{\Gamma_b}{2(\Gamma_b - \Gamma_{yg})}\right) \right] \quad (7.22)$$

This equation assumes for simplicity that $\Gamma_{by} = \Gamma_{bx}$, so that $\Gamma_b = 2\Gamma_{by}$. Note that the signal depends on the population difference ($\rho_{yy}^{(0)} - \rho_{bb}^{(0)}$) which is zero without the Π_y prepulse tuned to the excitonic transition.

The DT obtained in this way can be biexponential due to decay of both the $|b\rangle$ and $|y\rangle$ states. In the case that the biexciton relaxation rate is determined by the individual exciton relaxation rates, such that $\Gamma_{by} \approx \Gamma_{yg}$ and $\Gamma_{bx} \approx \Gamma_{xg}$, then $\Gamma_b \approx 2\Gamma_{yg}$. In this case, $\Gamma_b/2(\Gamma_b - \Gamma_{yg}) \approx 1$, and the decay of the DT is a single exponential with decay rate Γ_b.

Figure 7.20(b) shows the decay of the DT obtained from this three-beam technique. The data is fit to Eq. (7.22), where $1/\Gamma_{yg} = 41.2$ ps for state C (see Fig. 7.11(b)). The fit is shown in the plot and yields $1/\Gamma_b = 24 \pm 5$ ps. The ratio between the biexciton and exciton lifetimes for this particular dot is close to 0.5.

This result is different from that of Reference [197]. This difference could be due to the different types of QDs studied and the strength of the confinement. However, it agrees with Reference [203], which predicted a biexciton to exciton radiative lifetime ratio of 0.5 in self-assembled QDs. More precise measurements in various QD systems are therefore still needed for further understanding of the physics behind the radiative recombination of single QD biexcitons.

Biexcitonic transition dipole moment

Another important parameter is the exciton-to-biexciton transition dipole moment or oscillator strength. Takagahara [190] has calculated this transition strength using variational methods and found that both the excitonic and biexcitonic dipole moments increase as the QD size increases (but not deviating from the strong confinement regime) due to the increasing number of unit cells involved. The ratio between the biexcitonic and excitonic oscillator strength was also calculated. For a $GaAs/Al_{0.3}Ga_{0.7}As$ QD size of tens of nanometers, this ratio is between 1 and 2 [190].

Independent measurements based on linear absorption from single quantum dot excitons give a ground state to exciton dipole moment of order 50–100 Debye (compared to a few Debye for atomic systems) depending on dot size (see [102] and section 7.2). From the theory above for the biexciton, it is clear that with knowledge of the decay rates and by comparing the signal strength obtained at the biexciton to that obtained at the exciton in Fig. 7.19(c), it is possible to infer the ratio of the dipole moments.

Taking the decay parameters extracted from the previous section, the expected ratio of the resonant signal at ε_B and ε_X in this experiment is a function of $|\mu_{yb}/\mu_{gy}|$, which can then be determined from the experimental value. It is found that $|\mu_{yb}/\mu_{gy}| = 1.15 \pm 0.30$. This shows that the excitation of one exciton in a dot does not cause a significant change in the dipole of the other exciton despite the strong Coulomb correlation. The biexciton dipole is slightly larger than the exciton dipole, in agreement with calculations in Reference [190].

Optical selection rules

The predicted optical selection rules (OSRs) for biexcitons in quantum dot systems were discussed in section 7.1. For asymmetric QDs, the exciton and consequent biexciton transitions are colinearly polarized at zero magnetic field and cross-circularly polarized under nonzero magnetic field. It has been shown above that the GaAs interface fluctuation QD biexcitons based on Π_y–Π_y two-photon excitation agree with this prediction (without an external magnetic field). These optical selection rules were also confirmed in self-assembled $In_xGa_{1-x}As$/GaAs QDs [79,128] and CDSe/ZnSe QDs [77,78].

The optical selection rules for a typical GaAs interface fluctuation QD at zero magnetic field are illustrated by the DT measurements shown in Fig. 7.21(a) and (b). A QD with a relatively large exchange splitting is chosen. Top curves in Fig. 7.21(b) are the degenerate DT spectra showing the fine structure splitting and the linearly polarized optical selection rules for the exciton transitions. Using nondegenerate CW DT by fixing Ω_1 at the center of one of the exciton states (with a matching polarization) and monitoring E_2, the resonances due to the biexciton state are identified, as shown in Fig. 7.21(a). It is found that the excitonic and the subsequent biexcitonic transitions are co-linearly polarized. By making E_2 orthogonal to E_1, no signal is observed. The bottom curves of Fig. 7.21(b) are obtained by fixing E_2 at one of the biexciton resonances (again, with matching polarization) and tuning the co-linearly polarized E_1 around the exciton transitions.

Shown in Fig. 7.21(c) and (d) are similar experiments in the presence of the external magnetic field. It was found that the excitonic and the subsequent biexcitonic transitions become cross-circularly polarized. For (c), Ω_1 is fixed at the center of the exciton resonance with matching circular polarization, and E_2 is scanned around the biexciton transition with opposite circular polarization. The signal drops to zero by making the two fields co-circularly polarized. This is consistent with the model shown in the inset.

7.6 Optically induced two exciton-state entanglement

In the previous section, the optically induced two-photon coherence reprsents a coherent superposition state of the form: $|\psi\rangle = a|00\rangle + b|11\rangle$, a linear combination of the ground

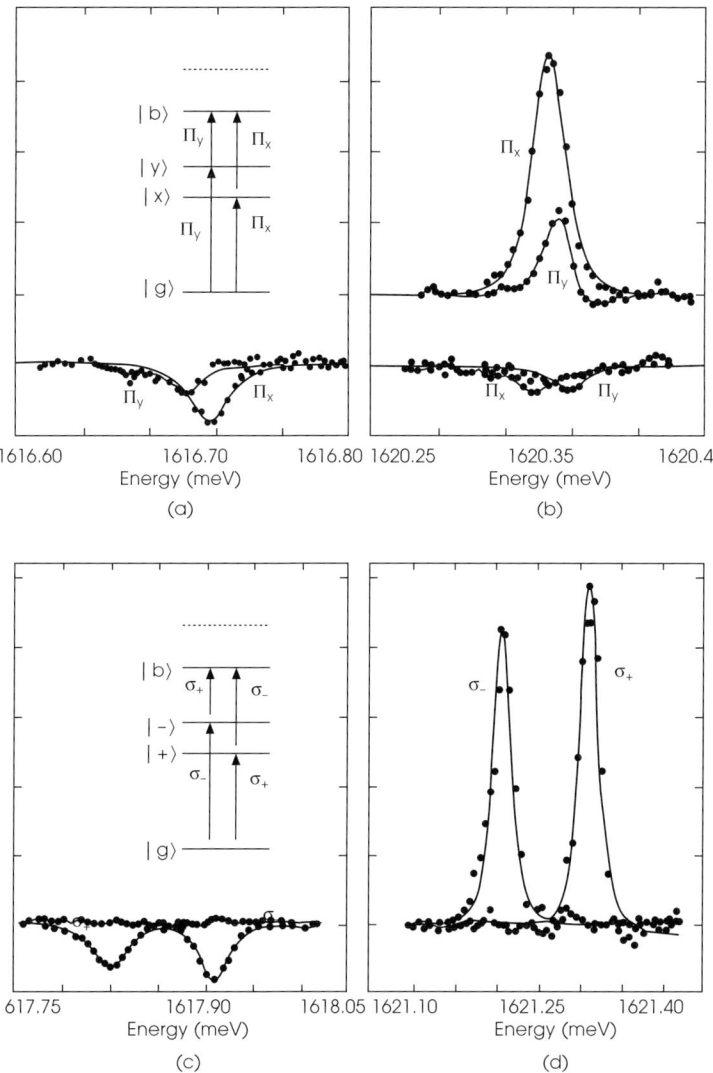

Fig. 7.21 Optical selection rules in typical QDs. Top curves of (b): degenerate spectra of exciton transitions showing the fine structure splitting without the external magnetic field. For the open (solid) circles, the two beams are Π_x (Π_y) polarized. (b) for the open (solid) circles, Ω_1 is fixed at the center of the Π_x (Π_y) exciton transition, Ω_2 is scanned around the exciton-to-biexciton transition and both beams are Π_x (Π_y) polarized. Bottom curves of (a): for the open (solid) circles, Ω_2 is fixed at the center of the Π_x (Π_y) biexciton transition, Ω_1 is tuned around the exciton resonances. (c) and (d) are similar experiments for another dot with the magnetic field turned on (1.1 T). The higher (lower) energy exciton is σ_+ (σ_-) polarized. By using E_1 to excite one of the exciton states, the exciton to biexciton transition is observed with E_2 orthogonally polarized with respect to E_1. The optical selection rules identified in these experiments are consistent with the model shown in the insets. Data are take from the 42 Å growth interrupt well.

state ($|00\rangle$, both electrons in the valance band, $|g\rangle$ of Fig. 7.2) and the biexciton state ($|11\rangle$, both electrons excited, $|b\rangle$ of Fig. 7.2). Considering that the absence and presence of an exciton can be regarded as a pseudo-spin [67], this coherence is analogous to a Bell state in a system consisting of two entangled spin-$\frac{1}{2}$ particles: $a|\downarrow\downarrow\rangle + b|\uparrow\uparrow\rangle$. In this section, a similar experiment based on nondegenerate DT is discussed in which a coherence of the form $|\psi\rangle = a|10\rangle + b|01\rangle$ is optically induced in GaAs interface fluctuation QDs [89]. $|01\rangle$ and $|10\rangle$ represent the $|-\rangle$ and $|+\rangle$ (or $|x\rangle$ and $|y\rangle$) single exciton states of Fig. 7.2 respectively. This is analogous to an entangled state $a|\downarrow\uparrow\rangle + b|\uparrow\downarrow\rangle$ of two spin-$\frac{1}{2}$ particles.

The ability to optically produce such entangled states is essential for the preparation of an arbitrary coherent superposition state of two qubits. It is also a prerequisite to a full demonstration of a quantum controlled-NOT gate. This can be understood by considering an initial state of the system being a product state of two qubits, $|10\rangle + |11\rangle$, where the first and second bits are the target and control bits respectively. By applying a quantum controlled-NOT operation, the state becomes an entangled state $|10\rangle + |01\rangle$. Thus, a successful controlled-NOT operation relies critically on the ability to entangle two qubits.

In the case of two excitons confined to a QD, the two qubits are coupled via the Coulomb interaction. As a result, the excitation of one exciton affects the transition energy of the other, as has been explicitly shown in the previous section. This allows for a controlled excitation of one exciton depending on which state the other exciton is in. This coupling allows for inducing entangled states of two excitons using coherent optical excitation.

To differentiate the two single exciton transitions not only by their orthogonal polarizations but also via transition energies, an external magnetic field is applied to fully split the two transitions. The degenerate DT spectra of the interface fluctuation QD to be studied in this section as a function of the magnetic field have been previously shown in Fig. 7.12. The exciton transitions become nondegenerate due to Zeeman splitting and they become σ_+ and σ_- polarized.

In the following measurements, the σ_- polarized $E_1(\Omega_1)$ and the σ_+ polarized $E_2(\Omega_2)$ are tuned to excite the σ_- and σ_+ exciton transitions respectively (see the inset of Fig. 7.23). Because of the large exciton–exciton binding energy, the excitation of the biexciton state can be effectively neglected. This excitation scheme therefore produces an interesting scenario: if the σ_- exciton is resonantly excited by the σ_- polarized E_1, then the excitation of the σ_+ exciton can be neglected due to the exciton–exciton Coulomb interaction (leading to a shift of ΔE in transition energy for the σ_+ exciton). Similarly, if the σ_+ exciton is resonantly excited, then the σ_- exciton cannot be effectively excited.

Coherent optical spectroscopy and manipulation of single quantum dots 333

Since E_1 and E_2 are mutually coherent, the system is expected to exist in a coherent superposition of the two possibilities. This is illustrated in Fig. 7.22. At any given time, at most one exciton exists because this experimental configuration prevents the excitation of the biexciton state.

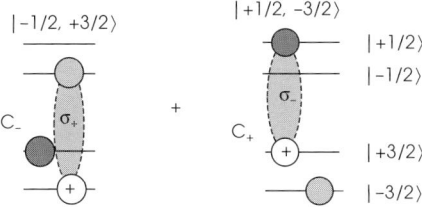

Fig. 7.22 A view of the exciton–exciton entanglement in the electron–hole picture.

In the two-exciton basis (see Fig. 7.2), this is equivalent to a coherent superposition within the exciton Zeeman doublet and is thus referred to as Zeeman coherence. In the weak field, the wavefunction of the system becomes

$$|\Psi\rangle = C_0|00\rangle + C_+|10\rangle + C_-|01\rangle \tag{9.23}$$

where the biexciton state does not contribute, making it nonfactorizable. In the following, it will be shown that a relatively long-lived Zeeman coherence would give rise to unique features in the CW DT spectrum and the measurements show that as a result of coherent excitation by E_1 and E_2, C_- and C_+ indeed have a well-defined phase relation within the exciton lifetime.

In these experiments, differential transmission of E_2 is monitored. By monitoring DT of E_1 instead of E_2, no new information can be obtained, unlike in the resonant excitation of biexciton. This is because states $|+\rangle$ and $|-\rangle$ are symmetric with regard to the ground state $|g\rangle$, whereas states $|+\rangle$ (or $|-\rangle$) and $|b\rangle$ are not.

Based on the equations of motion for the three level model containing the ground and the two single exciton states (the biexciton state is not excited), it is found that the Zeeman coherence contributes significantly to the DT signal provided that the pure dephasing of the Bell state $|\psi\rangle = a|10\rangle + b|01\rangle$ is not too large. This contribution is a result of a two-photon excitation between the σ_- and σ_+ exciton states using a virtual intermediate state near the ground state and follows

$$\rho_{gg}^{(0)} \begin{array}{c} \nearrow E_1^* \; \rho_{g-}^{(1)} \; E_2 \searrow \\ \searrow E_2 \; \rho_{+g}^{(1)} \; E_1^* \nearrow \end{array} \rho_{+-}^{(2)} \xrightarrow{E_1} \rho_{+g}^{(3)} \tag{7.24}$$

A second contribution to the coherent nonlinear optical signal comes from a real depletion of the ground state by E_1, which causes a reduced absorption of E_2. The perturbation path is

$$\rho_{gg}^{(0)} \begin{array}{c} E_1^* \rho_{g-}^{(1)} E_1 \\ \nearrow \searrow \\ \searrow \nearrow \\ E_1 \rho_{-g}^{(1)} E_1^* \end{array} \rho_{gg}^{(2)} \xrightarrow{E_2} \rho_{+g}^{(3)} \qquad (7.25)$$

Similar to the biexciton case, the first path is fully coherent and requires that E_1 and E_2 be mutually coherent within the decay of the Zeeman coherence. The second path, however, contributes even if the light sources are incoherent.

The DT signal (monitoring E_2) containing both components is calculated based on the above perturbation paths

$$I_2^{NL} = I_2^{\text{depletion}} + I_2^{\text{Zeeman}}$$

$$\propto \text{Im} \frac{i\beta}{(\gamma_{+g} - i\Delta_2)} \left[\begin{array}{c} \frac{1}{\Gamma_{-g}} \left(\frac{1}{\gamma_{-g} + i\Delta_1} + c.c \right) \\ + \frac{1}{\gamma_{+-} - i(\Delta_2 - \Delta_1)} \left(\frac{1}{\gamma_{-g} + i\Delta_1} + \frac{1}{\gamma_{+g} - i\Delta_2} \right) \end{array} \right] \qquad (7.26)$$

where $\beta = |\mu_{+g}\mu_{-g}E_1E_2|^2$, $\Delta_1 = \Omega_1 - \omega_{-g}$ and $\Delta_2 = \Omega_2 - \omega_{+g}$.

Based on Eq. (7.26), the Zeeman coherence contribution has a unique lineshape which is distinct from the depletion contribution and is used as a signature of its presence. This signature (theory) is shown in the spectra in Fig. 7.23 as a function of Ω_2, with Ω_1 fixed at various frequencies around the σ_- exciton transition.

There are two features that are unique for the signal caused by the Zeeman coherence, as shown in Fig. 7.23(a). First, as Ω_1 is moved off the center of the σ_- exciton resonance, a derivative-like lineshape develops. Secondly, the peak of the Zeeman coherence component moves in the same direction as the fixed Ω_1. This is because the σ_--to-σ_+ two-photon transition via a virtual intermediate state near the crystal ground is resonant when $\Omega_2 - \Omega_1$ matches $\omega_{+g} - \omega_{-g}$. Therefore, as Ω_1 is moved lower (higher) in energy, Ω_2 also needs to move to a lower (higher) position for maximum signal.

The $|g\rangle$ depletion component of the signal is shown in Fig. 7.23(b). It stays at the σ_+ resonance and only decreases rapidly in strength as Ω_1 is moved. The reason that the signal peak does not move is because the reduced absorption of E_2 is due to a real ground state population depeletion caused by $|E_1|^2$. Only the amount of depletion depends on Ω_1. As Ω_1 is moved off the center of the σ_- resonace, the strength of the $|g\rangle$ depletion of decreased rapidly. However, it is not important whether the depletion is created resonantly or not; E_2 still just maps out the σ_+ resonace.

Coherent optical spectroscopy and manipulation of single quantum dots 335

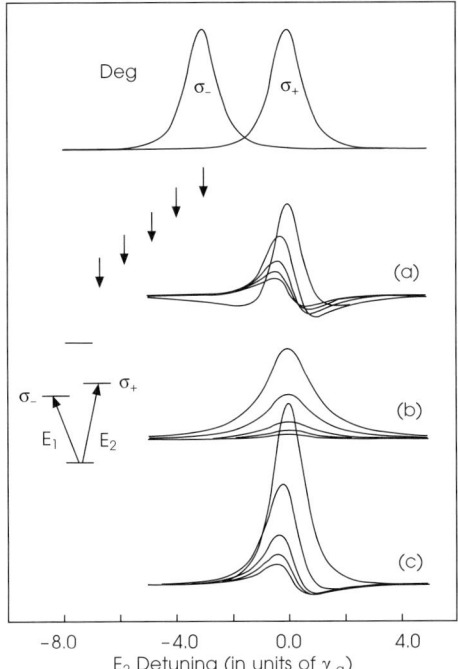

Fig. 7.23 Calculations for the contributions of the Zeeman coherence and the ground state depletion to the coherent nonlinear optical signal. Curves labeled as σ_- and σ_+ are the degenerate CW DT of the two exciton states. For all other curves, the σ_- polarized E_1 and σ_+ polarized E_2 excite the σ_- and σ_+ exciton transitions respectively, as shown in the inset. The biexciton level is far off resonance and does not contribute to the nonlinear optical response. E_1 is fixed in frequency for each spectrum but is tuned to various frequencies around the σ_- resonance, as indicated by the arrows. E_2 is tuned and the spectra are obtained as a function of Ω_2. The energy axis is in units of γ_{-g}. It is assumed that the pure dephasing is zero: $\gamma_{-g} = \gamma_{+g} = 1$, $\Gamma_{-g} = \Gamma_{+g} = 2$ and $\gamma_{-+} = 2$. (a) and (b) are the Zeeman coherence and incoherent ground state depletion contributions respectively. (c) is the total signal. Taken from Reference [111].

Similar to an exciton dipole coherence, the nonradiative Zeeman coherence could in principle decay rapidly due to pure dephasing even though the exciton lifetime is relatively long. Figure 7.23 has assumed zero pure dephasing of the Zeeman coherence. By adding pure dephasing, its contribution to the signal would broaden out and drop rapidly in strength. Figure 7.24 shows explicitly how pure dephasing of the Zeeman coherence affects the overall DT signal.

The above analysis is based on the three-level model consisting of the $|g\rangle$, $|+\rangle$ and $|-\rangle$ states. Without the exciton–exciton Coulomb interaction, the above approximation would be invalid. This is because in that case, E_1 and E_2 would also be resonant with the

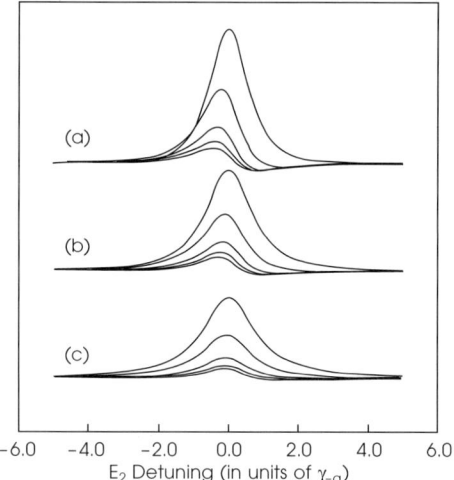

Fig. 7.24 Theory for the total DT signal under various Zeeman coherence dephasing rates. In each group of curves, E_1 is tuned to various positions, similar to Fig. 7.23. In (a), the decay parameters are the same as in Fig. 7.23(c) ($\gamma_{-+} = 2$). In (b), $\gamma_{-+} = 4$ and in (c) $\gamma_{-+} = 10$. The depletion contribution does not vary between cases. The differences in the spectra reflect the change of the Zeeman coherence contribution due to the change of γ_{-+}.

σ_+-to-biexciton and σ_--to-biexciton transitions, respectively. A full calculation including all four levels reveals a cancellation between different perturbation paths, resulting in a zero signal (for both the depletion and the Zeeman coherence components) when monitoring the differential transmission of E_2 (or E_1). This can be easily understood by noting that, in the absence of exciton–exciton Coulomb coupling, the two exciton transitions are essentially independent. The excitation of the σ_- exciton by E_1 would not have any effect on how E_2 interacts with the σ_+ exciton, leading to zero differential transmission of E_2.

Therefore, by studying the CW DT lineshape, we are able to not only identify the Zeeman coherence contribution and study its dephasing but also confirm the strong exciton–exciton Coulomb interaction in a QD.

Figure 7.25 shows the experimental results, which lead to the following conclusions:

1. The strong signal indicates that the involvement of state $|b\rangle$ is minimal. Thus, the two excitons must be strongly correlated by the Coulomb interaction, in agreement with the large biexciton binding energy measured in section 7.5.
2. The Zeeman coherence contributes significantly to the DT signal, suggesting that the pure dephasing of the Zeeman coherence is not significant. From the lineshape, the total Zeeman coherence dephasing rate is estimated to be around $(20 \text{ ps})^{-1}$.

Coherent optical spectroscopy and manipulation of single quantum dots   337

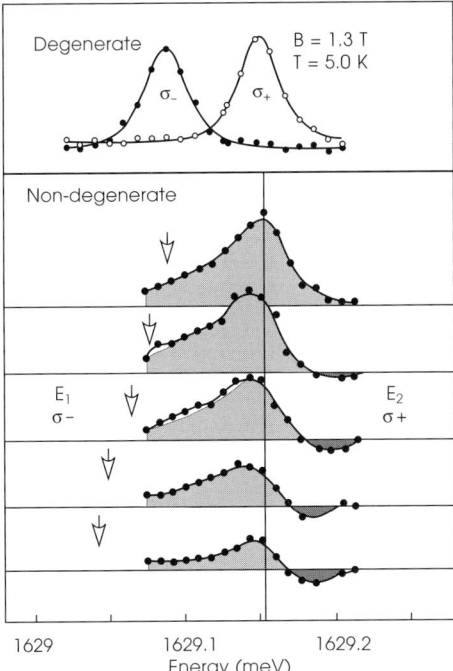

Fig. 7.25 Experimental results that correspond to the theory in Fig. 7.23. Taken from Reference [89].

3. The contribution from the Zeeman coherence indicates that due to the excitation by E_1 and E_2, the wavefunction of Eq. (7.23) is indeed a coherent superposition.

Due to the weak-field excitation condition to stay in the $\chi^{(3)}$ regime, as appropriate for spectroscopy studies where Eq. (7.26) is valid, the probability amplitude of the ground state is nonzero. The entanglement is therefore not maximal. We now quantify the optically induced two-exciton entanglement. In other words, we quantify all the amplitudes in Eq. (7.23), estimate the upper limit the biexciton level could have contributed and quantify how nonfactorizable the wavefunction is.

The signal strength of both the degenerate and nondegenerate response is used to validate the assumption that the two-exciton level does not contribute to our experiment. Any such contributions would give rise to a reduction in nondegenerate signal strength. A full cancellation of the signal takes place in the non-interacting case mentioned before, where the biexciton level would contribute with a state amplitude $C_b = C_-C_+/C_0$, leading to a factorizable wavefunction. By comparison with the theory, when Eq. (7.26) is modified to include the fourth level, the data then shows that the reduction is indeed negligible within an experimental error of 5%, meaning that the biexciton level contributes less than $C_b = 5\% \times C_-C_+/C_0$. From the detected signal strength, we are able to obtain $|C_+| = 0.30 \pm 0.05$ and $|C_-| = 0.30 \pm 0.05$. The state amplitude of state $|b\rangle$, $|C_b|$, is therefore less than 0.005 (0.1 would make the total wavefunction factorizable). The value of $|C_0|$ is 0.9. The density matrix for the two-exciton system is then established, from which the entropy of entanglement, E, is calculated to be $E = 0.08 \pm 0.02$, indicating quantum entanglement. The entropy of entanglement goes between zero and one, with zero for product states and one for maximally entangled states. For its definition, see Reference [205].

This number is relatively small compared to what has been achieved in an ion trap [206] (where E of 0.5 is reported) and reflects the weak-field condition of our experiment. For an entanglement with $C_0 = 0$ and the eventual application to a quantum logic device, the experiments would be done using high intensity coherent transient excitations with a pulse area of order π. This is feasible considering the successful demonstration of single QD Rabi oscillations discussed in section 7.4 despite the effects from delocalized excitons at high excitation intensity.

7.7 Single quantum dot as a prototype quantum computer

In the preceding sections, we discussed the laser manipulation of spin-polarized optical excitations in a semiconductor nanodot with the aim of its application to the realization of a quantum computer. In this section, we give a simple theoretical description of how a prototype quantum computer using a single dot would function. The basic element of a quantum computer is represented by a quantum unit of information, or qubit. In our case, the presence and absence of the exciton in a quantum dot constitutes a qubit and the up and down spin excitons in the same dot yield two qubits with a strong mutual interaction. With laser control of these two excitons in a dot, we have the makings of a two-qubit quantum computer. The examples of coherent control of these excitons, experimentally realized and explained in the previous sections, can already be seen as elementary quantum operations. The exciton Rabi rotations described in section 7.4, for instance, constitute the important single qubit manipulations. We shall show here how in principle the basic manipulations can be organized to run a quantum algorithm. A single quantum dot and a train of polarized pulses can in fact be used to implement an algorithm and solve efficiently a given problem. A numerical simulation of the time series of the exciton dynamics under optical control in a sequence of operations to solve the Deutsch–Jozsa problem [207] in a 2-qubit dot is given below to illustrate the basic physics of the prototype quantum computer and to furnish a test of the design.

Since the recent dramatic development of the theory of quantum computing, there have been a great variety of suggestions of implementation and a number of experimental demonstrations [81]. Among those based on semiconductors, quantum dots are favored for the confinement of single electron spin [208] and for the intersubband transitions with far infrared light [209,210] as the qubit. The use of optical control of excitons across the band gap in dots for quantum operations was suggested by Troiani *et al.* [211]. A theory of the spin excitons in quantum computing and of the physical implementation of quantum algorithms in a dot using ultrafast optical pulses was investigated in Ref. [4]. Several ideas for a scalable quantum computer involving excitons in different dots and optical quantum control were proposed [3,212,213]. In all of these schemes, each dot hosts a single qubit, and quantum gates are realized using interdot interactions such as the Förster mechanism [213], the exciton–exciton dipole interaction enhanced by an external electric field [3], or a cavity mode [212]. In some proposals the use of charged excitons, utilizing optical control of dots doped each with a single electron, are proposed [2,214]. For an illustration of basic quantum computing, it is sufficient to discuss here only the case of a single undoped dot hosting two qubits provided by two cross-polarized excitons.

The quality of the quantum information of the exciton depends on its coherence, i.e., the phase relation between the initial and final states of the light excitation. On one hand, the decoherence time of an exciton in a dot is of the order of tens of picoseconds. On the other hand, the resonant excitations require the operation time to be longer than the inverse of the transition energy difference (several meV in a dot). A solution to these two contradictory requirements can be given by the shaping of the optical pulses, thereby removing the second requirement of precise resonance. We address below the issue of how to design pulses for the realization of a given quantum transformation in the shortest time as possible.

Basic operations for quantum computation

The first requirement is two quantum states to represent a qubit. In a dot with two antiparallel spin excitons, the two qubit states $|x, y\rangle$, where $x, y = 0$ or 1, are mapped onto the four excitonic states as the ground state $|0\rangle \equiv |0, 0\rangle$, the σ_+ exciton state $|+\rangle \equiv |1, 0\rangle$, the σ_- exciton state $|-\rangle \equiv |0, 1\rangle$ and the lowest biexciton state $|+ -\rangle \equiv |1, 1\rangle$.

Any quantum computing algorithm can be constructed out of a number of single qubit operations and a logic operation between two qubits [81]. A commonly used single qubit operation to mix the two qubit states is the Hadamard transformation,

$$\frac{1}{\sqrt{2}}\begin{bmatrix} 1 & 1 \\ 1 & -1 \end{bmatrix}. \qquad (7.27)$$

A common quantum logic gate is the controlled-NOT gate (C-NOT) which changes, say, the first qubit (the target) depending on the state of the second qubit (the control). If the second qubit is 0, the first is unchanged and if the second is 1, the first bit is flipped. The matrix transformation in the basis $\{|00\rangle, |10\rangle, |01\rangle, |11\rangle\}$ is

$$\begin{bmatrix} 1 & 0 & 0 & 0 \\ 0 & 1 & 0 & 0 \\ 0 & 0 & 0 & 1 \\ 0 & 0 & 1 & 0 \end{bmatrix}. \qquad (7.28)$$

Optical pulses can directly perform Rabi rotations with generators σ_x and σ_y but rotations with generator σ_z need to be built as a combination of σ_y and σ_x. In our design of the laser implementation of a quantum algorithm we try to decompose the required global transformation directly in rotations generated by σ_y and σ_x for both single qubit and conditional operations without appealing to Hadmard or C-NOT. The general single qubit operation is implemented by two-color light driving combined Rabi rotations of both the exciton transition and the biexciton transition (say, between $|0\rangle$ and $|+\rangle$ and between $|-\rangle$ and $|+-\rangle$) through the same angle,

$$\begin{bmatrix} \cos(\alpha/2) & -\sin(\alpha/2) & 0 & 0 \\ \sin(\alpha/2) & \cos(\alpha/2) & 0 & 0 \\ 0 & 0 & \cos(\alpha/2) & -\sin(\alpha/2) \\ 0 & 0 & \sin(\alpha/2) & \cos(\alpha/2) \end{bmatrix}. \qquad (7.29)$$

The C-NOT gate is replaced by a controlled π rotation (C-ROT). If the σ_+ exciton is the target and σ_- exciton is the control, C-ROT is accomplished by a σ_+ polarized light driving a π Rabi rotation of the biexciton transition [209]. The matrix transformation is

$$\begin{bmatrix} 1 & 0 & 0 & 0 \\ 0 & 1 & 0 & 0 \\ 0 & 0 & 0 & -1 \\ 0 & 0 & 1 & 0 \end{bmatrix}. \qquad (7.30)$$

The Deutsch–Jozsa problem

The two-qubit Deutsch–Jozsa problem [207] is particularly instructive both for its simplicity and for the use of fundamental quantum operations. It has been implemented by nuclear magnetic resonance (NMR), which represented the first experimental realization of a two qubit quantum computer [215,216].

For a single bit variable $x = \{0,1\}$, there are four possible single-bit functions: $f_1(x) = 0$ and $f_2(x) = 1$ belong to the class of constant functions and $f_3(x) = x$ and $f_4(x) = 1 - x$ to the class of balanced functions. In the Deutsch–Jozsa problem we are given a closed box, which we can imagine, for instance, to be an unknown subroutine in a Fortran program able to calculate this function. The problem consists in determining whether the given subroutine calculates a constant or balanced function. Classically, the only way to find out if the function is constant or balanced is to calculate the function, for both the 0 and 1 input values, and then compare the results.

The solution of the Deutsch–Jozsa problem in a quantum computer illustrates the parallel processing of the qubits in the evaluation of the function. Here we are not demonstrating the power of the quantum computer over the classical one, which is meaningful only for a sufficiently large number of qubits [81]. First, the single bit variable x has to be changed to a quantum bit $|x\rangle$, which enables us to prepare arbitrary linear superpositions of 0 and 1. Our variable qubit is the presence or absence of a σ_- exciton. This qubit is manipulated by a Rabi rotation through a required angle using σ_- polarized light. Second, we replace the classical subroutine with a quantum subroutine given by a unitary transformation which parallel-processes all possible values of x. The solution of the problem involves the use of a second qubit $|y\rangle$, which is given in our single dot quantum computer by the presence or absence of a σ_+ exciton. The complete solution consists fo three steps:

(i) *Encoding.* It transforms the initial ground state dot into an input state:

$$|\text{in}\rangle = \frac{1}{2}(-|+-\rangle + |++\rangle - |--\rangle + |0\rangle) = \frac{1}{2}(-|11\rangle + |10\rangle - |01\rangle + |00\rangle).$$

This is accomplished by the single-qubit $-\pi/2\sigma_+$ rotation and $\pi/2\sigma_-$ rotation.

(ii) *Calling Quantum Subroutine.* We apply the unitary transformation U_{f_j} associated with the given function f_j. The explicit form of this unitary transformation acting on the two qubit $|y, x\rangle$ can be written as

$$U_{f_j}|y,x\rangle = R_x(\pi)^{1-2f_j(y)}|y,x\rangle.$$

This is a conditional π Rabi rotation acting on qubit x (σ_- exciton), conditioned on the effect of f_j on the qubit y (σ_+ exciton). For the constant functions this corresponds to an in-phase π pulse for the excitonic and biexcitonic transitions, while for the balanced functions the two π rotations are opposite in phase.

(iii) *Decoding.* By applying $\pi/2$ rotations of both circular polarizations, the final state is the ground state if the function is a constant and the $|+\rangle$ state if the function is balanced. The final measurement involves only the second qubit $|y\rangle$, i.e., the σ_+ exciton, and can be made with the cross-polarization probe or with the emitted photon polarization.

Figure 7.26 shows for a constant (f_2) and a balanced (f_3) function the same circuit diagram of the operations and the two desired sequence of states of the two excitons following each operation. The evolution of the qubits through the Deutsch–Jozsa algorithm is shown in Fig. 7.27 as the real parts of the coefficients $c_j(t)$ in the time evolution of the state

$$|\psi(t)\rangle = c_0(t)|0\rangle + c_-(t)|-\rangle + c_+(t)|+\rangle + c_{+-}(t)|-+\rangle.$$

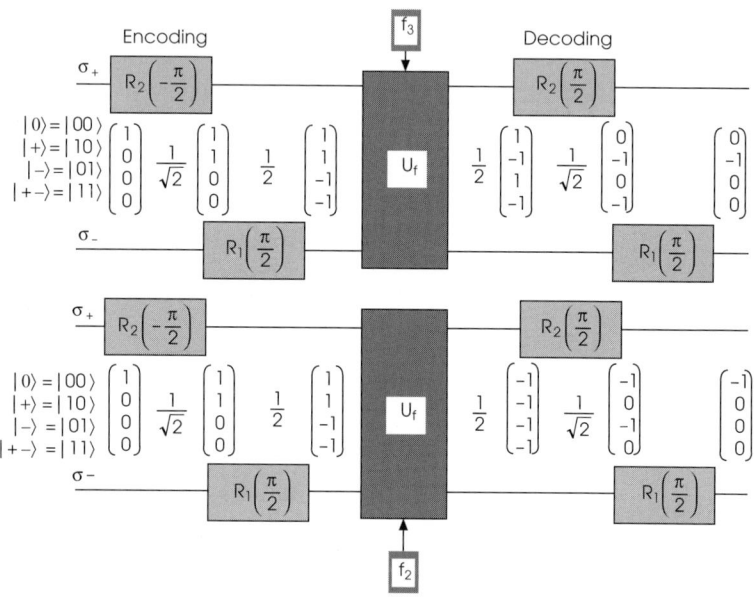

Fig. 7.26 Circuit diagram for the solution of the Deutsch–Jozsa problem for the balanced function f_3 (upper row) and for the constant function f_2 (lower row). Notice that in the final state the second qubit $|y\rangle$, associated with the σ_+ exciton, determines if the function is balanced or constant, the first qubit $|x\rangle$ being always $|0\rangle$. In the case of f_1 and f_4, the final states differ only by a global π phase.

In order to check the contribution to the unintended dynamics of the states out of the computational space we added two parallel-spin biexciton states $|++\rangle$ and $|--\rangle$,

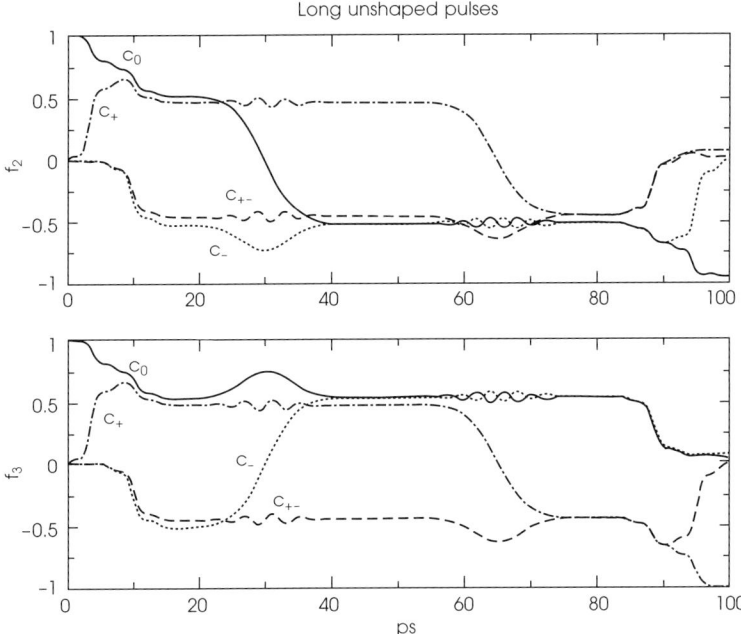

Fig. 7.27 Real parts of the coefficients $c_j(t)$ in $|\psi\rangle = c_0(t)|0\rangle + c_+(t)|+\rangle + c_-(t)|-\rangle + c_{+-}(t)|+-\rangle$ during the quatum computation. *Encoding* (0–20 ps), starting from the $|0\rangle$ we obtain $|in\rangle$. U_{f_i} (20–80 ps). Composed by two π pulses, one at the excitonic transition and one at the biexcitonic transition. For f_2 both the pulses correspond to $+\pi$. For f_3 the biexcitonic transition is $+\pi$ while the excitonic transition is $-\pi$. *Decoding* (80–100 ps), the output is $-|0\rangle$ for the constant function f_2 and $-|+\rangle$ for the balanced f_3. Taken from Reference [4].

calculated at 13 meV above the $|-+\rangle$. We used pulses with a peak electric field corresponding to a 0.2 meV Rabi energy, smaller than the separation between the exciton and the biexciton resonances which we assumed to be 1 meV. An ideal case with no dephasing is considered and the Rabi rotations are realized with spectrally narrow pulses. We pack the pulses with a maximum overlap at 10% of the peak electric field without substantial degeneration of the results.

Fast quantum computing by pulse shaping

The simulation given above assumes spectrally narrow pulses and, therefore, the total time needed to run the quantum computer of 100 ps is long compared to the measured dephasing time of the exciton. If the finite lifetime of the exciton due to spontaneous emission were included in the simulation of Fig. 7.27, it is evident that the quality of the operations would be so degraded as to make the results of the constant and balanced cases indistinguishable. The quantum operations of the state vector have to be made faster.

A faster control of the quantum system can be realized using somewhat more complex optical pulses. This idea, commonly referred to as pulse shaping, exploits the amplitude and phase spectral distributions of the pulses to realize efficient quantum control. We consider here a simple pulse shaping technique by the phase-locking of two subpicosecond laser pulses with different energies. This pulse design technique is experimentally realizable [217] and is particularly illustrative for its simplicity in the design of the fast control of excitons and biexcitons in quantum dots.

We can start by writing the Hamiltonian of the four-level systems coupled to an external electromagnetic pulse with σ^+ polarization, treated classically, in the form

$$H^+ = \begin{bmatrix} 0 & \Omega_+(t)/2 & 0 & 0 \\ \Omega_+^*(t)/2 & \varepsilon_+ & 0 & 0 \\ 0 & 0 & \varepsilon_- & f\Omega_+(t)/2 \\ 0 & 0 & f\Omega_+^*(t)/2 & \varepsilon_{-+} \end{bmatrix}, \quad (7.31)$$

where $\Omega_+(t) = d_+ E_+(t)$ represents a time dependent Rabi energy provided by an optical pulse. The dipole moment of the exciton $|+\rangle$ is denoted by d_+ and f is a correction factor to the dipole moment in the exciton-biexciton transition matrix element due to Coulomb interaction. The amplitude of the electric field $E_+(t) = \mathcal{E}_+(t)e^{-i\omega_+(t)}e^{i\phi}$ is assumed to be slowly varying. As in the atomic case, the condition on the frequency $\omega_+ \gg d_+\mathcal{E}_+$ enables the rotating wave approximation used in H^+ above. Thus, the counter-rotating terms, such as $H_{0,-}^+ = \Omega_+^*/2$, are set to zero. In the interaction representation, $\tilde{O} = \Lambda O \Lambda^\dagger$ denotes the transformed operator from O, with $\Lambda(t) = e^{iH_0 t}$, where H_0 is a diagonal matrix with elements $(0, \varepsilon_+, \varepsilon_-, \varepsilon_{-+})$. The time evolution operator at the end of the pulse U^σ can be written as

$$\tilde{U}^{\sigma+} = Te^{-i\frac{1}{2}\int_{-\infty}^{\infty} dt \tilde{V}^{\sigma+}(t)}, \quad (7.32)$$

where $\tilde{V}(t)^{\sigma+}$ is

$$\begin{bmatrix} 0 & \Omega_+(t)e^{i\varepsilon_+ t} & 0 & 0 \\ \Omega_+^*(t)e^{-i\varepsilon_+ t} & 0 & 0 & 0 \\ 0 & 0 & 0 & f\Omega_+(t)e^{i(\varepsilon_+ -\Delta)t} \\ 0 & 0 & f\Omega_+^*(t)e^{-i(\varepsilon_+ -\Delta)t} & 0 \end{bmatrix} \quad (7.33)$$

and $\Delta = \varepsilon_+ + \varepsilon_- - \varepsilon_{-+}$ is the biexciton binding energy. When only circularly polarized light is used, Eq. (7.33) shows that the four-level system behaves as a double two-level system, the first two-level transition (exciton transition) being represented by $|0\rangle \to |+\rangle$ and the second (biexciton transition) by $|-\rangle \to |-+\rangle$.

Consider now the desired operation where the exciton transition is a Rabi rotation through angle α and the biexciton transition a Rabi rotation through α',

$$\tilde{U}_j^{\sigma+} = \begin{bmatrix} \cos(\alpha/2) & -\sin(\alpha/2) & 0 & 0 \\ \sin(\alpha/2) & \cos(\alpha/2) & 0 & 0 \\ 0 & 0 & \cos(\alpha'/2) & -\sin(\alpha'/2) \\ 0 & 0 & \sin(\alpha'/2) & \cos(\alpha'/2) \end{bmatrix}. \quad (7.34)$$

The most direct solution for the realization of this transformation would be a two-pulse combination,

$$E_+(t) = \mathscr{E}_0 e^{-(t/s)^2} e^{-i\omega_{0+}t} + \mathscr{E}_1 e^{-(t/s_1)^2} e^{-i\omega_{1+}t + i\phi}. \quad (7.35)$$

If the two pulses are resonant respectively with the two transitions, i.e. $\omega_{0+} = \varepsilon_+$ and $\omega_{1+} = \varepsilon_{-+} - \varepsilon_-$, and are sufficiently narrow in frequency, the pulse resonant with the exciton transition would have a negligible effect on the biexciton transition and vice versa. However, this has been shown above to be costly in time. The problem is to find a composite pulse which would take much less time with tolerable deterioration of quality of the transformation.

For the quality of the transformation, we follow Ref. [218] in defining the fidelity of the transformation as

$$F = \overline{|\langle \psi_{in} | \tilde{U}^\dagger U_i | \psi_{in} \rangle|^2}, \quad (7.36)$$

where U_i is the ideal unitary operation, $\tilde{U}$ is the unitary transformation generated by the optical pulses, and the overline denotes the average over all the possible initial states. The operator $\tilde{U}^\dagger U_i$ is denoted by I for short. The average over all the possible states is done by considering an initial state with arbitrary complex coefficients $|\psi_{in}\rangle = \sum_j c_j |j\rangle$ with the normalization constraint $\sum_j |c_j|^2 = 1$. The fidelity can be then written in the form

$$F = \sum_{ijkl} \overline{c_i^* c_j c_k^* c_l} I_{ij} I_{lk}^* \quad (7.37)$$

and, in the four-level system considered here, the overline average is then on a hypersphere S in C^8 determined by the normalization conditon. This average $(1/S) \int_S d^2c_1 d^2c_2 d^2c_3 d^2c_4 c_i^* c_j c_k^* c_l$ is easily evaluated in polar coordinates and gives

$$F = 1/10 \sum_i |I_{ii}|^2 + 1/20 \sum_{i \neq j} (I_{ii} I_{jj}^* + I_{ij}^* I_{ij}). \quad (7.38)$$

The difference of the coefficients from those of Ref. [218] is due to their additional restrictions on the coefficients c_j. Our choice gives a more stringent estimate of the error in the operations.

There are three different approaches to pulse design for a quantum operation.

(i) *Approximation by the Area Theorem.* In the limit of very long pulses, the area theorem [108] determines the intensity of a Gaussian pulse that has to be used for a given rotation α

$$\mathscr{E}_0 = \frac{\alpha}{s\sqrt{\pi}d_+}. \tag{7.39}$$

For a single two-level system the pulse width s in Eq. (9.39) can be made arbitrarily small, but in the four-level case we are strongly limited by the resonance condition to $1/s$, $\mathscr{E}_0 d_+ \ll \Delta$. In order to shorten the time duration of the whole pulse, an intuitive approach would be to allow the two components of Eq. (7.35) to overlap in frequency but keep each satisfying the area theorem.

(ii) *The average Hamiltonian method.* The cumulant expansion (also known as the Magnus expansion [219]) of the evolution operator $\tilde{U}_j^{\sigma+}$ in Eq. (9.32) is given by [220]

$$\tilde{U}_j^{\sigma+} = e^{-\frac{i}{2}(\tilde{V}_1 + \tilde{V}_2 + \ldots)}. \tag{7.40}$$

The first term of the expansion corresponds to a time average of the interaction Hamiltonian,

$$\tilde{V}_1 = \int_0^\infty dt\, \tilde{V}(t). \tag{7.41}$$

The second term is given by

$$\tilde{V}_2 = \frac{-i}{4} \int_0^\infty dt \int_0^t dt'\, [\tilde{V}(t), \tilde{V}(t')]. \tag{7.42}$$

Keeping only the first term in the exponent constitutes the average Hamiltonian approximation. An estimation of the error in the truncation of the cumulant expansion is given by the second term.

(iii) *Numerical approach.* The parameters in Eq. (7.35) are varied to find the maximum fidelity. To lessen the numerical effort, physical considerations guide the reduction of the number of parameters varied. The first two approximation methods are also useful as starting points.

Examples of pulse design

We illustrate the above methods for a single qubit operation, i.e., a parallel rotation of both the exciton and biexciton transitions. For simplicity, let $f = 1$ and $s_1 = s$. Both theoretical estimates and experimental measurements have the f value not far from unity. In any case, the extension to $f \neq 1$ can be made in a similar manner to the treatment on the conditional rotation given below. We consider a composite pulse by superposing and phase-locking the two pulses as in Eq. (7.35) with $\mathscr{E}_0 = \mathscr{E}_1$ and $\omega_{0+} = \varepsilon_+$ and $\omega_{1+} = \varepsilon_+ - \Delta$. It remains to choose a value for $\mathscr{E}_0(s)$ by each of the three methods above and tests its efficacy by evaluating the fidelity of the operation.

In Fig. 7.28(a) the fidelity for $\alpha = \alpha' = \pi$ rotation is plotted as a function of the temporal width of the Gaussian pulse s. The corresponding value for the peak of the Rabi energy $\Omega_0 = d_+\mathscr{E}_0(s)$ is given in Fig. 7.28(b). The value of the biexcitonic binding energy Δ is 1 meV. The results by the area theorem approximation are shown as the dashed lines. The fidelity is close to unity only for $s \gg 1/\Delta$, corresponding to a region where the frequency selectivity is preserved. If for instance a 98% Fidelity is required, the area theorem approach will lead to optical pulses with $s > 4$ ps. The area theorem is not the best procedure of time optimization for single-qubit operations.

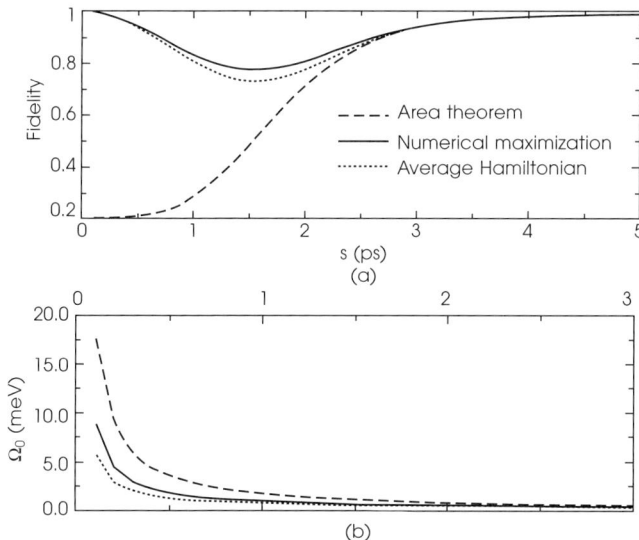

Fig. 7.28 (a) Fidelity as a function of the temporal width of the Gaussians s for a parallel rotation of $\alpha = \alpha' = \pi$. (b) Peak value of the Rabi energy $\Omega_0 = d_+\mathscr{E}_0(s)$. Dashed lines: the area theorem approximation. Dotted lines: the average Hamiltonian appproximation. Solid lines: numerical maximization of the fidelity.

Applying the average Hamiltonian approximation to the restricted pulse specified above leads to the single-qubit rotation $\tilde{U}^{\sigma_+}$ in the form of Eq. (7.34) with chosen values for $\alpha = \alpha'$ and for s, leading to $\mathcal{E}_0$ given by

$$\mathcal{E}_0 = \frac{\alpha}{s\sqrt{\pi}d_+ (1 + e^{-(\Delta s/2)^2})}. \tag{7.43}$$

The Gaussian term in the denominator on the right gives a correction to the area theorem, Eq. (7.39). The results are shown as dotted lines in Fig. 7.28(b). The resultant fidelity by the average Hamiltonian method is shown as dotted lines in Fig. 7.28(a). Note that it is possible to obtain a 98% Fidelity using much shorter pulses, of the order of 100 fs. In the limit of very short pulses this corresponds to spectrally very broad pulses which do not distinguish between the two transitions but yield a nearly parallel rotation.

The results of the numerical maximization using one variable $\mathcal{E}_0$ by Brent's method [221] are plotted as solid lines. They may be used as the standard to which the other two methods are compared. The optimal curve $\mathcal{E}_0(s)$ deviates considerably at short times from the area theorem approximation but is close to the average Hamiltonian approximation throughout the whole range of s.

The second example is a conditional operation for two qubits, viz., a σ_+ biexcitonic transition without affecting the excitonic $|+\rangle \to |0\rangle$, i.e., a rotation $\tilde{U}_j$ in Eq. (7.34) with $\alpha = 0$ and $\alpha' = \pi$. For the combined pulse in Eq. (7.35) we consider now $\phi = \pi$, and again $\mathcal{E}_0 = \mathcal{E}_1$ and $\omega_{0+} = \varepsilon_+$, $\omega_{1+} = \varepsilon_+ - \Delta$.

From the average Hamiltonian approximation (the first order term in the cluster expansion), we obtain relations for the three parameters of the pulse $\mathcal{E}_0$, s and s_1 for the desired rotations,

$$\alpha = d_+ \mathcal{E}_0 \sqrt{\pi}(s - s_1 e^{-(\Delta s_1/2)^2}), \tag{7.44}$$

$$\alpha' = d_+ \mathcal{E}_0 \sqrt{\pi}(s_1 - s e^{-(\Delta s/2)^2}). \tag{7.45}$$

For a given value of s_1, the other two parameters may be solved in the case with $\alpha = 0$ and $\alpha' = \pi$,

$$s = s_1 e^{-(\Delta s_1/2)^2}, \tag{7.46}$$

$$\mathcal{E}_0 = \sqrt{\pi}/d_+ (s_1 - s e^{-(\Delta s/2)^2}). \tag{7.47}$$

In the limit of large Δ the solution gives $s \to 0$ eliminating the term resonant with the excitonic transition and $\mathcal{E}_0 \to \sqrt{\pi}/s_1 d_+$ in accordance with the area theorem for the biexcitonic transition. For $\Delta \neq 0$ this system has always a solution for any $\alpha \neq \alpha'$.

In Fig. 7.29 we show (a) the fidelity and (b) the peak Rabi energy for the $\alpha = 0$ and $\alpha' = \pi$ transformation, for all three methods. The area theorem approximation amounts to taking a single pulse resonant with the biexciton transition. For the numerical maximization we maximize the fidelity for a given s_1 value as a function of s and $\mathcal{E}_0$ using the downhill simplex method [221]. We see clearly that the average Hamiltonian again gives a very good approximation: the deviations from the numerical maximization are negligible in most of the region. Also in this case we see that the use of a composite pulse provides a considerable saving in the time for the operation.

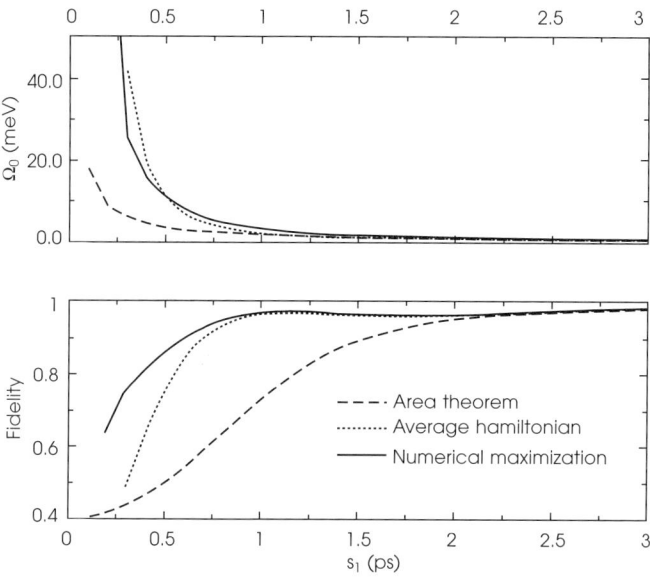

Fig. 7.29 (a) Fidelity as a function of the temporal width s_1 of the biexciton Gaussian component in the composite pulse for a rotation of $\alpha' = \pi$ only for the biexciton transition. (b) Peak value of the Rabi energy. Dashed lines: the area theorem approximation with a single pulse resonant with the biexciton transition. Dotted lines: the averaged Hamiltonian approximation. Solid lines: numerical maximization of the fidelity.

We can visualize the conditional dynamics of the four level system as two pseudospins evolving in the Bloch sphere. The two pseudospins $\vec{S}_X$ and $\vec{S}_{XX}$ are associated with the excitonic $|0\rangle \to |+\rangle$ and biexcitonic $|-\rangle \to |-+\rangle$ transitions, respectively. The initial state for the exciton pseudospin is taken to be the ground state, while for the biexciton pseudospin we assume the initial state $|-\rangle$. This choice is arbitrary, the calculation of the fidelity given above shows indeed that the transformation is realized independently of the initial state. A shaped pulse with $s = 0.56$ ps and $s_1 = 1.05$ ps, $\phi = \pi$ and $\Omega_0 = 2$ meV is used. We see that the pseudospin associated with the biexciton transition $\vec{S}_{XX}$ realizes the π rotation. The pseudospin associated with the exciton transition instead evolves in the Bloch sphere in a loop that brings it back to the initial state at the end of the pulse. This fulfills the requirement of the controlled π rotation of the σ_+ exciton only when the σ_- exciton is present, but leaving it back in its initial state when the σ_- exciton is absent. A simple intuitive picture is given by the frequency spectrum of the pulse in Eq. (7.35), shown in Fig. 7.30(c). The two phase-locked components have opposite phases and the chosen s and s_1 control the interference in the electric field giving a zero amplitude at the exciton resonance and a finite value at the biexciton resonance. We can therefore say that the ultrafast manipulation of the qubits is carried out using constructive and destructive light interference. This simple interpretation is equivalent to the average Hamiltonian approach.

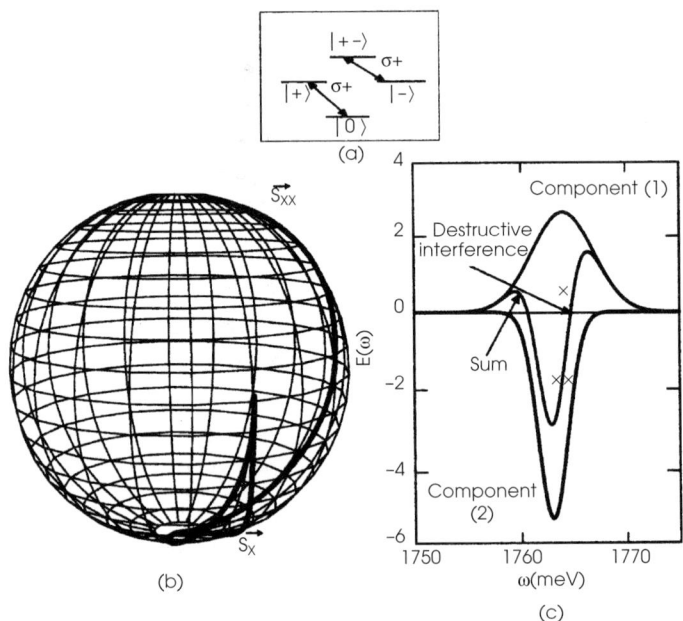

Fig. 7.30 Dynamics of spectral shaping. (a) Energy schematics for the transitions generated by a $\sigma+$ pulse. (b) Evolution of the exciton $\vec{S}_X$ and biexciton $\vec{S}_{XX}$ pseudospins on the Bloch sphere under a shaped pulse. (c) Fourier transform of the shaped pulse and its components.

Fast control applied to the Deutsch–Jozsa algorithm

We are now ready to check the effects of the pulse shaping remedy on the implementation of the Deutsch–Jozsa algorithm in the quantum dot. We plot in Fig. 7.31 the same coefficients as in Fig. 7.27 but we limit now the total time for the operation to be below 10 ps, i.e., the algorithm must be run one order of magnitude faster than in the ideal case of Fig. 7.27. The first column shows the effect on the dynamics when all the pulses' temporal widths are reduced by a factor ten and the Rabi energy consequently increased according to the prescriptions of the Area Theorem Approximation. As expected, the frequency selectivity is lost and the possibility of discriminating between constant and balanced function is definitively lost. The effect of the shaping in the computation is shown in Column 2. The possibility of distinguishing between the balanced and the constant function is recovered. The details of the pulse sequence for the shaped and unshaped cases are given in Fig. 7.32 and Table 7.1.

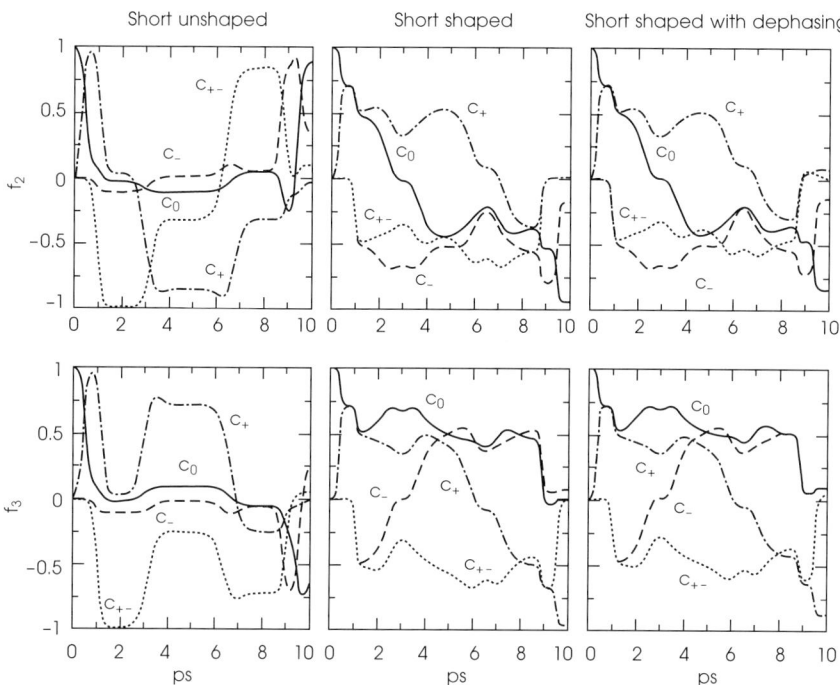

Fig. 7.31 Real parts of the coefficients $c_j(t)$ in $|\psi\rangle = c_0(t)|0\rangle + c_+(t)|+\rangle + c_-(t)|-\rangle + c_{+-}(t)|+-\rangle$ during the quantum computation. *Encoding* (0–20 ps), starting from the $|0\rangle$ we obtain $|in\rangle$. U_{f_i} (20–80 ps). *Decoding* (80–100 ps), the output is $-|0\rangle$ for the constant function f_2 and $-|+\rangle$ for the balanced f_3. Taken from Reference [4].

The first two columns of Fig. 7.31 do not include the effects of dephasing. In order to check the robustness of the use of composite pulses in the presence of dephasing, we include the spontaneous emission in the simulation by adding the Lindblad operators in the equation of motion for the density matrix [222]

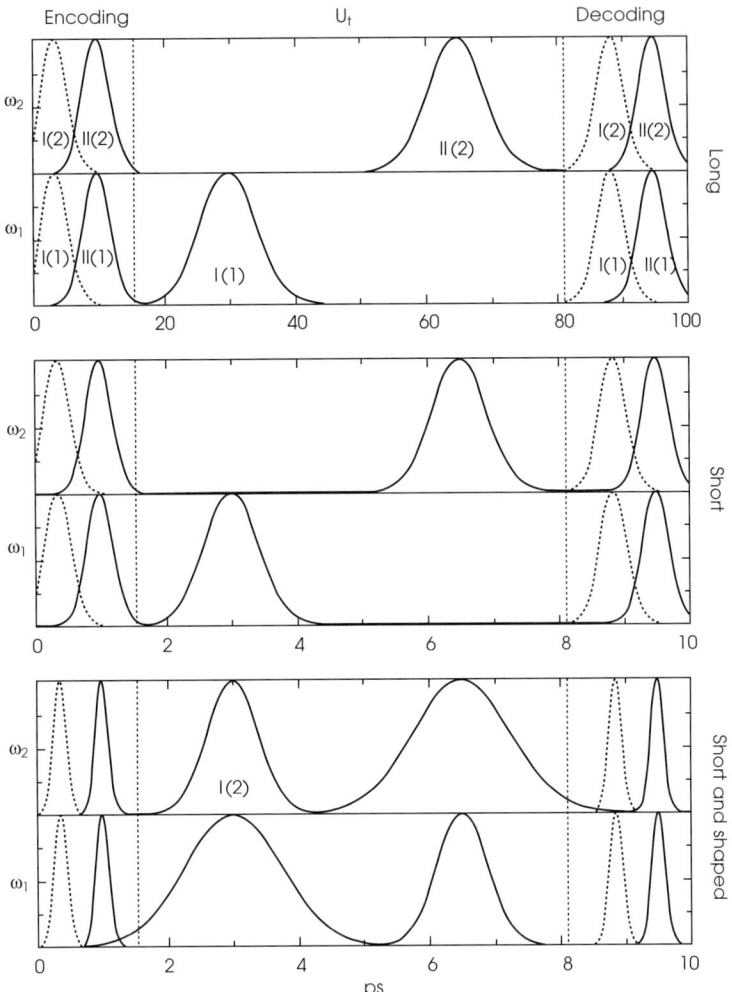

Fig. 7.32 Slowly varying envelopes of the pulse sequences used in the quantum computation for the solution of the DJ problem in Figs 7.27 and 7.31. The dotted (solid) line indicates $\sigma+$ ($\sigma-$) circular polarization. The two rows labeled by ω_i refers to the resonance frequency of the pulses: $\omega_1 = \varepsilon_+ = \varepsilon_-$, $\omega_2 = \varepsilon_+ - \Delta_{+-} = \varepsilon_- - \Delta_{+-}$.

$$\frac{d}{dt}\rho = -\frac{i}{\hbar}[H, \rho] + \sum_{j=1}^{4}\left(L_j\rho L_j^\dagger - \frac{1}{2}L_j^\dagger L_j\rho - \frac{1}{2}L_j^\dagger L_j\right). \tag{7.48}$$

where,

$$L_1 = \sqrt{\Gamma}\,|0\rangle\langle +|, \qquad L_2 = \sqrt{\Gamma}\,|0\rangle\langle -|,$$
$$L_3 = \sqrt{\Gamma}\,|+\rangle\langle -+|, \qquad L_4 = \sqrt{\Gamma}\,|-\rangle\langle -+|, \tag{7.49}$$

Table 7.1 Pulse sequences used in the computation, (1) and (2) indicate the two components of a shaped pulse of the form $\mathscr{E}(t) = \mathscr{E}_1(t) + \mathscr{E}_2(t) = \mathscr{E}_0 (e^{-(t/s)^2 - i\omega_1 t - i\phi_1} + e^{-(t/s_1)^2 - i\omega_2 t + i\phi_2})$. The area θ is defined as $\hbar\theta = \int d\mathscr{E}(t)\, dt$. $d\mathscr{E}_0$ is 0.2 meV in the long pulses case and 2 meV for short and shaped pulses. The pulse sequence of $U_{f_1}(U_{f_4})$ is identical to the $U_{f_2}(U_{f_3})$ except for a change of sign in the phase ϕ_j.

	Pol	ϕ_j	ω_j	Long pulses ps (θ)	Short pulses ps (θ)	Shaped pulses ps (θ)
				ENCODING		
I (1)	σ+	−π/2	ω_1	2.92 (π/2)	0.29 (π/2)	0.15 (0.81)
I (2)	σ+	−π/2	ω_2	2.92 (π/2)	0.29 (π/2)	0.15 (0.81)
II (1)	σ−	π/2	ω_1	2.92 (π/2)	0.29 (π/2)	0.15 (0.81)
II (2)	σ−	π/2	ω_2	2.92 (π/2)	0.29 (π/2)	0.15 (0.81)
				DECODING		
I (1)	σ+	π/2	ω_1	2.92 (π/2)	0.29 (π/2)	0.15 (0.81)
I (2)	σ+	π/2	ω_2	2.92 (π/2)	0.29 (π/2)	0.15 (0.81)
II (1)	σ−	π/2	ω_1	2.92 (π/2)	0.29 (π/2)	0.15 (0.81)
II (2)	σ−	π/2	ω_2	2.92 (π/2)	0.29 (π/2)	0.15 (0.81)
				U_{f_2}		
I (1)	σ−	−π/2	ω_1	5.83 (π)	0.58 (π)	1.05 (5.68)
I (2)	σ−	π/2	ω_2			0.56 (3.03)
II (2)	σ−	−π/2	ω_2	5.83 (π)	0.58 (π)	1.05 (5.68)
II (1)	σ−	π/2	ω_1			0.56 (3.03)
				U_{f_3}		
I (1)	σ−	−π/2	ω_1	5.83 (π)	0.58 (π)	1.05 (5.68)
I (2)	σ−	π/2	ω_2			0.56 (3.03)
II (2)	σ−	π/2	ω_2	5.83 (π)	0.58 (π)	1.05 (5.68)
II (1)	σ−	−π/2	ω_1			0.56 (3.03)

$\Gamma = 15$ μeV being chosen to approximate the measured dephasing time [88]. These operators represent all the possible spontaneous emission pathways in the four-level system. There are many equivalent ways to solve the master equation in terms of a nonlinear stochastic differential equation for a normalized state vector $|\psi\rangle$. We choose to use the quantum state diffusion (QSD) equation [223,224]

$$|d\psi\rangle = -\frac{i}{\hbar} H |\psi\rangle dt + \sum_j \left(\langle L_j^\dagger \rangle L_j - \frac{1}{2} L_j^\dagger L_j - \frac{1}{2} \langle L_j^\dagger \rangle \langle L_j \rangle \right) |\psi\rangle dt$$
$$+ \sum_j (L_j - \langle L_j \rangle) |\psi\rangle d\eta_j \qquad (7.50)$$

where $\langle L \rangle = \langle \psi | L | \psi \rangle$ and η_j are independent complex random variables. The density matrix can be expressed as $\rho = M | \psi \rangle \langle \psi |$ where M denotes ensemble average and the expectation value of any operator O is given by $M \langle \psi | O | \psi \rangle$. Inclusion of dephasing in this way reduces the fidelity of the operation but we can see from column 3 in Fig. 7.31 that it is still possible to distinguish between the constant and balanced functions within a margin of error less than 10%.

7.8 Summary

The optical response of single semiconductor QDs, particularly the GaAs interface fluctuation QDs, is reviewed in this chapter. The lowest exciton and biexciton states that are well localized in these QDs are coherently manipulated using both CW and transient optical fields. The Rabi oscillations of excitons are demonstrated. It is found that the decoherence of single QDs is dominated by energy relaxation and the pure-dephasing due to exciton-phonon and exciton–exciton elastic scattering is insignificant. The 3D confinement enhances the biexciton binding energy. Due to the exciton–exciton Coulomb correlation in a dot, the resonant energy of one exciton depends on whether the other exciton is excited or not, allowing for optically inducing two-exciton entangled states. The experiments discussed in this chapter have addressed the key issues related to quantum computing based on optically driven QDs.

The simplest prototype two-bit quantum computer based on laser manipulation of the excitons and the biexciton in QD is proposed. Shaping of femtosecond laser pulses keeps the computation within the time of environmental degradation of the quantum information. Numerical simulation of the complete solution of a simple basic problem demonstrates the feasibility of this primitive quantum computer. These results pave the way to resolving basic physics issues in the realization of larger quantum computers based on the strength of semiconductor and laser technologies.

To build a scalable computer based on this way of operation, it is necessary to be able to produce coupled QD arrays in the future. In addition, it is desired that the decoherence rate be further reduced. The studies discussed in this chapter show that the decoherence is limited by the energy relaxation. The double ground states of a charged QD have long energy relaxation time (lifetime) and can be used to achieve along decoherence time. In such proposals, the coherence between the two ground states (0 and 1 of a qubit) is manipulated optically via the trion states (charged exciton states) and the spectroscopic techniques presented in this chapter apply.

References

1. Ekert, A., and Jozsa, R. (1996). Quantum computing and Shor's algorithm. *Rev. Mod. Phys.* **68**, 733–753.
2. Imamoḡlu, A., *et al.* (1999). Quantum information processing using quantum dot spins and cavity QED. *Phys. Rev. Lett.* **83**, 4204–4207.
3. Biolatti, E., Iotti, R.C., Zanardi, P., and Rossi, F. (2000). Quantum information processing with semiconductor macroatoms. *Phys. Rev. Lett.* **85**, 5647–5650.
4. Chen, P., Piermarocchi, C., and Sham, L.J. (2001). Control of spin dynamics of excitons in nanodots for quantum operations. *Phys. Rev. Lett.* **87**, 067401.
5. Alivisatos, A.P. (1996). Semiconductor clusters, nanocrystals, and quantum dots. *Science* **271**, 933–937.
6. Eychmüller, A., Mews, A., and Weller, H. (1993). A quantum-dot quantum-well – CdS/HgS/CdS. *Chem. Phys. Lett.* **208**, 59–62.
7. Hines, M.A., and Guyot-Sionnest, P. (1996). Synthesis and characterization of strongly luminescing ZnS-capped CdSe nanocrystals. *J. Phys. Chem.* **100**, 468–471.
8. Peng, X.G., Schlamp, M.C., Kadavanich, A.V., and Alivisatos, A.P. (1997). Epitaxial growth of highly luminescent CdSe/GdS core/shell nanocrystals with photostability and electronic accessibility. *J. Am. Chem. Soc.* **119**, 7019–7029.
9. Peng, X., *et al.* (2000). Shape control of CdSe nanocrystals. *Nature* **404**, 59–61.
10. Puntes, V.F., Krishnan, K.M., and Alivisatos, A.P. (2001). Colloidal nanocrystal shape and size control: The case of cobalt. *Science* **291**, 2115–2117.
11. Hu, J., *et al.* (2001). Linearly polarized emission from colloidal semiconductor quantum rods. *Science* **292**, 2060–2063.
12. Manna, L., Scher, E.C., and Alivisatos, A.P. (2000). Synthesis of soluble and processable rod-, arrow-, teardrop-, and tetrapod-shaped CdSe nanocrystals. *J. Am. Chem. Soc.* **122**, 12700–12706.
13. Danek, M., Jensen, K.F., Murray, C.B., and Bawendi, M.G. (1996). Synthesis of luminescent thin-film CdSe/ZnSe quantum dot composites using CdSe quantum dots passivated with an overlayer of ZnSe. *Chem. Mater.* **8**, 173–180.
14. Murray, C.B., and Bawendi, M.G. (1995). Self-organization of CdSe nanocrystallites into three-dimensional superlattices. *Science* **270**, 1335–1338.
15. Hess, H.F., Betzig, E., and Harris, T.D. (1994). Near-field spectroscopy of the quantum constituents of a luminescent system. *Science* **264**, 1740–1745.
16. Brunner, K., Abstreiter, G., Böhm, G., Tränkle, G., and Weimann, G. (1994). Sharp-line photoluminescence and two-photon absorption of zero-dimensional biexcitons in a GaAs/AlGaAs structure. *Phys. Rev. Lett.* **73**, 1138–1141.
17. Zrenner, A., *et al.* (1994). Quantum dots formed by interface fluctuations in AlAs/GaAs coupled quantum well structures. *Phys. Rev. Lett.* **72**, 3382–3385.
18. Gammon, D., Snow, E.S., Shanabrook, B.V., Katzer, D.S., and Park, D. (1996). Homogeneous linewidths in the optical spectrum of a single gallium arsenide quantum dot. *Science* **273**, 87–90.
19. Petroff, P.M., and DenBaars, S.P. (1994). MBE and MOCVD growth and properties of self-

assembing quantum dot arrays in III-V semiconductor structures. *Superlattices and Mcrostructures* **15**, 15–21.
20. Eaglesham, D.J., and Cerullo, M. (1990). Dislocation-free Stranski-Krastanow growth of Ge on Si (100). *Phy. Rev. Lett.* **64**, 1943–1946.
21. Richard, N. (1996). Self-organized growth of quantum-dot structures. *Semicond. Sci. Tech.* **11**, 1365–1379.
22. Petroff, P.M., Lorke, A., and Imamoglu, A. (2001). Epitaxially self-assembled quantum dots. *Phys. Today* **54**, 46–52.
23. Brunner, K., et al. (1992). Photoluminescence from a single GaAs/AlGaAs quantum dot. *Phys. Rev. Lett.* **76**, 3216–3219.
24. Steffen, R., Koch, T., Oshinowo, J., Faller, F., and Forchel, A. (1995). Photoluminescence study of deep etched InGaAs/GaAs quantum wires and dots defined by low-voltage electron beam lithography. *Appl. Phys. Lett.* **68**, 223–225.
25. Bockelmann, U., et al. (1996). Time resolved spectroscopy of single quantum dots: Fermi gas of excitons? *Phys. Rev. Lett.* **76**, 3622–3625.
26. Wegscheider, W., Schedelbeck, G., Abstreiter, G., Rother, M., and Bichler, M. (1997). Atomically precise GaAs/AlGaAs quantum dots fabricated by twofold cleaved edge overgrowth. *Phys. Rev. Lett.* **79**, 1917–1920.
27. Bockelmann, U., Heller, W., Filoramo, A., and Roussignol, P. (1997). Microphotoluminescence studies of single quantum dots. I. Time-resolved experiments. *Phys. Rev. B* **55**, 4456–4468.
28. Bockelmann, U. and Heller, Abstreiter, G. (1997). Microphotoluminescence studies of single quantum dots. II. Magnetic-field experiments. *Phys. Rev. B* **55**, 4469–4472.
29. Xie, Q., Madhukar, A., Chen, P., and Kobayshi, N.P. (1995). Vertically self-organized InAs quantum box islands on GaAs (100). *Phys. Rev. Lett.* **75**, 2542–2545.
30. Solomon, G.S., Trezza, J.A., Marshall, A.F., and Harris Jr., J.S. (1996). Vertically aligned and electronically coupled growth induced InAs islands in GaAs. *Phys. Rev. Lett.* **76**, 952–955.
31. Tersoff, J., Teichert, C., and Lagally, M.G. (1996). Self-organization in growth of quantum dot superlattices. *Phys. Rev. Lett.* **76**, 1675–1678.
32. Schedelbeck, G., Wegschelder, W., Bichler, M., and Abstreiter, G. (1997). Coupled quantum dots fabricated by cleaved edge overgrowth: From artificial atoms to molecules. *Science* **278**, 1792–1795.
33. Springholz, G., Holy, V., Pinczolits, M., and Bauer, G. (1998). Self-organized growth of three-dimensional quantum-dot crystals with fcc-like stacking and a tunable lattice constant. *Science* **282**, 734–737.
34. Holy, V., Springholz, G., Pinczolits, M., and Bauer, G. (1999). Strain induced vertical and lateral correlations in quantum dot superlattices. *Phys. Rev. Lett.* **83**, 356–359.
35. Springholz, G., et al. (2000). Tuning of vertical and lateral correlations in self-organized PbSe/Pb$_{1-x}$Eu$_x$Te quantum dot superlattices. *Phys. Rev. Lett.* **84**, 4669–4672.
36. Bayer, M., et al. (2001). Coupling and entangling of quantum states in quantum dot molecules. *Science* **291**, 451–453.
37. Kamath, K., Bhattacharya, P., Sosnowski, T., Norris, T., and Phillips, J. (1996). Room-temperature operation of In$_{0.4}$Ga$_{0.6}$As/GaAs self-organized quantum dot lasers. *Electronic Lett.* **32**, 1374–1375.

38. Fafard, S., *et al.* (1996). Red-emitting semiconductor quantum dot lasers. *Science* **274**, 1350–1353.
39. Saito, H., Nishi, K., Ogura, I., Sugou, S., and Sugimoto, Y. (1996). Room-temperature lasing operation of a quantum-dot vertical-cavity surface-emitting laser. *Appl. Phys. Lett.* **69**, 3140–3142.
40. Klimov, V.I., *et al.* (2000). Optical gain and stimulated emission in nanocrystal quantum dots. *Science* **290**, 314–317.
41. Grundmann, M. (2000). The present status of quantum dot lasers. *Physica. E* **5**, 167–184.
42. Bhattacharya, P., Krishna, S., Phillips, J., McCann, P.J., and Namjou, K. (2000). Carrier dynamics in self-organized quantum dots and their application to long-wavelength sources and detectors. *J. Crystal Growth* **84**, 2513–2516.
43. Huang, X., *et al.* (2000). Passive mode-locking in 1.3 μm two-section InAs quantum dot lasers. *Appl. Phys. Lett.* **78**, 2825–2827.
44. Colvin, V., Schlamp, M., and Alivisatos, A.P. (1994). Light-emitting-diodes made from cadmium selenide nanocrystals and a semiconducting polymer. *Nature* **370**, 354–357.
45. Dabbousi, B.O., Bawendi, M.G., Onotsuka, O., and Rubner, M.F. (1995). Electroluminescence from cdse quantum-dot polymer composites. *Appl. Phys. Lett.* **66**, 1316–1318.
46. Michler, P., *et al.* (2000). A quantum dot single-photon turnstile device. *Science* **290**, 2282–2285.
47. Imamoğlu, A. and Yamamoto, Y. (1994). Turnstile device for heralded single photons: Coulomb blockade of electron and hole tunneling in quantum confined p-i-n heterojunctions. *Phys. Rev. Lett.* **72**, 210–213.
48. Santori, C., Pelton, M., Solomon, G., Dale, Y., and Yamamoto, E. (2001). Triggered single photons from a quantum dot. *Phys. Rev. Lett.* **86**, 1502–1505.
49. Flissikowski, T., Hundt, A., Lowisch, M., Rabe, M., and Henneberger, F. (2001). Photon beats from a single semiconductor quantum dot. *Phys. Rev. Lett.* **86**, 3172–3175.
50. Becher, C., *et al.* (2001). Nonclassical radiation from a single self-assembled InAs quantum dot. *Phys. Rev. B* **63**, 121312(R).
51. Zwiller, V., *et al.* (2001). Single quantum dots emit single photons at a time: antibunching experiments. *Appl. Phys. Lett.* **78**, 2476–2478.
52. Benson, O., Santori, C., Pelton, M., and Yamamoto, Y. (2001). Regulated and entangled photons from a single quantum dot. *Phys. Rev. Lett.* **84**, 2513–2516.
53. Bruchez Jr., M., Moronne, M., Gin, P., Weiss, S., and Alivisatos, A.P. (1998). Semiconductor nanocrystals as fluorescent biological labels. *Science* **281,** 2013–2016.
54. Chan, W.C.W. and Nie, S. (1998). Quantum dot bioconjugates for ultrasensitive nonisotopic detection. *Science* **281**, 2016–2018.
55. Lundstrom, T., Schoenfeld, W., Lee, H., and Petroff, P.M. (1999). Exciton storage in semiconductor self-assembled quantum dots. *Science* **286**, 2312–2314.
56. Recher, P., Sukhorukov, V., and Loss, D. (2000). Quantum dot as spin filter and spin memory. *Phys. Rev. Lett.* **85**, 1962–1965.
57. O'Regan, B., and Graätzel, M. (1991). A low-cost, high-effeciency solar-cell based on dye-sensitized colloidal TiO_2 films. *Nature* **353**, 737–740.
58. Pan, D., Towe, E., and Kennerly, S. (2000). Photovoltaic quantum-dot infrared detectors. *Appl. Phys. Lett.* **76**, 3301–3303.

59. Aguado, R. and Kouwenhoven, L.P. (2000). Double quantum dots as detectors of high-frequency quantum noise in mesoscopic conductors. *Phys. Rev. Lett.* **84**, 1986–1989.
60. Gammon, D., *et al.* (1997). Nuclear spectroscopy in single quantum dots: Nanoscopic Raman scattering and nuclear magnetic resonance with single quantum dots. *Science* **277**, 85–88.
61. Brown, S.W., Kennedy, T.A., and Gammon, D. (1998). Optical NMR from single quantum dots. *Sol. State Nuc. Mag. Res.* **11**, 49–58.
62. Bányai, L. and Koch, S.W. (1993). Semiconductor Quantum Dots. World Scientific.
63. Bimberg, D. (1999). Quantum Dot Heterostructures. John Wiley, Chichester, New York.
64. Jacak, L. (1998). Quantum dots. Springer, New York.
65. Wannier, G.H. (1937). The structure of electronic excitation levels in insulating crystals. *Phys. Rev.* **52**, 191–197.
66. Kohn, W. (1957). Shallow impurity states in silicon and germanium. *Adv. Solid State Phys.* **5**, 257–320.
67. Sham, L.J. and Rice, T.M. (1966). Many-particle derivation of the effective-mass equation for the Wannier exciton. *Phys. Rev.* **144**, 708–714.
68. Frenkel, J. (1931). On the transformation of light into heat in solids. *Phys. Rev.* **37**, 17–44.
69. Östreich, T., Schönhammer, K., and Sham, L.J. (1995). Exciton–exciton correlation in the nonlinear optical regime. *Phys. Rev. Lett.* **74**, 4698–4701.
70. Gammon, D., Snow, E.S., Shanabrook, B.V., Katzer, D.S., and Park, D. (1996). Fine structure splitting in the optical spectra of single GaAs quantum dots. *Phys. Rev. Lett.* **76**, 3005–3008.
71. Wu, Q., Grober, R.D., Gammon, D., and Katzer, D.S. (1999). Imaging spectroscopy of two-dimensional excitons in a narrow GaAs/AlGaAs quantum well. *Phys. Rev. Lett.* **83**, 2652–2655.
72. Guest, J.R., *et al.* (2001). Near-field coherent spectroscopy and microscopy of a nanoscopic quantum system. *Science* **293**, 2224–2227.
73. Bayer, M., *et al.* (1998). Exciton complexes in $In_xGa_{1-x}As/GaAs$ quantum dots. *Phys. Rev. B* **58**, 4740–4753.
74. Dekel, E., *et al.* (1998). Multiexciton spectroscopy of a single self-assembled quantum dot. *Phys. Rev. Lett.* **80**, 4991–4994.
75. Tanaka, T., Singh, J., Arakawa, Y., and Bhattacharya, P. (1993). Near band edge polarization dependence as a probe of structural symmetry in GaAs/AlGaAs quantum dot structures. *Appl. Phys. Lett.* **62**, 756–758.
76. Takagahara, T. (2000). Theory of exciton fine structures and extremely slow spin relaxation in single quantum dots. *J. Lumi.* **87–89**, 308–311.
77. Besombes, L., Kheng, K., and Martrou, D. (2000). Exciton and biexciton fine structure in single elongated islands grown on a vicinal surface. *Phys. Rev. Lett.* **85**, 425–428.
78. Kulakovskii, V.D., *et al.* (1999). Fine structure of biexciton emission in symmetric and asymmetric CdSe/ZnSe quantum dots. *Phys. Rev. Lett.* **82**, 1780–1783.
79. Bayer, M., *et al.* (1999). Electron and hole g factors and exchange interaction from studies of the exciton fine structure in $In_{0.60}Ga_{0.40}As$ quantum dots. *Phys. Rev. Lett.* **82**, 1748–1751.
80. DiVincenzo, D.P. and Loss, D. (1998). Quantum information is physical. *Superlatt. and Microstruct.* **23**, 419–432.

81. Neilson, M.A. and Chuang, I.L. (2000). Quantum Computation and Quantum Information. Cambridge University Press, Cambridge.
82. Macchiavello, C., Palma, G.M., and Zeilinger, A. (2000). Quantum Computation and Quantum Information Theory. World Scientific, Singapore.
83. Berman, P.R. (1977). Effects of collisions on linear and nonlinear spectroscopic lineshapes. *Phys. Report* **43**, 101–149.
84. Liao, P.F., Bjorkholm, J.E., and Berman, P.R. (1979). Study of collisional redistribution using two-photon absorption with a nearly-resonant intermediate state. *Phys. Rev. A* **20**, 1489–1494.
85. Meystre, P. and Sargent, M. (1998). Elements of Quantum Optics. Springer, New York.
86. Vinattieri, A., *et al.* (1994). Exciton dynamics in GaAs quantum wells under resonant excitation. *Phys. Rev. B* **50**, 10868–10879.
87. Wang, H., *et al.* (1993). Transient nonlinear optical response from excitation induced dephasing in GaAs. *Phys. Rev. Lett.* **71**, 1261–1264.
88. Bonadeo, N.H., *et al.* (1998). Nonlinear nano-optics: Probing one exciton at a time. *Phys. Rev. Lett.* **81**, 2759–2762.
89. Chen, G., *et al.* (2000). Optically induced entanglement of excitons in a single quantum dot. *Science* **289**, 1906–1909.
90. Toda, Y., Moriwaki, O., Nishioka, M.S., and Arakawa, Y. (1999). Efficient carrier relaxation mechanism in InGaAs/GaAs self-assembled quantum dots based on the existence of continuum states. *Phys. Rev. Lett.* **82**, 4114–4117.
91. Landin, L., Miller, M.S., Pistol, M.E., Pryor, C.E., and Samuelson, L. (1998), Optical studies of individual inas quantum dots in GaAs: Few-particle effects. *Science* **280**, 262–264.
92. Marzin, J., Gérard, J., Izraël, A., Barrier, D., and Bastard, G. (1994). Photoluminescence of single InAs quantum dots obtained by self-organized growth on GaAs. *Phys. Rev. Lett.* **73**, 716–719.
93. Hinzer, K., *et al.* (2001). Optical spectroscopy of a single $Al_{0.36}In_{0.64}As/As_{0.33}Ga_{0.67}As$ quantum dot. *Phys. Rev. B* **63**, 075314.
94. Empedocles, S.A., Norris, D.J., and Bawendi, M.G. (1996). Photoluminescence spectroscopy of single CdSe nanocrystallite quantum dots. *Phys. Rev. Lett.* **77**, 3873–3876.
95. Empedocles, S.A., and Bawendi, M.G. (1997). Quantum-confined stark effect in single CdSe nanocrystallite quantum dots. *Science* **278**, 2114–2117.
96. Grundmann, M., *et al.* (1995). Ultranarrow luminescence lines from single quantum dots. *Phys. Rev. Lett.* **74**, 4043–4046.
97. Wagner, H.P., *et al.* (1999). Exciton dephasing and biexciton binding in CdSe/ZnSe islands. *Phys. Rev. B* **60**, 10640–10643.
98. Kuther, A., *et al.* (1998). Zeeman splitting of excitons and biexcitons in single $In_{0.6}Ga_{0.4}As$/GaAs self-assembled quantum dots. *Phys. Rev. B* **58**, R7508–R7511.
99. Heller, O., Lelong, P., and Bastard, G. (1997). Biexcitons bound to single-island interface defects. *Phys. Rev. B* **56**, 4702–4709.
100. Kingshirn, C.F. (1997). Semiconductor Optics. Springer, New York.
101. Hawrylak, P., Narvaez, G.A., Bayer, M., and Forchel, A. (2000). Excitonic absorption in a quantum dot. *Phys. Rev. Lett.* **85**, 389–392.

102. Guest, J.R., et al. (2002). Measurement of optical absorption by a single quantum dot exciton. *Phys. Rev. B* **65**, 241310(R).
103. Stievater, T.H., et al. (2002). Wavelength modulation spectroscopy of single quantum dot states. *Appl. Phys. Lett.* **80**, 1876–1878.
104. Peyghambarian, N., Koch, S.W., and Mysyrowicz, A. (1993). Introduction to Semiconductor Optics. Solid State Physical Electronics. Prentice Hall, Englewood Cliffs. NJ.
105. Bonadeo, N.H., et al. (1998). Coherent optical control of the quantum state of a single quantum dot. *Science* **282**, 1473–1476.
106. Stievater, T.H., et al. (2001). Rabi oscillations of excitons in single quantum dots. *Phys. Rev. Lett.* **87**, 133603.
107. Stievater, T.H., et al. (2001). Transient nonlinear spectroscopy of excitons and biexcitons in single quantum dots. To be published.
108. Allen, L., and Eberly, J.H. (1987). Optical Resonance and Two-Level Atoms. Dover Publications, Inc. NY.
109. Mollow, B.R. (1969). Power spectrum of light scattered by two-level systems. *Phys. Rev.* **188**, 1969–1975.
110. Ciuti, C., and Quochi, F. (1998). Theory of the excitonic mollow spectrum in semiconductors. *Sol. State Comm.* **107**, 715–718.
111. Chen, G., et al. (2001). Zeeman coherence in single quantum dots. *Solid State Commun.* **119**, 199–205.
112. Chen, G., et al. (2002). Biexciton quantum coherence in a single quantum dot: Towards quantum information processing. *Phys. Rev. Lett.* **88**, 117901.
113. Toda, Y., Shinomori, S., Suzuki, K., and Arakawa, Y. (1998). Polarized photoluminescence spectroscopy of single self-assembled InAs quantum dots. *Phys. Rev. B* **58**, R10147–R10150.
114. Andreani, L.C., Panzarini, G., and Gérard, J.M. (1999). Strong-coupling regime for quantum boxes in pillar microcavities: Theory. *Phys. Rev. B* **60**, 13276–13279.
115. Takagahara, T. (2000). Personal communication.
116. Khitrova, G., Gibbs, H.M., Jahnke, F., Kira, M., and Koch, S.W. (1999). Nonlinear optics of normal-mode-coupling semiconductor microcavities. *Rev. Mod. Phys.* **71**, 1591–1639.
117. Bonadeo, N.H. (1999). Nano-optics: Coherent optical spectroscopy of single semiconductor quantum dots. PhD thesis, The University of Michigan, The Harrison M. Randall Laboratory of Physics, Ann Arbor MI 48109–1120.
118. Runge, E. and Zimmermann, R. (1998). Level repulsion in excitonic spectra of disordered systems and local relaxation kinetics. *Ann. Phys.* **7**, 417–426.
119. Guest, J.R., et al. (2000). Nonlinear near-field spectroscopy and microscopy of single excitons in a disordered quantum well. *In* "QELS 2000 Tech. Dig.", pp. 6–7.
120. Intonti, F., et al. (2001). Quantum mechanical repulsion of exciton levels in a disordered quantum well. *Phys. Rev. Lett.* **87**, 076801.
121. von Freymann, G., et al. (2001). Autocorrelation spectroscopy on single ultrathin layers of CdSe/ZnSe: hints for a non-thermal distribution of excitons in quantum islands. *J. Microsc-Oxford* **202**, 218–222.
122. Citrin, D.S. (1993). Radiative lifetimes of excitons in quantum wells: Localization and phase-coherence effects. *Phys. Rev. B* **47**, 3832–3841.

123. Bockelmann, U. (1993). Exciton relaxation and radiative recombination in semiconductor quantum dots. *Phys. Rev. B* **48**, 17637–17640.
124. Gotoh, H., Ando, H., and Takagahara, T. (1997). Radiative recombination lifetime of excitons in thin quantum boxes. *J. Appl. Phys.* **81**, 1785–1789.
125. Takagahara, T. (1999). Theory of exciton dephasing in semiconductor quantum dots. *Phys. Rev. B* **60**, 2638–2652.
126. Brown, S.W., Kennedy, T.A., Gammon, D., and Snow, E.S. (1996). Spectrally resolved overhauser shifts in single GaAs/Al$_x$Ga$_{1-x}$As quantum dots. *Phys. Rev. B* **54**, R17339–R17342.
127. Rinaldi, R., *et al.* (1996). Zeeman effect in parabolic quantum dots. *Phys. Rev. Lett.* **77**, 342–345.
128. Toda, Y., Shinomori, S., Suzuki, K., and Arakawa, Y. (1998). Near-field magneto-optical spectroscopy of single self-assembled InAs quantum dots. *Appl. Phys. Lett.* **73**, 517–519.
129. Bayer, M., Stern, O., and Forchel (2000). Spectroscopic study of dark excitons in In$_x$Ga$_{1-x}$As self-assembled quantum dots by magnetic-field-induced symmetry breaking. *Phys. Rev. B* **61**, 7273–7276.
130. Overhauser, A.W. (1958). Polarization of nuclei in metals. *Phys. Rev.* **92**, 411–415.
131. Meier, F., and Zakharchenya, B. (Eds.) (1984). Optical Orientation. Modern Problems in Condensed Matter Sciences Vol. 8. North-Holland, Amsterdam.
132. Levis, R.J., Menkir, Getahun, M., and Rabitz, H. (2001). Selective bond dissociation and rearrangement with optimally tailored, strong-field laser pulses. *Science* **292**, 709–713.
133. Rice, S.A. (2001). Interfering for the good of a chemical reaction. *Nature* **409**, 422–426.
134. Weinacht, T.C. (1998). Measurement of the amplitude and phase of a sculpted Rydberg wave packet. *Phys. Rev. Lett.* **80**, 5508–5511.
135. Zare, R.N. (1998). Laser control of chemical reactions. *Science* **279**, 1875–1879.
136. Wang, F., Chen, C., and Elliot, D.S. (1996). Product state control through interfering excitation routes. *Phys. Rev. Lett.* **77**, 2416–2419.
137. Meekhof, D.M., Monroe, C., King, B.E., Itano, W.M., and Wineland, D.J. (1996). Generation of nonclassical motional states of a trapped atom. *Phys. Rev. Lett.* **76**, 1796–1799.
138. Warren, W.S., Rabitz, H., and Dahleh, M. (1993). Coherent control of quantum dynamics: The dream is alive. *Science* **259**, 1581–1589.
139. Yeazell, J.A., and Stroud Jr. C.R. (1988). Observation of spatially localized atomic electron wave packets. *Phys. Rev. Lett.* **60**, 1494–1497.
140. Haché, A., *et al.* (1997). Observation of coherently controlled photocurrent in unbiased, bulk GaAs. *Phys. Rev. Lett.* **78**, 306–309.
141. Wehner, M.U., Ulm, M.H., Chemla, D.S., and Wegener, M. (1998). Coherent control of electron-lo-phonon scattering in bulk GaAs. *Phys. Rev. Lett.* **80**, 1992–1995.
142. Planken, P.C.M., Brener, I., Nuss, M.C., Luo, M.S.C., and Chuang, S.L. (1993). Coherent control of terahertz charge oscillations in a coupled quantum well using phase-locked optical pulses. *Phys. Rev. B* **48**, 4903–4906.
143. Huggard, P.G., *et al.* (1997). Coherent control of cyclotron emission from a semiconductor using sub-picosecond electric field transients. *Appl. Phys. Lett.* **71**, 2647–2649.
144. Heberle, A.P., Baumberg, J.J., and Kohler, K. (1995). Ultrafast coherent control and destruction on excitons in quantum wells. *Phys. Rev. Lett.* **75**, 2598–2601.

145. Yee, D.S., Yee, K.J., Hohng, S.C., and S., K.D. (2000). Coherent control of absorption and polarization decay in a GaAs quantum well: time and spectral domain studies. *Phys. Rev. Lett.* **84**, 3474–3477.
146. Erland, J., Lyssenko, G., and Hvam, J.M. (2001). Optical coherent control in semiconductors: Fringe contrast and inhomogeneous broadening. *Phys. Rev. B* **63**, 155317.
147. Özgür, U., Lee, C.-W., and Everitt, H.O. (2001). Control of coherent acoustic phonons in semiconductor quantum wells. *Phys. Rev. Lett.* **86**, 5604–5607.
148. Marie, X., *et al.* (1999). Coherent control of exciton polaritons in a semiconductor microcavity. *Phys. Rev. Lett.* **59**, R2494–R2497.
149. Vu, Q.T, Haug, H., Hügel, W.A., Chatterjee, S., and Wegener, M. (2000). Signature of electron-plasmon quantum kinetics in GaAs. *Phys. Rev. Lett.* **85**, 3508–3511.
150. Lenihan, A.S., Dutt, M.V.G., Steel, D.G., Schoenfeld, W., and Petroff, P.M. (2000). Transient optical excitation and control in self-assembled quantum dots. *In* "QELS 2000 Tech. Dig.", p. 38.
151. Toda, Y., Sugimoto, T., Nishioka, M., and Arakawa, Y. (2000). Time correlation measurement on single self-assembled quantum dots. *In* "QELS 2000 Tech. Dig.", pp. 36–37.
152. Htoon, H., *et al.* (2001). Carrier relaxation and quantum decoherence of excited states in self-assembled quantum dots. *Phys. Rev. B* **63**, 241303 (R).
153. Scherer, N.F., Carson, R.J., Matro, A., Du, M., and Ruggiero, A.J. (1991). Fluorescence-detected wave packet interferometry: time resolved molecular spectroscopy with sequence of femtosecond phase-locked pulses. *J. Chem. Phys.* **95**, 1487–1511.
154. Metiu, H., and Engel, V. (1990). Coherence, transients, and interference in photodissociation with ultrashort pulses. *J. Opt. Soc. Am. B* **7**, 1709–1726.
155. Fourkas, J.T., Wilson, W.L., and Wackerle, G. (1989), Picosecond time-scale phase-related optical pulses – measurement of sodium optical coherence decay by observation of incoherent fluorescence. *J. Opt. Soc. Am. B* **6**, 1905–1910.
156. Spano, F., Haner, M., and Warren, W.S. (1987). Spectroscopic demonstration of picosecond, phase-shifted laser multiple-pulse sequence. *Chem. Phys. Lett.* **135**, 97–102.
157. Blanchet, V., Nicole, C., Bouchene, M.A., and Girard, B. (1997). Temporal coherent control in two-photon transitions: From optical interferences to quantum interferences. *Phys. Rev. Lett.* **78**, 2716–2719.
158. Allen, L., and Eberly, J.H. (1975). Optical resonance and two-level atoms. Dover Publications, Inc., NY.
159. Rabi, I.I. (1937). Space quantization in a gyrating magnetic field. *Phys. Rev.* **51**, 652–654.
160. Rabi, I.I., Milliman, S., Kusch, P., and Zacharias, J.R. (1939). The molecular beam resonance method for measuring nuclear magnetic moments: The magnetic moments of $_3Li^6$, $_3Li^7$ and $_9F^{19}$. *Phys. Rev.* **55**, 526–535.
161. Hocker, G.B., and Tang, C.L. (1968). Observation of the optical transient nutation effect. *Phys. Rev. Lett.* **21**, 591–594.
162. Gibbs, H.M. (1973). Incoherent resonance fluorescence from a Rb atomic beam excited by a short coherent optical pulse. *Phys. Rev. A* **8**, 446–455.
163. Binder, R., Koch, S.W., Lindberg, M., Peyghambarian, N., and Schäfer, W. (1990). Ultrafast adiabatic following in semiconductors. *Phys. Rev. Lett.* **65**, 899–902.

164. Fürst, C., Leitenstorfer, A., Nutch, A., Tränkle, G., and Zrenner, A. (1997). Ultrafast Rabi oscillations of free-carrier transitions in InP. *Phys. Stat. Sol. B* **204**, 20–22.
165. Cundiff, S.T., *et al.* (1994). Rabi flopping in semiconductors. *Phys. Rev. Lett.* **73**, 1178–1181.
166. Schülzgen, A., *et al.* (1999). Direct observation of excitonic Rabi oscillations in semiconductors. *Phys. Rev. Lett.* **82**, 2346–2349.
167. Quochi, F., *et al.* (1998). Strongly driven semiconductor microcavities: From the polariton doublet to an AC Stark triplet. *Phys. Rev. Lett.* **80**, 4733–4736.
168. Saba, M., *et al.* (2000). Direct observation of the excitonic AC Stark splitting in a quantum well. *Phys. Rev. B* **62**, R16322–R16325.
169. Cole, B.E., Williams, J.B., King, B.T., Sherwin, M.S., and Stanley, C.R. (2001). Coherent manipulation of semiconductor quantum bits with terahertz radiation. *Nature* **410**, 60–63.
170. Chen, P., Piermarocchi, C., and Sham, L.J. (2001). Theory of coherent optical control of exciton spin dynamics in a semiconductor quantum dot. *Physica E* **10**, 7–12.
171. Bott, K., *et al.* (1993). Influence of exciton–exciton interactions on the coherent optical response in GaAs quantum wells. *Phys. Rev. B* **48**, 17418–17426.
172. Erland, J., Kim, J.C., Bonadeo, N.H., Katzer, D.G.D.S., and Steel, D.G. (1999). Nonexponential photon echo decays from nanostructures: Strongly and weakly localized degenerate exciton states. *Phys. Rev. B* **60**, R8497–R8500.
173. Wu, Q., Grober, R.D., Gammon, D., and Katzer, D.S. (2000). Excitons, biexcitons, and electron–hole plasma in a narrow 2.8-nm $GaAs/Al_xGa_{1-x}As$ quantum well. *Phys. Rev. B* **62**, 13022–13027.
174. Deveaud, B., *et al.* (1991). Enhanced radiative recombination of free excitons in GaAs quantum wells. *Phys. Rev. Lett.* **67**, 2355–2358.
175. Andreani, L.C. (1995). Confined electrons and photons. Plenum Press, New York.
176. Ciuti, C., Savona, V., Piermarocchi, C., Quattropani, A., and Schwendimann, P. (1998). Role of the exchange of carriers in elastic exciton–exciton scattering in quantum wells. *Phys. Rev. B* **58**, 7926–7933.
177. Honold, A., Schultheis, L., Kuhl, J., and Tu, C.W. (1989). Collision broadening of two-dimensional excitons in a GaAs single quantum well. *Phys. Rev. Lett.* **40**, 6442–6445.
178. Hallerass, E.A. and Ore, A. (1947). Binding energy of the positronium molecule. *Phys. Rev.* **71**, 493–496.
179. Ore, A. (1947). Structure of the quadrielectron. *Phys. Rev.* **71**, 913–914.
180. Lampert, M.A. (1958). Mobile and immobile effective-mass-particle complexes in nonmetallic solids. *Phys. Rev. Lett.* **1**, 450–453.
181. Moskalenko, S.A. (1958). *Opt. Spekt.* **5**, 147.
182. Sharma, R.R. (1968) Binding energy of the excitonic molecule. *Phys. Rev.* **170**, 770–772.
183. Akimoto, O., and Hanamura, E. (1972). Binding-energy of excitonic molecule. *Solid State Commun.* **10**, 253–255.
184. Wehner, R.K. (1969). On exciton molecule. *Solid State Commun.* **7**, 457–458.
185. Adamowski, J., Bednarck, S., and Suffczyn, M. (1971). Binding energy of biexcitons. *Solid State Commun.* **9**, 2037–2038.
186. Brinkman, W.E., Rice, T.M., and Bell, J.B. (1973). The excitonic molecule. *Phys. Rev. B* **8**, 1570–1580.

187. Akimoto, A., and Hanamura, E. (1972). Excitonic molecule I. Calculation of the binding energy. *J. Phys. Soc. Jpn.* **33**, 1537.
188. Haynes, J.R. (1966). Experimental observation of the excitonic molecule. *Phys. Rev. Lett.* **17**, 860–863.
189. Kleinman, D.A. (1983). Binding energy of biexcitons and bound excitons in quantum wells. *Phys. Rev. B* **28**, 871–879.
190. Takagahara, T. (1989). Biexciton states in semiconductor quantum dots and their nonlinear optical properties. *Phys. Rev. B* **39**, 10206–10231.
191. Hu, Y.Z., et al. (1990). Biexcitons in semiconductor quantum dots. *Phys. Rev. Lett.* **64**, 1805–1807.
192. Bryant, G.W. (1990). Biexciton binding in quantum boxes. *Phys. Rev. B* **41**, 1243–1246.
193. Singh, J., Birkedal, D., Lyssenko, V.G., and Hvam, J.M. (1996). Binding energy of two-dimensional biexcitons. *Phys. Rev. B* **53**, 15909–15913.
194. Banyai, L., Galbraith, I., Ell, C., and Haug, H. (1987). Excitons and biexcitons in semiconductor quantum wires. *Phys. Rev. B* **36**, 6099–6104.
195. Madarasz, F.L., Szmulowicz, F., and Hopkins, F.K. (1995). Exciton/biexciton energies in rectangular GaAs/Al$_x$Ga$_{1-x}$As quantum-well wires. *Phys. Rev. B* **52**, 8964–8973.
196. Xie, W. (2000). Binding energy of biexcitons in a quantum disk. *Mod. Phys. Lett. B* **14**, 701–707.
197. Bacher, G., et al. (1999). Biexciton versus exciton lifetime in a single semiconductor quantum dot. *Phys. Rev. Lett.* **83**, 4417–4420.
198. Beterov, I.M., and Chebotaev, V.P. (1974). Three-level gas systems and their interaction with radiation. *Prog. Quantum Electron.* **3**, 1–106.
199. Ferrio, K., and Steel, D.G. (1996). Observation of the ultrafast two-photon coherent biexciton oscillation in a GaAs/Al$_x$Ga$_{1-x}$As multiple quantum well. *Phys. Rev. B* **54**, R5231–R5234.
200. Ferrio, K.B., and Steel, D.G. (1998). Raman quantum beats of interacting excitons. *Phys. Rev. Lett.* **80**, 786–789.
201. Langbein, W., and Havam, J.M. (2000). Dephasing in the quasi-two-dimensional exciton biexciton system. *Phys. Rev. B* **61**, 1692–1695.
202. Citrin, D.S. (1994). Long radiative lifetimes of biexcitons in GaAs/Al$_x$Ga$_{1-x}$As quantum wells. *Phys. Rev. B* **50**, 17655–17658.
203. Dekel, E., et al. (2000). Cascade evolution and radiative recombination of quantum dot multiexcitons studied by time-resolved spectroscopy. *Phys. Rev. B* **62**, 11038–11045.
204. Kim, J.C., Wake, D.R., and Wolfe, J.P. (1994). Thermodynamics of biexcitons in a GaAs quantum well. *Phys. Rev. B* **50**, 15099–15107.
205. Bennett, C.H., DiVincenzo, D.P., Smolin, J.A., and Wootters, W.K. (1996). Mixed-state entanglement and quantum error correction. *Phys. Rev. A* **54**, 3824–3851.
206. Sackett, C.A., et al. (1993). Experimental entanglement of four particles. *Nature* **404**, 256–259.
207. Deutsch, D. and Jozsa, R. (1992). Rapid solution of problems by quantum computation. *Proc. R. Soc. London Ser. A* **439**, 553–558.
208. Loss, D. and DiVincenzo, D.P. (1998). Quantum computation with quantum dots. *Phys. Rev. A* **57**, 120–126.

209. Barenco, A., Deutsch, D., Ekert, A., and Jozsa, R. (1995). Conditional quantum dynamics and logic gates. *Phys. Rev. Lett.* **74**, 4083–4086.
210. Sanders, G.D., Kim, K.W., and Holton, W.C. (2000). Scalable solid-state quantum computer based on quantum dot pillar structures. *Phys. Rev. B* **61**, 7526–7535.
211. Troiani, F., Hohenester, U., and Molinari, E. (2000). Exploiting exciton–exciton interactions in semiconductor quantum dots for quantum-information processing. *Phys. Rev. B* **62**, R2263–R2266.
212. Brun, T.A., and Wang, H. (2000). Coupling nanocrystals to a high-q silica microsphere: Entanglement in quantum dots via photon exchange. *Phys. Rev. A* **61**, 032307.
213. Quiroga, L. and Johnson, N.F. (1999). Entangled bell and Greenberger–Horne–Zeilinger states of excitons in coupled quantum dots. *Phys. Rev. Lett.* **83**, 2270–2273.
214. Pazy, E., *et al.* (2001). Spin-based optical quantum gates via Pauli blocking in semiconductor quantum dots. *cond-mat/0109337*.
215. Chuang, I.L., Vandersypen, L.M.K., Zhou, X., Leung, D.W., and Lloyd, S. (1998). Experimental realization of a quantum algorithm. *Nature* **393**, 143–146.
216. Jones, J.A., and Mosca, M. (1998). Implementation of a quantum algorithm on a nuclear magnetic resonance quantum computer. *J. Chem. Phys.* **109**, 1648–1653.
217. Shelton, R.K., *et al.* (2001). Phase-coherent optical pulse synthesis from separate femtosecond laser. *Science* **293**, 1286–1289.
218. Poyatos, J.F., Cirac, J.I., and Zoller, P. (1997). Complete characterization of a quantum process: the two-bit quantum gate. *Phys. Rev. Lett.* **390**, 78–81.
219. Wilcox, R.M. (1967). Exponential operators and parameter differentiation in quantum physics. *J. Math. Phys.* **8**, 962.
220. Ernst, R.R., Bodenhausen, G., and Wokaun, A. (1987). Principles on Nuclear Magnetic Resonace in One and Two Dimensions, p. 72. Clarendon Press, Oxford.
221. Press, W.H., Teulolsky, S.A., Vetterling, W.T., and Flannery, B.P. (1986). Numerical Recipes. Cambridge University Press, Cambridge.
222. Carmichael, H. (1993). An Open Systems Approach to Quantum Optics. Springer, Berlin.
223. Gisin, N. and Percival, I.C. (1992). The quantum-state diffusion model applied to open systems. *J. Phys. A* **25**, 5677–5691.
224. Schack, R. and Brun, T.A. (1997). A C++ library using quantum trajectories to solve quantum master equations. *Computer Physics Communications* **102**, 210–228.

Quantum Coherence, Correlation and Decoherence in
Semiconductor Nanostructures
T. Takagahara (Ed.)
Copyright © 2003 Elsevier Science (USA). All rights reserved.

Chapter 8
Cavity QED of quantum dots with dielectric microspheres

Hailin Wang

Department of Physics and Oregon Center for Optics, University of Oregon, Eugene, OR 97403, USA

Abstract

Coupling nanocrystals synthesized through colloidal chemistry or nanostructures grown by molecular beam epitaxy to whispering gallery modes in a dielectric microsphere forms a unique composite semiconductor-microcavity system. In a low-Q regime, enhanced spontaneous emission has been demonstrated in CdSe/ZnS core/shell nanocrystals embedded in the interior surface of a polystyrene microsphere, providing important information on radiative dynamics in these quantum dots. In a high-Q regime, the good cavity limit in cavity QED, in which absorption or emission occurring in a single quantum dot can significantly affect the dynamics of the composite quantum dot-microcavity system, has been reached by coupling the nanocrystals to whispering gallery modes in a fused silica microsphere. Studies of homogeneous linewidth of the CdSe/ZnS core/shell nanocrystals using high resolution spectral hole burning has also shown the suppression of dephasing associated with electron–phonon interactions due to complete quantization of acoustic phonon modes in the colloidal quantum dots. While excessive pure dephasing in CdSe/ZnS core/shell nanocrystals has thus far prevented us from achieving the strong-coupling limit in cavity QED, this limit should be readily achievable with a composite nanocrystal-microsphere system where nanocrystals are nearly lifetime broadened.

8.1 Introduction

Spontaneous emission represents one of the most visible and fundamental manifestations of the dynamical interaction between matter and vacuum. The possibility of modifying spontaneous emission with an optical microcavity was first pointed out by E. Purcell [1]. Studies in this area, broadly referred to as cavity QED, have traditionally focused on atomic systems. Experimental investigations using composite atom-cavity systems have led to beautiful demonstration of effects such as enhanced or inhibited spontaneous emission, Jaynes–Cummings energy ladder, single-atom masers and lasers, and have led to applications in quantum measurement, quantum entanglement, and more recently quantum information processing [2–7]. The tremendous progresses achieved in atomic cavity QED have also stimulated considerable interest in cavity QED in quantum dot (QD) systems that feature atomic-like discrete energy levels. In addition to applications such as microlasers, QD-microcavity systems avoid the difficulty of center-of-mass motion inherent in atomic cavity QED and can thus scale up to a relatively large mesoscopic system, for example, with an array of QDs coupling strongly to a cavity mode.

For a simple system where a two-level dipole transition couples to a resonant optical microcavity, dynamics of spontaneous emission can be categorized into two different regimes. In the limit that $g < (\kappa, \gamma)$ where κ is the cavity decay rate, γ is the dephasing rate of the optical transition, and g is the coherent dipole coupling rate between the optical transition and the resonant cavity mode, enhanced or inhibited spontaneous emission is expected and the system remains in a weak-coupling regime. In comparison, when g is large compared to both κ and γ, the system reaches a strong-coupling regime. In this regime, the interaction between the optical transition and the cavity mode becomes so strong that spontaneous emission becomes reversible and a single photon can drastically change the dynamics of the system. The strong coupling regime, which is easy to specify but is extremely difficult to achieve, is essential for many applications of cavity QED systems, especially for quantum information processing and for generating quantum entanglement [4–9].

Two important cavity parameters for cavity QED are the cavity Q-factor and the effective cavity mode volume. The Q-factor determines the cavity decay rate. The effective mode volume, V_0, defined as the spatial integral of the field intensity, normalized to unity at the maximum, determines the strength of the vacuum electric field (or the electric field generated by a single photon in the cavity mode) given by [10]

$$E_{vac} = \sqrt{\hbar\omega/2\varepsilon_0 V_0} .\tag{8.1}$$

Since $g = \mu E_{vac}/\hbar$ where μ is the dipole moment for the relevant optical transition, cavities with high Q-factor and small mode volume are necessary for achieving the strong-coupling regime.

A variety of monolithic semiconductor microcavities, including micro-pillar Fabry-Perot cavities, microdiscs, and photonic crystals, have been used for cavity QED studies in a QD system [11–16]. While enhanced spontaneous emission has been demonstrated in these systems, experimental efforts toward achieving the strong coupling regime using monolithic semiconductor microcavities have thus far been hindered by inadequate cavity Q-factors.

Dielectric microspheres, especially fused silica microspheres, are uniquely suitable to applications in cavity QED. In a dielectric microsphere, light can circulate along the curved surface through total internal reflection to form a whispering gallery mode (WGM). Lower order WGMs in these microspheres can feature both small mode volume and unprecedented high cavity finesse [17]. A composite semiconductor-microsphere system, where semiconductor nanostructures couple to WGMs in a dielectric microsphere, can take advantage of the extreme photonic confinement in the dielectric microsphere and also allows the flexibility of separate engineering of photonic and electronic confinement. Semiconductor nanocrystals, such as core/shell CdSe/ZnS nanocrystals synthesized through colloidal chemistry, as well as nanostructures grown by molecular beam epitaxy (MBE) can be incorporated in these composite microcavity systems.

In this chapter, we will discuss our recent experimental efforts in using the composite semiconductor-microsphere system for cavity QED of QDs. Basic properties of WGMs in a dielectric microsphere will be briefly reviewed in section 8.2. Composite systems combining a fused silica microsphere and a MBE-grown nanostructure will be discussed in section 8.3. Many of the difficulties encountered in these systems can be avoided when a composite nanocrystal-microsphere system is used. Cavity QED of core/shell CdSe/ZnS nanocrystals in both low-Q and high-Q regimes will be discussed in detail in section 8.4. Since excessive pure dephasing in nanocrystals is presently the main obstacle for achieving the strong coupling regime, high-resolution hole burning studies on dephasing processes in these nanocrystals will also be discussed in this section. A brief summary will be presented in section 8.5.

8.2 Whispering gallery modes in a dielectric microsphere

In this section we review briefly basic properties of WGMs in a dielectric microsphere, including mode structures and Q-factors of fused silica microspheres. More detailed discussions on properties of WGMs can be found in [18,19].

Whispering gallery modes in a dielectric sphere are solutions to Maxwell's equations

$$\nabla \times (\nabla \times \mathbf{E}) - (\omega/c)^2 \, \varepsilon(r)\mathbf{E} = 0 \qquad (8.2)$$

where $\varepsilon(r)$ is the dielectric constant. We assume $\varepsilon = 1$ outside the sphere and $\varepsilon > 1$ but constant inside the sphere. The transverse electric (TE) modes are described by

$$E(\mathbf{r}) = \phi(r) \mathbf{X}_{lm}(\theta, \phi) \tag{8.3}$$

where $\mathbf{X}_{lm} = [l(l+1)]^{-1/2}\mathbf{L}Y_{lm}$ is the vector spherical harmonic with $\mathbf{L} = \mathbf{r} \times i\nabla$. The radial distribution is determined by

$$\frac{d^2\Phi}{d^2r} + \left[(\omega/c)^2 \varepsilon(r) - \frac{l(l+1)}{r^2}\right]\Phi = 0 \tag{8.4}$$

with $\Phi = r\phi(r)$. The transverse magnetic (TM) field modes are described by

$$\mathbf{E}(\mathbf{r}) = \frac{1}{\varepsilon(r)} \nabla \times [\phi(r)\mathbf{X}_{lm}(\theta, \phi)] \tag{8.5}$$

where the radial distribution is determined by

$$\frac{d}{dr}\left[\frac{1}{\varepsilon(r)}\frac{d\Phi}{dr}\right] + \frac{1}{\varepsilon(r)}\left[(\omega/c)^2 \varepsilon(r) - \frac{l(l+1)}{r^2}\right]\Phi = 0 \tag{8.6}$$

again with $\Phi = r\phi(r)$. Inside the sphere $\phi(r)$ is a spherical Bessel function and outside the sphere $\phi(r)$ is an outgoing spherical Hankel function. For each l, the electromagnetic boundary condition leads to a characteristic equation whose roots determine the discrete resonant frequency ω_{plm} of the WGMs where p is the principal or the radial mode number.

Whispering gallery modes can also be understood by using an analogy between Maxwell's equations and the Schrödinger equation. If one views $[1-\varepsilon(r)]$ as an analogue of a potential, the region with $\varepsilon > 1$ then behaves like an attractive potential well. For the electric field distribution along the radial direction, the effective potential is the sum of the attractive well due to the dielectric sphere and a repulsive angular momentum barrier as shown in Eq. (8.4). Whispering gallery modes are then the quasi-bound solutions confined near the surface of the sphere with the confinement potential due to the discontinuity in the dielectric constant across the surface of the sphere. This also naturally leads to the analogy between WGMs and atomic wave functions: $p-1$ corresponds to the number of nodes in the electric field along the radial direction, l corresponds to the angular momentum quantum number, and m corresponds to the azimuthal angular momentum quantum number. One can also estimate l by assuming that the round-trip optical path length equals $l\lambda$ where λ is the wavelength in vacuum. For light rays with near glancing incidence, this yields

$$l \approx r_0 n\omega/c \tag{8.7}$$

where r_0 and n are the radius and refractive index of the sphere, respectively. In this regard, we also refer l as the orbital mode number.

Whispering gallery modes with $p = 1$ and $m = \pm l$ are essentially a ring circling on the equator ($\theta = \pi/2$) and are confined within a wavelength of the sphere surface. For these WGMs the effective mode volume scales approximately with $r_0^2 \lambda$ and can be much smaller than the volume of the sphere. Vacuum field strength of order 300 V/cm can be achieved at the surface of a fused silica microsphere with $r_0 = 5$ μm and with λ near 600 nm.

Whispering gallery modes with $p = 1$ and $|m| < l$ are also confined near the sphere surface but has $l - |m| + 1$ major lobes in its angular intensity dependence, centered at $\theta = \pi/2$ and extending to $\theta = \pi/2 \pm \cos^{-1}(m/l)$. For a prefect sphere, TE or TM modes with the same p and l but different m are degenerate. A small ellipticity in typical fused silica microspheres, however, removes this azimuthal degeneracy.

We now discuss in more detail WGMs in fused silica microspheres. Fused silica microspheres can be fabricated by fusing the tip of an optical fiber or a silica wire with a focused CO_2 laser beam. The size of the sphere can range from a few to a few hundred μm. Figure 8.1 shows the optical image of a fused silica microsphere. The sphere remains attached to a fiber stem for easy manipulation and is slightly deformed due to the presence of the fiber stem. The ellipticity can range from 0.1% to more than 3%. In addition to CO_2 laser beams, microtorches have also been used to fabricate fused silica microspheres.

Fig. 8.1 A fused silica microsphere with a radius near 50 μm attached to a fiber stem. The microsphere is fabricated by fusing the end of a fiber tip with a focused CO_2 laser beam.

Whispering gallery modes in a fused silica microsphere can be conveniently excited through frustrated total internal reflection. Efficient excitation of WGMs with lower p can be achieved by using an angle of incidence near the critical angle between a high index optical prism and the sphere. Figure 8.2 shows the mode structure of a sphere with $r_0 = 70$ µm where we measured the intensity of the scattering light from the excited WGMs as a function of the excitation laser wavelength. The free spectral range (FSR) of the WGMs for this sphere is 1 nm and is determined by FSR $= \lambda^2/2\pi n\, r_0$ with $n = 1.452$. The two peaks within one FSR shown in Fig. 8.2 correspond to WGMs with $p = 1$ and 2, respectively. Each peak in Fig. 8.2(a) also contains a number of much sharper resonances as shown in an expanded plot in Fig. 8.2(b). These resonances correspond to WGMs that have the same p and l but different m. The frequency separation between two neighboring azimuthal modes can range from 0.001 nm to 0.5 nm depending on the ellipticity and the size of the sphere. The total number of p modes and m modes that can be excited in a given excitation configuration can be controlled by varying the angle of incidence and the laser beam profile. The spectral linewidth of the WGM resonance shown in Fig. 8.2(b) corresponds to a Q-factor of 10^6 and is limited by the output coupling loss through the high index prism. Q-factors near 10^9 were obtained when we introduced a greater gap between the sphere and the prism.

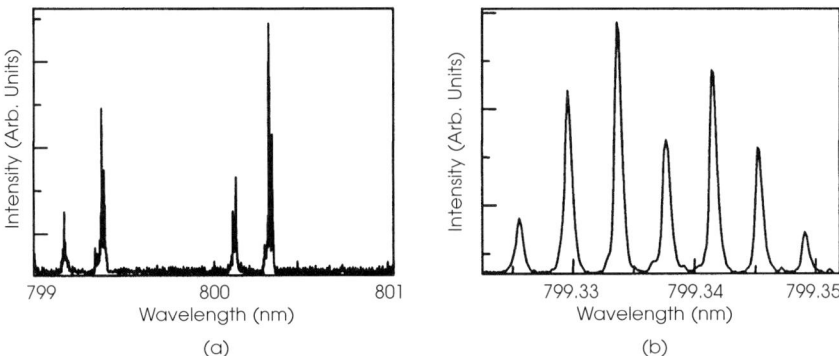

Fig. 8.2 Whispering gallery modes excited through frustrated total internal reflection for a fused silica microsphere with a radius of 70 µm. An expanded plot of the peak near 799.3 nm in (a) is displayed in (b).

There are several mechanisms that limit the finesse of a fused silica microsphere. If one considers a fused silica microsphere with $r_0 = 100$ µm and λ near 600 nm, the absorption and scattering loss of high purity fused silica limits the Q-factor to near 10^{10}. Examination of the surface of fused silica microspheres using atomic force microscopy revealed nearly atomically smooth surface with surface fluctuations on the order of a few tenth of nm in height and with a correlation length of several nm. The resulting surface scattering also limits the Q-factor to near 10^{10} [20,21]. Another important limiting factor for the cavity finesse is surface adsorption of water vapors, which can reduce the Q-factor to below 10^9 [20,21]. Intrinsic diffraction or tunneling loss, which depends exponentially on the sphere size, does not become important as long as $r_0/\lambda > 5$. The unprecedented high cavity finesse along with the extremely small mode volume makes fused silica microspheres a highly promising optical microresonator for use in cavity QED of QDs as well as atoms and molecules [22–27].

8.3 Composite system of dielectric microsphere and MBE-grown nanostructure

In this section we discuss a composite semiconductor-microsphere system where optical excitations in a MBE-grown semiconductor nanostructure are coupled to WGMs in a fused silica microsphere. This system takes advantage of the high-Q WGMs in a fused silica microsphere and at the same time can use a variety of well-developed and well-characterized MBE-grown semiconductor nanostructures.

The experimental setup is shown schematically in Fig. 8.3. The composite system consists of a microsphere in contact with a semiconductor sample [26,27]. A high index prism is placed on the other side of the microsphere for excitation and output coupling of WGMs. The prism is mounted on a piezoelectric translation stage that can adjust the gap between the prism and sphere. The entire setup is attached to the cold finger of a cryostat. To characterize the mode structure of the composite system, we have used in our initial experiments an inhomogeneously broadened GaAs/AlGaAs quantum well

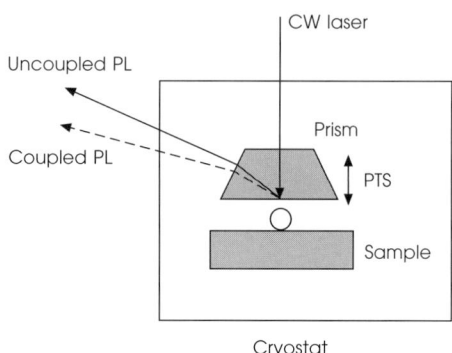

Fig. 8.3 Schematics of the experimental setup for coupling optical excitations in a MBE-grown nanostructure to WGMs in a fused silica microsphere. The optical prism is mounted on a piezoelectric translation stage (PTS). (From [27] with permission.)

(QW) sample that contains 8 periods of QWs with well width of 13 nm and barrier width of 15 nm.

For photoluminescence (PL) measurements, the QW sample was excited above the band gap with a continuous wave (CW) laser. Excitonic PL coupled into the WGMs (referred to as the coupled PL) were collected via output coupling through the high index prism. The coupled PL came out of the prism with a refraction angle greater than that of the uncoupled PL (PL not coupled into the WGMs), as indicated in Fig. 8.3. Figure 8.4 shows spectra of both uncoupled and coupled PL obtained at 10 K. The coupled PL spectra with TE and TM polarization feature periodic mode structures with a FSR of 1.5 nm, determined by the radius ($r_0 = 47$ μm) of the microsphere.

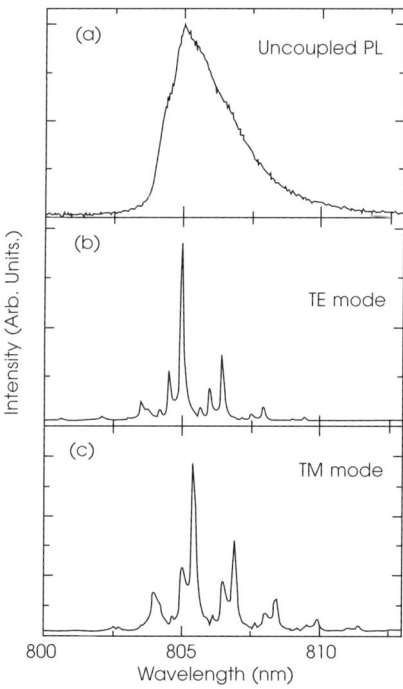

Fig. 8.4 Uncoupled PL spectrum (a) and coupled PL spectra (b and c) from a composite system of a GaAs/AlGaAs QW and a fused silica microsphere. The data were obtained at 10 K. (From [27] with permission.)

As we have discussed in section 8.2, WGMs in a dielectric sphere are characterized by radial mode number p, orbital mode number l, and azimuthal mode number m. As shown in Fig. 8.4, within each FSR there are several modes with different p, with the $p = 1$ mode featuring the greatest coupling efficiency. Additional PL studies performed at higher spectral resolution also showed that each peak in the coupled PL spectra in Fig. 8.4 consists of a number of sharper resonances corresponding to WGMs with different m, similar to that shown in Fig. 8.2 [26]. The spectral linewidth of these WGM resonances, however, is limited by the spectrometer resolution (0.01 nm).

To achieve adequate spectral resolution and especially to determine the Q-factor of the combined system, we have used the resonant light scattering technique, in which WGMs near the equator of the microsphere were launched directly via frustrated total internal reflection, as discussed in section 8.2. Emissions or scatterings from these WGMs were collected away from the reflection direction of the incident laser beam. Whispering gallery resonances with spectral linewidth of order 0.004 nm were observed for the combined semiconductor-microsphere system, corresponding to a Q-factor of 2 $\times 10^5$, compared to a Q-factor of 10^8 that is easily attainable for the fused silica microsphere even after repeated measurements [27]. The significant Q-spoiling observed is in part due to the refractive loss in the contact area between the QW sample and the sphere since the QW sample has an index of refraction greater than that of the fused silica microsphere and is also in part due to the scattering loss occurring in the contact area. Both of these loss mechanisms can, however, be avoided or significantly reduced if one uses as the semiconductor sample a tip with a dimension small compared with the optical wavelength. In spite of the Q-spoiling, the Q-factor of the combined system is still significantly greater than that of monolithic planar semiconductor microcavities or semiconductor microdisks.

A major difficulty in using the above composite microcavity system for cavity QED is that for typical MBE-grown nanostructures, the sample is at the very tail of the evanescent field of the WGMs such that the dipole coupling rate between the optical transition and the resonant WGM is relatively small. To determine the dipole coupling rate, we have attempted to measure Q-spoiling of the combined QW-microsphere system induced by absorption of excitons in the QW sample. For these measurements, we have selected microspheres that have WGMs resonant with the heavy hole (hh) excitonic transition in a QW sample containing a single 17 nm GaAs well. The absorption linewidth of the hh exciton resonance at 10 K is 0.3 nm. Since evanescent fields of WGMs decay exponentially away from the sphere surface, we chemically etched the GaAs capping layer of the sample down to approximately 25 nm thick to position the GaAs well nearly 32 nm away from the sample surface (there is a 7 nm AlGaAs barrier between the GaAs capping layer and the GaAs well). Spectral linewidth of 0.004 nm was observed for WGMs resonant with the hh excitonic transition. The same spectral linewidth, however, was also observed for WGMs below, at, or above the hh exciton absorption resonance [27]. The absence of absorption-induced Q-spoiling implies that the absorption loss due to the dipole coupling between the exciton and the resonant WGM is small compared with refractive and scattering losses occurring in the contact area.

For a microsphere in vacuum and for WGMs that are near glancing incidence, the field penetration depth into the vacuum is given by $\lambda/2\pi\sqrt{n^2-1}$. While the actual penetration depth when the microsphere is in contact with a semiconductor sample requires a detailed calculation of the WGM field distribution including the proper boundary condition, for a simple estimate we take the penetration depth in the sample to be $\lambda_s/2\pi\sqrt{n^2-1}$ where λ_s is the wavelength in the sample (the refractive index of the sample is 3.5). At $\lambda = 800$ nm this yields a penetration depth of only 35 nm. For the sample used in our study, the distance between the center of the QW and the sample surface is nearly 40 nm, leading to a reduction of a factor of 3 in the dipole coupling rate.

While the above experimental studies clearly indicate that PL from a semiconductor sample can couple efficiently into the WGMs of the composite semiconductor-microsphere system, to further increase the dipole coupling between excitons and relevant WGMs, one needs to use nanostructures where the active medium such as a QW or a QD can be positioned within a few nm from the microsphere surface. For GaAs-based nanostructures, a capping layer at least 25 nm to 40 nm thick is necessary in order to avoid severe nonradiative surface effects that can result in a drastic broadening of the exciton resonance [28]. However, for nanostructures such as II-VI nanocrystals, efficient surface passivation can be achieved by a capping layer of only 1 nm thick. Furthermore, a composite nanocrystal-microsphere system, in which chemically-synthesized colloidal QDs are coupled to WGMs of a dielectric microsphere, also avoids the refractive loss occurring in the contact area between a microsphere and a planar MBE-grown nanostructure. Properties of composite nanocrystal-microsphere systems will be discussed in detail in the next section.

8.4 Composite system of dielectric microsphere and semiconductor nanocrystals

Recent advances in colloidal chemistry have led to the fabrication of high quality semiconductor core/shell nanocrystals [29–31]. In these nanocrystals, a nearly defect free core such as CdSe is capped by a thin layer (~1 nm) of a semiconductor with greater bandgap such as ZnS. These core/shell nanocrystals feature much improved quantum yield compared with earlier generations of semiconductor nanocrystals.

A composite nanocrystal-microsphere system overcomes many of the difficulties encountered in the composite system of microsphere and MBE-grown nanostructure discussed in the previous section and provides a nearly ideal model system for cavity QED of QDs. On one hand, extremely high-Q factor and small cavity mode volume can be easily obtained. On the other hand, discrete acoustic phonon modes in colloidal QDs can lead to suppression of dephasing associated with electron–phonon interactions. In this section we will discuss in detail cavity QED studies in a low-Q regime where polystyrene spheres were used and in a high-Q regime where fused silica microspheres were used. This is followed by discussions on dephasing processes in CdSe/ZnS core/shell nanocrystals since dephasing in these QDs is presently the primary limiting factor for achieving the strong-coupling regime.

Coupling nanocrystals to a dielectric microsphere: low-Q regime

In a low-Q regime, the cavity decay rate is large compared with both the dipole coupling rate and the dephasing rate of the relevant optical transition. Enhanced spontaneous emission is expected in this regime when the cavity is resonant with the relevant optical transition. The enhancement in the spontaneous emission rate or the radiative decay rate is characterized by the Purcell factor given by $F_p = \Gamma_c/\Gamma_r$, where Γ_c is the radiative decay rate into a cavity mode and Γ_r is the radiative decay rate in a homogeneous dielectric medium. The cavity-induced relative change in the total decay rate, $\Gamma_t = \Gamma_{nonr} + \Gamma_r$, where Γ_{nonr} is the nonradiative decay rate, however, also depends on the relative contribution of Γ_r and Γ_{nonr} and is given by

$$\xi = \frac{\Gamma_{nonr} + (1 + F_p)\Gamma_r}{\Gamma_{nonr} + \Gamma_r} - 1 = F_p \frac{1}{1 + \Gamma_{nonr}/\Gamma_r}, \tag{8.8}$$

since Γ_{nonr} is not affected by the cavity. The manifestation of enhanced spontaneous emission depends on the ratio Γ_{nonr}/Γ_r, thus providing valuable information on the underlying radiative dynamics that is otherwise not available from conventional time-resolved PL studies.

For cavity QED studies in the low-Q regime, CdSe/ZnS nanocrystals were doped in the interior surface of a polystyrene sphere [32]. No indications of degradation due to the doping process have been observed. An inset in Plate 8 shows the optical image of a typical nanocrystal-doped polystyrene sphere. For studies presented in this chapter polystyrene spheres with a diameter of 15 μm were used. Separate control experiments also used spheres with a diameter of 100 μm (no enhanced spontaneous emission was observed with these large spheres). CdSe/ZnS core/shell nanocrystals used in our study were fabricated by using a high temperature organometallic synthesis developed earlier [29,30]. The nanocrystals feature an average room-temperature quantum yield of 40%~50%. Three groups of nanocrystals with average core radius R of 2 nm, 2.7 nm, and 4.5 nm were used. The average core radii were determined from PL spectra and are consistent with results of transmission electron microscopy. Plate 8 shows the PL spectrum obtained at 10 K from nanocrystals with $R = 2.7$ nm embedded in a polystyrene sphere. The linewidth of the narrowest WGMs, as shown in the inset, is 0.2 nm, corresponding to a Q-factor of 3000.

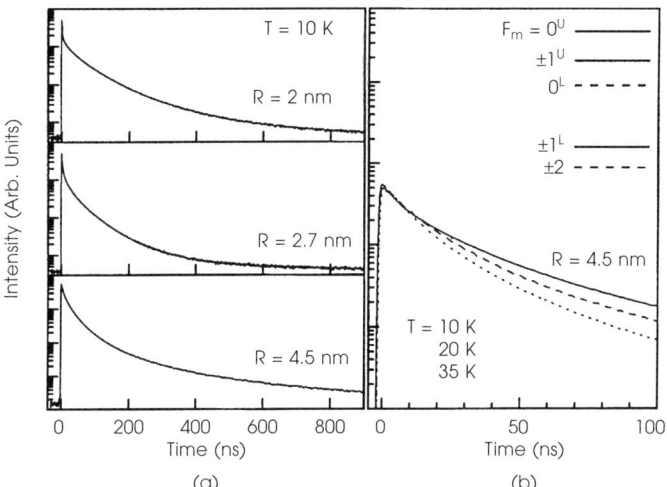

Fig. 8.5 Time-resolved PL from CdSe/ZnS core/shell nanocrystals in free space. (a) At 10 K. (b) Temperature dependence for nanocrystals with $R = 4.5$ nm. Solid, dashed, and dotted curves are obtained at 10 K, 20 K, and 35 K, respectively, with the amplitude normalized to the same peak intensity. The inset in (b) shows schematically the excitonic energy structure for CdSe nanocrystals near the band edge. (From [32] with permission.)

If one assumes that optical dipoles are randomly oriented, resonant with the cavity mode, and positioned at the maximum of the vacuum electric field, and that the homogeneous linewidth of the relevant optical transition is smaller than the cavity linewidth, the Purcell factor is then given by [1]

$$F_p = QD\lambda^3/4\pi^2 n^3 V_0, \quad (8.9)$$

where D is the mode degeneracy. For a 15 μm sphere, using $Q = 3000$, $\lambda = 620$ nm, $n = 1.5$, $D = 2$, and $V_0 = 35$ μm³, we obtain $F_p = 0.3$. The actual effective Purcell factor is smaller due to the spatial and spectral distributions of nanocrystals in the measurement. Using statistical average similar to that used in Ref. 9, we obtain an effective Purcell factor of 0.2.

Time-resolved PL was carried out by using correlated photon counting. The excitation pulses, obtained by frequency-doubling a mode-locked Ti: Sapphire laser, were centered near λ = 400 nm with a reduced repetition rate of 500 kHz. A photomultiplier tube was used as the detector with a system response near 1.5 ns. Experimental results presented were obtained at excitation levels where behaviors of time-resolved PL are independent of the input pulse energy.

While high quality colloidal QDs feature atomic-like discrete energy structures, radiative dynamics, especially the physical nature of band edge PL, in these QDs is very complex [33,34], which can greatly complicate the manifestation of cavity QED processes in these nanocrystals. Figure 8.5 shows time-resolved PL obtained in free space. At 10 K decay times of PL range from of order 1 ns to a few hundred ns.

The complex behavior of multiple decay components shown in Fig. 8.5 is in part due to the exciton energy structure in nanocrystals. The energy structure of band edge excitons in CdSe nanocrystals, drawn schematically in the inset of Plate 8, is characterized by five levels with angular momentum projection $F_m = \pm 2, \pm 1^L, 0^L \pm 1^U, 0^U$, where U and L denote upper and lower states with the same F_m [33]. Within the effective mass approximation (EMA), transitions from the crystal ground state to states ± 2 and 0^L (dashed lines) are dipole-forbidden, while transitions to states $\pm 1^L$, $\pm 1^U$ and 0^U (solid lines) are dipole-allowed. In addition to direct radiative recombination, phonon-assisted optical transitions from both dipole-allowed and dipole-forbidden transitions can also contribute to the PL. In principle, each direct or phonon-assisted transition can lead to a single or multiple decay components in the time-resolved PL with the decay including both radiative and nonradiative processes.

For a collection of nanocrystals with a large inhomogeneous linewidth, PL at a given wavelength can result from a number of direct and indirect transitions near the band edge. Little information on details of the underlying radiative dynamics can be extracted from the conventional time-resolved PL since it is difficult to correlate decay components with a specific optical transition. In order to reduce contributions from higher excited states, time-resolved PL presented here was obtained at the lower energy end of the PL spectrum unless otherwise specified.

Figure 8.6 shows time-resolved PL of nanocrystals embedded in polystyrene spheres. In each sub-figure, the lower curve is obtained at a wavelength resonant with a WGM while the upper curve is obtained at a wavelength near but is off-resonant (~3 Å away) with the given WGM. A spectral bandwidth of 1 Å is used for these studies in order to separate the resonant and off-resonant contributions. Note that in free space, behaviors of time-resolved PL within a narrow spectral range of 5 Å are nearly identical. An example of the relevant wavelength positions is also shown in an inset in Plate 8. The enhancement in PL decay rates when the PL is resonant with a given WGM occurs for only one PL component.

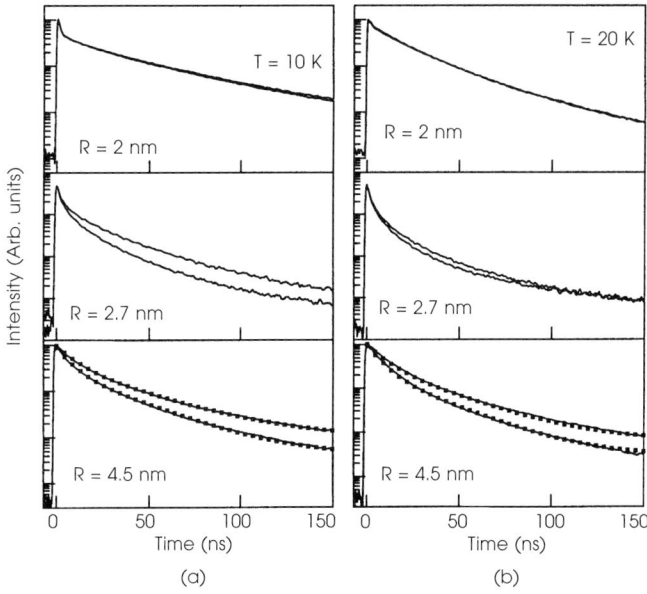

Fig. 8.6 Time-resolved PL from CdSe/ZnS core/shell nanocrystals embedded in a polystyrene sphere at spectral positions resonant (the lower curve in each figure) or off-resonant (the upper curve in each figure) with given WGMs. The measurements were carried out at the lower energy side of the respective PL spectra. For each figure, the amplitude is normalized to the same peak intensity. (a) At 10 K. (b) At 20 K. Results of numerical fits to PL from nanocrystals with $R = 4.5$ nm are shown as squares and are discussed in the text. (From [32] with permission.)

The relative enhancement in the PL decay rate shown in Fig. 8.6 depends sensitively on temperature and especially nanocrystal sizes. This is not due to changes in the underlying Purcell factor. As long as the homogeneous linewidth is small compared with the cavity linewidth, F_p is determined by the properties of the cavity and remains nearly the same for all measurements in Fig. 8.6. For a relatively small F_p (~0.2), a clear signature of relative changes in the total decay rate can be observed only when Γ_r is greater than or at least comparable to Γ_{nonr}, as indicated by Eq. (8.8).

We attribute the PL component that exhibits pronounced cavity-induced enhancement in the total decay rate to optical emissions from states $\pm 1^L$, the lowest dipole-allowed transition. For these states, Γ_{nonr} includes contributions from thermal activation to higher excited states, decay to states ± 2, and possibly relaxation to surface states. The decay into the ± 2 states requires spin flipping of excitons, which is shown to be extremely slow in recent studies using time-resolved Faraday rotation [35]. At low temperature, thermal activation from the $\pm 1^L$ to $\pm 1^U$ states is also slow because of the large energy separation (>10 meV) between these states. In comparison, for states with energies higher than the $\pm 1^L$ states, rapid relaxation into lower energy states occurs since in this case the relaxation does not require spin flipping. For these higher excited states, Γ_{nonr} is much greater than Γ_r. No significant cavity-induced relative increase in the decay rates is expected for PL from these higher excited states. This is also confirmed by the observation that for all the nanocrystals we have used, no pronounced enhancement in PL decay rates was seen for PL obtained at the center or at the higher energy side of the PL spectra (see Fig. 8.7) where contributions from higher excited states become much more important.

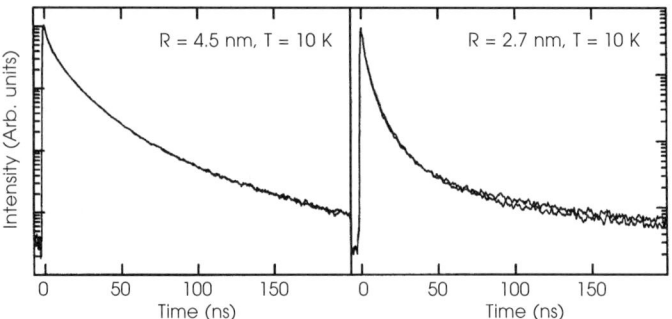

Fig. 8.7 Time-resolved PL from CdSe/ZnS core/shell nanocrystals embedded in a polystyrene sphere obtained near the center of the respective PL spectra and with the amplitude normalized to the same peak intensity. Results obtained at a wavelength resonant with a given WGM and at a nearby wavelength but off-resonant with the WGM are shown in each figure and are nearly the same. (From [32] with permission.)

The size and the temperature dependence of the enhanced spontaneous emission process shown in Fig. 8.6 further supports the above assignment on enhanced spontaneous emission from the lowest dipole-allowed transition. The enhancement in PL decay rates can only be observed in relatively large nanocrystals ($R = 2.7$ and 4.5 nm but not $R = 2$ nm). Earlier theoretical calculations have shown that the relative oscillator strength for the $\pm 1^L$ states increases with increasing nanocrystal size whereas the relative oscillator strength for the $\pm 1^U$ states decreases with increasing nanocrystal size [33]. When the nanocrystal size decreases from more than 3 nm to 1.5 nm in radius, radiative lifetime $1/\Gamma_r$ for the $\pm 1^L$ states increases from 10 ns to 100 ns [34]. The ratio Γ_r/Γ_{nonr} for the $\pm 1^L$ states is thus expected to decrease with decreasing nanocrystal size. Hence, smaller nanocrystals should feature a smaller ξ as well as a smaller relative contribution from the $\pm 1^L$ states to the overall PL, leading to negligible enhancement in total PL decay rates as observed for nanocrystals with $R = 2$ nm.

Figure 8.6 also shows that at low temperature ξ for nanocrystals with $R = 4.5$ nm exhibits only a weak temperature dependence while much stronger temperature dependence of ξ is observed for nanocrystals with $R = 2.7$ nm. This size variation in the temperature dependence of ξ is expected since larger nanocrystals are expected to feature a greater ratio of Γ_r/Γ_{nonr}. For very large nanocrystals, the contribution to the total decay rate from temperature dependent nonradiative decay processes is relatively small. This is in agreement with the temperature dependence of the time resolved PL obtained in free space. As shown in Fig. 8.5(b), within the first 20 ns, relatively weak temperature dependence was observed in the time resolved PL in free space for nanocrystals with $R = 4.5$ nm. In comparison, much stronger temperature dependence was observed in similar measurements for nanocrystals with $R = 2.7$ nm (not shown).

The observation of the pronounced cavity-induced enhancement in the total decay rate along with the weak temperature dependence in both ξ and Γ_t for nanocrystals with $R \approx 4.5$ nm indicates that in these large nanocrystals and for the $\pm1^L$ transition, Γ_r is considerably greater than Γ_{nonr}. For a more quantitative analysis, we have used two exponential components to fit the PL in the first 150 ns of the time resolved PL. For nanocrystals off-resonant with the relevant WGM, the decay time for the first decay component is $1/\Gamma_t = 8.7$ ns and 8.1 ns at 10 K and 20 K respectively (the decay time for the second and slower component is of order 30 ns at 10 K). Whereas for nanocrystals resonant with the relevant WGM, the decay time for the first decay component is $1/\Gamma_t = 7.3$ ns and 6.8 ns at 10 K and 20 K respectively. These results yield $\xi = 0.2$ and 0.19 at 10 K and 20 K, respectively, approaching the theoretically expected F_p and also indicating that the underlying decay process is primarily radiative in origin. For very large nanocrystals, radiative decay time, $1/\Gamma_r$, for the lowest dipole-allowed transition is theoretically expected to be of order 10 ns, in general agreement with the experimental result.

The above experimental studies on cavity QED of nanocrystals in the low-Q regime have shown enhanced spontaneous emission from CdSe/ZnS core/shell nanocrystals and have demonstrated the feasibility of using whispering gallery optical microcavities to manipulate and control spontaneous emission in these nanocrystals. These studies also provided valuable and much needed information on radiative dynamics in these remarkable QDs. For large nanocrystals ($R\sim4.5$ nm), population decay of the underlying optical transition is primarily radiative in origin in spite of the complex energy structure and relaxation processes, which is important for the use of these QDs for cavity QED in the strong coupling regime.

Coupling nanocrystals to a dielectric microsphere: high-Q regime

While the composite system based on polystyrene microspheres is suitable for the demonstration of enhanced spontaneous emission, the Q-factor of this system is relatively small. To achieve much greater Q-factors, we have carried out experimental studies using composite systems based on fused silica microspheres. The high-Q microcavity system has enabled us to reach the good cavity limit in cavity QED, in which absorption or emission occurring in a single nanocrystal can significantly affect the dynamics of the composite nanocrystal-microsphere system [36].

Plate 9 shows the image of a fused silica microsphere with CdSe/ZnS core/shell nanocrystals deposited on the surface of the microsphere through solution deposition, in which nanocrystals suspended in a solution of chloroform were used. The bright red PL from nanocrystals on the sphere surface can be seen clearly. A portion of the PL spectrum from a composite nanocrystal-microsphere system is shown in Fig. 8.8. The spectrum was obtained at room temperature and the nanocrystals were excited above the band gap with an excitation wavelength of 532 nm. A high-index optical prism was also used to couple WGMs out of the microsphere. The PL spectrum shows a FSR of 0.7 nm. For the two peaks within one FSR in Fig. 8.8, we assign the stronger and the weaker peaks to the $p = 1$ and $p = 2$ modes, respectively, since the $p = 1$ mode has the best mode matching with nanocrystals on the sphere surface. Each peak in Fig. 8.8 also contains several WGMs with different m. These modes, however, are not resolved in the figure due to the limited spectrometer resolution.

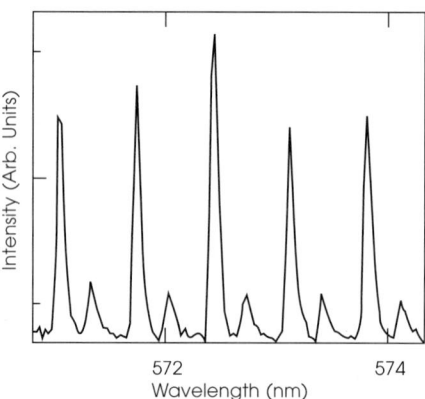

Fig. 8.8 Photoluminescence spectrum of a composite nanocrystal-microsphere system where the sphere radius is near 50 μm and the average core radius of the CdS/ZnS core/shell nanocrystals is 2 nm. (From [36] with permission.)

Resonant light scattering techniques similar to that discussed in section 8.2 and section 8.3 were used to characterize the Q-factor of the composite nanocrystal-fused silica microsphere system. Figures 8.9(a) and 8.9(b) show the resonant scattering spectra obtained far below the band gap and near the absorption line center of the nanocrystals, respectively. The mode spacing in Fig. 8.9(a) corresponds to separation between two adjacent azimuthal modes. Near the absorption line center, the spectral linewidth of the WGM is 4×10^{-4} nm, corresponding to a Q-factor of 1.6×10^6. In comparison, below the band gap, the linewidth decreases to 5×10^{-6} nm, corresponding to a Q-factor of 1.6×10^8, as shown in the inset in Fig. 8.9(b). This linewidth is limited by the scanning step size of the tunable diode laser used in the measurement. The difference in the Q-factors obtained far below and at the absorption line center is primarily due to absorption of nanocrystals coupling to the relevant WGM. Separate time-domain ring-down spectroscopy, which can provide a more accurate measure of Q-factors, has further shown that the Q-factor of the composite nanocrystal-fused silica microsphere system is limited to a few times of 10^8 by surface adsorption of chloroform used in the nanocrystal deposition process [36]. Note that CdSe nanocrystals can also be chemically linked to the surface of glass spheres via mercaptosilances [37,38]. However, compared with the solution deposition, chemical processing involved in the linking process results in much lower cavity Q-factors.

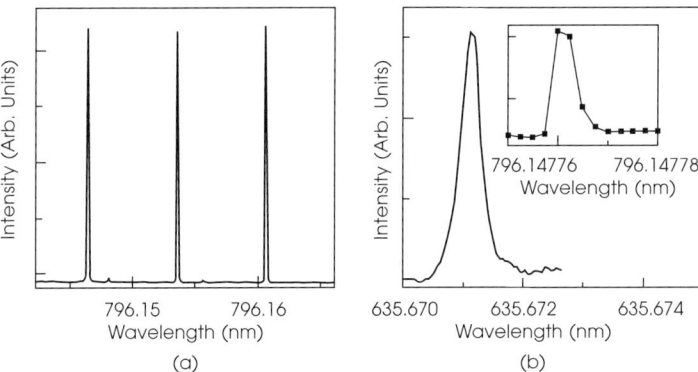

Fig. 8.9 Resonant scattering spectra from a composite nanocrystal-microsphere system where the sphere radius is near 50 μm and the average core radius of the CdS/ZnS core/shell nanocrystals is 4.5 nm. (a) Below the band gap of the nanocrystals. (b) Near the exciton absorption line center. The inset in (b) also shows a resonance in (a) obtained with greater spectral resolution. (From [36] with permission.)

While theoretical models that include relevant electron–electron interactions are needed in order to provide a satisfactory description of interactions between a QD and a resonant cavity mode, at the low excitation limit a simple coupled oscillator model widely used for composite atom-cavity systems can provide an adequate description.

The equation of motion for the coupled oscillator model in the low excitation limit is given by [2]:

$$\dot{\alpha} = -(i\omega_c + \kappa)\alpha + g\beta \qquad (8.10a)$$

$$\dot{\beta} = -(i\omega_0 + \gamma)\beta - g\alpha \qquad (8.10b)$$

where α is the expectation value of the annihilation field operator for the cavity mode at the position of the QD, β is the expectation value of the Pauli lowering operator σ_- for the QD transition, ω_c and ω_0 are the resonant frequencies of the cavity and the relevant dipole optical transition in the QD, respectively. Excitations of the coupled oscillator system can be characterized by the two normal modes of the system.

We discuss here two different limits in the weak coupling regime. In the low-Q limit, i.e. $\kappa \gg (g, \gamma)$, the coupling between a QD and a resonant cavity mode leads to enhanced decay of the QD-like normal mode. The decay rate of the QD-like mode is given by

$$\gamma' = \gamma + \frac{g^2}{\kappa} \qquad (8.11)$$

where $2g^2/\kappa$ is the enhancement in the spontaneous emission rate of the QD. This result is the same as that obtained by Purcell but is expressed in a different form. In comparison, in the limit that. $\gamma \gg (g, \kappa)$, the coupling between a QD and a resonant cavity mode modifies the decay rate of the cavity-like normal mode. The spectral linewidth of the cavity-like mode in the weak excitation limit is given by:

$$\kappa' = \kappa + \frac{g^2}{\gamma}. \qquad (8.12)$$

Spectral broadening of the WGM due to the QD absorption is thus given by g^2/γ. In atomic cavity QED, $g^2/\gamma\kappa$ is defined as the critical atom number since it determines the number of atoms needed in the cavity in order to significantly affect the dynamics of the composite system. For a QD system, we define $g^2/\gamma\kappa$ as the critical QD number.

For WGMs with $p = 1$ and $m = \pm l$, the effective mode volume at $\lambda = 600$ nm is of order 1000 μm^3 for a sphere with $r_0 = 50$ μm. The dipole coupling rate $g/2\pi$ is then of order 100 MHz for a nanocrystal on the sphere surface where we have used a dipole moment of 3×10^{-19} C·Å for the relevant optical transition. To account for the broadening of the WGM due to nanocrystal absorption shown in Fig. 8.9(b), we estimate the number of nanocrystals resonantly coupling to the WGM to be of order 2×10^4 where we have assumed a homogeneous linewidth of order 10 meV for CdSe/ZnS core/shell nanocrystals at room temperature.

At very low temperature, the extremely high-Q factor along with the small effective mode volume of lower order WGMs allows us to reach the good cavity limit where the critical QD number becomes smaller than 1. For a simple estimate, we take the homogeneous linewidth at low temperature to be 0.04 meV (corresponding to $\gamma/2\pi = 5$ GHz). The increase in the WGM spectral linewidth due to absorption from a single nanocrystal is then 2 MHz, compared with the bare cavity linewidth $\kappa/2\pi \approx 1$ MHz. In this case, the critical QD number is 0.5. Note that much greater dipole coupling rates and thus much smaller critical QD number can be achieved by using spheres with smaller radii since g is inversely proportional to $\sqrt{V_0}$. The extreme sensitivity of WGMs to effects of single nanocrystals should open up a new avenue for probing dynamics, decoherence, and individual quantum transitions in a single QD. In the limit that the bare cavity decay rate is small compared with the spontaneous emission rate into the relevant cavity mode, laser emission from a single QD also becomes possible.

For nanocrystals whose dephasing is entirely due to radiative decay, $\gamma/2\pi$ is of order 10 MHz where we have assumed a radiative lifetime of order 10 ns. In this limit, $g > (\gamma, \kappa)$ can be easily achieved with microspheres with a relatively large size. Since cavity QED studies in the low-Q regime in the preceding sections have already indicated that population relaxation in large CdSe/ZnS nanocrystals for the relevant optical transition is primarily due to radiative decay, whether the strong coupling regime can be achieved with the composite nanocrystal-microsphere system discussed here depends on the details of pure dephasing processes in these nanocrystals.

Dephasing in semiconductor nanocrystals

A resonant optical excitation creates an excited state population and also induces an optical polarization. Dynamics of this optical excitation is characterized by relaxation of the population as well as decay of the induced optical polarization, i.e. dephasing. Population relaxation processes are expected to contribute to dephasing with a dephasing rate given by $\Gamma/2$ where Γ is the population decay rate. Pure dephasing processes that do not involve population or energy relaxation can also contribute to dephasing and can be a dominant contribution to dephasing.

Pure dephasing can arise from coupling of excitonic states with a continuum of acoustic phonons [39]. To elucidate this, we show schematically in Fig. 8.10 eigen states of a coupled exciton–phonon system where we assumed that mixing of electronic states due to electron–phonon coupling can be ignored. In this limit, the exciton–phonon coupling leads to a shift in the equilibrium position of lattice vibrations and a temperature-independent polaron shift of the exciton energy [40]. The ground and excited states of the coupled exciton–phonon system can be described (in terms of eigen functions of the uncoupled system) by $|\phi_g> |\varphi_m(x)>$ and $|\phi_e> |\varphi_n(x-a)>$, where $\varphi_m(x)$ is the wave function for a phonon state with m phonons, x is the phonon coordinate with a being a relative shift in the equilibrium position induced by the exciton–phonon coupling, and ϕ_g and ϕ_e are the wave functions of the electronic ground and excited states, respectively. The dipole matrix element between the ground and excited states of the coupled system is then $<\phi_g| \mathbf{r} |\phi_e><\varphi_m(x)| \varphi_n(x-a)>$. The optical transition can now take place between states involving different phonon numbers since $<\varphi_m(x)| \varphi_n(x-a)> \neq 0$ even when $m \neq n$ and can be viewed as transitions between two quasi-continuous manifolds, as illustrated in Fig. 8.10. For discrete acoustic phonon modes, this will lead to phonon side bands in absorption and emission spectra similar to well-known longitudinal optical phonon side bands [40]. For a continuum of acoustic phonon modes, these side bands will overlap and merge together. The resulting spectral broadening of the optical transition depends on the relative strength of relevant transitions but does not involve population relaxation of the excitonic states.

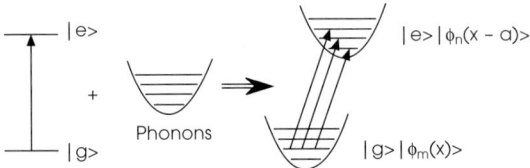

Fig. 8.10 A schematic showing that electron–phonon coupling can lead to acoustic phonon sidebands in absorption spectra.

Pure dephasing associated with the electron-phonon interactions cannot be suppressed by the discrete electronic energy levels in a QD. Recent experimental studies using stimulated photon echoes have shown that pure dephasing due to electron phonon interactions can be a dominant contribution to dephasing in MBE-grown QDs [41,42]. This pure dephasing process, however, can be suppressed if the phonon modes are completely discrete. In this regard, high quality semiconductor nanocrystals can in principle feature dephasing rates that are limited by the radiative lifetime and can be much smaller than that of the corresponding MBE-grown QDs.

While extensive studies of semiconductor nanocrystals have led to considerable understanding of the electronic and optical properties of these nanocrystals, dephasing in these nanocrystals still remains poorly understood. Two different experimental approaches have been used so far in order to determine the homogeneous linewidth of semiconductor nanocrystals: single-nanocrystal measurements on spatially isolated nanocrystals or nonlinear optical measurements such as photon echoes and spectral hole burning in a collection of inhomogeneously broadened nanocrystals. Resolution limited homogeneous linewidth of 120 μeV in CdSe nanocrystals has been observed by using single-nanocrystal PL spectroscopy [43]. Earlier studies of stimulated photon echoes have suggested a linewidth of order 10 meV for CdSe nanocrystals [44]. More recent accumulated photon echo studies in CdSe nanocrystals embedded in glass matrix have revealed size-dependent linewidth between 100 and 200 μeV, limited by surface relaxation processes [45].

To probe homogeneous linewidth of a collection of inhomogeneously broadened nanocrystals, we have used the technique of high resolution spectral hole burning [46]. The frequency domain hole burning measurement can be carried out at very low excitation level (~1 W/cm^2) and can also avoid the problem of spectral diffusion encountered in earlier single-nanocrystal PL studies by using a relatively fast modulation rate.

Two tunable diode lasers were used for the hole burning study. The pump-beam was modulated with an acousto-optic modulator (AOM) at frequencies up to 100 kHz. A lock-in amplifier was used to detect the change in the transmission of the sample induced by the pump beam. The spectral resolution of the measurement is limited by the spectral linewidth of one of the diode lasers (~100 MHz). Due to the limited tuning range of the diode lasers, large nanocrystals with $R = 4.5$ nm were used. The nanocrystals were dispersed in a thin polystyrene film, which was then deposited on a sapphire disk. The sample was mounted on the cold finger of a continuous flow cryostat. For measurements below 4 K, a liquid helium immersion cryostat was used.

Figure 8.11 shows the differential transmission spectrum of the sample at 10 K as

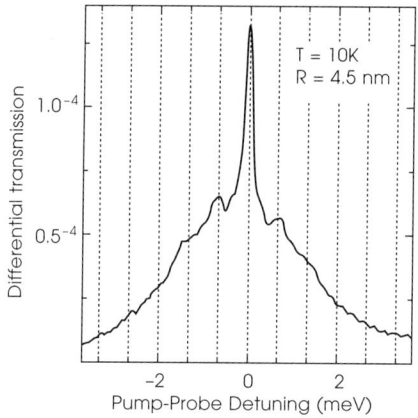

Fig. 8.11 Hole burning spectrum of CdSe/ZnS core/shell nanocrystals with $R = 4.5$ nm obtained at 10 K. Two acoustic phonon sidebands are located 0.67 meV and 1.5 meV away from the hole burning resonance to both the lower and higher energy side. (From [46] with permission.)

a function of the detuning, δ, between the pump and probe beams with the two beams collinearly polarized. A pronounced resonance appears at the zero detuning, indicating that the resonance is due to the spectral hole burning created by the pump beam. Similar results were also obtained at other pump energies with the spectral position of the resonance occurring at the zero detuning. In addition to the spectral hole, the differential transmission spectrum also shows two pronounced sidebands at an energy $E_a = 0.67$ meV away from the spectral hole and at both lower and higher energy side of the spectral hole burning resonance. Barely resolved but clearly visible are also two broader sidebands at roughly $E_b = 1.5$ meV away from the spectral hole.

There can be two contributions to the differential transmission: a coherent contribution and an incoherent contribution [47]. The coherent contribution arises from coherent wave mixing between the pump and probe beam. The amplitude of the coherent response scales with $1/(\delta + i\Gamma)$ where Γ is the population decay rate of the excitation. This coherent response is not observed in our measurements since the spectral linewidth of the pump laser used in our study far exceeds Γ. The incoherent contribution arises from the bleaching of the nanocrystal absorption induced by the pump beam, which gives rise to the hole burning resonance as well as the sidebands in the different transmission spectrum.

The hole burning response shown in Fig. 8.11 is due to the lowest dipole-allowed excitonic transition in the nanocrystals. In steady state the amplitude of the nonlinear response scales with inverse of both the population decay rate and dephasing rate of the optical excitation. Since higher energy optical transitions are expected to feature greater dephasing and population decay rates, contributions to the nonlinear response from these optical transitions should be negligible.

The sidebands in Fig. 8.11, whose energy positions are symmetric with the spectral hole burning resonance, arise from acoustic phonon sidebands involving absorption or emission of discrete acoustic phonons in nanocrystals. The energy positions of these phonon sidebands are in good agreement with the energies of spherical and torsional acoustic phonon modes determined in earlier studies based on Raman scattering [48,49]. The acoustic phonon sidebands, especially the sideband with 1.5 meV away from the hole burning resonance shown in Fig. 8.11, are much broader than the spectral hole burning resonance. The phonon sidebands in this case are more likely the results of multiple phonon modes instead of single phonon modes. For nanocrystals with a radius of 4.5 nm, energy spacing between different acoustic phonon modes can be as small as 0.1 meV.

Figure 8.12 shows the temperature dependence of the homogeneous linewidth derived from the hole burning response. For an inhomogeneously broadened system and in the limit of third order nonlinear response, the width of the spectral hole burning is twice the homogeneous linewidth of the corresponding optical transition. The data shown in Fig. 8.12 were obtained with a 100 kHz modulation frequency for the pump beam to eliminate effects of spectral diffusion of optical excitations at relatively slow time scales. Very low incident pump and probe intensities ($I_{pump} < 1.5$ W/cm^2 and $I_{probe} < 0.15$ W/cm^2) were also used in order to avoid power broadening of the spectral hole burning response.

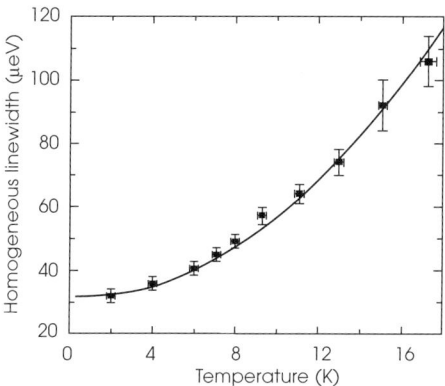

Fig. 8.12 Temperature dependence of the homogeneous linewidth of CdSe/ZnS core/shell nanocrystals with $R = 4.5$ nm. The linewidth was obtained from a Lorentzian fit to the hole burning resonance. (From [46] with permission.)

Figure 8.12 reveals a nonlinear temperature dependence of the homogeneous linewidth, in contrast to the usual linear dependence for excitons in a QW or a bulk system. This nonlinear temperature dependence reflects effects of the quantization of acoustic phonon modes. For a discrete acoustic phonon mode, the lowest order electron–phonon interaction that can result in dephasing is a two-phonon process that involves both absorption and emission of a phonon. We have attempted to fit the temperature dependence by including the contributions of two acoustic phonon modes at $E_a = 0.67$ meV and $E_b = 1.5$ meV. The numerical fit shown in Fig. 8.12 is described by [50]

$$\gamma = \gamma_0 + A \sinh^{-2}(E_a/2k_BT) + B \sinh^{-2}(E_b/2k_BT) \qquad (8.13)$$

where $\gamma_0 = 16$ μeV, $A = 1.5$ μeV and $B = 2.35$ μeV. The phonon energy used in the above numerical fit should be viewed as the average energy of a group of acoustic phonon modes. Note that homogeneous linewidth obtained above is considerably smaller than that reported in previous single-nanocrystal PL studies. The PL studies are resolution-limited and are complicated by spectral diffusion processes.

The homogeneous linewidth obtained from the hole burning study corresponds to the linewidth of the zero acoustic phonon line and demonstrates the suppression of the pure dephasing process associated with electron–phonon interactions. In the absence of the phonon quantization, pure dephasing associated with a continuum of acoustic phonon sidebands can be a dominant contribution to the intrinsic dephasing rate. For CdSe quantum dots, theoretical studies indicate that these phonon sidebands lead to an absorption linewidth on the order of a few meV at low temperature [51,52]. As shown in Fig. 8.11, the spectral linewidth of the overall phonon sidebands in the differential transmission response is in good agreement with the above theoretical expectation. Contributions from these phonon sidebands also lead to sub-picosecond dephasing time observed in time domain measurements, as shown in previous photon echo studies. The time domain studies, however, were not able to clearly resolve the discrete acoustic phonon sidebands.

The homogeneous linewidth of the zero acoustic phonon line, however, is still much greater than that expected from the population relaxation time. Time-resolved PL studies shown in Fig. 8.6 have shown that the population decay time for the relevant optical transition is of order 10 ns, corresponding to a homogeneous linewidth of order 0.1 μeV (or 20 MHz). This large difference between γ and $\Gamma/2$ reveals significant contributions from pure dephasing processes to the homogeneous linewidth even when the acoustic phonons are completely discrete. The underlying physical mechanism for the excessive pure dephasing still remains unclear at this point. The excessive pure dephasing has thus far prevented us from achieving the strong coupling regime in cavity QED with CdSe/ZnS nanocrystals.

8.5 Summary

In this chapter we have discussed cavity QED studies using a composite semiconductor-microsphere system. We found that the use of semiconductor nanocrystals in the composite system can overcome many of the difficulties encountered in composite systems that use MBE-grown nanostructures. Enhanced spontaneous emission has been demonstrated in CdSe/ZnS core/shell nanocrystals embedded in a polystyrene sphere. The manifestation of the enhanced spontaneous emission process in the low-Q regime has also provided important information on radiative dynamics in these nanocrystals. Coupling nanocrystals to a high-Q fused silica microsphere system, we have reached the good cavity limit where the critical QD number is smaller than 1. In this limit, dynamics of a QD-microcavity system is sensitive to absorption or emission occurring in a single QD.

Using high-resolution spectral hole burning, we have shown that dephasing associated with electron–phonon interactions can be suppressed by the complete quantization of acoustic phonon modes in colloidal QDs. Residual excessive pure dephasing processes in CdSe/ZnS nanocrystals, however, have thus far prevented us from achieving the strong coupling regime of cavity QED. Since the composite nanocrystal-microsphere system can feature cavity finesse approaching 10^6, the strong coupling regime should be readily achievable with a composite nanocrystal-microsphere system when nanocrystals that are nearly lifetime broadened become available.

Finally, we note that the strong coupling regime in cavity QED of QDs, while extremely difficult to achieve, will open up a new frontier for manipulating and controlling optical interactions in a mesoscopic quantum system. As an example, quantum entanglement of two colloidal QDs can be created through the exchange of a photon in a WGM between the two QDs. Successive photon exchange processes can also lead to entanglement of an array of QDs, providing a potential model systems for quantum information processing [9].

Acknowledgments

The author gratefully acknowledges contributions of Xudong Fan, Scott Lacey, Phedon Palinginis, Sash Tavenner, and Mark Lonergan. This work was partially supported by National Science Foundation under grant No. DMR-9733230 and No. ECS-9988542 and by Army Research Office under grant No. DAAD 19-00-01-0393.

Some of the figures in this chapter appeared in the following papers. The author is grateful to the copyright holders for granting permission to reproduce these figures:

Fan, X., Lonergan, Y., and Wang, H. (2001). Enhanced spontaneous emission from semiconductor nanocrystals embedded in whispering gallery optical microcavities. *Phys. Rev. B* **64**, 1153101–1153105. Copyright (2001) by the American Physical Society.

Reprinted with permission from Fan, X., Doran, A., and Wang, H. (1998). High-Q whispering gallery modes from a composite system of GaAs quantum well and fused silica microsphere. *Appl. Phys. Lett.* **73**, 3190–3192. Copyright 1998, American Institute of Physics.

Reprinted with permission from Palinginis, P. and Wang, H. (2001). High resolution spectral hole burning in Cd/ZnS core/shell nanocrystals. *Appl. Phys. Lett.* **78**, 1541–1543 (2001). Copyright 2001, American Institute of Physics.

Fan, X., Lacey, S., and Wang, H. (1999). Microcavities combining a semiconductor heterostructure with a fused silica microsphere. *Opt. Lett.* **24**, 771–773. Copyright 1999, Optical Society of America.

Fan, X., Lacey, S., Palinginis, P., Wang, H., and Lonergan, M. (2000). Coupling semiconductor nanocrystals to a fused silica microsphere: a quantum dot microcavity with extremely high Q-factors. *Opt. Lett.* **25**, 1600–1602. Copyright 2000, Optical Society of America.

References

1. Purcell, E.M. (1946). Spontaneous emission probabilities at radio frequencies. *Phys. Rev.* **69**, 681.
2. For a review, see Berman, P.R., ed. (1994). Cavity Quantum Electrodynamics, Academic Press, Boston.
3. For recent development, see, for example, Hood, C.J., Lynn, T.W., Doherty, A.C., Papkins, A.S., and Kimble, H.J. (2000). The atom-cavity microscope: single atoms bound in orbit by single photons. *Science* **287**, 1447–1453.
4. Pellizzari, T., Gardiner, S.A., Cirac, J.I., and Zoller, P. (1995). Decoherence, continuous observation, and quantum computing: a cavity QED model. *Phys. Rev. Lett.* **75**, 3788–3791.
5. Turchette, Q.A., Hood, C.J., Lange, W., Mabuchi, H., and Kimble, H.J. (1995). Measurement of conditional phase shifts for quantum logic. *Phys. Rev. Lett.* **75**, 4710–4713.
6. Hagley, E., Maitre, X., Nogues, G., Wunderlich, C., Brune, M., Raimond, J.M., and Haroche, S. (1997). Generation of Einstein–Podolsky–Rosen pairs of atoms. *Phys. Rev. Lett.* **79**, 1–5.
7. Rauschenbeutel, A., Nogues, G., Osnaghi, S., Bertet, P., Brune, M., Raimond, J.M., and Haroche, S. (1999). Coherent operation of a tunable quantum phase gate in cavity QED. *Phys. Rev. Lett.* **83**, 5166–5169.
8. Imamoglu, A., Awschalom, D.D., Burkard, G., DiVincenzo, D.P., Loss, D., Sherwin, M., and Small, A. (1999). Quantum information processing using quantum dot spins and cavity QED. *Phys. Rev. Lett.* **83**, 4204–4207.
9. Brun, T.A., and Wang, H. (2000). Coupling nanocrystals to a high-Q silica microsphere: Entanglement in quantum dots via photon exchange. *Phys. Rev. A* **61**, 032307/1–5.
10. See for example, Scully, M.O., and Zubairy, M.S. (1997). Quantum Optics, Cambridge University Press.
11. Gerard, J.M., Sermage, B., Gayral, B., Legrand, B., Costard, E., and Thierry-Meig, V. (1998). Enhanced spontaneous emission by quantum boxes in a monolithic optical microcavity. *Phys. Rev. Lett.* **81**, 1110–1113.
12. Graham, L.A., Huffaker, D.L., and Deppe, D.G. (1999). Spontaneous lifetime control in a native-oxide-apertured microcavity. *Appl. Phys. Lett.* **74**, 2408–2410.
13. Solomon, G.S., Pelton, M., and Yamamoto, Y. (2001). Single-mode spontaneous emission from a single quantum dot in a three-dimensional microcavity. *Phys. Rev. Lett.* **86**, 3903–3906.
14. Kiraz, A., Michler, P., Becher, C., Gayral, B., Imamoglu, A., Lidong Zhang, Hu, E., Schoenfeld, W.V., and Petroff, P.M. (2001). Cavity-quantum electrodynamics using a single InAs quantum dot in a microdisk structure. *Appl. Phys. Lett.* **78**, 3932–3934.
15. Yoshie, T., Scherer, A., Chen, H., Huffaker, D., and Deppe, D. (2001). Optical characterization of two-dimensional photonic crystal cavities with indium arsenide quantum dot emitters. *Appl. Phys. Lett.* **79**, 114–116.

16. Reese, C., Becher, C., Imamoglu, A., Hu, E., Gerardot, B.D., and Petroff, P.M. (2001). Photonic crystal microcavities with self-assembled InAs quantum dots as active emitters. *Appl. Phys. Lett.* **78**, 2279–2281.
17. Braginsky, V.B., Gorodetsky, M.L., and Ilchenko, V.S. (1989). Quality-factor and nonlinear properties of optical whispering-gallery modes. *Phys. Lett. A* **137**, 393–397.
18. For a review, see for example, Chang, R.K., and Campillo, A.J., ed. (1996). Optical Processes in Microcavities, World Scientific, Singapore.
19. For a textbook discussion on WGM's, see for example, Nussenzveig, H.M. (1992). Diffraction Effects in Semiclassical Scattering, Cambridge University Press.
20. Gorodetsky, M.L., Savchenkov, A.A., and Ilchenko, V.S. (1996). Ultimate Q of optical microsphere resonators. *Opt. Lett.* **21**, 453–455.
21. Vernooy, D.W., Ilchenko, V.S., Mabuchi, H., Streed, E.W., and Kimble, H.J. (1998). High-Q measurements of fused-silica microspheres in the near infrared. *Opt. Lett.* **23**, 247–249.
22. Collot, L., Lefevre-Seguin, V., Brune, M., Raimond, J.M., and Haroche, S. (1993). Very high-Q whispering-gallery mode resonances observed on fused silica microspheres. *Europhy. Lett.* **23**, 327–334.
23. Vernooy, D.W., Furusawa, A., Georgiades, N.P., Ilchenko, V.S., and Kimble, H.J. (1998). Cavity QED with high-Q whispering gallery modes. *Phys. Rev. A* **57**, R2293–2296.
24. Norris, D.J., Kuwata-Gonokami, M., and Moerner, W.E. (1997). Excitation of a single molecule on the surface of a spherical microcavity. *Appl. Phys. Lett.* **71**, 297–299.
25. Pelton, M., and Yamamoto, Y. (1999). Ultralow threshold laser using a single quantum dot and a microsphere cavity. *Phys. Rev. A* **59**, 2418–2421.
26. Fan, X., Doran, A., and Wang, H. (1998). High-Q whispering gallery modes from a composite system of GaAs quantum well and fused silica microsphere. *Appl. Phys. Lett.* **73**, 3190–3192.
27. Fan, X., Lacey. S., and Wang, H. (1999). Microcavities combining a semiconductor with a fused-silica microsphere. *Opt. Lett.* **24**, 771–773.
28. Hess, H.F., Betzig, E., Harris, T.D., Pfeiffer, L.N., and West, K.W. (1994). Near-field spectroscopy of the quantum constituents of a luminescent system. *Science* **264**, 1740–1745.
29. Hines, M.A., and Guyot-Sionnest, P. (1996). Synthesis and characterization of strongly luminescing ZnS-capped CdSe nanocrystals. *J. Phys. Chem.* **100**, 468–471.
30. Dabbousi, B.O., Rodriguez-Viejo, J., Mikulec, F.V., Heine, J.R., Mattoussi, H., Ober, R., Jensen, K.F., and Bawendi, M.G. (1997). (CdSe)ZnS core-shell quantum dots: synthesis and characterization of a size series of highly luminescent nanocrystallites. *J. Phys. Chem. B* **101**, 9463–9475.
31. Peng, X., Schlamp, M.C., Kadavanich, A.V., and Alivisatos, A.P. (1997). Epitaxial growth of highly luminescent CdSe/CdS core/shell nanocrystals with photostability and electronic accessibility. *J. Am. Chem. Soc.* **119**, 7019–7029.
32. Fan, X., Lonergan, M.C., Zhang, Y., and Wang, H. (2001). Enhanced spontaneous emission from semiconductor nanocrystals embedded in whispering gallery optical microcavities. *Phys. Rev. B*, to appear in Sept. 15 issue.
33. Efros, A.L., Rosen, M., Kuno, M., Nirmal, M., Norris, D.J., and Bawendi, M. (1996). Bandedge exciton in quantum dots of semiconductors with a degenerate valence band: dark and bright exciton states. *Phys. Rev. B* **54**, 4843–4856.

34. Leung, K., Pokrant, S., and Whaley, K.B. (1998). Exciton fine structure in CdSe nanoclusters. *Phys. Rev. B* **57**, 12291–12301.
35. Gupta, J.A., Awschalom, D.D., Peng, X., and Alivisatos, A.P. (1999). Spin coherence in semiconductor quantum dots. *Phy. Rev. B* **59**, R10421–10424.
36. Fan, X., Palinginis, P., Lacey, S., Wang, H., and Lonergan, M.C. (2000). Coupling semiconductor nanocrystals to a fused-silica microsphere: a quantum-dot microcavity with extremely high Q factors. *Opt. Lett.* **25**, 1600–1602.
37. Artemyev, M.V., Woggon, U., Wannemacher, R., Jaschinski, H., and Langbein, W. (2001). Light trapped in a photonic dot: microspheres as a cavity for quantum dot emission. *Nano Letters* **1**, 309.
38. Tevanner, S., Palinginis, P., and Wang, H. (2001). Unpublished.
39. Duke, C.B., and Mahan, G.D. (1965). Phonon-broadened impurity spectra I – Density of states. *Phys. Rev.* **139**, A 1965–1982.
40. Huang, K., and Rhys, A. (1950). Theory of light absorption and nonradiative transition in F-centers. *Proc. Roy. Soc. (London) A* **204**, 406–423.
41. Fan, X., Takagahara, T., Cunningham, J.E., and Wang, H. (1998). Pure dephasing induced by exciton–phonon interactions in narrow GaAs quantum wells. *Solid State Comm.* **108**, 857–861.
42. Takagahara, T. (1999). Theory of exciton dephasing in semiconductor quantum dots. *Phys. Rev. B* **60**, 2638–2652.
43. Empedocles, S.A., Norris, D.J., and Bawendi, M.G. (1996). Photoluminescence spectroscopy of single CdSe nanocrystallite quantum dots. *Phys. Rev. Lett.* **77**, 3879–3882.
44. Schoenlein, R.W., Mittleman, D.M., Shiang, J.J., Alivisatos, A.P., and Shank, C.V. (1993). Investigation of femtosecond electronic dephasing in CdSe nanocrystals using quantum-beat-suppressed photon echoes. *Phys. Rev. Lett.* **70**, 1014–1017.
45. Takemoto, K., Hyun, B.-R., and Masumoto, Y. (2000). Heterodyne-detected accumulated photon echo in CdSe quantum dots. *Solid State Comm.* **114**, 521–525.
46. Palinginis, P., and Wang, H. (2001). High-resolution spectral hole burning in CdSe/ZnS core/shell nanocrystals. *Appl. Phys. Lett.* **78**, 1541–1543.
47. See for example, Meystre, P., and Sargent III, M. (1991). Elements of Quantum Optics, 2nd ed., Springer, Berlin.
48. Saviot, L., Champagnon, B., Duval, E., Kudriavtsev, I.A., and Ekimov, A.I. (1996). Size dependence of acoustic and optical vibrational modes of CdSe nanocrystals in glasses. *J. Non-Cryst. Solids* **197**, 238–246.
49. Woggon, U., Gindele, F., Wind, O., and Klingshirn, C. (1996). Exchange interaction and phonon confinement in CdSe quantum dots. *Phys. Rev. B* **54**, 1506–1509.
50. Ikezawa, M., and Masumoto, Y. (2000). Ultranarrow homogeneous broadening of confined excitons in quantum dots: Effect of the surrounding matrix. *Phys. Rev. B* **61**, 12662–12665.
51. Takagahara, T. (1993). Electron-phonon interactions and excitonic dephasing in semiconductor nanocrystals. *Phys. Rev. Lett.* **71**, 3577–3580.
52. Li, X., and Arakawa, Y. (1999). Optical linewidths in an individual quantum dot. *Phys. Rev. B* **60**, 1915–1920.

Quantum Coherence, Correlation and Decoherence in
Semiconductor Nanostructures
T. Takagahara (Ed.)
Copyright © 2003 Elsevier Science (USA). All rights reserved.

Chapter 9
Theory of exciton coherence and decoherence in semiconductor quantum dots

T. Takagahara

Department of Electronics and Information Science, Kyoto Institute of Technology, Matsugasaki, Sakyo-ku, Kyoto 606-8585, Japan

Abstract

Semiconductor quantum dots are considered as a promising candidate to implement the quantum state control and the quantum information processing. The discrete energy level structures due to the three-dimensional confinement is favorable to realize an ideal two-level system, namely, the ground state and an excited state, whose superposition states can be manipulated by optical means. Also the exciton states have a huge transition dipole moment (~ 100 Debye) for typical size of quantum dots. The most prominent manifestation of the quantum coherence is the Rabi oscillation which represents the coherent evolution of the excitonic dipole moment. This is a key operation in the quantum state manipulation and has been achieved successfully by several groups. Here the excitonic Rabi splitting and Rabi oscillation in a single quantum dot are discussed theoretically and their new features due to the quantum interference are clarified. The mechanisms of the population relaxation and dephasing (decoherence) of excitons are discussed. A quantitative theory of the exciton dephasing due to the electron-phonon interaction is developed based on the Green function formalism and is extended for application to the case of dephasing of the general non-radiative coherence.

9.1 Introduction

Coherent manipulation of quantum states is a critical step toward many novel technological applications ranging from manipulation of qubits in quantum logic gates [1–5] to controlling the reaction pathways of molecules [6]. Both Rabi oscillation and quantum interference play central roles in coherent control. In strong excitation regime, Rabi oscillation provides a direct control to the excited state population of a quantum system through the input area (i.e., time integrated Rabi frequency) of a single excitation pulse [7]. On the other hand, in weak excitation regime, the quantum interference of probability waves excited with phase-tailored pulse pairs is utilized for wavefunction manipulation in atoms [8,9], molecules [6], and semiconductor heterostructures [10,11]. Atomic like states of semiconductor quantum dots have been envisioned to be the building blocks of solid-state quantum computer [1]. This technological motivation intensifies research efforts in coherent phenomena of semiconductor quantum dots. Numerous fundamental coherent properties common to ideal quantum systems such as atomic-like spectra [12–16], entanglement [17,18], photon anti-bunching [19,20], modification of spontaneous emission rate by a micro-cavity [21] and coherent wavefunction manipulation by using quantum interference [22,16] have been recently demonstrated in various type of semiconductor quantum dots. However, there have been very few investigations on Rabi oscillation in semiconductor quantum dots. This fundamental coherent phenomenon stands as a principal criterion in establishing analogy among the fundamentally different quantum states of atomic/molecular systems [23–29] and semiconductors [30–32]. Furthermore, a recent theoretical investigation [33] predicted a dramatic change of quantum interference patterns when the excitation strengths of the phase-tailored pulse pairs become strong enough to induce nonlinear effects of Rabi oscillation in excited state population. Very recently, the exciton Rabi splitting has been observed in the luminescence spectrum of a single InGaAs quantum dot [34] and the exciton Rabi oscillation has also been observed in the pump-probe spectroscopy of a single GaAs quantum dot [35]. Also the predicted new quantum interference phenomena have been observed successfully in a single InGaAs quantum dot [36].

In this Chapter, we will discuss the theoretical aspects of the Rabi splitting and Rabi oscillation of excitons in semiconductor quantum dots, clarify their characteristic features and predict new aspects in conjunction with recent experiments. We will also discuss the mechanisms of the exciton decoherence in semiconductor quantum dots and clarify the ultimate limit of the exciton decoherence time.

9.2 Exciton Rabi splitting in a single quantum dot

Now we consider a two-level system under a strong resonant excitation and study its emission spectrum. This was studied by Mollow [37] for the first time and a triplet structure was discovered in the emission spectrum which is now interpreted in terms of the dressed state (see section 9.3). The relevant Hamiltonian is given by

$$H = |2\rangle E_2 \langle 2| + |1\rangle E_1 \langle 1| - \mu \mathscr{E} \cos \overline{\omega} t = H_0 - \mu \mathscr{E} \cos \overline{\omega} t, \tag{9.1}$$

where $|2\rangle(|1\rangle)$ denotes the upper (lower) level, $E_2(E_1)$ the energy of the upper (lower) level, μ is the dipole moment operator between $|1\rangle$ and $|2\rangle$, $\mathscr{E}(\overline{\omega})$ the electric field amplitude (frequency) of the excitation light and H_0 is defined by the first two terms. The equation of motion of the density matrix is given by

$$\frac{d}{dt}\rho = -\frac{i}{\hbar}[H, \rho] = \Gamma\rho, \tag{9.2}$$

where $\Gamma\rho$ represents the relaxation terms symbolically. More explicitly, they are written as

$$\dot{\rho}_{11} = i\Omega(e^{i\overline{\omega}t}\rho_{21} - e^{-i\overline{\omega}t}\rho_{12}) + \kappa\rho_{22}, \tag{9.3}$$

$$\dot{\rho}_{22} = -i\Omega(e^{i\overline{\omega}t}\rho_{21} - e^{-i\overline{\omega}t}\rho_{12}) - \kappa\rho_{22}, \tag{9.4}$$

$$\dot{\rho}_{12} = i\Omega e^{i\overline{\omega}t}(\rho_{22} - \rho_{11}) + (i\omega_0 - \kappa/2)\rho_{12}, \tag{9.5}$$

$$\dot{\rho}_{21} = -i\Omega e^{-i\overline{\omega}t}(\rho_{22} - \rho_{11}) + (-i\omega_0 - \kappa/2)\rho_{21}, \tag{9.6}$$

where the rotating wave approximation is used, κ denotes the radiative decay rate of the upper level and $\kappa/2$ is the decay rate of the coherence between the upper and lower levels, $\hbar\omega_0 = E_2 - E_1$, and $\Omega = \langle 1|\mu|2\rangle \mathscr{E}/(2\hbar)$ is assumed as a real number. Separating out the oscillating part as

$$\rho_{12}(t) = e^{i\overline{\omega}t}\tilde{\rho}_{12}, \quad \rho_{21}(t) = e^{-i\overline{\omega}t}\tilde{\rho}_{21}, \tag{9.7}$$

we have

$$\dot{\rho}_{11} = i\Omega(\tilde{\rho}_{21} - \tilde{\rho}_{12}) + \kappa\rho_{22}, \tag{9.8}$$

$$\dot{\rho}_{22} = -i\Omega(\tilde{\rho}_{21} - \tilde{\rho}_{12}) - \kappa\rho_{22}, \tag{9.9}$$

$$\dot{\tilde{\rho}}_{12} = i\Omega(\rho_{22} - \rho_{11}) + (-i\Delta\overline{\omega} - \kappa/2)\tilde{\rho}_{12}, \tag{9.10}$$

$$\dot{\tilde{\rho}}_{21} = -i\Omega(\rho_{22} - \rho_{11}) + (i\Delta\overline{\omega} - \kappa/2)\tilde{\rho}_{21}, \tag{9.11}$$

with $\Delta\overline{\omega} = \overline{\omega} - \omega_0$. The emission spectrum under the cw excitation is calculated as follows: the stationary state of the system is determined and the emission spectrum from the stationary state is calculated. By noting the conservation of the probability, namely, $\rho_{11} + \rho_{22} = 1$, we choose $\rho_{22}, \tilde{\rho}_{12}, \tilde{\rho}_{21}$ as independent variables and we have

$$\dot{\rho}_{22} = -i\Omega(\tilde{\rho}_{21} - \tilde{\rho}_{12}) - \kappa\rho_{22}, \tag{9.12}$$

$$\dot{\tilde{\rho}}_{12} = -i\Omega + i2\Omega\rho_{22} + (-i\Delta\overline{\omega} - \kappa/2)\tilde{\rho}_{12}, \tag{9.13}$$

$$\dot{\tilde{\rho}}_{21} = i\Omega - i2\Omega\rho_{22} + (i\Delta\overline{\omega} - \kappa/2)\tilde{\rho}_{21}. \tag{9.14}$$

The stationary state is obtained by putting $\dot{\rho}_{22} = \dot{\tilde{\rho}}_{12} = \dot{\tilde{\rho}}_{21} = 0$ and will be denoted by

ρ^{st}. Now we derive the expression of the emission spectrum from the stationary state. First of all, the emission spectrum from the $|2\rangle$ level is calculated as

$$I(\omega) = |\langle 1|\mu|2\rangle|^2 \delta(\omega - \omega_0) \propto |\langle 1|\mu|2\rangle|^2 \frac{1}{(\omega - \omega_0)^2 + \gamma^2}$$

$$\propto \text{Re}\left[\int_0^\infty dt\, \langle 1|\mu e^{-\frac{i}{\hbar}H_0 t}|2\rangle\langle 2|\mu e^{\frac{i}{\hbar}H_0 t}|1\rangle e^{(i\omega-\gamma)t}\right]$$

$$= \text{Re}\left[\int_0^\infty dt\, \langle 1|\mu e^{-\frac{i}{\hbar}H_0^\times t}(|2\rangle\langle 2|\mu)|1\rangle e^{(i\omega-\gamma)t}\right], \quad (9.15)$$

where $H_0^\times$ represents the motion as the density matrix and is defined by $H_0^\times \rho = [H_0, \rho]$ and γ is is the population decay rate of the upper level $|2\rangle$. Equation (9.15) gives the emission spectrum from the pure state $|2\rangle\langle 2|$. Extending the above formalism, the emission spectrum from the stationary state ρ^{st} is calculated by replacing $|2\rangle\langle 2|$ by ρ^{st} as

$$\text{Re}\left[\int_0^\infty dt\, \langle 1|\mu e^{-\frac{i}{\hbar}H_0^\times t}(\rho^{st}\mu)|1\rangle e^{(i\omega-\gamma)t}\right]$$

$$= \text{Re}\left[\langle 1|\mu \frac{1}{\gamma - i\omega + \frac{i}{\hbar}H_0^\times}(\rho^{st}\mu)|1\rangle\right]$$

$$= |\langle 1|\mu|2\rangle|^2 \text{Re}\left[\langle 2|\frac{1}{\gamma - i\omega + \frac{i}{\hbar}H_0^\times}(\rho^{st}|2\rangle\langle 1|)|1\rangle\right] \quad (9.16)$$

with

$$\rho^{st}|2\rangle\langle 1| = \rho_{12}^{st}|1\rangle\langle 1| + \rho_{22}^{st}|2\rangle\langle 1|. \quad (9.17)$$

This emission spectrum shows a triplet structure as demonstrated by Mollow [37] for the first time.

The above formulation is given for a two-level system. In recent experiments on a single semiconductor quantum dot [34], the strong excitation is resonant with an excited exciton state and the emission from the exciton ground state is monitored. Here we have to consider at least three levels, namely, the ground state denoted by $|1\rangle$, the excited exciton state $|2\rangle$ and the exciton ground state $|3\rangle$. The level scheme and the relaxation pathways of population are shown in Fig. 9.1. The corresponding equations of motion for the density matrix are given by

$$\dot{\rho}_{11} = i\Omega(\tilde{\rho}_{21} - \tilde{\rho}_{12}) + \kappa_{21}\rho_{22} + \kappa_{31}\rho_{33}, \quad (9.18)$$

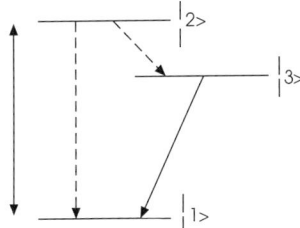

Fig. 9.1 An energy level scheme. The non-radiative relaxation pathways of population are depicted by dashed lines. Solid lines represent optical transitions.

$$\dot{\rho}_{22} = -(\kappa_{21} + \kappa_{23})\rho_{22} + i\Omega(\tilde{\rho}_{12} - \tilde{\rho}_{21}), \quad (9.19)$$

$$\dot{\rho}_{33} = -\kappa_{31}\rho_{33} + \kappa_{23}\rho_{22}, \quad (9.20)$$

$$\dot{\tilde{\rho}}_{12} = i\Omega(\rho_{22} - \rho_{11}) + (-i\Delta\overline{\omega} - \gamma_{12})\tilde{\rho}_{12}, \quad (9.21)$$

$$\dot{\tilde{\rho}}_{21} = -i\Omega(\rho_{22} - \rho_{11}) + (i\Delta\overline{\omega} - \gamma_{12})\tilde{\rho}_{21}, \quad (9.22)$$

$$\dot{\tilde{\rho}}_{13} = (-i(\overline{\omega} - \omega_3) - \gamma_{13})\tilde{\rho}_{13} + i\Omega\rho_{23}, \quad (9.23)$$

$$\dot{\tilde{\rho}}_{31} = (i(\overline{\omega} - \omega_3) - \gamma_{13})\tilde{\rho}_{31} - i\Omega\rho_{32}, \quad (9.24)$$

$$\dot{\rho}_{23} = (i(\omega_3 - \omega_0) - \gamma_{23})\rho_{23} + i\Omega\tilde{\rho}_{13}, \quad (9.25)$$

$$\dot{\rho}_{32} = (-i(\omega_3 - \omega_0) - \gamma_{23})\rho_{32} - i\Omega\tilde{\rho}_{31}, \quad (9.26)$$

where the energies of $|1\rangle$, $|2\rangle$, and $|3\rangle$ levels are taken as 0, $\hbar\omega_0$ and $\hbar\omega_3$, respectively, γ_{ij} is the dephasing rate of the off-diagonal coherence ρ_{ij}, Ω has the same meaning as before, κ_{ij} is the population decay rate from the level $|i\rangle$ to the level $|j\rangle$ and the oscillating part is separated out as

$$\rho_{12}(t) = e^{i\overline{\omega}t}\tilde{\rho}_{12}, \quad \rho_{21}(t) = e^{-i\overline{\omega}t}\tilde{\rho}_{21},$$

$$\rho_{13}(t) = e^{i\overline{\omega}t}\tilde{\rho}_{13}, \quad \rho_{31}(t) = e^{-i\overline{\omega}t}\tilde{\rho}_{31}. \quad (9.27)$$

It can be seen that the components $\tilde{\rho}_{13}$, $\tilde{\rho}_{31}$, ρ_{23} and ρ_{32} are completely decoupled from ρ_{11}, ρ_{22}, ρ_{33}, $\tilde{\rho}_{12}$ and $\tilde{\rho}_{21}$ because there is no optical field connecting the levels $|1\rangle$ and $|2\rangle$ with the level $|3\rangle$. Thus to obtain the stationary density matrix, we need to take into account only ρ_{11}, ρ_{22}, ρ_{33}, $\tilde{\rho}_{12}$ and $\tilde{\rho}_{21}$. Because of the conservation of probability, namely, $\rho_{11} + \rho_{22} + \rho_{33} = 1$, the stationary density matrix ρ^{st} can be obtained from

$$\begin{pmatrix} \kappa_{21}+\kappa_{23} & 0 & -i\Omega & i\Omega \\ -\kappa_{23} & \kappa_{31} & 0 & 0 \\ -2i\Omega & -i\Omega & i\Delta\overline{\omega}+\gamma_{12} & 0 \\ 2i\Omega & i\Omega & 0 & -i\Delta\overline{\omega}+\gamma_{12} \end{pmatrix} \begin{pmatrix} \rho_{22}^{st} \\ \rho_{33}^{st} \\ \tilde{\rho}_{12}^{st} \\ \tilde{\rho}_{21}^{st} \end{pmatrix} = \begin{pmatrix} 0 \\ 0 \\ -i\Omega \\ i\Omega \end{pmatrix}. \quad (9.28)$$

The emission spectrum from the stationary state is given by (16). Since the transition dipole moment μ has matrix elements $\langle 1 | \mu | 2 \rangle$, $\langle 1 | \mu | 3 \rangle$ and their conjugates, there can be an interference between the emission from the level $|2\rangle$ and that from the level $|3\rangle$. However, when the transition energies $\hbar\omega_0$ and $\hbar\omega_3$ are much different, the interference can be neglected and one of the transitions can be treated separately. In the following, the emission spectrum from the level $|3\rangle$ will be studied. The spectrum is given by

$$I(\omega) = \text{Re}\left[\int_0^\infty dt\, \langle 1 | \mu e^{-\frac{i}{\hbar}H_0^\times t}(\rho^{st}\mu)|1\rangle\, e^{(i\omega-\gamma)t}\right]$$

$$\rightarrow |\langle 1|\mu|3\rangle|^2 \text{Re}\left[\int_0^\infty dt\, \langle 3| e^{-\frac{i}{\hbar}H_0^\times t}(\rho^{st}|3\rangle\langle 1|)|1\rangle\, e^{(i\omega-\gamma)t}\right]$$

$$= |\langle 1|\mu|3\rangle|^2 \text{Re}\left[\left\langle 3 \left| \frac{1}{\gamma - i\omega + \frac{i}{\hbar}H_0^\times}(\rho^{st}|3\rangle\langle 1|)\right|1\right\rangle\right] \tag{9.29}$$

with

$$\rho^{st}|3\rangle\langle 1| = \rho_{13}^{st}|1\rangle\langle 1| + \rho_{23}^{st}|2\rangle\langle 1| + \rho_{33}^{st}|3\rangle\langle 1| = \rho_{33}^{st}|3\rangle\langle 1|, \tag{9.30}$$

where $\rho_{13}^{st} = \rho_{23}^{st} = 0$ as mentioned above. Equation (9.29) indicates that the spectrum can be obtained as the Fourier–Laplace transform of the component $\rho_{31}(t)$ of the density matrix which evolves from the initial state $\rho^{st}|3\rangle\langle 1|$. $1/(s + \frac{i}{\hbar}H_0^\times)$ can be calculated by noting that $\tilde{\rho}_{31}$ and ρ_{32} are coupled together in (24) and (26). Restricting to these two components, we have

$$\frac{1}{s + \frac{i}{\hbar}H_0^\times} = \begin{pmatrix} s + \gamma_{13} - i(\overline{\omega} - \omega_3) & i\Omega \\ i\Omega & s + \gamma_{23} - i(\omega_0 - \omega_3) \end{pmatrix}^{-1}$$

$$= \frac{1}{D(s)}\begin{pmatrix} s + \gamma_{23} - i(\omega_0 - \omega_3) & -i\Omega \\ -i\Omega & s + \gamma_{13} - i(\overline{\omega} - \omega_3) \end{pmatrix} \tag{9.31}$$

with

$$D(s) = (s + \gamma_{13} - i(\overline{\omega} - \omega_3))(s + \gamma_{23} - i(\omega_0 - \omega_3)) + \Omega^2, \tag{9.32}$$

where the bases of the 2×2 representation are $\tilde{\rho}_{31}$ and ρ_{32}. Then the emission spectrum in Eq. (9.29) can be written as

$$|\langle 1|\mu|3\rangle|^2 \rho_{33}^{st} \text{Re}\, \frac{\gamma_{23} + i(\overline{\omega} - \omega - \omega_0 + \omega_3)}{D(i(\overline{\omega} - \omega))}. \tag{9.33}$$

Theory of exciton coherence and decoherence 401

The evolution of the emission spectrum with the detuning $\bar{\omega} - \omega_0$ of the excitation light is exhibited in Fig. 9.2. The relevant parameters are chosen as

$$\hbar\kappa_{21} = \hbar\kappa_{23} = 10 \text{ μeV}, \hbar\kappa_{31} = 5 \text{ μeV}, \hbar(\omega_0 - \omega_3) = 20 \text{ meV},$$

$$\hbar\gamma_{12} = \hbar\gamma_{13} = \hbar\gamma_{23} = 50 \text{ μeV}, \hbar\Omega = 100 \text{ μeV},$$

$$\Delta\omega = \bar{\omega} - \omega_0 = -0.4 \text{ meV (a)}, -0.2 \text{ meV (b)}, 0 \text{ meV (c)}, 0.2 \text{ meV (d)}, 0.4 \text{ meV (e)}.$$

(9.34)

At the resonance ($\bar{\omega} - \omega_0 = 0$, Fig. 9.2(c)) double peaks with equal intensity appear at $\omega = \omega \pm \Omega$, when $\gamma_{13}, \gamma_{23} \ll \Omega$. This feature can be seen readily from the denominator in (9.33). For the positive detuning ($\bar{\omega} > \omega_0$, Fig. 9.2(d), (e)), the lower energy peak has a larger intensity compared to the higher energy peak. The situation is reversed for the negative detuning (Fig. 9.2(a), (b)). These features can be understood in more detail in terms of the dressed exciton states.

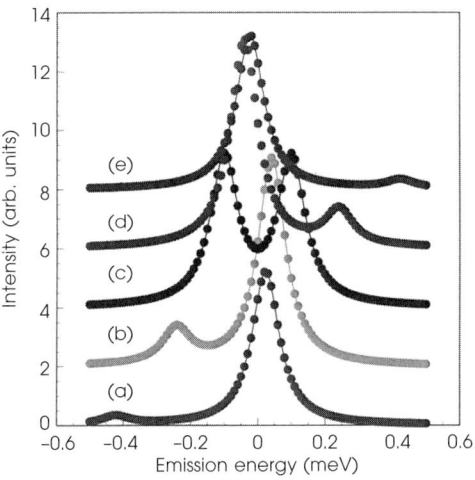

Fig. 9.2 The evolution of the emission spectrum of a three-level system in Fig. 9.1 is plotted for several values of the detuning $\bar{\omega} - \omega_0$: (a) – 0.4 meV (b) – 0.2 meV (c) 0 meV (d) 0.2 meV (e) 0.4 meV.

9.3 Dressed exciton state

In the photon number picture where the number of photons in the excitation light is denoted by n, $|1, n\rangle$ and $|2, n-1\rangle$ have almost the same energy in the case of resonant excitation, i.e., $\bar{\omega} \cong \omega_0$. In this representation the two states are coupled together and the relevant Hamiltonian is written as

$$\begin{pmatrix} \hbar\overline{\omega} & -\hbar\Omega \\ -\hbar\Omega & \hbar\omega_0 \end{pmatrix}. \tag{9.35}$$

The energy eigenvalues and the corresponding eigenstates are given by

$$E_\pm = \frac{\hbar}{2}[\overline{\omega} + \omega_0 \pm \sqrt{(\overline{\omega} - \omega_0)^2 + 4\Omega^2}], \tag{9.36}$$

$$\phi_\pm = c_1^\pm |1, n\rangle + c_2^\pm |2, n-1\rangle, \tag{9.37}$$

$$c_1^\pm = \frac{\Omega}{\sqrt{\frac{1}{2}\delta(\delta \mp (\overline{\omega} - \omega_0))}}, \quad c_2^\pm = \mp\sqrt{\frac{\delta \mp (\overline{\omega} - \omega_0)}{2\delta}} \tag{9.38}$$

with

$$\delta = \sqrt{(\overline{\omega} - \omega_0)^2 + 4\Omega^2}. \tag{9.39}$$

These states in (9.37) are usually called the dressed exciton state, namely the exciton state dressed with photons [38]. The emission from the state $|3, n-1\rangle$ occurs to the final states $|\phi_\pm\rangle$, leading to a doublet structure in the spectrum. The emission intensity is proportional to $|c_1^\pm|^2$ because the component of the level $|1\rangle$ in the final states is relevant. From the expression in (9.38), we see that for the positive detuning ($\overline{\omega} > \omega_0$), $|c_1^+|^2 > |c_1^-|^2$ and that the intensity of the lower energy peak is stronger than that of the higher energy peak. For the negative detuning the situation is reversed. These arguments explain the numerical results in Fig. 9.2. Recently, a systematic study on the detuning dependence of the emission spectrum from an InGaAs single quantum dot was reported by Kamada *et al.* [34] and the above features were successfully confirmed.

9.4 Exciton Rabi oscillation in a single quantum dot

It is important to examine the quantum mechanical interference between two perturbations induced by external stimuli such as optical pulses. In order to observe the quantum mechanical interference, the two external perturbations should be mutually coherent. For example, in the usual pump–probe experiments, the incoherent change in the system induced by the pump pulse is probed by the change in the transmission spectrum of a weak probe pulse. In this case the mutual coherence between the pump and probe pulses is not relevant. However, the importance of the relative coherence of the excitation pulses was demonstrated for the first time in the coherent destruction of the excited state population [10]. In the latter case, a phase-locked pulse pair was used to excite and deexcite the system. The dynamical evolution in the system depends sensitively on the relative phase of two excitation pulses. This sensitivity can be manifested even when we observe an incoherent quantity such as the time-integrated luminescence intensity, as will be discussed later. These coherent optical phenomena can be observed most clearly in semiconductor quantum dots compared with other higher-dimensional structures such as quantum wells and bulk materials, because the energy levels are completely discrete and the contribution from nearby levels and continuum states can be reduced much in quantum dots.

In a semiconductor quantum dot the exciton dephasing time is about several tens of picoseconds [12] and the population lifetime is of the order of several hundreds of picoseconds [39]. Thus the coherent dynamics can be observed by using picosecond pulses. If we can observe directly the temporal variation of the excited state population, it can be claimed that the Rabi oscillation is observed in the time-domain. However, this is difficult especially for the case of a single quantum dot because of the weakness of the signal. Instead of this direct measurement, the time-integrated emission intensity can be measured as a function of the delay time between two phase-locked pulses. Although this is an indirect measurement, the results of this measurement reveal clearly the quantum mechanical interference arising from the Rabi oscillation. In the following we discuss the coherent optical phenomena in a quantum dot extending the concept of the wave-packet interferometry initiated by Bonadeo *et al.* [22] to the strong excitation regime. Here we consider a three-level system in Fig. 9.1 and the semi-classical radiation fields which are almost resonant with the transition between $|1\rangle$ and $|2\rangle$. For a phase-locked pulse pair with the same amplitude, the electric field can be written as

$$E(t) = \varepsilon(t) \cos \omega t + \varepsilon(t - t_d) \cos \omega(t - t_d)$$

$$= \frac{1}{2} \left(\varepsilon(t) + \varepsilon(t - t_d) e^{-i\omega t_d} \right) e^{i\omega t} + c.c., \quad (9.40)$$

where $\varepsilon(t)$ is the pulse envelope and is assumed in a hyperbolic secant form with a pulse width parameter t_p:

$$\varepsilon(t) = \varepsilon_0 \operatorname{sech}\left(\frac{t}{t_p}\right). \quad (9.41)$$

The basic equations of motion for the density matrix are given by

$$\dot{\rho}_{11} = \kappa_{21}\rho_{22} + \kappa_{31}\rho_{33} - i\Omega^*(t)\tilde{\rho}_{12} + i\Omega(t)\tilde{\rho}_{21}, \quad (9.42)$$

$$\dot{\rho}_{22} = -(\kappa_{21} + \kappa_{23})\rho_{22} + i\Omega^*(t)\tilde{\rho}_{12} - i\Omega(t)\tilde{\rho}_{21}, \quad (9.43)$$

$$\dot{\rho}_{33} = \kappa_{23}\rho_{22} - \kappa_{31}\rho_{33}, \quad (9.44)$$

$$\dot{\tilde{\rho}}_{12} = -i\Omega(t)(\rho_{11} - \rho_{22}) + (-i\Delta\omega - \gamma_{12})\tilde{\rho}_{12}, \quad (9.45)$$

$$\dot{\tilde{\rho}}_{21} = i\Omega^*(t)(\rho_{11} - \rho_{22}) + (i\Delta\omega - \gamma_{12})\tilde{\rho}_{21}, \tag{4.46}$$

$$\dot{\tilde{\rho}}_{13} = i\Omega(t)\rho_{23} + (-i(\omega - \omega_3) - \gamma_{13})\tilde{\rho}_{13}, \tag{9.47}$$

$$\dot{\tilde{\rho}}_{31} = -i\Omega^*(t)\rho_{32} + (i(\omega - \omega_3) - \gamma_{13})\tilde{\rho}_{31}, \tag{9.48}$$

$$\dot{\rho}_{23} = i\Omega^*(t)\tilde{\rho}_{13} + (i(\omega_3 - \omega_0) - \gamma_{23})\rho_{23}, \tag{9.49}$$

$$\dot{\rho}_{32} = -i\Omega(t)\tilde{\rho}_{31} + (-i(\omega_3 - \omega_0) - \gamma_{23})\rho_{32}, \tag{9.50}$$

with

$$\Delta\omega = \omega - \omega_0, \quad \Omega(t) = \frac{\langle 1|\mu|2\rangle}{2\hbar}(\varepsilon(t) + \varepsilon(t - t_d)e^{-i\omega t_d}), \tag{9.51}$$

where the notations are the same as in (9.18)–(9.26), the oscillating part is separated out as in (9.27) with replacing $\bar{\omega}$ by ω and $\Omega(t)$ is a complex quantity even when $\langle 1|\mu|2\rangle$ is a real number. Here ρ_{13}, ρ_{23} and their complex conjugates are not excited and remain zero when starting from the initial state such as $\rho_{11} = 1$ and $\rho_{ij} = 0$ for other combinations of i, j and thus can be discarded in the following calculations. The population ρ_{33} of the state $|3\rangle$ is fed by the population ρ_{22} of the state $|2\rangle$ through the non-radiative processes. The measured quantity is the time-integrated intensity of luminescence from the exciton ground state $|3\rangle$, namely

$$\Gamma_{31}\int_0^\infty dt\, \rho_{33}(t), \tag{9.52}$$

where Γ_{31} is the radiative decay rate of the state $|3\rangle$. This quantity can be shown to be proportional to

$$\int_0^\infty dt\, \rho_{22}(t). \tag{9.53}$$

The proof goes as follows. With the notation of the Laplace transform defined by

$$f[s] = \int_0^\infty dt\, e^{-st} f(t), \tag{9.54}$$

the quantity in (9.52) is written as $\Gamma_{31}\rho_{33}[0]$. From the Laplace transform of (9.44) we find that

$$\rho_{33}[0] = \frac{\kappa_{23}}{\kappa_{31}}\rho_{22}[0]. \tag{9.55}$$

This equation shows that the quantity in (9.52) is proportional to (9.53). It is advantageous for the numerical calculation to replace (9.52) by (9.53) because the lifetime of the state $|3\rangle$ is in general longer than that of the state $|2\rangle$ and the time-integration of (9.53) is much time-saving.

Before exhibiting numerical results, we examine analytical expressions for the case of weak excitation where the perturbational calculation is applicable. First of all, we obtain

$$\tilde{\rho}_{12}(t) = -i \int_0^t d\tau\, e^{-(\gamma_{12}+i\Delta\omega)(t-\tau)} \Omega(\tau), \tag{9.56}$$

$$\tilde{\rho}_{21}(t) = i \int_0^t d\tau\, e^{-(\gamma_{12}-i\Delta\omega)(t-\tau)} \Omega^*(\tau). \tag{9.57}$$

Substituting these into (9.43), we have

$$\dot{\rho}_{22}(t) = -\kappa_2 \rho_{22}(t) + \Omega^*(t) \int_0^t d\tau\, e^{-(\gamma_{12}+i\Delta\omega)(t-\tau)} \Omega(\tau)$$

$$+ \Omega(t) \int_0^t d\tau\, e^{-(\gamma_{12}-i\Delta\omega)(t-\tau)} \Omega^*(\tau), \tag{9.58}$$

and then

$$\rho_{22}(t) = \int_0^t dt_1 e^{-\kappa_2(t-t_1)} \left[\Omega^*(t_1) \int_0^{t_1} dt_2 e^{-(\gamma_{12}+i\Delta\omega)(t_1-t_2)} \Omega(t_2) \right.$$

$$\left. + \Omega(t_1) \int_0^{t_1} dt_2 e^{-(\gamma_{12}-i\Delta\omega)(t_1-t_2)} \Omega^*(t_2) \right], \tag{9.59}$$

where $\kappa_2 = \kappa_{21} + \kappa_{23}$. The time-integrated luminescence intensity is proportional to the Laplace transform $\rho_{22}[0]$ and is given by

$$\rho_{22}[0] = \frac{2}{\kappa_2} \operatorname{Re}\left[\int_0^\infty dt_1 \int_0^{t_1} dt_2 e^{-(\gamma_{12}+i\Delta\omega)(t_1-t_2)} \Omega^*(t_1) \Omega(t_2) \right]. \tag{9.60}$$

In the limit of a short pulse, i.e., $t_p \ll \gamma_{12}^{-1}, \Delta\omega^{-1}$, we can approximate as $\varepsilon(t) \propto \delta(t)$ and we have

$$\rho_{22}[0] \propto 1 + \exp[-\gamma_{12} t_d] \cos \omega_0 t_d. \tag{9.61}$$

This indicates that the time-integrated luminescence intensity shows a fine oscillation with the optical frequency ω_0 and the envelopes of its maximums and minimums decay with the exciton dephasing rate γ_{12}. In the following, the time-integrated luminescence intensity as a function of the delay time t_d will be called simply *the correlation trace*. In order to calculate the correlation trace in the case of stronger excitation, we have to solve numerically the equations of motion for the density matrix:

$$\frac{d}{dt}\begin{pmatrix} \rho_{11} \\ \rho_{22} \\ \rho_{33} \\ \tilde{\rho}_{12} \\ \tilde{\rho}_{21} \end{pmatrix} = \begin{pmatrix} 0 & 0 & \kappa_{21} & \kappa_{31} & -i\Omega^*(t) & i\Omega(t) \\ 0 & -(\kappa_{21}+\kappa_{23}) & 0 & i\Omega^*(t) & -i\Omega(t) \\ 0 & \kappa_{23} & -\kappa_{31} & 0 & 0 \\ -i\Omega(t) & i\Omega(t) & 0 & -i\Delta\omega-\gamma_{12} & 0 \\ i\Omega^*(t) & -i\Omega^*(t) & 0 & 0 & i\Delta\omega-\gamma_{12} \end{pmatrix}\begin{pmatrix} \rho_{11} \\ \rho_{22} \\ \rho_{33} \\ \tilde{\rho}_{12} \\ \tilde{\rho}_{21} \end{pmatrix}.$$

(9.62)

9.5 Bloch vector model

First of all, we notice a simple model which enables us to have simple physical interpretations of numerical results. As mentioned before, in semiconductor quantum dots, the exciton dephasing time is about several tens of picoseconds (ps) and the population lifetime is about several hundreds of ps. In this situation, the relaxation processes can be neglected for typical values of the pulse width about a few ps and the delay time between two phase-locked pulses less than a few tens of ps and we can introduce a simple Bloch vector model [40]. When only the two levels $|1\rangle$ and $|2\rangle$ are included and the relaxation terms are dropped, the equations of motion in (9.62) can be reduced to

$$\dot{\rho}_{11} = -i\Omega^*(t)\tilde{\rho}_{12} + i\Omega(t)\tilde{\rho}_{21},\tag{9.63}$$

$$\dot{\rho}_{22} = i\Omega^*(t)\tilde{\rho}_{12} - i\Omega(t)\tilde{\rho}_{21},\tag{9.64}$$

$$\dot{\tilde{\rho}}_{12} = -i\Delta\omega\tilde{\rho}_{12} - i\Omega(t)(\rho_{11}-\rho_{22}),\tag{9.65}$$

$$\dot{\tilde{\rho}}_{21} = i\Delta\omega\tilde{\rho}_{21} + i\Omega^*(t)(\rho_{11}-\rho_{22}).\tag{9.66}$$

Then introducing the Bloch vector defined by

$$\vec{\rho} = \begin{pmatrix} \rho_x \\ \rho_y \\ \rho_z \end{pmatrix} \tag{9.67}$$

with

$$\rho_x = \tilde{\rho}_{12} + \tilde{\rho}_{21}, \quad \rho_y = i(\tilde{\rho}_{21} - \tilde{\rho}_{12}), \quad \rho_z = \rho_{22} - \rho_{11},\tag{9.68}$$

we have an equation of motion given by

$$\frac{d}{dt}\vec{\rho} = \vec{\rho} \times \vec{G} \tag{9.69}$$

with the gyration vector defined by

$$\vec{G} = \begin{pmatrix} \Omega(t) + \Omega^*(t) \\ i(\Omega^*(t) - \Omega(t)) \\ \Delta\omega \end{pmatrix}. \tag{9.70}$$

For the pulse shape in (9.41), we have

$$\vec{G} = \begin{pmatrix} \Omega_0 [\operatorname{sech}\frac{t}{t_p} + \operatorname{sech}\frac{t-t_d}{t_p}\cos\omega t_d] \\ -\Omega_0 \operatorname{sech}\frac{t-t_d}{t_p}\sin\omega t_d \\ \Delta\omega \end{pmatrix} \tag{9.71}$$

with $\Omega_0 = \langle 1|\mu|2\rangle\varepsilon_0/\hbar$, where $\langle 1|\mu|2\rangle$ is assumed to be real. Thus the magnitude and the direction of the gyration vector are sensitively dependent on the delay time t_d of the second pulse.

In order to introduce the concept of the pulse area, we deal with a case of constant $\vec{G}$ vector. Then the equations of motion are given by

$$\dot{\rho}_x = G_z\rho_y - G_y\rho_z, \tag{9.72}$$

$$\dot{\rho}_y = G_x\rho_z - G_z\rho_x, \tag{9.73}$$

$$\dot{\rho}_z = G_y\rho_x - G_x\rho_y. \tag{9.74}$$

This set of equations can be solved by the Laplace transform with an initial condition that $\rho_x(0) = \rho_y(0) = 0$ and $\rho_z(0) = -1$ and we have

$$\rho_{22}(t) = \frac{G_y^2 + G_z^2}{G^2}\sin^2\left[\frac{1}{2}|G|t\right] \tag{9.75}$$

with

$$G^2 = G_x^2 + G_y^2 + G_z^2, \quad |G| = \sqrt{G^2}. \tag{9.76}$$

Thus the population of the excited state $|2\rangle$ oscillates with a period of $2\pi/|G|$. When $\Delta\omega = 0$ or $G_z = 0$, namely the excitation light is resonant with the transition between $|1\rangle$ and $|2\rangle$, the system is completely excited to the level $|2\rangle$ at the time of $|G|t = \pi, 3\pi, 5\pi, \ldots$. From this we can infer that the quantity $|G|t$ can be interpreted as the pulse area which indicates the rotation angle of the Bloch vector [7].

Now we note that the population lifetime of the excited exciton state is much longer than the typical pulse width. Then for small values of t_d, the population of the excited exciton state and the time-integrated luminescence intensity can be approximated as

$$\rho_{22}(t) \propto e^{-\gamma t} \rho_{22} \text{ (just after pulses)} \tag{9.77}$$

and

$$\int_0^\infty dt\, \rho_{22}(t) \propto \rho_{22} \text{ (just after pulses)}, \tag{9.78}$$

where γ is the population decay rate. Thus the time-integrated luminescence intensity is proportional to ρ_{22} (just after pulses) and can be approximated by using (9.75) as

$$\propto \sin^2\left[\frac{1}{2}\int_{-\infty}^\infty d\tau\, |G(\tau)|\right]. \tag{9.79}$$

This dependence is plotted in Fig. 9.3 as a function of the total pulse area defined by

$$\int_{-\infty}^\infty d\tau\, |G(\tau)|. \tag{9.80}$$

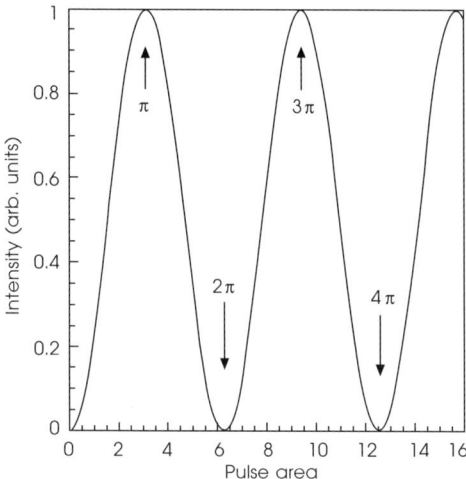

Fig. 9.3 The time-integrated luminescence intensity as a function of the pulse area.

The pulse-area dependence of the luminescence intensity was observed for the first time by Gibbs [26] for Rb atoms and the oscillatory behaviors were confirmed, although the actual luminescence intensity does not vanish at the pulse area of integer multiples of

2π due to the presence of relaxation channels. Under the resonant excitation ($\Delta\omega = 0$) and for $\omega t_d = 2n\pi$

$$\vec{G} = \begin{pmatrix} \Omega_0 [\mathrm{sech}\frac{t}{t_p} + \mathrm{sech}\frac{t-t_d}{t_p}] \\ 0 \\ 0 \end{pmatrix} \tag{9.81}$$

and we have

$$\int_{-\infty}^{\infty} dt\, |G(t)| = 2\pi\Omega_0 t_p. \tag{9.82}$$

On the other hand, for $\omega t_d = (2n+1)\pi$

$$\vec{G} = \begin{pmatrix} \Omega_0 [\mathrm{sech}\frac{t}{t_p} - \mathrm{sech}\frac{t-t_d}{t_p}] \\ 0 \\ 0 \end{pmatrix} \tag{9.83}$$

and we have

$$\int_{-\infty}^{\infty} dt\, |G(t)| \cong 0 \tag{9.84}$$

for $t_d \ll t_p$. Thus the pulse area increases from 0 to $2\pi\Omega_0 t_p$ and decreases from $2\pi\Omega_0 t_p$ to 0 continuously with the change of 2π in ωt_d. When $0 < 2\Omega_0 t_p \leq 1$, the total pulse area increases from 0 to a value less than π and then decreases to 0 with the change of 2π in ωt_d and the corresponding change of the luminescence intensity shows a single peak according to Fig. 9.3. The change of 2π in ωt_d corresponds to a change of $2\pi/\omega$ in t_d which is nothing but one optical cycle and is of the order of a few femtoseconds. When $1 < 2\Omega_0 t_p \leq 2$, the total pulse area increases from 0 to a value more than π but less than 2π and then decreases to 0 with the change of 2π in ωt_d and the corresponding change of the luminescence intensity shows double peaks. In a similar way, we find that when $n - 1 < 2\Omega_0 t_p \leq n$, the luminescence intensity shows n peaks with the change of 2π in ωt_d. Thus the time-integrated luminescence intensity shows a fine oscillation in the femtosecond range. With increasing time delay t_d between two phase-locked pulses, the temporal overlap between two pulses becomes incomplete, leading to a smaller number of peaks within the change of 2π in ωt_d. But more details cannot be discussed within the qualitative arguments based on the approximate expression in (9.79).

9.6 Numerical results and discussion

With increasing excitation intensity, there appear two characteristic features: one is a modulation of the envelopes of the maximums and minimums and the other is a change in the fine oscillation in the femtosecond range. Typical numerical results of the correlation trace are shown in Fig. 9.4 for increasing excitation intensities. The relevant parameters in (9.62) are chosen as

$$t_p = 6 \text{ ps}, \quad \kappa_{21} t_p = \kappa_{23} t_p = 0.01, \quad \kappa_{31} t_p = 0.005,$$

$$\gamma_{12} t_p = \gamma_{13} t_p = \gamma_{23} t_p = 0.06, \quad \omega = \omega_0,$$

$$\Omega_0 t_p = 0.2(a), 0.6(b), \ldots \tag{9.85}$$

One of the most prominent features in Fig. 9.4 is the sudden shrinkage of the interval between the maximum and the minimum envelopes at $\Omega_0 t_p = 1.0$, which corresponds to the case that the pulse area of each pulse is π. The first pulse turns around the Bloch vector from the south pole to the north pole and the second pulse comes in after relaxation processes. In the case that the second pulse is in phase with the first pulse, the Bloch vector rotates in the same direction as for the first pulse and returns to the neighborhood of the south pole because the second pulse has also a pulse area of π. In the case that the second pulse is out of phase relative to the first pulse, the Bloch vector rotates in the backward direction, returning to the vicinity of the south pole. Then the time-integrated luminescence intensity is almost the same for both cases, leading to the shrinkage of the interval between the maximum and the minimum envelopes as in Fig. 9.4(d). When the pulse area deviates from π, the incidental coincidence of the Bloch vector position after the passage of a pulse pair does not occur between the in-phase and

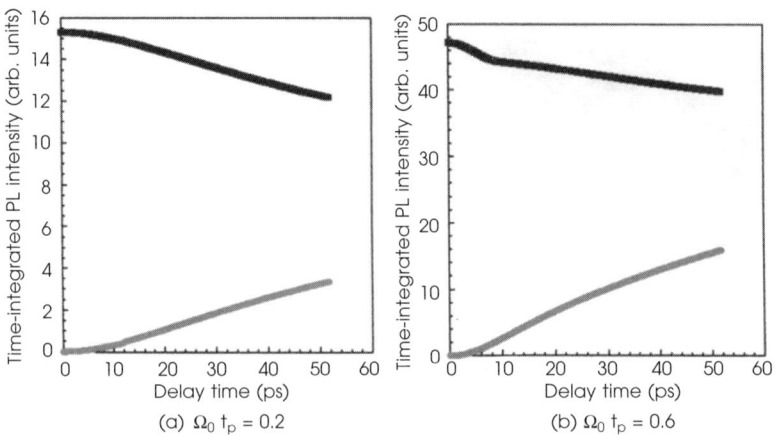

(a) $\Omega_0 t_p = 0.2$

(b) $\Omega_0 t_p = 0.6$

Theory of exciton coherence and decoherence 411

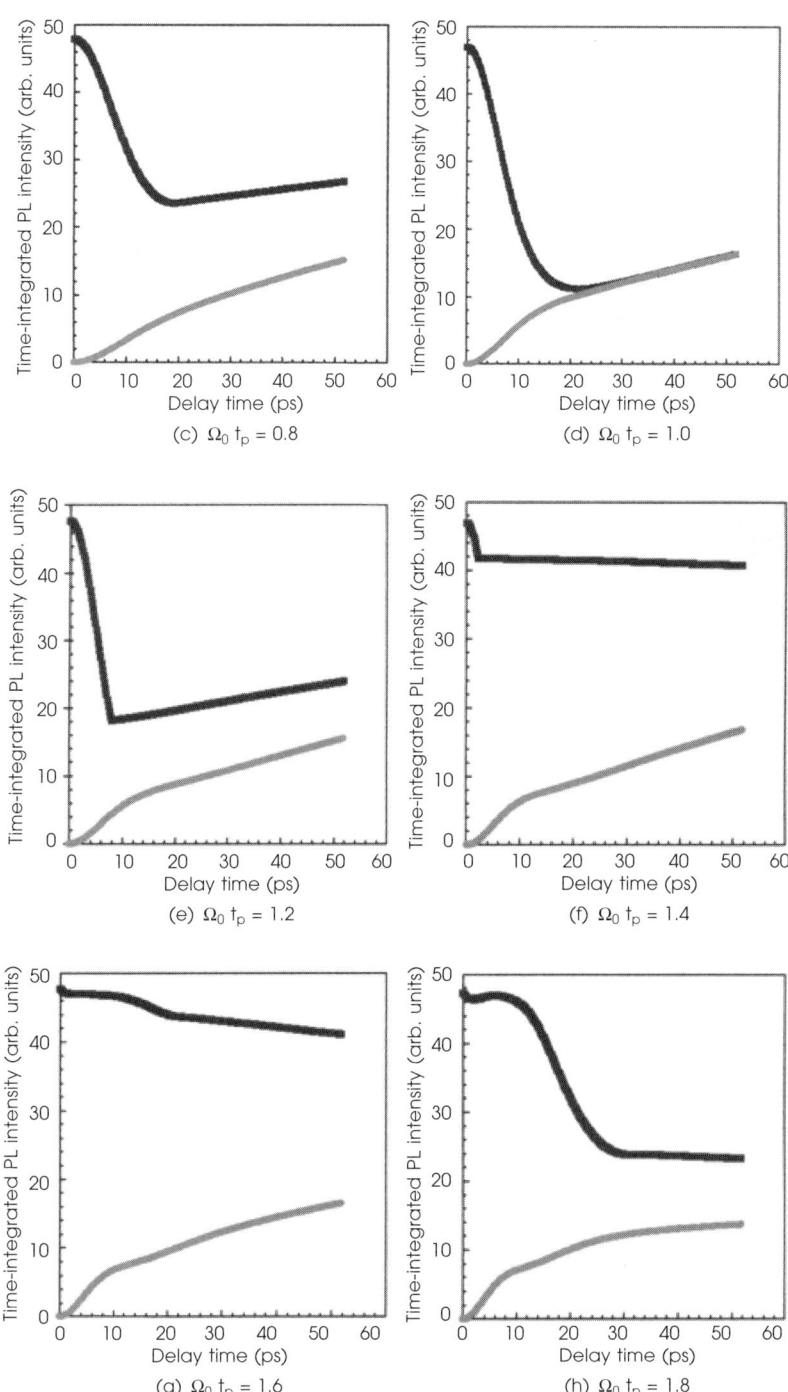

(c) $\Omega_0 t_p = 0.8$

(d) $\Omega_0 t_p = 1.0$

(e) $\Omega_0 t_p = 1.2$

(f) $\Omega_0 t_p = 1.4$

(g) $\Omega_0 t_p = 1.6$

(h) $\Omega_0 t_p = 1.8$

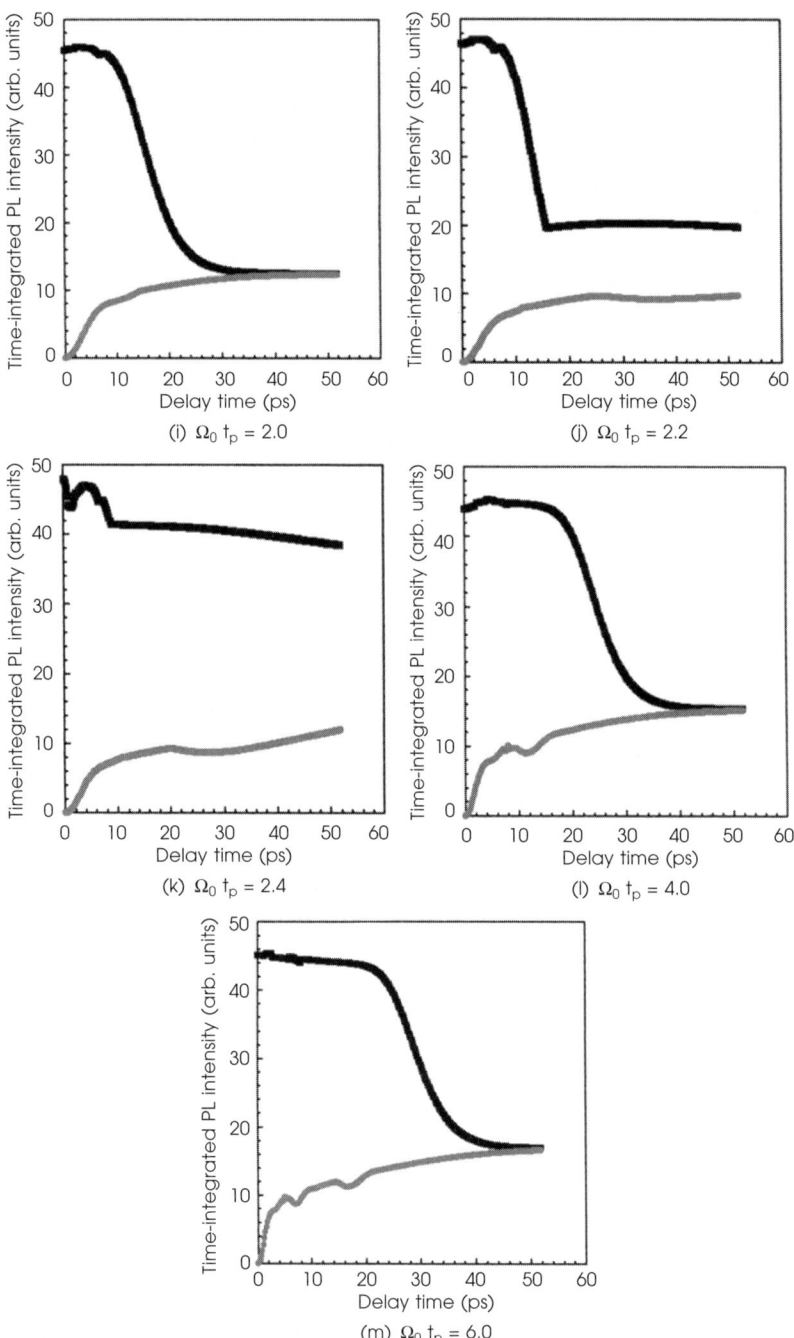

Fig. 9.4 The correlation traces for the three-level system in Fig. 9.1 for several values of the excitation intensity. Only the envelopes of maximums and minimums are plotted.

the out-of-phase cases. From the above reasoning, the collapse of the maximum and the minimum envelopes is expected to occur periodically at $\Omega_0 t_p = 1.0, 2.0, \ldots$ These features can be confirmed in the systematic changes in Fig. 9.4, although at higher excitation intensities the excited state population accumulates due to the presence of relaxation processes and the above effect becomes smeared.

The modulation of the envelopes, which is seen clearly in Figs 9.4(l) and (m), can be interpreted as a kind of beat oscillation. Under the strong excitation, the dressed exciton state is formed and the doublet splitting occurs in the exciton state and also in the ground state. Thus the situation is similar to that discussed in section 9.7 where closely separated two levels are simultaneously excited and the luminescence from the exciton ground state is time-integrated. The difference is that the doublet splitting is induced here by the excitation pulse itself and its magnitude is time-dependent. As a consequence, the period of the beat-like oscillation in Figs 9.4(l) and (m) cannot be identified definitely.

The changes of the correlation trace in the femtosecond range are exhibited in Fig. 9.5 for various excitation intensities and for two typical delay times. As discussed above, there appear n peaks within one optical cycle for the range of $n - 1 < 2\Omega_0 t_p \leq n$ and for $t_d < t_p$ such that the overlap between two phase-locked pulses is significant. For larger values of t_d, the number of peaks is reduced within the change of t_d of one optical cycle.

Another interesting feature is related to the detuning of the excitation light denoted by $\Delta\omega = \omega - \omega_0$, where $\hbar\omega_0$ is the energy difference between the excited state and the ground state. Figures 9.6 and 9.7 exhibit the detuning dependence of the envelopes of the maximums and minimums of the correlation trace in the coarse time range. The relevant parameters are chosen as

$t_p = 6$ ps, $\kappa_{21} t_p = \kappa_{23} t_p = 0.01$, $\kappa_{31} t_p = 0.005$,

$$\gamma_{12} t_p = 0.08 \tag{9.86}$$

and for Fig. 9.6

$$(\omega - \omega_0) t_p = 0.1, 0.4, 0.8, 1.2, \quad \Omega_0 t_p = 0.2 \tag{9.87}$$

and for Fig. 9.7

$$(\omega - \omega_0) t_p = 0, 0.1, 0.4, 0.8, \quad \Omega_0 t_p = 0.8. \tag{9.88}$$

In the case of weak excitation (Fig. 9.6), a modulation in the envelopes appears at large detuning but the peak of the maximum envelope remains at $t_d = 0$ ps. On the other hand, in the case of stronger excitation (Fig. 9.7), with increasing detuning, there appears a

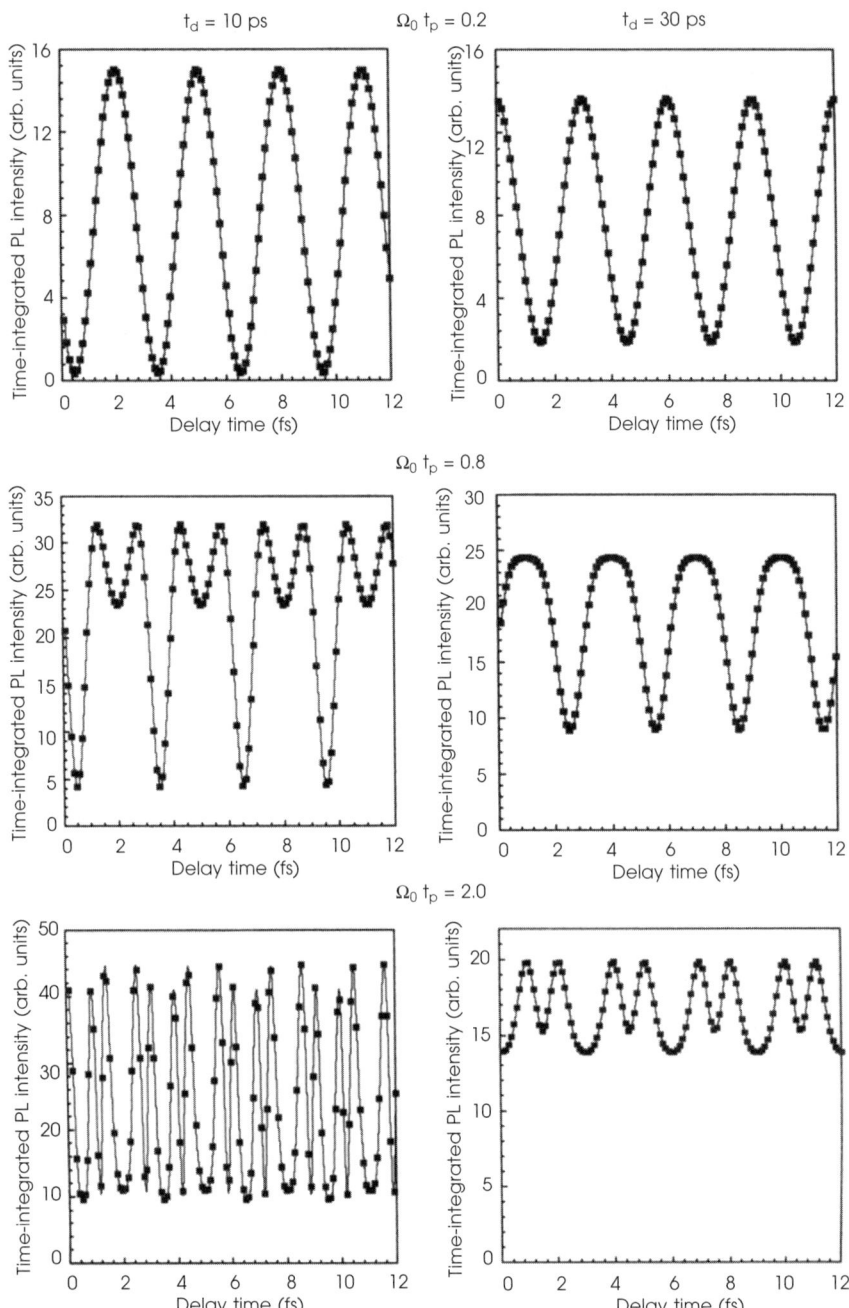

Fig. 9.5 The correlation traces in the femtosecond range around $t_d = 10$ ps and 30 ps for $\Omega_0 t_p = 0.2$, 0.8 and 2.0.

Theory of exciton coherence and decoherence   415

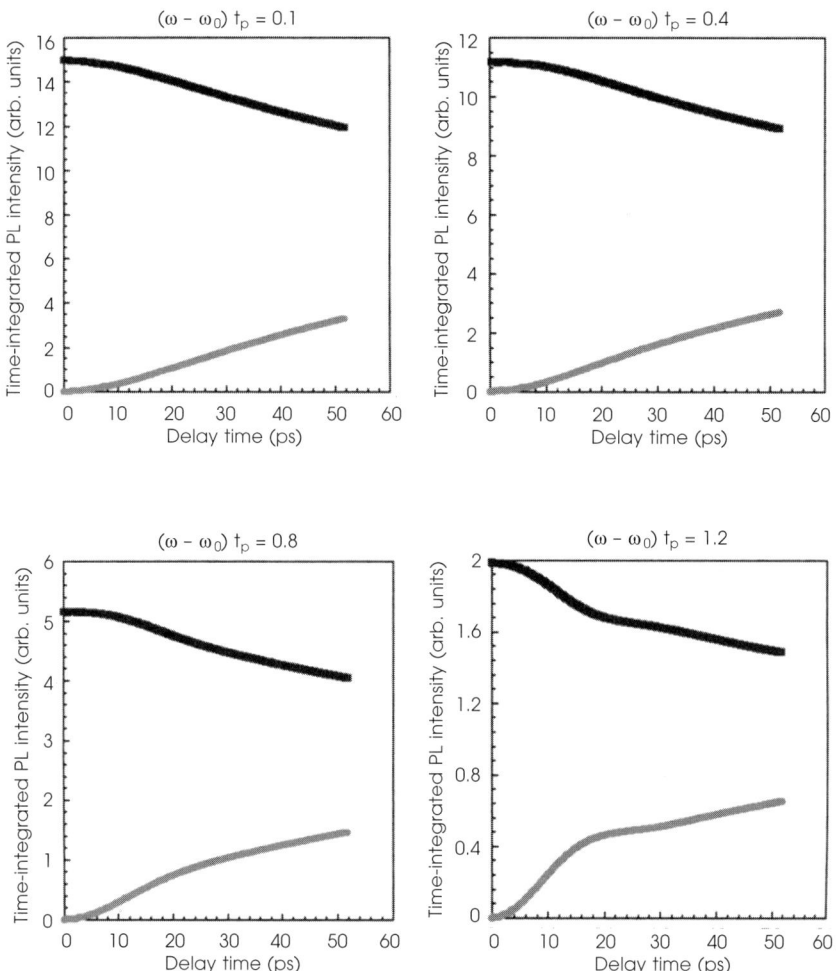

Fig. 9.6 Detuning dependence of the envelopes of maximums and minimums of the correlation trace for $\Omega_0 t_p = 0.2$.

dip around $t_d = 0$ ps. This feature can be understood most clearly in terms of the simple Bloch vector model. The Bloch vector rotates around the vector $\vec{G}$, starting from the south pole. But the gyration vector $\vec{G}$ is not lying in the horizontal plane due to the presence of detuning. The main concern is the z-component of the Bloch vector after the passage of a phase-locked pulse pair which determines the time-integrated luminescence intensity. The motion of the Bloch vector is exhibited in Fig. 9.8 for the case of detuning $(\omega - \omega_0)t_p = 0.8$ and for the maximum points at $t_d = 0$ ps and 14 ps, as indicated by circles in Fig. 9.7. The z-component of the Bloch vector after the passage of the pulse

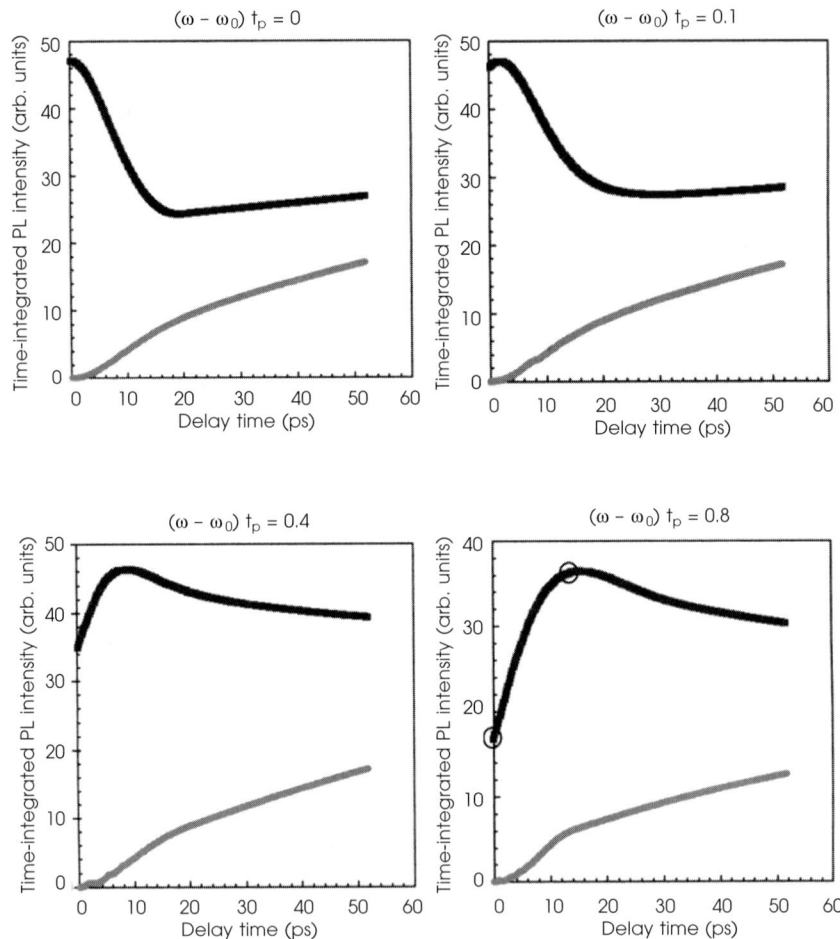

Fig. 9.7 Detuning dependence of the envelopes of maximums and minimums of the correlation trace for $\Omega_0 t_p = 0.8$.

pair takes the maximum value at $t_d = 14$ ps instead at $t_d = 0$ ps, leading to the larger time-integrated luminescence intensity at $t_d = 14$ ps. The situation is the same also for the negative detuning.

As mentioned before, when $1 < 2\Omega_0 t_p \leq 2$, there appear two peaks in the correlation trace within the change of $2\pi/\omega$ in t_d. When the excitation light is exactly resonant with the exciton transition, the double peaks are symmetric with equal intensity irrespective of the value of t_d (not shown). However, in the presence of detuning, the double peaks appear asymmetric and its shape changes with increasing coarse delay time t_d, eventually merging into a single peak as shown in Fig. 9.9, where the simple Bloch vector model in (9.69) is used instead of (9.62) and the relevant parameters are chosen as

Theory of exciton coherence and decoherence 417

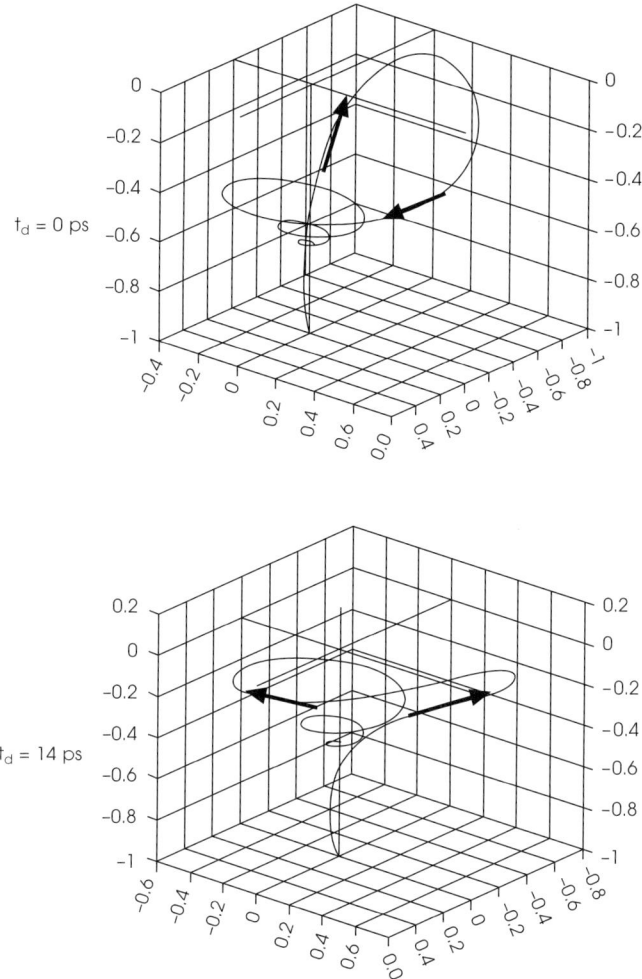

Fig. 9.8 The time-evolution of the Bloch vector at the detuning of $(\omega - \omega_0)t_p = 0.8$ and corresponding to the maximum points at $t_d = 0$ ps and 14 ps as marked by circles in Fig. 9.7.

$$t_p = 6 \text{ ps}, \ \Omega_0 t_p = 0.8, \ (\omega - \omega_0)t_p = \pm 0.8. \tag{9.89}$$

For the positive detuning, the left peak diminishes with increasing delay time, whereas for the negative detuning the right peak loses its intensity. These features can be understood in terms of the motion of the Bloch vector which is sensitively dependent on the temporal evolution of the magnitude and direction of the gyration vector $\vec{G}$. The evolution of the Bloch vector for the case of positive detuning $(\omega - \omega_0)t_p = 0.8$ is shown in Fig. 9.10. The left column is the same as in Fig. 9.9. The middle column corresponds to the

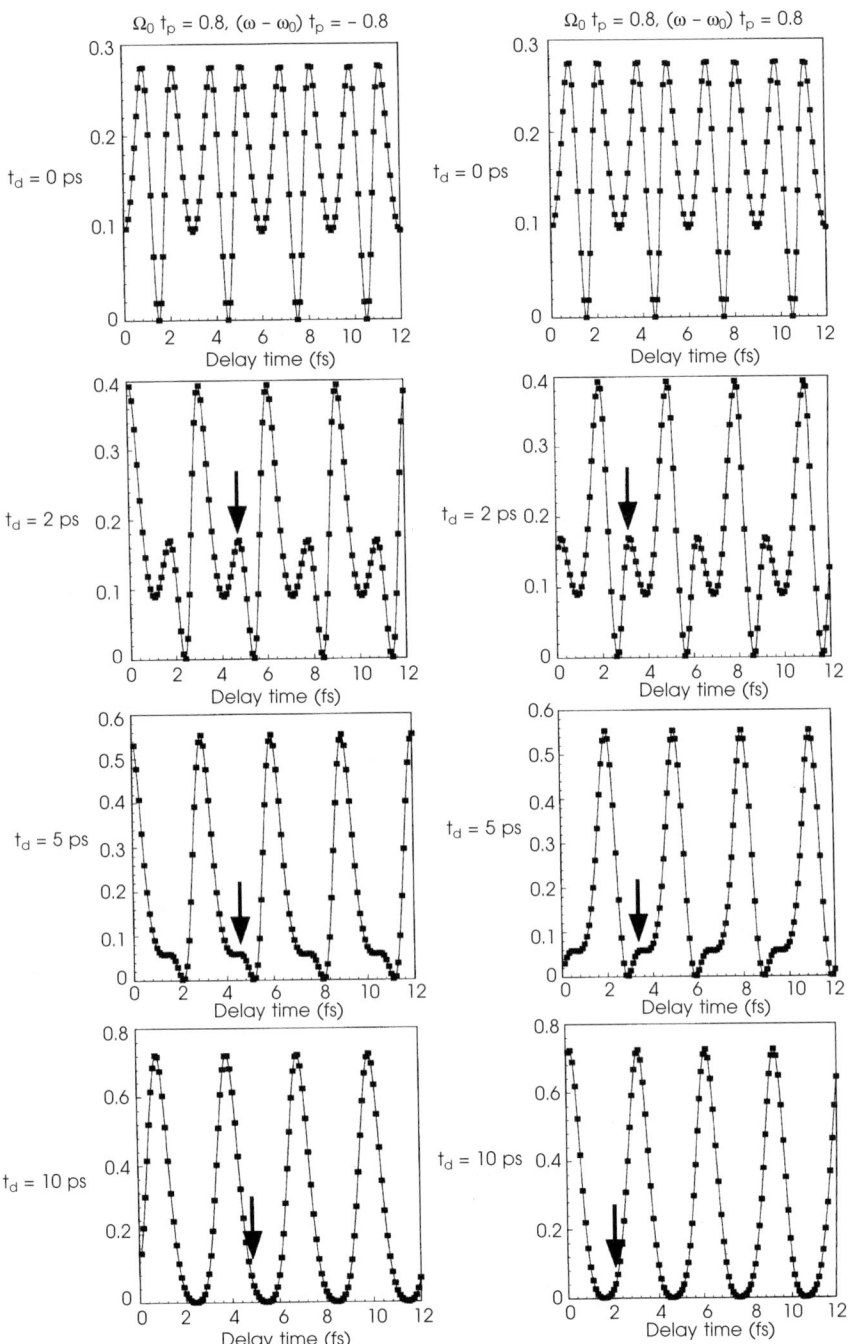

Fig. 9.9 Detuning dependence of the correlation trace in the femtosecond range for $\Omega_0 t_p = 0.8$ and $(\omega - \omega_0)t_p = \pm 0.8$ around the delay time t_d of 0 ps, 2 ps, 5 ps and 10 ps. The arrows indicate the diminishing peaks.

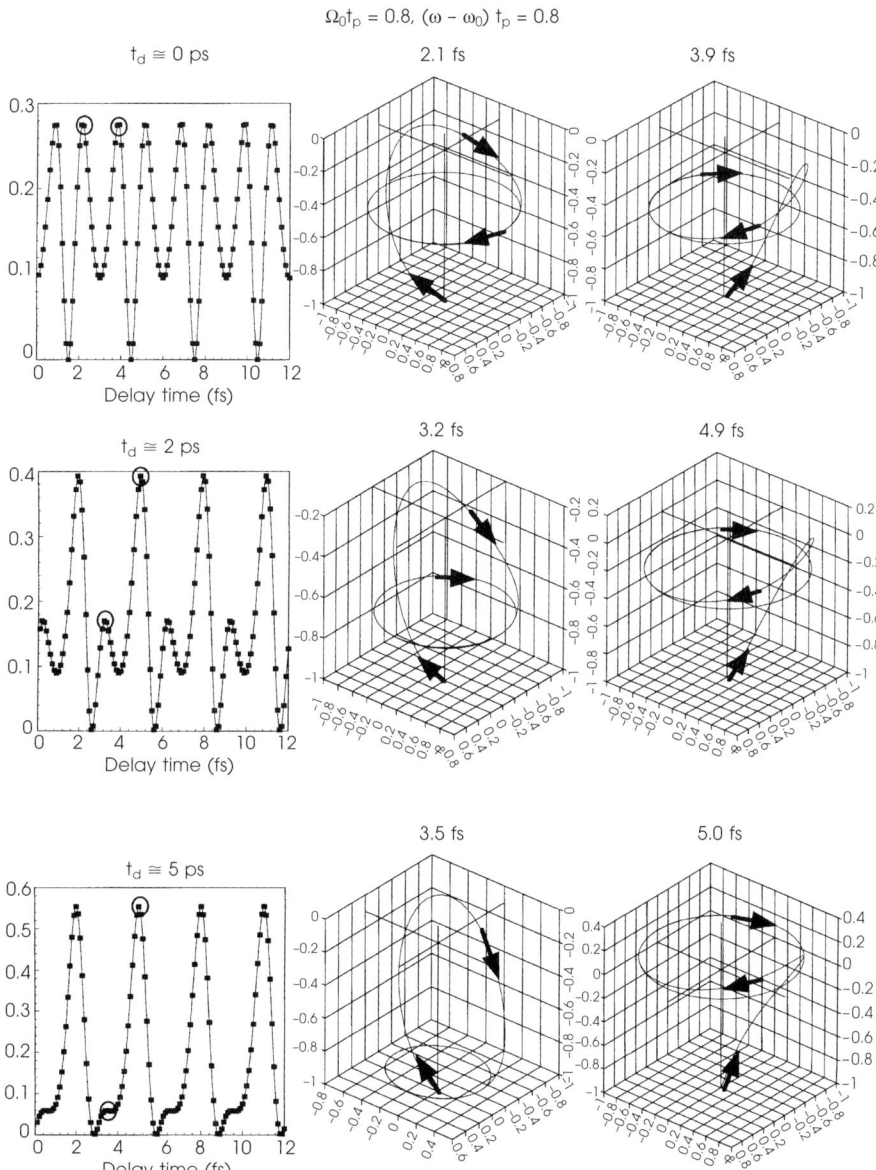

Fig. 9.10 The time-evolution of the Bloch vector corresponding to the points indicated by circles in the left column. The middle (right) column corresponds to the left (right) peak indicated by circles. The correlation traces in the left column are the same as in Fig. 9.9 for $\Omega_0 t_p = 0.8$ and $(\omega - \omega_0)t_p = 0.8$.

left peak in the left column, while the right column to the right peak in the left column. In the middle column, the Bloch vector starts from the south pole and turns around, returning to the vicinity of the south pole. This means that the population of the excited exciton state is rather small after the passage of the phase-locked pulse pair and correspondingly the time-integrated luminescence intensity is rather small. On the other hand, in the right column for the right peaks the Bloch vector does not return to the neighborhood of the south pole and its z-component retains a larger value after the passage of the pulse pair, indicating a larger value of the time-integrated luminescence intensity. Thus the asymmetry of the double peaks in the femtosecond time range can be understood in terms of the motion of the Bloch vector. The situation is the same for the negative detuning except that the relative weight of the left and right peaks is interchanged.

9.7 Wave packet interferometry

Coherent optical control of the wavefunction in a single quantum dot was initiated by Bonadeo et al. [22]. In their experiments, an island-like structure in a single quantum well (QW) due to the fluctuation of the QW thickness was used and was regarded effectively as a quantum dot. That island-like structure is usually elongated along the [$\bar{1}$10] direction [41]. As a consequence of the long-range part of the electron-hole exchange interaction [42], the exciton states show the doublet splitting. Each doublet consists of orthogonally polarized exciton states with an energy splitting of the order of several tens of μeV. In the experiments an exciton doublet was excited by a phase-locked pulse pair and the time-integrated intensity of luminescence from the exciton ground state was measured. The typical energy level structure is depicted in Fig. 9.11. Now we present a theoretical framework to describe these experiments. The polarization

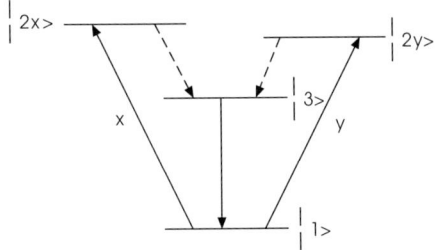

Fig. 9.11 An energy level scheme. The non-radiative relaxation pathways of population are depicted by dashed lines. Solid lines represent optical transitions. $|2x\rangle$ ($|2y\rangle$) denotes the exciton state polarized in the x-(y-)direction.

direction of the orthogonally polarized exciton states will be denoted as x- or y-direction. The electric field of the phase-locked pulse pair is written as

$$\vec{E}(t) = \vec{\varepsilon}_1(t)\cos\omega t + \vec{\varepsilon}_2(t-t_d)\cos\omega(t-t_d)$$

$$= \frac{1}{2}(\vec{\varepsilon}_1(t) + \vec{\varepsilon}_2(t-t_d)e^{-i\omega t_d})e^{i\omega t} + \frac{1}{2}(\vec{\varepsilon}_1(t) + \vec{\varepsilon}_2(t-t_d)e^{i\omega t_d})e^{-i\omega t}$$

with

$$\vec{\varepsilon}_1(t) = \vec{e}_1\varepsilon(t), \quad \vec{\varepsilon}_2(t-t_d) = \vec{e}_2\varepsilon(t-t_d), \tag{9.90}$$

where $\vec{e}_1$ and $\vec{e}_2$ are the polarization unit vectors and $\varepsilon(t)$ is the common pulse envelope assumed in a hyperbolic secant form as in (9.41). The electric dipole interaction term can be written as

$$-\vec{\mu}\cdot\vec{E}(t) = -\hbar\Omega_x(t)e^{i\omega t} - \hbar\Omega_y(t)e^{i\omega t} + \text{c.c.}$$

with

$$\Omega_x(t) = \frac{\mu_x}{2\hbar}(e_{1x}\varepsilon(t) + e_{2x}\varepsilon(t-t_d)e^{-i\omega t_d}), \tag{9.91}$$

$$\Omega_y(t) = \frac{\mu_y}{2\hbar}(e_{1y}\varepsilon(t) + e_{2y}\varepsilon(t-t_d)e^{-i\omega t_d}), \tag{9.92}$$

where μ_i, e_{1i} and e_{2i} ($i = x, y$) are the Cartesian components of the dipole moment operator and the polarization unit vectors. With these notations the equations of motion of the density matrix for the four levels $|1\rangle, |2x\rangle, |2y\rangle$ and $|3\rangle$ in Fig. 9.11 are given by

$$\dot{\rho}_{11} = -i\Omega_x^*(t)e^{-i\omega t}\rho_{1,2x} + i\Omega_x(t)e^{i\omega t}\rho_{2x,1} - i\Omega_y^*(t)e^{-i\omega t}\rho_{1,2y}$$
$$+ i\Omega_y(t)e^{i\omega t}\rho_{2y,1} + \kappa_{21}(\rho_{2x,2x} + \rho_{2y,2y}) + \kappa_{31}\rho_{33}, \tag{9.93}$$

$$\dot{\rho}_{2x,2x} = i\Omega_x^*(t)e^{-i\omega t}\rho_{1,2x} - i\Omega_x(t)e^{i\omega t}\rho_{2x,1} - (\kappa_{21} + \kappa_{23})\rho_{2x,2x}, \tag{9.94}$$

$$\dot{\rho}_{2y,2y} = i\Omega_y^*(t)e^{-i\omega t}\rho_{1,2y} - i\Omega_y(t)e^{i\omega t}\rho_{2y,1} - (\kappa_{21} + \kappa_{23})\rho_{2y,2y}, \tag{9.95}$$

$$\dot{\rho}_{33} = -\kappa_{31}\rho_{33} + \kappa_{23}(\rho_{2x,2x} + \rho_{2y,2y}), \tag{9.96}$$

$$\dot{\rho}_{1,2x} = -i\Omega_x(t)e^{i\omega t}(\rho_{11} - \rho_{2x,2x}) + (i\omega_{2x} - \gamma_{12})\rho_{1,2x} + i\Omega_y(t)e^{i\omega t}\rho_{2y,2x}, \tag{9.97}$$

$$\dot{\rho}_{1,2y} = -i\Omega_y(t)e^{i\omega t}(\rho_{11} - \rho_{2y,2y}) + (i\omega_{2y} - \gamma_{12})\rho_{1,2y} + i\Omega_x(t)e^{i\omega t}\rho_{2x,2y}, \tag{9.98}$$

$$\dot{\rho}_{2x,1} = i\Omega_x^*(t)e^{-i\omega t}(\rho_{11} - \rho_{2x,2x}) + (-i\omega_{2x} - \gamma_{12})\rho_{2x,1} - i\Omega_y^*(t)e^{-i\omega t}\rho_{2x,2y}, \tag{9.99}$$

$$\dot{\rho}_{2y,1} = i\Omega_y^*(t)e^{-i\omega t}(\rho_{11} - \rho_{2y,2y}) + (-i\omega_{2y} - \gamma_{12})\rho_{2y,1} - i\Omega_x^*(t)e^{-i\omega t}\rho_{2y,2x}, \tag{9.100}$$

$$\dot{\rho}_{2x,2y} = i\Omega_x^*(t)e^{-i\omega t}\rho_{1,2y} - i\Omega_y(t)e^{i\omega t}\rho_{2x,1} + (-i(\omega_{2x} - \omega_{2y}) - \gamma_{22})\rho_{2x,2y}, \tag{9.101}$$

$$\dot{\rho}_{2y,2x} = i\Omega_y^*(t)e^{-i\omega t}\rho_{1,2x} - i\Omega_x(t)e^{i\omega t}\rho_{2y,1} + (i(\omega_{2x} - \omega_{2y}) - \gamma_{22})\rho_{2y,2x}, \tag{9.102}$$

$$\dot{\rho}_{13} = i\Omega_x(t)e^{i\omega t}\rho_{2x,3} + i\Omega_y(t)e^{i\omega t}\rho_{2y,3} + (i\omega_3 - \gamma_{13})\rho_{13}, \tag{9.103}$$

$$\dot{\rho}_{31} = -i\Omega_x^*(t)e^{-i\omega t}\rho_{3,2x} - i\Omega_y^*(t)e^{-i\omega t}\rho_{3,2y} + (-i\omega_3 - \gamma_{13})\rho_{31}, \tag{9.104}$$

$$\dot{\rho}_{2x,3} = i\Omega_x^*(t)e^{-i\omega t}\rho_{13} + (-i(\omega_{2x} - \omega_3) - \gamma_{23})\rho_{2x,3}, \tag{9.105}$$

$$\dot{\rho}_{3,2x} = -i\Omega_x(t)e^{i\omega t}\rho_{31} + (i(\omega_{2x} - \omega_3) - \gamma_{23})\rho_{3,2x}, \tag{9.106}$$

$$\dot{\rho}_{2y,3} = i\Omega_y^*(t)e^{-i\omega t}\rho_{13} + (-i(\omega_{2y} - \omega_3) - \gamma_{23})\rho_{2y,3}, \tag{9.107}$$

$$\dot{\rho}_{3,2y} = -i\Omega_y(t)e^{i\omega t}\rho_{31} + (i(\omega_{2y} - \omega_3) - \gamma_{23})\rho_{3,2y}, \tag{9.108}$$

where the same notations are used as in (9.18)–(9.26) and $\hbar\omega_i$ ($i = 2x, 2y, 3$) is the energy of each level relative to the ground state $|1\rangle$. It is easily seen that the components $\rho_{13}, \rho_{31}, \rho_{2x,3}, \rho_{3,2x}, \rho_{2y,3}$ and $\rho_{3,2y}$ are completely decoupled from other components and they do not have non-zero values for the initial condition that $\rho_{11}(0) = 1$ and $\rho_{ij}(0) = 0$ for other combinations of i and j. Thus those components will be dropped in the following. It is convenient to separate out the major oscillating factor as

$$\rho_{1,2x} = e^{i\omega t}\tilde{\rho}_{1,2x}, \quad \rho_{1,2y} = e^{i\omega t}\tilde{\rho}_{1,2y}, \tag{9.109}$$

$$\rho_{2x,1} = e^{-i\omega t}\tilde{\rho}_{2x,1}, \quad \rho_{2y,1} = e^{-i\omega t}\tilde{\rho}_{2y,1}. \tag{9.110}$$

Then the relevant equations of motion for ten components are given as

$$\frac{d}{dt}\rho = M\rho \tag{9.111}$$

with

$$\rho = {}^t(\rho_{11}, \rho_{2x,2x}, \rho_{2y,2y}, \rho_{33}, \rho_{1,2x}, \rho_{1,2y}, \rho_{2x,1}, \rho_{2y,1}, \rho_{2x,2y}, \rho_{2y,2x}), \tag{9.112}$$

$$M = \begin{pmatrix} 0 & \kappa_{21} & \kappa_{21} & \kappa_{31} & -i\Omega_x^*(t) & -i\Omega_y^*(t) & i\Omega_x(t) & i\Omega_y(t) & 0 & 0 \\ 0 & -\kappa_{23}-\kappa_{21} & 0 & 0 & i\Omega_x^*(t) & 0 & -i\Omega_x(t) & 0 & 0 & 0 \\ 0 & 0 & -\kappa_{23}-\kappa_{21} & 0 & 0 & i\Omega_y^*(t) & 0 & -i\Omega_y(t) & 0 & 0 \\ 0 & \kappa_{23} & \kappa_{23} & -\kappa_{31} & 0 & 0 & 0 & 0 & 0 & 0 \\ -i\Omega_x(t) & i\Omega_x(t) & 0 & 0 & \alpha_{12} & 0 & 0 & 0 & 0 & i\Omega_y(t) \\ -i\Omega_y(t) & 0 & i\Omega_y(t) & 0 & 0 & \beta_{12} & 0 & 0 & i\Omega_x(t) & 0 \\ i\Omega_x^*(t) & -i\Omega_x^*(t) & 0 & 0 & 0 & 0 & \alpha_{12}^* & 0 & -i\Omega_y^*(t) & 0 \\ i\Omega_y^*(t) & 0 & -i\Omega_y^*(t) & 0 & 0 & 0 & 0 & \beta_{12}^* & 0 & -i\Omega_x^*(t) \\ 0 & 0 & 0 & 0 & i\Omega_x^*(t) & -i\Omega_y(t) & 0 & 0 & \delta_{22} & 0 \\ 0 & 0 & 0 & 0 & i\Omega_y^*(t) & 0 & 0 & -i\Omega_x(t) & 0 & \delta_{22}^* \end{pmatrix}$$

$$\tag{9.113}$$

$$\alpha_{12} = -i\Delta\omega - \gamma_{12}, \quad \beta_{12} = -i(\Delta\omega + \Delta) - \gamma_{12}, \quad \delta_{22} = -i\Delta - \gamma_{22}, \tag{9.114}$$

$$\Delta = \omega_{2x} - \omega_{2y}, \quad \Delta\omega = \omega - \omega_{2x}. \tag{9.115}$$

Now in order to examine the correlation trace in the weak excitation case, the perturbative calculation will be carried out. Starting from the initial state that $\rho_{11} = 1$ and other $\rho_{ij} = 0$, we have

$$\tilde{\rho}^{(1)}_{1,2x}(t) = -i \int_0^t d\tau\, e^{-(i\Delta\omega + \gamma_{12})(t-\tau)} \Omega_x(\tau), \tag{9.116}$$

$$\tilde{\rho}^{(1)}_{1,2y}(t) = -i \int_0^t d\tau\, e^{-(i(\Delta\omega + \Delta) + \gamma_{12})(t-\tau)} \Omega_y(\tau). \tag{9.117}$$

In the second-order perturbation, we obtain

$$\rho^{(2)}_{2x,2x}(t) = \int_0^t dt_1\, e^{-\kappa_2(t-t_1)} \Omega_x^*(t_1) \int_0^{t_1} dt_2\, e^{-(i\Delta\omega + \gamma_{12})(t_1-t_2)} \Omega_x(t_2) + c.c., \tag{9.118}$$

$$\rho^{(2)}_{2y,2y}(t) = \int_0^t dt_1\, e^{-\kappa_2(t-t_1)} \Omega_y^*(t_1) \int_0^{t_1} dt_2\, e^{-(i(\Delta\omega + \Delta) + \gamma_{12})(t_1-t_2)} \Omega_y(t_2) + c.c. \tag{9.119}$$

with $\kappa_2 = \kappa_{21} + \kappa_{23}$. In order to see the qualitative features, we consider optical pulses with short duration compared with relevant relaxation times and assume as $E(t) \propto \delta(t)$. Then the time-integrated luminescence intensity is proportional to

$$\int_0^\infty dt\, (\rho^{(2)}_{2x,2x}(t) + \rho^{(2)}_{2y,2y}(t)) \tag{9.120}$$

and is calculated as

$$|\langle 1|\mu_x|2x\rangle|^2 (e_{1x}^2 + e_{2x}^2 + 2e_{1x}e_{2x} e^{-\gamma_{12}t_d} \cos\omega_{2x} t_d)$$
$$+ |\langle 1|\mu_y|2y\rangle|^2 (e_{1y}^2 + e_{2y}^2 + 2e_{1y}e_{2y} e^{-\gamma_{12}t_d} \cos\omega_{2y} t_d). \tag{9.121}$$

Since for the doublet states $|2x\rangle$ and $|2y\rangle$ the magnitude of $|\langle 1|\mu_x|2x\rangle|^2$ and $|\langle 1|\mu_y|2y\rangle|^2$ is almost equal to each other, the above expression is proportional to

$$I(t_d) = 1 + e_{1x}e_{2x} e^{-\gamma_{12}t_d} \cos\omega_{2x} t_d + e_{1y}e_{2y} e^{-\gamma_{12}t_d} \cos\omega_{2y} t_d. \tag{9.122}$$

When the pulse pair is polarized in the same direction along the half way between the x-axis and the y-axis, namely

$$e_{1x} = e_{1y} = e_{2x} = e_{2y} = \frac{1}{\sqrt{2}}, \tag{9.123}$$

we have

$$I(t_d) = 1 + e^{-\gamma_{12}t_d} \cos \frac{\omega_{2x} + \omega_{2y}}{2} t_d \cos \frac{\Delta}{2} t_d. \tag{9.124}$$

Thus the time-integrated luminescence intensity shows a beating with a period of $2\pi/\Delta$ due to the simultaneous excitation of the doublet with energy separation of Δ. On the other hand, when the pulse pair is orthogonally polarized, namely

$$e_{1x} = e_{1y} = e_{2x} = -e_{2y} = \frac{1}{\sqrt{2}}, \tag{9.125}$$

we have

$$I(t_d) = 1 - e^{-\gamma_{12}t_d} \sin \frac{\omega_{2x} + \omega_{2y}}{2} t_d \sin \frac{\Delta}{2} t_d. \tag{9.126}$$

Thus the beating structure shows a node at $t_d = 0$ in the case of orthogonal polarization, whereas it shows a peak in the case of parallel polarization.

For stronger excitation, we have to calculate numerically the luminescence intensity based on (9.111). Results are shown in Figs 9.12 and 9.13 for respective cases of the parallel and orthogonal polarization of the phase-locked pulse pair. The relevant parameters are chosen as

$$t_p = 6 \text{ ps}, \kappa_{21}t_p = \kappa_{23}t_p = 0.01, \kappa_{31}t_p = 0.005,$$

$$\gamma_{12}t_p = 0.08, \gamma_{22}t_p = 0.03, \Delta t_p = (\omega_{2x} - \omega_{2y})t_p = 2.0,$$

$$\Delta\omega t_p = (\omega - \omega_{2x})t_p = -1.0, \langle 1 | \mu_x | 2x \rangle = \langle 1 | \mu_y | 2y \rangle = \mu,$$

$$\Omega_0 t_p = \frac{\mu \varepsilon_0}{\hbar} t_p = 0.2, 1.0, 2.0, 4.0. \tag{9.127}$$

With increasing excitation intensity, the sinusoidal beating structure deforms due to incoherent processes via accumulated population in the excited states. Since both $|2x\rangle$ and $|2y\rangle$ levels show the Rabi splitting, the interference pattern in the correlation trace becomes much involved.

9.8 Effect of two-photon coherence

So far we have considered single exciton states in a quantum dot. However, with increasing excitation intensity, we have to take into account two-exciton states. It is easily expected that the behavior of the quantum interference will be modified significantly by the two-

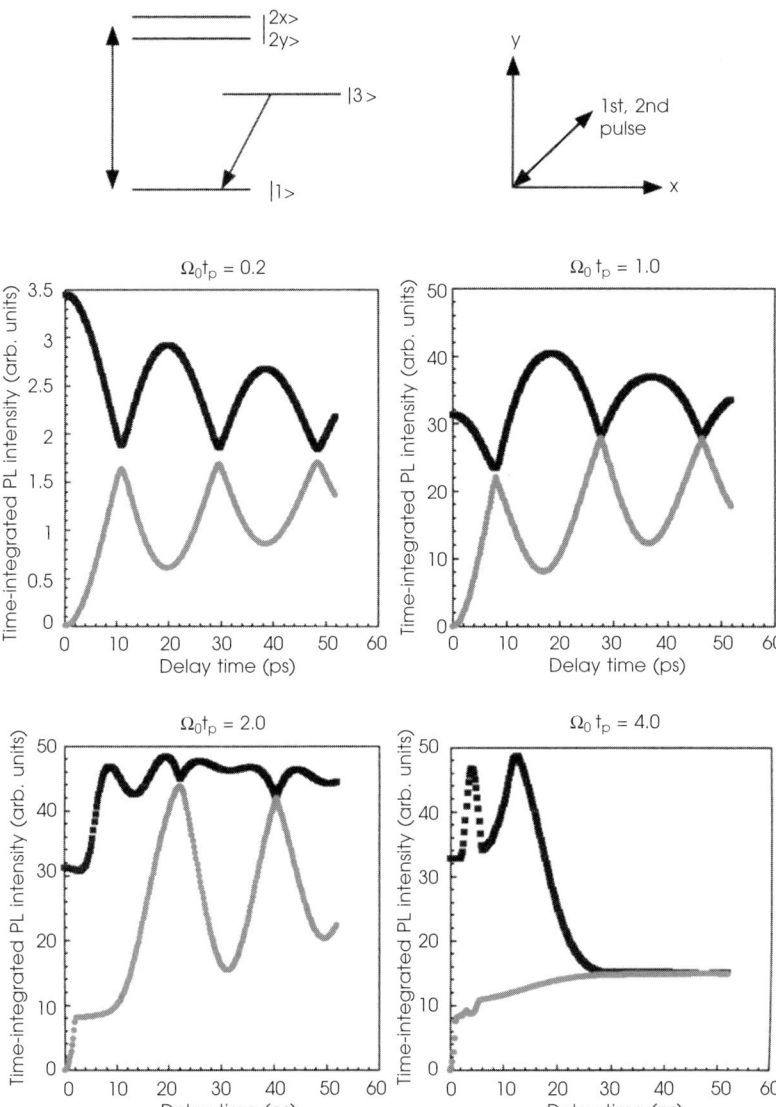

Fig. 9.12 Correlation traces for a four-level system in Fig. 9.11. The phase-locked pulse pair is polarized in the same direction along the half way between the x-axis and the y-axis.

photon coherence via two-exciton states. This feature is important in achieving the exciton entangled state [17] because the contribution from two-exciton states cannot be precluded only by the large detuning. In fact, under the pulse excitation, the contribution from off-resonant levels is much emphasized and cannot be neglected since the intrinsic

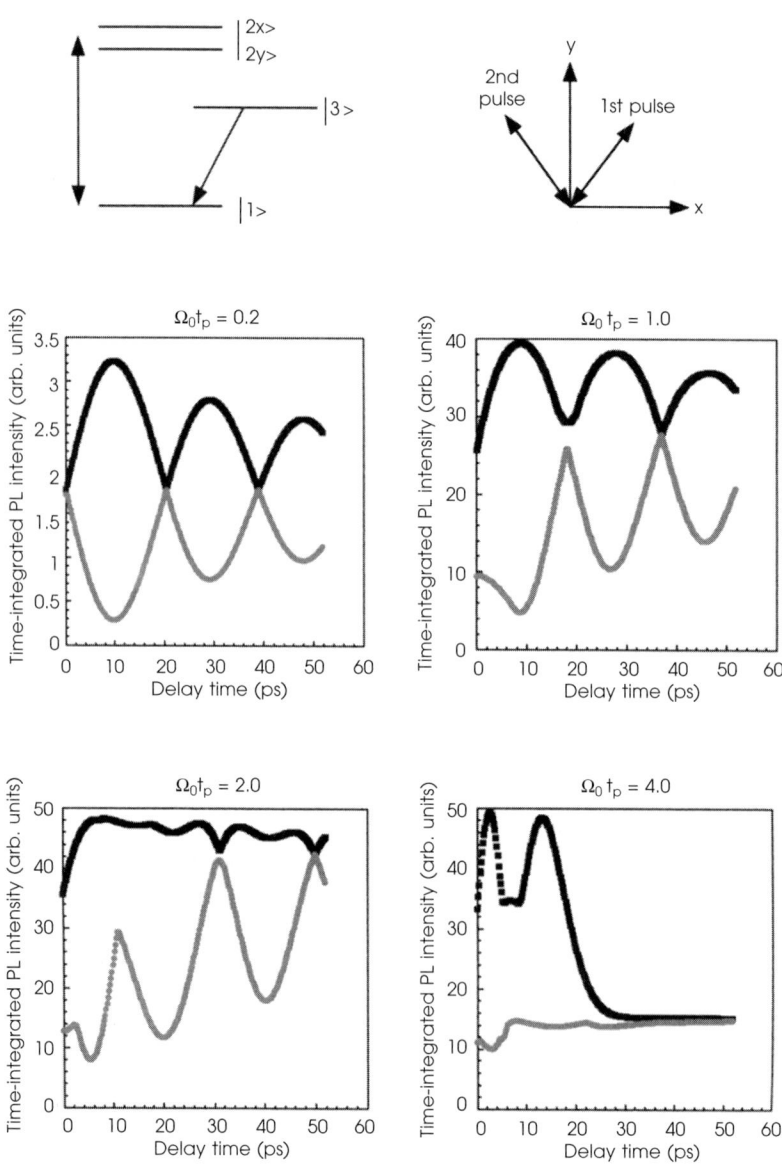

Fig. 9.13 Correlation traces for a four-level system in Fig. 9.11. The phase-locked pulse pair is polarized orthogonally along the half way between the x-axis and the y-axis.

spectral broadening of the pulse induces the excitation of those states. In order to see typical behaviors, we add a two-exciton state denoted by m in Fig. 9.14. The state m is assumed to be excited by the combination of two photons with x- or y-polarization. This selection rule corresponds to the two-exciton state with the total angular momentum $J = 0$ [43]. The level $|3\rangle$, namely, the exciton ground state is not coupled with other levels coherently via optical fields. By the same reasoning as in section 9.7, the components of the density matrix of the form ρ_{3j} or ρ_{j3} ($j = 1, 2x, 2y, m$) are completely decoupled from other components. Thus 17 members out of 25 components of the density matrix are relevant in the equations of motion. The explicit expression of the equations of motion is rather lengthy and is omitted. In this case also, the calculated quantity is the time-integrated luminescence intensity from the exciton ground state.

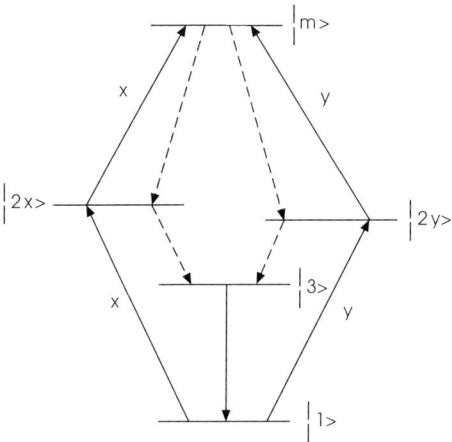

Fig. 9.14 An energy level scheme. The state $|m\rangle$ is added to the level scheme in Fig. 9.11. The non-radiative relaxation pathways of population are depicted by dashed lines. Solid lines represent optical transitions.

A model calculation is carried out to see the excitation intensity dependence of the correlation trace as shown in Fig. 9.15. The relevant parameters are chosen as

$t_p = 6$ ps, $\kappa_{21}t_p = \kappa_{23}t_p = 0.01$, $\kappa_{31}t_p = 0.005$, $\kappa_{m2}t_p = 0.01$,

$\gamma_{12}t_p = 0.08$, $\gamma_{22}t_p = 0.03$, $\gamma_{m1}t_p = 0.08$, $\gamma_{m2}t_p = 0.1$,

$\Delta t_p = (\omega_{2x} - \omega_{2y})t_p = 2.0$, $\Delta\omega t_p = (\omega - \omega_{2x})t_p = -1.0$, $\dfrac{(E_m - 2E_{2x})t_p}{\hbar} = -1.0$,

$\langle 1 | \mu_x | 2x \rangle = \langle 1 | \mu_y | 2y \rangle = \langle 2x | \mu_x | m \rangle = \langle 2y | \mu_y | m \rangle = \mu$,

$\Omega_0 t_p = \dfrac{\mu\varepsilon_0}{\hbar} t_p = 0.2, 1.0, \ldots,$ \hfill (9.128)

where κ_{m2} denotes the population relaxation rate from the state $|m\rangle$ to the exciton state

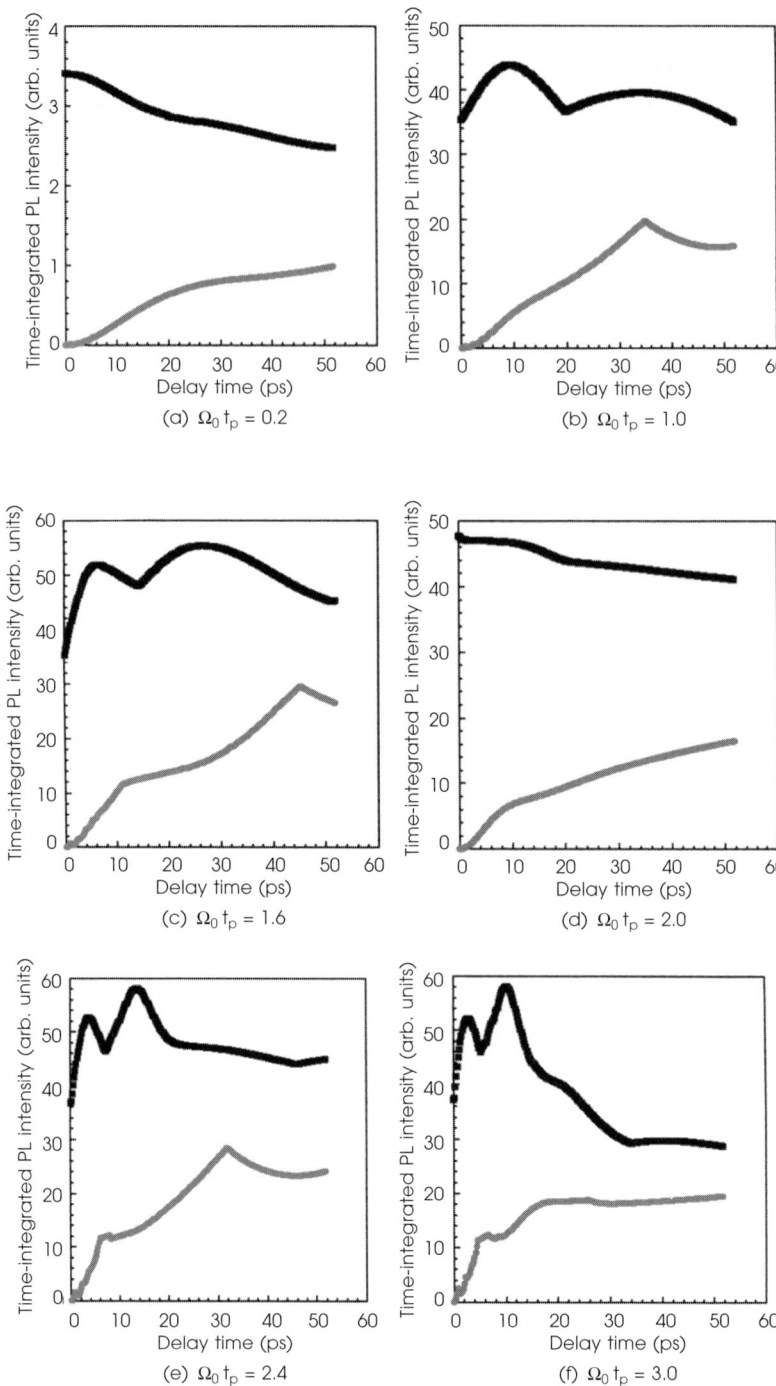

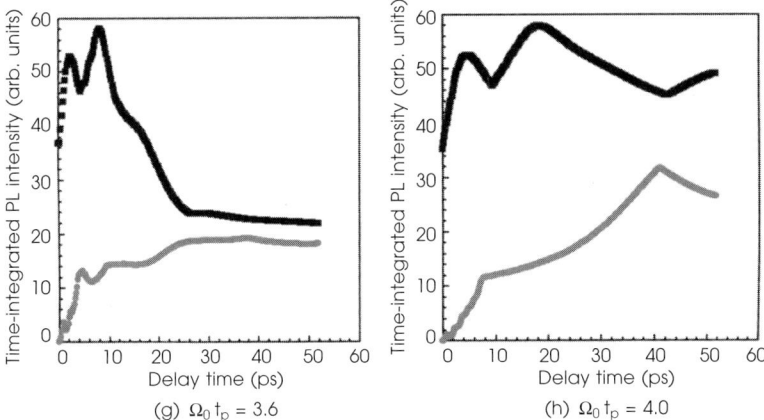

Fig. 9.15 Correlation traces for a five-level system in Fig. 9.14. The phase-locked pulse pair is polarized in the same direction along the x-axis.

$|2x\rangle$ or $|2y\rangle$ ($\kappa_{m,2x} = \kappa_{m,2y} = \kappa_{m2}$) and γ_{m2} is the dephasing rate of the coherence between $|m\rangle$ and $|2x\rangle$ or $|2y\rangle$ ($\gamma_{m,2x} = \gamma_{m,2y} = \gamma_{m2}$). A remarkable feature is that a quantum-beat like behavior appears even when the single polarization component is excited, namely, the phase-locked pulse pair is polarized in the same direction, e.g., x- or y-direction. The Rabi oscillation is induced not only between the ground state and the exciton state but also between the exciton state and the two-exciton state with respective oscillation frequencies. As a result of interference between two Rabi oscillations, the exciton population shows a beating modulation and the time-integrated intensity of luminescence from the exciton ground state exhibits a beating feature as a function of the delay time between two phase-locked pulses. In the weak excitation case, the excitonic transition is mainly excited and a simple behavior of exponential decay is apparent. With increasing excitation intensity, the transition between the exciton state and the two-exciton state is also induced and the interference mentioned above tends to manifest in the form of a beating structure in the correlation trace. The frequency difference between two Rabi oscillations is determined mainly by the biexciton binding energy $B_{XX} = E_m - 2E_{2x}$ (or E_{2y}). The period of the beating is given by $2\pi\hbar/B_{XX}$, as can be guessed from the arguments in (9.124). The actual beating behavior in the correlation trace is sensitively dependent on the relative phase of the Rabi oscillation between the ground state and the exciton state and that between the exciton state and the two-exciton state. It is interesting to note that in the case of $\Omega_0 t_p = 2$, which corresponds to a pulse area of 2π, a simple behavior of exponential decay is recovered as shown in Fig. 9.15(d). This is understood as follows.

The system evolves from the ground state, namely the Bloch vector starts from the south pole and turns around by 2π radian, returning to the vicinity of the south pole. This situation is similar to the case of small angle rotation of the Bloch vector under the weak excitation in which only the excitonic transition is excited. Thus the correlation trace exhibits an exponential decay as in the weak excitation case. On the basis of the same arguments, we expect to observe an exponential decay of the correlation trace in the case of $\Omega_0 t_p = 4$, which corresponds to a pulse area of 4π, since the Bloch vector returns to the neighborhood of the south pole as in the case of $\Omega_0 t_p = 2$. Contrary to this expectation, however, the correlation trace exhibits a complicated beating behavior as shown in Fig. 9.15(h). In this case the exciton population accumulates within the pulse duration due to the presence of relaxation channels and the coherence induced by the second pulse via this exciton population gives rise to a complicated interference pattern.

Now we consider the case in which the first pulse is polarized along the half way between the x-axis and the y-axis and the second pulse is polarized perpendicular to that. In the weak excitation case, the correlation trace shows a typical beating feature with exponential decay and is almost the same as in Fig. 9.13, since the two-exciton state $|m\rangle$ is not excited. With increasing excitation intensity, the envelope of the correlation trace begins to deviate from a sinusoidal beating behavior and exhibits a seemingly irregular pattern as shown in Fig. 9.16. In a typical case of $\Omega_0 t_p = 2$, the fine time behavior in the femtosecond range of the correlation trace is quite interesting as exhibited in Fig. 9.17. Around the coarse time delay of 0 ps, there appear four peaks within one optical cycle (~3 fs) corresponding to the interference among four optical transitions, namely the $2x$ and $2y$ exciton transitions and the $2x$-m and $2y$-m transitions between the exciton state and the two-exciton state. With increasing delay time t_d, the interference fades away and the four peaks within one optical cycle of t_d merge into two peaks.

In the case of parallel polarization in which the phase-locked pulse pair is polarized in the same direction along the half way between the x-axis and the y-axis, the overall beating features of the coarse time envelope and the fine time behaviors are exhibited in Figs 9.18 and 9.19, respectively. The seemingly irregular pattern in the case of higher excitation intensity are similar to those in Fig. 9.16. These studies will provide important informations on physical conditions to realize the exciton entanglement in a single quantum dot. Actually there are several bound and even continuum two-exciton states and these would contribute significantly to the quantum interference phenomena. Studies on the effects of these states are left for the future.

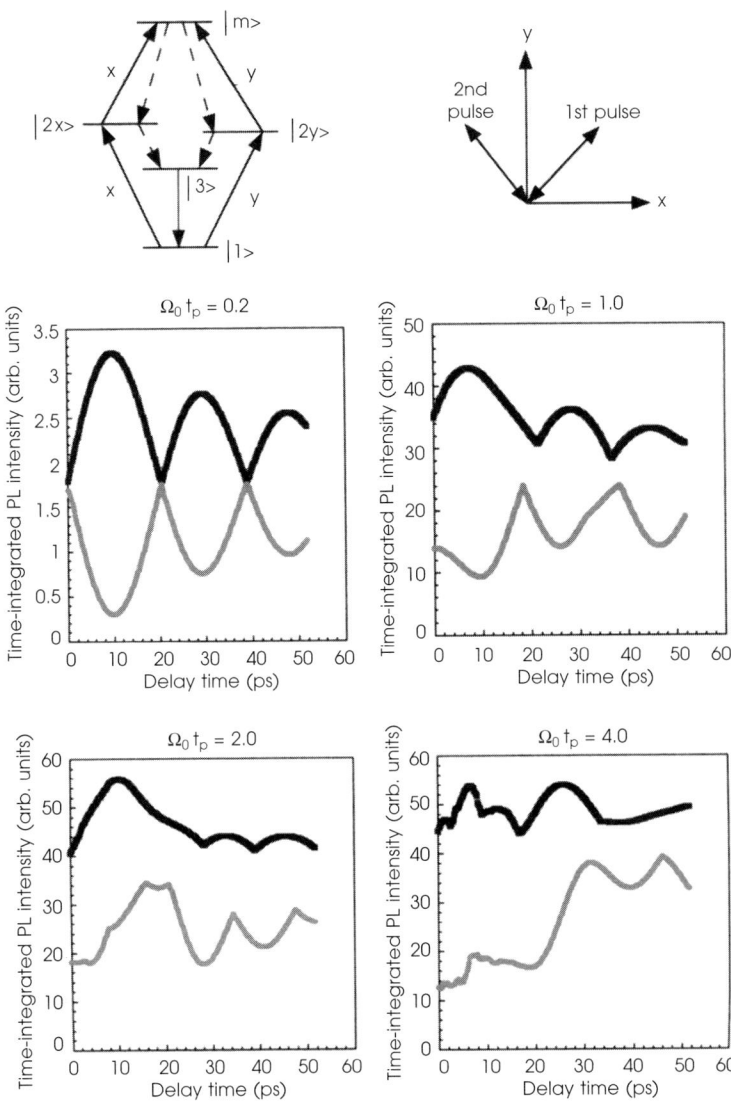

Fig. 9.16 Correlation traces for a five-level system in Fig. 9.14. The phase-locked pulse pair is polarized orthogonally along the half way between the x-axis and the y-axis.

9.9 Exciton dephasing in semiconductor quantum dots

A resonant optical excitation creates an excited state population and also induces an optical polarization. Dynamics of this optical excitation is characterized by relaxation

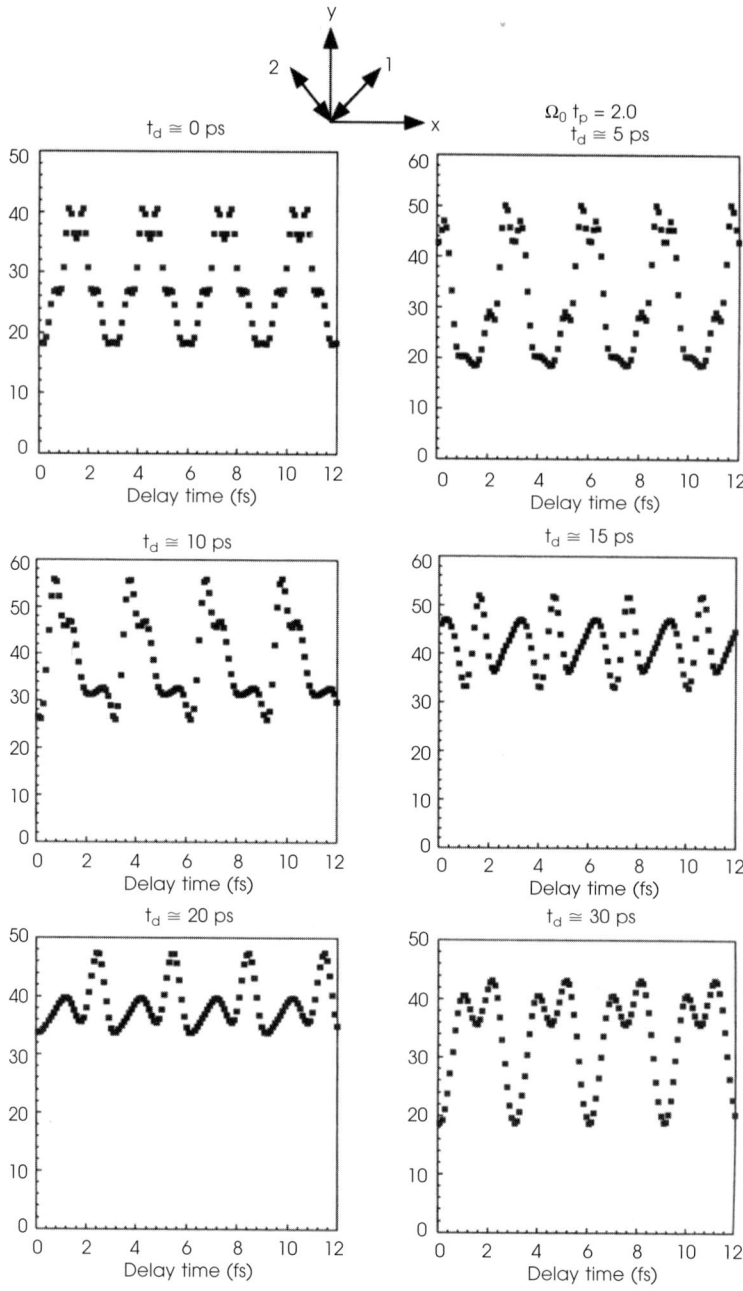

Fig. 9.17 Correlation traces in the femtosecond range around t_d = 0 ps, 5 ps, 10 ps, 15 ps, 20 ps and 30 ps corresponding to the case of $\Omega_0 t_p$ = 2.0 in Fig. 9.16.

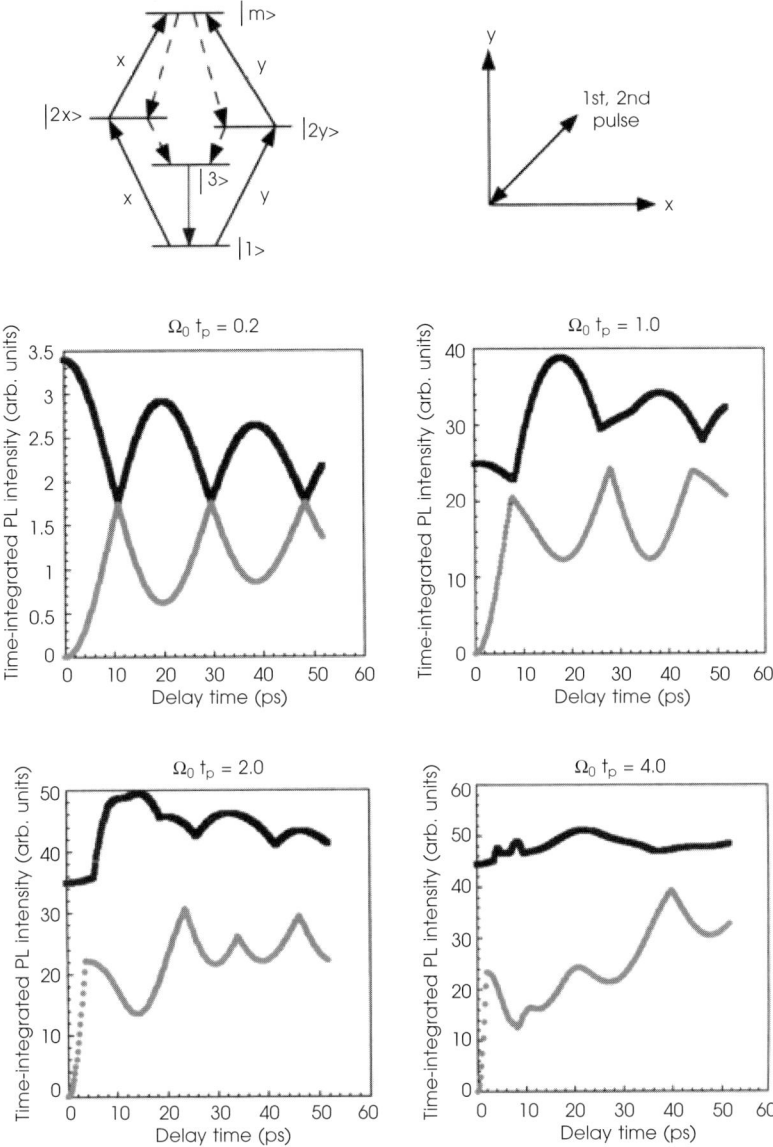

Fig. 9.18 Correlation traces for a five-level system in Fig. 9.14. The phase-locked pulse pair is polarized in the same direction along the half way between the x-axis and the y-axis.

of the population as well as decay of the induced optical polarization. In lower dimensional semiconductors, electronic confinement leads to qualitative changes in population relaxation including spontaneous emission and exciton-phonon scattering, as shown in

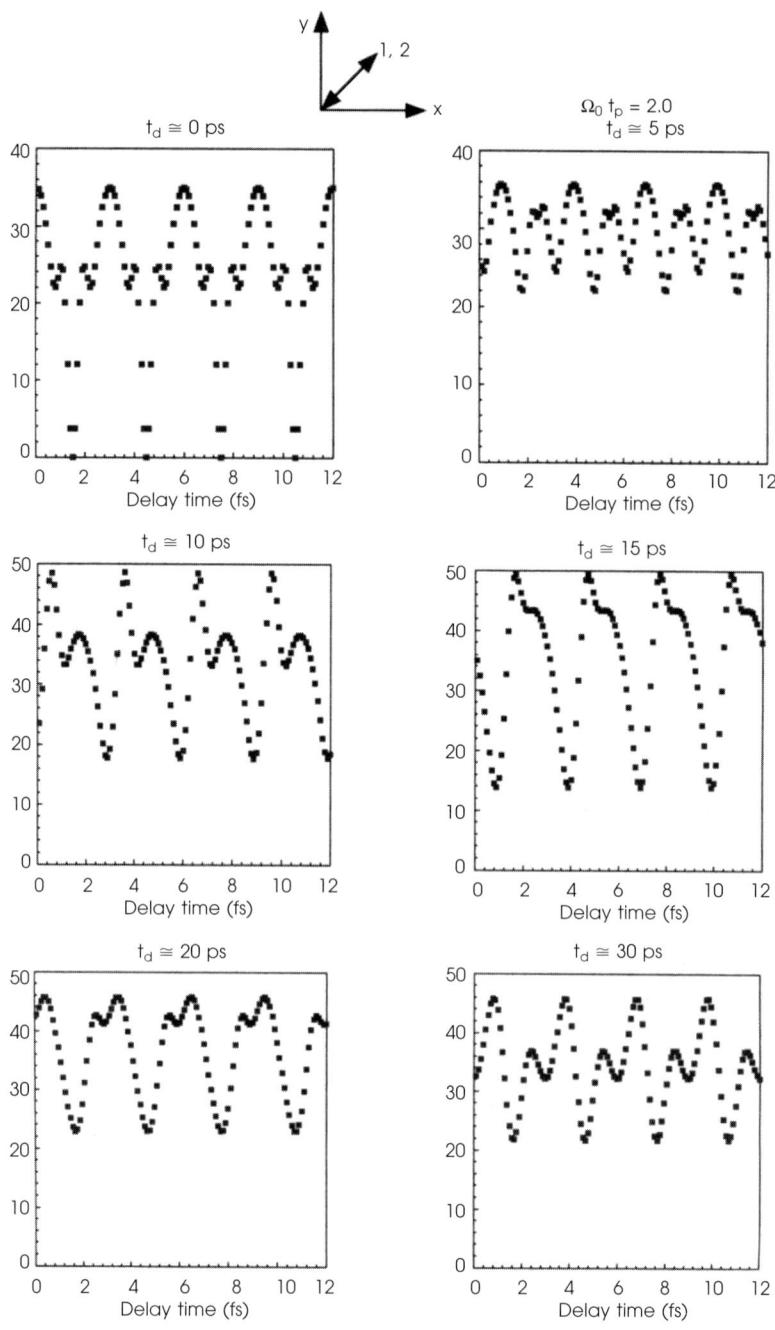

Fig. 9.19 Correlation traces in the femtosecond range around $t_d = 0$ ps, 5 ps, 10 ps, 15 ps, 20 ps and 30 ps corresponding to the case of $\Omega_0 t_p = 2.0$ in Fig. 9.18.

extensive recent studies [44]. These population relaxation processes are expected to contribute to dephasing with a dephasing rate given by half the population decay rate. Pure dephasing processes that do not involve population or energy relaxation of excitons can also contribute to dephasing. Pure dephasing, which is a well-established concept for atomic systems, remains yet to be investigated in lower dimensional semiconductors due to a lack of direct comparison between dephasing and population relaxation and between theory and experiment. Studies of pure dephasing processes in lower dimensional semiconductors will renew and deepen our understanding of dephasing of collective excitations in solids, although several seminal studies were done on the exciton dephasing in quantum well (QW) structures [45–48].

Narrow GaAs QWs grown by molecular beam epitaxy (MBE) and with growth interruptions have provided a model system for investigating dephasing processes in lower dimensional semiconductors. In these narrow QWs, fluctuations at the interface between GaAs and AlGaAs lead to localization of excitons at monolayer-high islands. These localized states can be regarded effectively as weakly-confined quantum dot (QD)-like states. One dimension of the confinement is defined by the width of the QW, while the other two lateral dimensions are defined by the effective size of the islands. To avoid inhomogeneous broadening due to well-width and island-size fluctuations, earlier studies have used photoluminescence (PL) and PL excitation with high spatial resolution to probe excitons in individual islands [49,12,13]. As a result, a very narrow linewidth of about several tens of μeV was observed. Without additional information on population relaxation, it was suggested that at very low temperature dephasing of excitons in these structures is due to radiative recombination, while at elevated temperatures dephasing is mainly caused by thermal activation of excitons to higher excited states [12]. However, this interpretation is not complete since both of the suggested processes belong to the longitudinal decay processes and the dephasing rate is in general composed of half the longitudinal decay rate and the pure dephasing rate. In order to examine the presence of pure dephasing in this system, we carried out nonlinear optical measurements of the exciton dephasing in GaAs QD-like islands based on the three-pulse stimulated photon echo method [50]. This method enables the simultaneous measurement of the dephasing rate and the population decay rate. At very low temperatures the observed dephasing rate $\Gamma_\perp$ is very close to half the population decay rate $\Gamma_\parallel/2$, suggesting that dephasing is caused mainly by the population decay. With increasing temperature the dephasing rate increases much faster than the population decay rate. At elevated temperatures (>30 K), dephasing rates become much greater than $\Gamma_\parallel/2$, indicating a dominant contribution of pure dephasing. Thus our measurements revealed convincingly the presence of pure dephasing process that dominates excitonic dephasing at elevated temperatures.

436 T. Takagahara

The observed strong increase of the pure dephasing rate above 30 K suggests that interactions between excitons and low-energy acoustic phonons play an essential role in the pure dephasing process. In this chapter, we present a theoretical model that takes into account the interaction between excitons and acoustic phonons and can explain satisfactorily the magnitude as well as the temperature dependence of the dephasing rate. Our model generalizes the Huang–Rhys theory of F-centers [51–53] to include mixing among the ground and excited exciton states through exciton–phonon interactions [54] and as a result enables us to identify the elementary processes of exciton pure dephasing.

9.10 Green function formalism of exciton dephasing rate

As discussed in the last section, the strong increase of the exciton dephasing rate above 30 K suggests the important role of the low-energy acoustic phonons in determining the dephasing rate. The exciton dephasing rate can be estimated most directly from the half-width at half maximum (HWHM) of the absorption spectrum which can be calculated from the Fermi golden rule as

$$I(\hbar\omega) = \frac{2\pi}{\hbar} \text{Av.} \sum_f |\langle f | V_R | g \rangle|^2 \delta(\hbar\omega + E_g - E_f), \tag{9.129}$$

where V_R is the electromagnetic interaction, $|f\rangle$ and $|g\rangle$ denote the final exciton state and the initial ground state, respectively, including the phonon degrees of freedom and Av. means the average over the thermal equilibrium state of phonons. This expression can be rewritten as

$$I(\hbar\omega) \propto \text{Re}\left[\int_0^\infty dt\, e^{-i\omega t} \text{Av.}\left\langle g \left| V_R \exp\left[\frac{i}{\hbar} H_e t\right] V_R \exp\left[-\frac{i}{\hbar} H_g t\right] \right| g \right\rangle\right], \tag{9.130}$$

where H_e and H_g are the Hamiltonians in the excited state and the ground state, respectively. This can be confirmed by inserting a closure relation between two V_R's. Equation (9.130) is a Fourier–Laplace transform of a correlation function. In order to proceed further, the Hamiltonians will be specified as

$$H_g = \sum_\alpha \hbar\omega_\alpha b_\alpha^\dagger b_\alpha, \tag{9.131}$$

$$H_e = \sum_i E_i |i\rangle\langle i| + \sum_\alpha \hbar\omega_\alpha b_\alpha^\dagger b_\alpha + \sum_\alpha M_\alpha (b_\alpha + b_\alpha^\dagger), \tag{9.132}$$

where the index α denotes the acoustic phonon mode, E_i the energy of the exciton states and M_α is the exciton–phonon coupling matrix within the exciton state manifold. This is a generalization of the Huang–Rhys model of F-centers [51] to include mixing among the exciton state manifold which is reflected in the matrix form of M_α. In the following, we take into account only the exciton–phonon coupling which is linear with respect to the phonon coordinates. Even within this range, however, the well-known deformation potential coupling and the piezoelectric coupling are included. Thus our Hamiltonian is sufficiently general. We note that in the elementary processes of the exciton–phonon interaction the crystal momentum conservation needs to be satisfied in directions where the translational invariance holds. The dephasing process becomes prominent in systems with three-dimensional (3D) electronic confinement because the 3D confinement relaxes the crystal momentum conservation and also suppresses exciton population relaxation [55,56] due to the exciton–phonon interactions.

Hereafter the three terms of H_e in (9.132) will be denoted, respectively as

$$H_e = H_e^0 + H_g + V. \tag{9.133}$$

Then the Laplace transform of $\exp[(i/\hbar)H_e t]$ can be expanded as

$$\int_0^\infty dt\, e^{-st} \exp[(i/\hbar)H_e t]$$

$$= \frac{1}{s - \frac{i}{\hbar}H_e} = \frac{1}{s - \frac{i}{\hbar}(H_e^0 + H_g)} + \frac{i}{\hbar}\frac{1}{s - \frac{i}{\hbar}(H_e^0 + H_g)} V \frac{1}{s - \frac{i}{\hbar}(H_e^0 + H_g)}$$

$$+ \left(\frac{i}{\hbar}\right)^2 \frac{1}{s - \frac{i}{\hbar}(H_e^0 + H_g)} V \frac{1}{s - \frac{i}{\hbar}(H_e^0 + H_g)} V \frac{1}{s - \frac{i}{\hbar}(H_e^0 + H_g)} + \cdots \tag{9.134}$$

For example, the third term on the right hand side of (9.134) can be expressed in the convolution form as

$$\left(\frac{i}{\hbar}\right)^2 \int_0^t dt_1 \int_0^{t_1} dt_2\, e^{\frac{i}{\hbar}(H_e^0 + H_g)(t - t_1)} V e^{\frac{i}{\hbar}(H_e^0 + H_g)(t_1 - t_2)} V e^{\frac{i}{\hbar}(H_e^0 + H_g)t_2}. \tag{9.135}$$

Substituting this term for $\exp[(i/\hbar)H_e t]$ in (9.130) and noting the commutability between H_e^0 and H_g and between M_α and H_g, we have

$$\left\langle g \left| V_R \exp\left[\frac{i}{\hbar}H_e t\right] V_R \exp\left[-\frac{i}{\hbar}H_g t\right] \right| g \right\rangle$$

$$\rightarrow \left(\frac{i}{\hbar}\right)^2 \int_0^t dt_1 \int_0^{t_1} dt_2 \sum_{\alpha,\beta} \left\langle g \left| V_R \exp\left[\frac{i}{\hbar}H_e^0(t - t_1)\right] M_\alpha \exp\left[\frac{i}{\hbar}H_e^0(t_1 - t_2)\right] M_\beta \right.\right.$$

$$\times \exp\left[\frac{i}{\hbar}H_e^0 t_2\right] V_R \left| g \right\rangle \times \left\langle 0 \left| \exp\left[\frac{i}{\hbar}H_g(t - t_1)\right] (b_\alpha + b_\alpha^\dagger) \exp\left[\frac{i}{\hbar}H_g(t_1 - t_2)\right]\right.\right.$$

$$\times (b_\beta + b_\beta^\dagger) \exp\left[\frac{i}{\hbar}H_g t_2\right] \exp\left[-\frac{i}{\hbar}H_g t\right] \left| 0 \right\rangle, \tag{9.136}$$

where $|0\rangle$ denotes symbolically the thermal equilibrium state of phonons. The phonon part of (9.136) can be written as

$$\left\langle 0 \left| \exp\left[\frac{i}{\hbar}H_g(t-t_1)\right](b_\alpha + b_\alpha^\dagger)\exp\left[\frac{i}{\hbar}H_g(t_1-t_2)\right](b_\beta + b_\beta^\dagger) \right.\right.$$

$$\left.\left. \times \exp\left[\frac{i}{\hbar}H_g t_2\right]\exp\left[-\frac{i}{\hbar}H_g t\right] \right| 0 \right\rangle$$

$$= \mathrm{T_r}\,\rho_0 \exp\left[-\frac{i}{\hbar}H_g t_1\right](b_\alpha + b_\alpha^\dagger)\exp\left[\frac{i}{\hbar}H_g(t_1-t_2)\right](b_\beta + b_\beta^\dagger)\exp\left[\frac{i}{\hbar}H_g t_2\right]$$

$$= \mathrm{Tr}\rho_0\, (b_\alpha(-t_1) + b_\alpha^\dagger(-t_1))(b_\beta(-t_2) + b_\beta^\dagger(-t_2))$$

$$= \delta_{\alpha\beta}[N_\alpha e^{-i\omega_\alpha(t_1-t_2)} + (1+N_\alpha)e^{i\omega_\alpha(t_1-t_2)}] \tag{9.137}$$

with

$$\rho_0 = e^{-\beta H_g}/\mathrm{Tr}e^{-\beta H_g},$$

$$b_\alpha(t) = \exp\left[\frac{i}{\hbar}H_g t\right]b_\alpha \exp\left[-\frac{i}{\hbar}H_g t\right] = e^{-i\omega_\alpha t}b_\alpha,$$

and

$N_\alpha = 1/[\exp(\beta\hbar\omega_\alpha) - 1]$ with $\beta = 1/(k_B T)$.

Then again making a Laplace transform, we have

$$\left(\frac{i}{\hbar}\right)^2 \left\langle g \left| V_R \frac{1}{s-\frac{i}{\hbar}H_e^0} \sum_\alpha M_\alpha \left\{ \frac{N_\alpha}{s-\frac{i}{\hbar}H_e^0 + i\omega_\alpha} + \frac{1+N_\alpha}{s-\frac{i}{\hbar}H_e^0 - i\omega_\alpha} \right\} M_\alpha \frac{1}{s-\frac{i}{\hbar}H_e^0} V_R \right| g \right\rangle$$

$$= \left\langle g \left| V_R \frac{1}{s-\frac{i}{\hbar}H_e^0} \Sigma_0^{(2)}(s) \frac{1}{s-\frac{i}{\hbar}H_e^0} V_R \right| g \right\rangle \tag{9.138}$$

with

$$\Sigma_0^{(2)}(s) = \left(\frac{i}{\hbar}\right)^2 \sum_\alpha M_\alpha \left\{ \frac{N_\alpha}{s-\frac{i}{\hbar}H_e^0 + i\omega_\alpha} + \frac{1+N_\alpha}{s-\frac{i}{\hbar}H_e^0 - i\omega_\alpha} \right\} M_\alpha. \tag{9.139}$$

Since the exciton–phonon interaction Hamiltonian V is linear with respect to the phonon coordinates, the terms of odd powers in V in (9.134) vanish in the final expression. The next non-vanishing term is the fourth order term in V and this has three contraction diagrams as depicted in Fig. 9.20. For example, the term corresponding to Fig. 9.20(c) can be written as

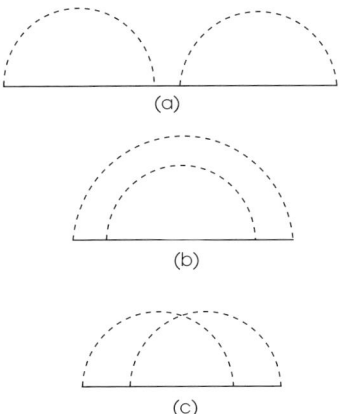

Fig. 9.20 Possible diagrams of the fourth order term in (9.134). A solid (dashed) line denotes the exciton (phonon) propagator (from Takagahara [54]).

$$\Sigma_0^{(4)}(s)$$

$$= \left(\frac{i}{\hbar}\right)^4 \sum_{\alpha,\beta} \left\{ M_\alpha \frac{N_\alpha}{s - \frac{i}{\hbar}H_e^0 + i\omega_\alpha} M_\beta \frac{1}{s - \frac{i}{\hbar}H_e^0 + i(\omega_\alpha + \omega_\beta)} M_\alpha \frac{N_\beta}{s - \frac{i}{\hbar}H_e^0 + i\omega_\beta} M_\beta \right.$$

$$+ M_\alpha \frac{N_\alpha}{s - \frac{i}{\hbar}H_e^0 + i\omega_\alpha} M_\beta \frac{1}{s - \frac{i}{\hbar}H_e^0 + i(\omega_\alpha - \omega_\beta)} M_\alpha \frac{1 + N_\beta}{s - \frac{i}{\hbar}H_e^0 - i\omega_\beta} M_\beta$$

$$+ M_\alpha \frac{1 + N_\alpha}{s - \frac{i}{\hbar}H_e^0 - i\omega_\alpha} M_\beta \frac{1}{s - \frac{i}{\hbar}H_e^0 - i(\omega_\alpha - \omega_\beta)} M_\alpha \frac{N_\beta}{s - \frac{i}{\hbar}H_e^0 + i\omega_\beta} M_\beta$$

$$+ M_\alpha \frac{1 + N_\alpha}{s - \frac{i}{\hbar}H_e^0 - i\omega_\alpha} M_\beta \frac{1}{s - \frac{i}{\hbar}H_e^0 - i(\omega_\alpha + \omega_\beta)} M_\alpha \frac{1 + N_\beta}{s - \frac{i}{\hbar}H_e^0 - i\omega_\beta} M_\beta \right\}.$$

(9.140)

On the other hand, the diagrams in Figs 9.20(a) and (b) can be incorporated into the zeroth order and second order terms, respectively, by renormalizing the exciton propagator and the self-energy, as will be shown below.

First of all, we introduce the Green function defined by

$$G_0^{(2)}(s) = \frac{1}{s - \frac{i}{\hbar}H_e^0 - \Sigma_0^{(2)}(s)}$$

$$= \frac{1}{s - \frac{i}{\hbar}H_e^0} + \frac{1}{s - \frac{i}{\hbar}H_e^0} \Sigma_0^{(2)}(s) \frac{1}{s - \frac{i}{\hbar}H_e^0}$$

$$+ \frac{1}{s - \frac{i}{\hbar}H_e^0} \Sigma_0^{(2)}(s) \frac{1}{s - \frac{i}{\hbar}H_e^0} \Sigma_0^{(2)}(s) \frac{1}{s - \frac{i}{\hbar}H_e^0} + \ldots, \quad (9.141)$$

where $\Sigma_0^{(2)}$ has a meaning of the second order self-energy and $G_0^{(2)}$ represents the exciton propagator in the phonon field. The diagram in Fig. 9.20(a) is included in this Green function. Using this propagator, we improve the second order self-energy as

$$\Sigma^{(2)}(s) = \left(\frac{i}{\hbar}\right)^2 \sum_\alpha M_\alpha [N_\alpha G_0^{(2)}(s + i\omega_\alpha) + (1 + N_\alpha) G_0^{(2)}(s - i\omega_\alpha)] M_\alpha \quad (9.142)$$

and including this self-energy into the denominator, we obtain the improved Green function as

$$G^{(2)}(s) = \frac{1}{s - \frac{i}{\hbar}H_e^0 - \Sigma^{(2)}(s)}. \quad (9.143)$$

This Green function incorporates the diagram in Fig. 9.20(b). Thus only the diagram in Fig. 9.20(c), namely an irreducible diagram, should be included in the fourth order.

Now, using the exciton propagator in (9.143), we calculate the self-energy including the fourth order irreducible diagram, namely

$$\Sigma^{(2)}(s) + \Sigma^{(4)}(s) = \left(\frac{i}{\hbar}\right)^2 \sum_\alpha M_\alpha [N_\alpha G^{(2)}(s + i\omega_\alpha) + (1 + N_\alpha) G^{(2)}(s - i\omega_\alpha)] M_\alpha$$

$$+ \left(\frac{i}{\hbar}\right)^4 \sum_{\alpha,\beta} \{N_\alpha N_\beta M_\alpha G^{(2)}(s + i\omega_\alpha) M_\beta G^{(2)}(s + i(\omega_\alpha + \omega_\beta)) M_\alpha G^{(2)}(s + i\omega_\beta) M_\beta$$

$$+ N_\alpha(1 + N_\beta) M_\alpha G^{(2)}(s + i\omega_\alpha) M_\beta G^{(2)}(s + i(\omega_\alpha - \omega_\beta)) M_\alpha G^{(2)}(s - i\omega_\beta) M_\beta$$

$$+ (1 + N_\alpha) N_\beta M_\alpha G^{(2)}(s - i\omega_\alpha) M_\beta G^{(2)}(s - i(\omega_\alpha - \omega_\beta)) M_\alpha G^{(2)}(s + i\omega_\beta) M_\beta$$

$$+ (1 + N_\alpha)(1 + N_\beta) M_\alpha G^{(2)}(s - i\omega_\alpha) M_\beta G^{(2)}(s - i(\omega_\alpha + \omega_\beta)) M_\alpha G^{(2)}(s - i\omega_\beta) M_\beta\}$$

$$(9.144)$$

and incorporating this self-energy into the denominator, we have the improved Green function as

$$G^{(4)}(s) = \frac{1}{s - \frac{i}{\hbar}H_e^0 - \Sigma^{(2)}(s) - \Sigma^{(4)}(s)}. \quad (9.145)$$

We can extend this procedure up to the higher order iteratively.

As a consequence, we find

$$\int_0^\infty dt\, e^{-st} \text{Av.} \left\langle g \left| V_R \exp\left[\frac{i}{\hbar} H_e t\right] V_R \exp\left[-\frac{i}{\hbar} H_g t\right] \right| g \right\rangle$$

$$= \left\langle g \left| V_R \frac{1}{s - \frac{i}{\hbar} H_e^0 - \Sigma(s)} V_R \right| g \right\rangle \tag{9.146}$$

with

$$\Sigma(s) = \Sigma^{(2)}(s) + \Sigma^{(4)}(s) + \Sigma^{(6)}(s) + \ldots,$$

where the self-energy parts can be represented by irreducible diagrams as shown in Fig. 9.21.

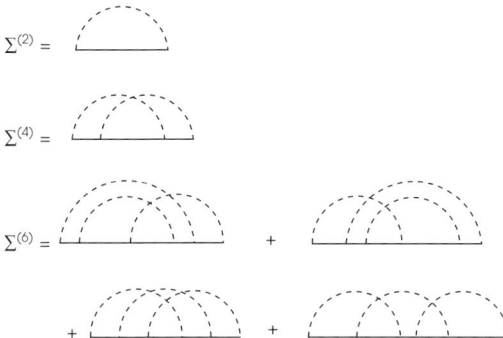

Fig. 9.21 Irreducible diagrams corresponding to the self-energy terms of the second, fourth and sixth order with respect to the exciton–phonon interaction (from Takagahara [54]).

The optical absorption spectrum is calculated from (9.146) by putting s as

$$s \to i\omega + \delta,$$

where δ is a half of the population decay rate of the exciton state excluding the contribution from the acoustic phonon-mediated relaxation, because such contribution is automatically included in the self-energy part $\Sigma(s)$. More concretely, δ should include the radiative decay rate, the trapping rate to some defects and the rate of exciton migration to neighboring islands. The latter two processes are phonon-mediated but should be included in δ because they are not taken into account in the present Green function formalism. δ can be estimated from the observed population decay rate subtracting the phonon-assisted population decay rate within an island which can be calculated theoretically as given in section 9.15.

In order to examine the convergence of the above procedure, we have estimated the Green function up to the sixth order and compared the optical absorption spectrum and its HWHM by calculating

$$I^{(2)}(\omega) = \text{Re}[\text{Av.} \langle g \mid V_R G^{(2)}(s = i\omega + \delta) V_R \mid g \rangle],$$

$$I^{(4)}(\omega) = \text{Re}[\text{Av.} \langle g \mid V_R G^{(4)}(s = i\omega + \delta) V_R \mid g \rangle],$$

$$I^{(6)}(\omega) = \text{Re}[\text{Av.} \langle g \mid V_R G^{(6)}(s = i\omega + \delta) V_R \mid g \rangle]. \tag{9.147}$$

Typical results are shown in Fig. 9.22. The relevant parameters are explained in section 9.12. The size parameters of a quantum disk for Fig. 9.22 are $a = 20$ nm and $b = 15$ nm. It is seen that the percentage difference of the dephasing rates calculated from $G^{(2)}$ and $G^{(4)}$ is about 15–20%, whereas that calculated from $G^{(4)}$ and $G^{(6)}$ is about several %. Hence we carry out the calculation up to the fourth order and estimate the dephasing rate from the HWHM of $I^{(4)}(\omega)$.

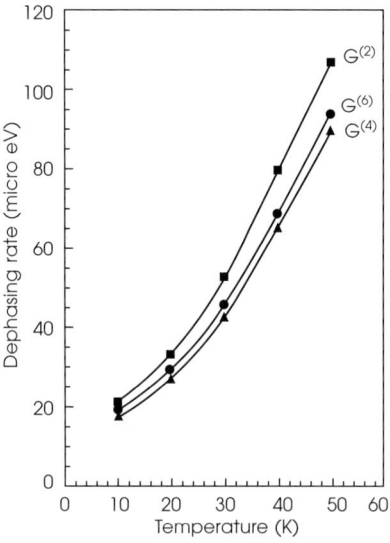

Fig. 9.22 Temperature dependence of exciton dephasing rates calculated by including the self-energy terms up to the second, fourth and sixth order denoted as $G^{(2)}$, $G^{(4)}$ and $G^{(6)}$, respectively. The size parameters of the quantum disk are the same as in Fig. 9.23 (from Takagahara [54]).

9.11 Exciton–phonon interactions

The microscopic details of the interaction Hamiltonian between the exciton and the acoustic phonons will be described. In GaAs/AlGaAs QWs, the elastic properties of both materials are not much different and thus the bulk-like acoustic phonon modes can be assumed as the zeroth order approximation. Hereafter the phonon modes will be specified by the wavevector $\vec{q}$. The interaction between electrons and acoustic phonons arises from the deformation potential coupling and the piezoelectric coupling [57].

The dominant interaction term of the deformation potential coupling is given as

$$H_{DF} = \sum_{r,q} \sqrt{\frac{\hbar |q|}{2\rho u V}} (D_c a_{cr}^\dagger a_{cr} + D_v a_{vr}^\dagger a_{vr}) e^{iqr} (b_q + b_{-q}^\dagger), \qquad (9.148)$$

where $a(a^\dagger)$ denotes the annihilation (creation) operator of the electron in the conduction (c) or valence (v) band, $b(b^\dagger)$ is the annihilation (creation) operator of the acoustic phonon, $D_c(D_v)$ the deformation potential of the conduction (valence) band, u the sound velocity of the longitudinal acoustic (LA) mode, V the quantization volume, ρ the mass density and the vector symbols of $\vec{r}$ and $\vec{q}$ are dropped.

The piezoelectric coupling is given as

$$H_{PZ} = -\sum_{r,q} \frac{8\pi e e_{14}}{\varepsilon q^2} \sqrt{\frac{\hbar}{2\rho u |q| V}} (\xi_x q_y q_z + \xi_y q_z q_x + \xi_z q_x q_y) e^{iqr}$$

$$\times (a_{cr}^\dagger a_{cr} + a_{vr}^\dagger a_{vr})(b_q + b_{-q}^\dagger), \qquad (9.149)$$

where $\varepsilon(e_{14})$ is the dielectric (piezoelectric) constant and $\vec{\xi}$ is the polarization vector of the acoustic phonon modes. In this case the transverse acoustic (TA) mode as well as the longitudinal acoustic (LA) mode contribute to the coupling. The polarization vectors for the LA mode and the two TA modes with a wavevector $\vec{q}$ are given as

$$\vec{\xi}(LA) = (q_x, q_y, q_z)/|\vec{q}|,$$

$$\vec{\xi}(TA1) = (q_y, -q_x, 0)/\sqrt{q_x^2 + q_y^2},$$

$$\vec{\xi}(TA2) = (-q_x q_z, -q_y q_z, q_x^2 + q_y^2)/|\vec{q}|\sqrt{q_x^2 + q_y^2}. \qquad (9.150)$$

Then the matrix element of the interaction Hamiltonian between two exciton states given by

$$|X_i\rangle = \sum_{r_e, r_h} F_i(r_e, r_h) a_{cr_e}^\dagger a_{vr_h} |0\rangle \qquad (9.151)$$

and

$$|X_f\rangle = \sum_{r_e, r_h} F_f(r_e, r_h) a_{cr_e}^\dagger a_{vr_h} |0\rangle \qquad (9.152)$$

is calculated as

$$\langle X_f | H_{DF} | X_i \rangle = \sum_q \sqrt{\frac{\hbar |q|}{2\rho u_{LA} V}} (D_c \langle X_f | e^{iqr_e} | X_i \rangle$$

$$- D_v \langle X_f | e^{iqr_h} | X_i \rangle)(\sqrt{N_q} \text{ or } \sqrt{N_q + 1}),$$

$$\langle X_f | H_{PZ} | X_i \rangle = -\sum_{\sigma,q} \frac{8\pi e e_{14}}{\varepsilon q^2} \sqrt{\frac{\hbar}{2\rho u_\sigma |q| V}} (\xi_x q_y q_z + \xi_y q_z q_x$$
$$+ \xi_z q_x q_y)(\langle X_f | e^{iqr_e} | X_i \rangle - \langle X_f | e^{iqr_h} | X_i \rangle) \cdot (\sqrt{N_q} \text{ or } \sqrt{N_q + 1})$$
(9.153)

with

$$\langle X_f | e^{iqr_e(r_h)} | X_i \rangle = \int d^3 r_e \int d^3 r_h F_f^*(r_e, r_h) e^{iqr_e(r_h)} F_i(r_e, r_h),$$
(9.154)

where $\sigma = LA$, $TA1$ and $TA2$ and the factor $\sqrt{N_q}$ ($\sqrt{N_q + 1}$) corresponds to the phonon absorption (emission) process.

The parameter values employed for GaAs are $D_c = -14.6$ eV, $D_v = -4.8$ eV [58], $u_{LA} = 4.81 \cdot 10^5$ cm/s, $u_{TA} = 3.34 \cdot 10^5$ cm/s, $e_{14} = 1.6 \cdot 10^{-5}$ C/cm^2 and $\varepsilon = 12.56$ [59].

9.12 Excitons in anisotropic quantum disk

Now the theoretical formulation has been completed. In a more concrete calculation, we have to specify a model for the QD-like island structures. The extremely narrow linewidth of exciton emission was observed for the first time in QW samples [49,13,12]. The lateral fluctuation of the QW thickness gives rise to an island-like structure. The localized excitons at such structures can be viewed as the zero-dimensional excitons. In these samples, the confinement in the direction of the crystal growth is strong, whereas the confinement in the lateral direction is rather weak. Furthermore, the island structures were found to be elongated along the $[\bar{1}10]$ direction [41]. Thus these island structures can be modeled by an anisotropic quantum disk. In order to facilitate the calculation, the lateral confinement potential in the x and y directions is assumed to be Gaussian as

$$V_e(r) = V_e^0 \exp\left[-\left(\frac{x}{a}\right)^2 - \left(\frac{y}{b}\right)^2\right], \quad V_h(r) = V_h^0 \exp\left[-\left(\frac{x}{a}\right)^2 - \left(\frac{y}{b}\right)^2\right], \quad (9.155)$$

where the lateral size parameters a and b can be fixed in principle from the measurement of morphology by e.g., STM but are left as adjustable parameters. The exciton wavefunction in such an anisotropic quantum disk can be approximated as

$$F(r_e, r_h) = \sum_{l_e, l_h, m_e, m_h} C(l_e, l_h, m_e, m_h) \left(\frac{x_e}{a}\right)^{l_e} \left(\frac{x_h}{a}\right)^{l_h} \left(\frac{y_e}{b}\right)^{m_e} \left(\frac{y_h}{b}\right)^{m_h}$$

$$\times \exp\left[-\frac{1}{2}\left\{\left(\frac{x_e}{a}\right)^2 + \left(\frac{x_h}{a}\right)^2 + \left(\frac{y_e}{b}\right)^2 + \left(\frac{y_h}{b}\right)^2\right\}\right.$$

$$\left. - \alpha_x (x_e - x_h)^2 - \alpha_y (y_e - y_h)^2 \right] \varphi_0(z_e) \varphi_0(z_h) \quad (9.156)$$

Theory of exciton coherence and decoherence   445

with

$$\varphi_0(z) = \sqrt{\frac{2}{L_z}} \cos\left(\frac{\pi z}{L_z}\right), \tag{9.157}$$

where $C(l_e, l_h, m_e, m_h)$ is the expansion coefficient, L_z is the QW thickness, the factor $1/2$ in the exponent is attached to make the probability distribution $|F(r_e, r_h)|^2$ to follow the functional form of the confining potential and α_x and α_y indicate the degree of the electron-hole correlation and are determined variationally. The electron-hole relative motion within the exciton state is not much different from that in the bulk because the lateral confinement is rather weak. As a result, the parameters α_x and α_y are weakly dependent on the lateral size. Because of the inversion symmetry of the confining potential, the parity is a good quantum number and the wavefunction can be classified in terms of the combination of parities of $x^{l_e+l_h}$ and $y^{m_e+m_h}$. The exciton ground state belongs to the (even, even) series, where the first (second) index indicates the parity with respect to the $x(y)$ coordinate. As can be seen easily, the optically allowed exciton states belong to the (even, even) series and other exciton states associated with (even, odd), (odd, even) and (odd, odd) series are dark states. In actual calculations, terms up to the sixth power are included, namely, $0 \leq l_e + l_h,\ m_e + m_h \leq 6$ to ensure the convergence of the calculation.

The potential depth for the exciton lateral motion can be guessed from the splitting energy of the heavy hole excitons due to the monolayer fluctuation of the QW thickness. The value of $|V_e^0 + V_h^0|$ is typically about 10 meV for the nominal QW thickness about 3 nm [41,50]. Of course, even if $V_e^0 + V_h^0$ is fixed, each value of V_e^0 and V_h^0 cannot be determined uniquely. Here we employ $V_e^0 = -6$ meV and $V_h^0 = -3$ meV throughout this paper, referring to the experimental results and assuming $V_e^0 : V_h^0 = 2:1$.

A typical example of the exciton level structure is shown in Fig. 9.23 for $a = 20$ nm and $b = 15$ nm. The disk height, namely the QW thickness, is fixed at 3 nm throughout this chapter. The transition intensities of optically active exciton states are plotted by solid lines and the corresponding radiative lifetime is also given alongside. The dark exciton states are exhibited by triangles slightly above the horizontal axis to indicate the energy positions. In the calculation of the optical absorption spectrum in (9.130), the lowest 13 exciton levels are taken into account including the dark exciton states because this number of levels is sufficient for converged results.

9.13   Temperature-dependence of the exciton dephasing rate

First of all, we are interested in the lineshape of the calculated absorption spectrum. The lineshape of the exciton ground state is plotted in Fig. 9.24 for a quantum disk model

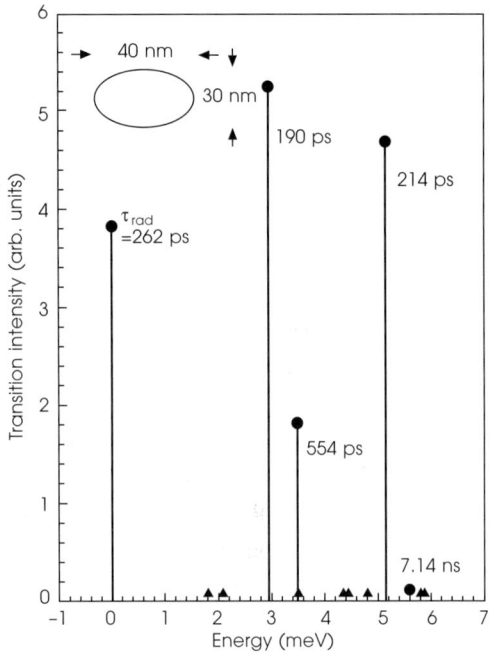

Fig. 9.23 Exciton energy levels in a GaAs quantum disk with parameters of $a = 20$ nm, $b = 15$ nm, $L_z = 3$ nm, $V_e^0 = 6$ meV and $V_e^0 = 3$ meV (see the text). The origin of energy is taken at the exciton ground state. The dark exciton states are denoted by triangles slightly above the horizontal axis (from Takagahara [54]).

in section 9.12 at 10K and 50K. The squares show the calculated spectra and the circles are the Lorentzian fit. At low temperatures, the spectra can be fitted very well by the Lorentzian as expected. At elevated temperatures, however, the lineshape deviates from the Lorentzian and shows additional broadening. In any case, the dephasing rate, i.e., the HWHM of the absorption spectrum can be estimated unambiguously. In addition, it is interesting to note the red shift of the exciton peak position about several tens of µeV relative to the purely electronic transition energy indicated by the origin of energy. This is caused by the lattice relaxation energy given by (4.5) in Ref. [57].

The size dependence of the dephasing rate is shown in Fig. 9.25 for the size parameters of $(a, b) = $ (12 nm, 10 nm), (20 nm, 15 nm) and (30 nm, 20 nm). In this size range the dephasing rate is larger for smaller sizes. This can be considered to be caused by the enhanced coupling strength between the exciton and acoustic phonons since the confinement effect on the spectral density of acoustic phonon modes is not significant in this size range. From the comparison of these results with experimental data, we can guess the likely size of the quantum disk. Hereafter we employ the size parameters of $(a, b) = $ (20 nm, 15 nm).

Theory of exciton coherence and decoherence 447

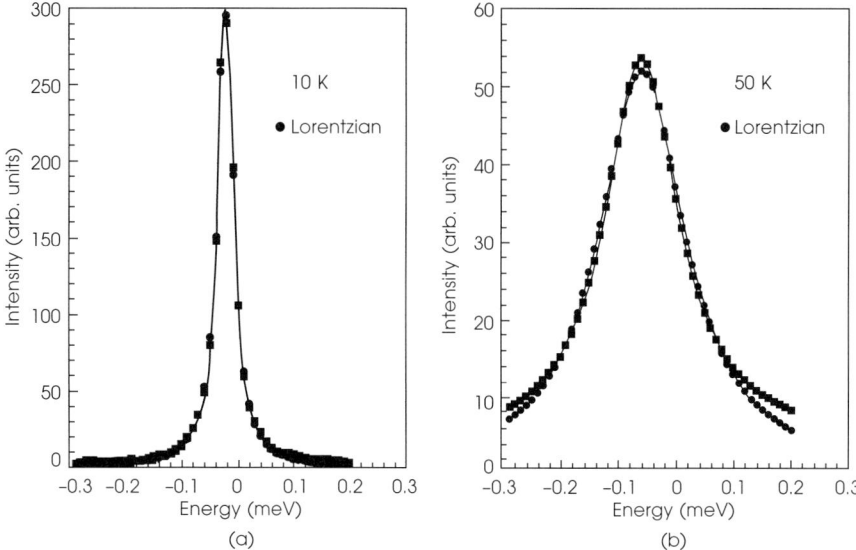

Fig. 9.24 Absorption spectra of the exciton ground state at (a) 10K and (b) 50K for the quantum disk in Fig. 9.23 (from Takagahara [54]).

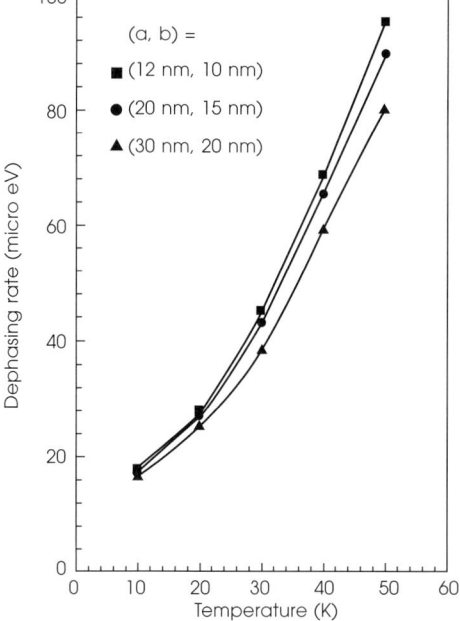

Fig. 9.25 Dephasing rates are plotted as a function of temperature for the exciton ground state in three quantum disks with size parameters of $(a, b) = (12$ nm, 10 nm$)$, $(20$ nm, 15 nm$)$ and $(30$ nm, 20 nm$)$ and the disk height of 3 nm (from Takagahara [54]).

The calculated dephasing rate is plotted in Fig. 9.26 as a function of temperature with experimental data [50]. As mentioned in the Introduction, the difference between the dephasing rate $\Gamma_\perp$ and half the population decay rate $\Gamma_\parallel/2$ indicates the pure dephasing rate. It is seen that the overall agreement between the theory and experiments is satisfactory concerning both the absolute magnitude and the temperature dependence. Furthermore, in the theory, we can separate out the contribution of the deformation potential coupling to the exciton dephasing and this part is shown by the arrow denoted as "Def. pot.". The remaining part is coming from the piezoelectric coupling and the interference term between the two couplings. But this part is simply denoted as "Piezo." in Fig. 9.26. It is seen that the deformation potential coupling is dominantly contributing to the pure dephasing.

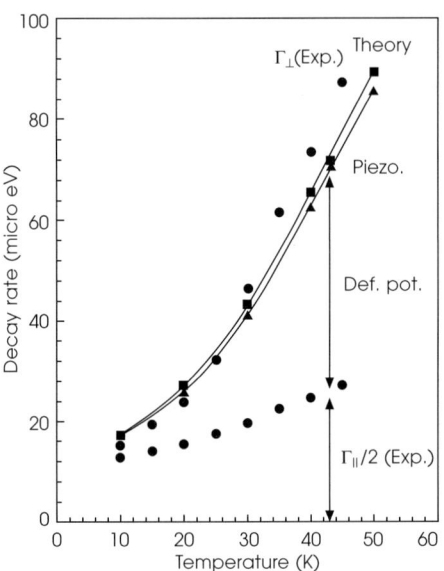

Fig. 9.26 Calculated dephasing rates of the exciton ground state are shown with experimental data (Ref. [50]) as a function of temperature. A quantum disk model is employed with the same parameters as in Fig. 9.23. The pure dephasing rate is decomposed into the contribution from the deformation potential coupling denoted as Def. pot. and that from the piezoelectric coupling and the interference term denoted as Piezo (from Takagahara [54]).

In order to see the reason in more detail, we look into the matrix elements of the exciton–phonon interactions. In Fig. 9.27, we plot the angular average of squared matrix element of the exciton–phonon interaction defined by

$$f_{ij}^\alpha(|q|) = \int d\Omega |\langle i|V_\alpha(q)|j\rangle|^2, \tag{9.158}$$

where $\alpha = LA$ or TA, the indices i and j denote the exciton state and for the case of $\alpha = TA$ the contribution from both $TA1$ and $TA2$ modes in (9.150) is combined. V_{LA} includes the contribution from both the deformation potential coupling and the piezoelectric coupling, whereas V_{TA} contains only the contribution from the piezoelectric coupling. In the inset, the energy level scheme is shown for the lowest four exciton states including the dark exciton states. From the comparison between Fig. 9.27(a) and Fig. 9.27(b), we see that the contribution from LA phonons is more than one order of magnitude larger than that from TA phonons. Furthermore, it is important to note that the wavevector of the most efficiently coupled phonons is roughly determined by max. $(1/a_B, 1/L)$, where a_B is the exciton Bohr radius and L is the typical size of the lateral confinement. The vanishing z-component of the wavevector is favored because the common envelope function in (9.157) is assumed for both the electron and the hole and they are uncorrelated in the z-direction. The relevant wavevector is about 10^6 cm^{-1} for GaAs islands. Hence the corresponding phonon energy is rather small (<1 meV), since the phonon energy versus wavevector ($|q|$) relation is $\hbar\omega = 0.32(0.22) |q|$ (meV) for the LA (TA) modes

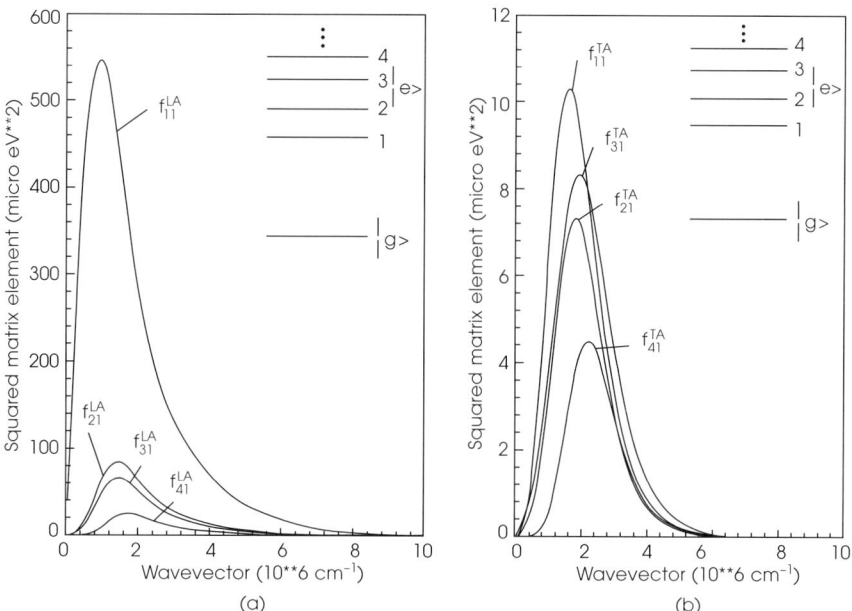

Fig. 9.27 Angularly averaged squared matrix elements of the exciton–phonon interaction are plotted as a function of the phonon wavevector for (a) the LA phonons and (b) the TA phonons. The horizontal (vertical) axis is scaled by 10^6 cm^{-1} ((μeV)^2). The employed GaAs quantum disk is the same as in Fig. 9.23. The lowest four exciton states are numbered consecutively including the dark exciton states (from Takagahara [54]).

with $|q|$ scaled in units of 10^6 cm^{-1}. This property will be invoked later in discussing the correlation between the temperature dependence of the exciton dephasing rate and the strength of the quantum confinement.

We have also calculated the dephasing rate of the excited exciton states. The results are shown in Fig. 9.28 for the lowest four optically active exciton states. In the calculation for the excited exciton states, the value of δ in (9.147) is assumed to be the same as for the exciton ground state because the relevant relaxation processes may be dependent on the exciton state but the absolute magnitude of δ is rather small. The dephasing rate is in general larger for the higher-lying exciton states. But this tendency is not monotonic as seen by the reversed order of magnitude between the second and third exciton states.

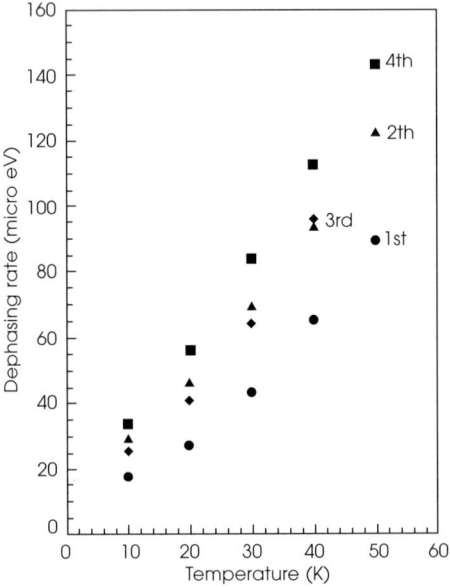

Fig. 9.28 Calculated dephasing rates of the lowest four optically active exciton states are shown as a function of temperature of the same quantum disk as in Fig. 9.23 (from Takagahara [54]).

9.14 Elementary processes of exciton pure dephasing

Now we discuss the mechanism of exciton pure dephasing. Generally speaking, pure dephasing means the decay of the dipole coherence without change in the state of the system. Any real transition to other states leads to the population decay. Thus the pure dephasing is caused by virtual processes which start from a relevant state and through some excursion in the intermediate states return to the same initial state. These virtual processes give rise to the temporal fluctuation of the phase of wavefunction. Previously this kind of temporal phase fluctuation was treated by a stochastic model of random frequency modulation [60] and the resulting pure dephasing was discussed in the context of resonant secondary emission [61].

Here we treat these processes microscopically. There are two kinds of such virtual processes which contribute to the pure dephasing. The first kind of process is induced by the off-diagonal exciton–phonon interaction. Those processes start from the exciton ground state, pass through excited exciton states and return to the exciton ground state. The second kind of process is induced by the diagonal exciton–phonon interaction and the relevant state remains always within the exciton ground state. These processes are shown schematically in the inset of Fig. 9.29. The contribution to the pure dephasing from the second kind of process can be singled out theoretically by carrying out the calculation which includes only the exciton ground state. That contribution is denoted

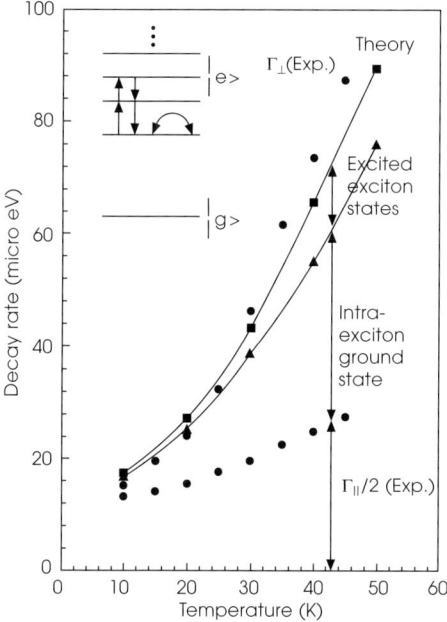

Fig. 9.29 Calculated dephasing rates of the exciton ground state are shown with experimental data (Ref. [50]) as a function of temperature for the same quantum disk as in Fig. 9.23. The pure dephasing rate is decomposed into the contribution from the diagonal exciton–phonon interaction denoted as Intra-exciton ground state and that from the off-diagonal interaction and the interference term denoted as Excited exciton states (from Takagahara [54]).

by "Intra-exciton ground state" in Fig. 9.29. The remaining part denoted as "Excited exciton states" comes from the first kind of process and the interference between the two kinds of processes. It is seen that the "intra-exciton ground state" (diagonal) processes contribute substantially to the pure dephasing but the contribution from the off-diagonal processes is not negligible. This feature can be understood from Figs 9.27(a) and 9.27(b) since the squared matrix element within the exciton ground state denoted by f_{11}^α is much larger than other squared matrix elements. As a result, the "intra-exciton ground state" processes contribute significantly to the pure dephasing.

9.15 Mechanisms of population decay of excitons

The possible mechanisms of the population decay will be discussed. Experimentally, two decay time constants were observed [50]. The slow time constant (~200 ps) is almost independent of temperature suggesting the radiative decay as its mechanism. In fact, the calculated radiative lifetime of the exciton ground state is around 200 ps as shown in Fig. 9.23. On the other hand, the fast time constant (~30 ps) is weakly dependent on temperature. The likely mechanisms are the thermal activation to excited exciton states and the phonon-assisted migration to neighboring islands. In this section we present detailed calculation of these relaxation rates and examine the significance of these mechanisms.

Phonon-assisted population relaxation

The phonon-assisted transition rate to other exciton states is calculated as

$$P_i = \sum_{j \neq i} w_{ij}, \tag{9.159}$$

where w_{ij} is the transition rate from the exciton state i to other exciton state j. Here P_i's are calculated for the lowest four optically active exciton states and are plotted in Fig. 9.30 as a function of temperature. The same quantum disk model as in Fig. 9.23 is employed and the same 13 exciton levels are included in the calculation. Since the energy difference between exciton levels is less than several meV, it is sufficient to take into account only the one-phonon processes. For the exciton ground state, the transition rate is about several µeV. For the excited exciton states, the transition rates are about one order of magnitude larger than that of the exciton ground state. In general, the higher-lying exciton states have a larger population decay rate. But this trend is not monotonic as seen by the reversed order of magnitude between the second and third exciton states.

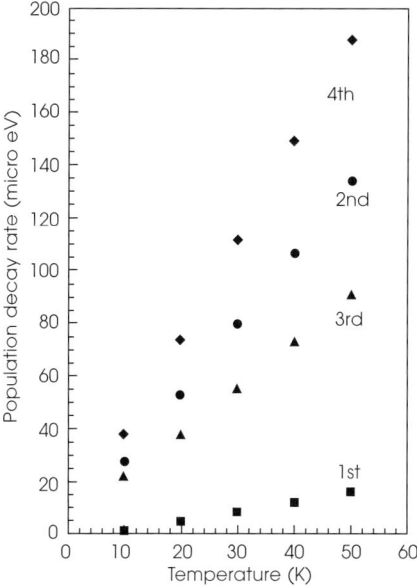

Fig. 9.30 Phonon-assisted population decay rates of the lowest four optically active exciton states are plotted as a function of temperature for the same quantum disk as in Fig. 9.23 (from Takagahara [54]).

It is interesting to note that the linear temperature dependence is clearly seen. This indicates that the energy of relevant acoustic phonons is rather small as shown in Figs 9.27(a) and 9.27(b) and the high temperature approximation holds as

$$\frac{1}{e^{\hbar\omega/k_BT} - 1} \approx \frac{k_BT}{\hbar\omega}. \tag{9.160}$$

This is the origin of the linear temperature dependence.

Phonon-assisted exciton migration

The excitons localized at island structures can migrate among them accompanying phonon absorption or emission to compensate for the energy mismatch. Since the energy mismatch is typically about a few meV, the acoustic phonons are dominantly contributing to the exciton migration process. Now let us consider two island sites at R_a and R_b and assume that the island at R_b is larger in size and has a localized exciton state of lower energy than in the island at R_a. Then we consider the phonon-assisted exciton migration from the site R_a to the site R_b, namely a transition of

$| R_a; n_q \rangle \to | R_b: n_q + 1 \rangle,$

where n_q indicates the occupation number of an acoustic phonon mode with wavevector $\vec{q}$. As discussed previously [62], there are three elementary processes of this transition;

1. $| R_a; n_q \rangle \xrightarrow{H_{ep}} | R_b; n_q + 1 \rangle,$
2. $| R_a; n_q \rangle \xrightarrow{H_{ep}} | R_b; n_q + 1 \rangle \xrightarrow{H_{ss}} | R_b; n_q + 1 \rangle,$
3. $| R_a; n_q \rangle \xrightarrow{H_{ss}} | R_b; n_q \rangle \xrightarrow{H_{ep}} | R_b; n_q + 1 \rangle,$

where $H_{ep}(H_{ss})$ represents the exciton–phonon interaction (inter-site transfer) Hamiltonian. The first process is a direct process through the overlap between exciton wavefunctions at two island sites which is strongly dependent on the distance between two islands. Thus this process contributes only for the case of short distance. On the other hand, the second and third processes are indirect ones which can contribute to the exciton transfer even for the case of long distance.

The inter-site transfer Hamiltonian H_{ss} is caused by the electron–electron interaction and is calculated as

$$J(R_a, R_b) = \langle R_a | H_{ss} | R_b \rangle$$

$$= \sum_{r_e, r_h} \sum_{r'_e, r'_h} F^*_{c\tau,v\sigma}(r_e, r_h; R_a) F_{c\tau',v\sigma'}(r'_e, r'_h; R_b) [V(c\tau r_e, v\sigma' r'_h; c\tau' r'_e, v\sigma r_h)$$

$$- V(c\tau r_e, v\sigma' r'_h; v\sigma r_h, c\tau' r'_e)], \tag{9.161}$$

where the localized exciton at the site R_i is described as

$$| R_i \rangle = \sum_{\tau, r_e; \sigma, r_h} F_{c\tau,v\sigma}(r_e, r_h; R_i) a^\dagger_{c\tau r_e} a_{v\sigma r_h} |0\rangle \tag{9.162}$$

with an envelope function F corresponding to (9.156), and the suffix $\tau(\sigma)$ denotes the Wannier function index of the conduction (valence) band, e.g., the total angular momentum. The first term in (9.161) represents the electron–hole exchange interaction and the second term corresponds to the Coulomb interaction. Because of the localized nature of the Wannier functions, we can approximate as

$$V(c\tau r_e, v\sigma' r'_h; v\sigma r_h, c\tau' r'_e) \approx \delta_{r_e r'_e} \delta_{r_h r'_h} \delta_{\tau\tau'} \delta_{\sigma\sigma'} \frac{e^2}{\varepsilon |r_e - r_h|},$$

$$V(c\tau r_e, v\sigma' r'_h; c\tau' r'_e, v\sigma r_h) \approx \delta_{r_e r_h} \delta_{r'_e r'_h} \Big[\delta_{r_e r'_e} V(c\tau r_e, v\sigma' r_e; c\tau' r_e, v\sigma r_e)$$

$$+ (1 - \delta_{r_e r'_e}) \vec{\mu}_{c\tau v\sigma} \frac{[1 - 3\vec{n} \cdot {}^t\vec{n}]}{|r_e - r'_e|^3} \vec{\mu}_{v\sigma' c\tau'} \Big] \tag{9.163}$$

with

$$\vec{n} = \frac{\vec{r}_e - \vec{r}'_e}{|r_e - r'_e|}, \tag{9.164}$$

$$\vec{\mu}_{c\tau v\sigma} = \int d^3 r\, \phi^*_{c\tau R}(r)(\vec{r} - \vec{R})\, \phi_{v\sigma R}(r), \tag{9.165}$$

where $\phi_{c\tau(v\sigma)R}(r)$ is a Wannier function localized at the site R. Hereafter the vector symbols will be dropped because of no fear of confusion. The Coulomb term decreases rapidly when $|R_a - R_b|$ exceeds the lateral size of the exciton wavefunction. On the other hand, the exchange term contains the dipole–dipole interaction and has a long-range character decreasing as $|R_a - R_b|^{-3}$. Thus the exchange term contributes dominantly to the inter-site exciton transfer when $|R_a - R_b|$ is larger than the lateral size of the confining potential.

The exciton transfer probability is calculated as

$$w(R_a \to R_b) = \frac{2\pi}{\hbar} \sum_\sigma \Big| \langle R_b; n_q + 1 | H^\sigma_{ep} | R_a; n_q \rangle$$

$$+ \frac{\langle R_b; n_q + 1 | H_{ss} | R_a; n_q + 1\rangle \langle R_a; n_q + 1 | H^\sigma_{ep} | R_a; n_q \rangle}{-\hbar\omega^\sigma_q}$$

$$+ \frac{\langle R_b; n_q + 1 | H^\sigma_{ep} | R_b; n_q \rangle \langle R_b; n_q | H_{ss} | R_a; n_q \rangle}{E_X(R_a) - E_X(R_b)} \Big|^2 \delta(E_X(R_a) - E_X(R_b) - \hbar\omega^\sigma_q), \tag{9.166}$$

where $E_X(R_i)$ is the energy of a localized exciton at the site R_i and the summation concerning the acoustic phonon mode σ is taken over the LA, $TA1$ and $TA2$ modes in (9.150).

In the exchange matrix element in (9.163), the first (second) term is usually called the short- (long-)range part of the exchange interaction. The contribution from the long-range part can be rewritten into a more tractable form as [63]

$$\sum_{r_e \ne r'_e} F^*_{c\tau,v\sigma}(r_e, r_e; R_a) F_{c\tau',v\sigma'}(r'_e, r'_e; R_b) \mu_{c\tau,v\sigma} \frac{[1 - 3n \cdot {}^t n]}{|r_e - r'_e|^3} \mu_{v\sigma',c\tau'}$$

$$= -\sum_{r_e \ne r'_e} F^*_{c\tau,v\sigma}(r_e, r_e; R_a) \mu_{c\tau,v\sigma} \operatorname{grad}_{r_e} \operatorname{div}_{r_e} \left[\frac{\mu_{v\sigma',c\tau'}}{|r_e - r'_e|} F_{c\tau',v\sigma'}(r'_e, r'_e; R_b) \right],$$

$$= -\int d^3 r_e \int' d^3 r'_e F^*_{c\tau,v\sigma}(r_e, r_e; R_a) \mu_{c\tau,v\sigma} \operatorname{grad}_{r_e} \operatorname{div}_{r_e} \left[\frac{\mu_{v\sigma',c\tau'}}{|r_e - r'_e|} F_{c\tau',v\sigma'}(r'_e, r'_e; R_b) \right], \tag{9.167}$$

where the integration with respect to r'_e is carried out over the whole space excluding a small sphere around the point r_e and this is indicated by a primed integral symbol. Then making use of a relation for an arbitrary vector field $\vec{Q}(r)$

$$\int' d^3r' \, \text{grad}_r \text{div}_r \frac{\vec{Q}(r')}{|r-r'|} = \frac{4\pi}{3} \vec{Q}(r) + \text{grad}_r \text{div}_r \int d^3r' \frac{\vec{Q}(r')}{|r-r'|}, \quad (9.168)$$

we can rewrite (9.167) as

$$-\int d^3r_e \, F^*_{c\tau,\upsilon\sigma}(r_e, r_e; R_a) \mu_{c\tau,\upsilon\sigma} \left[\frac{4\pi}{3} \mu_{\upsilon\sigma',c\tau'} F_{c\tau',\upsilon\sigma'}(r_e, r_e; R_b) \right.$$

$$\left. + \text{grad}_{r_e} \text{div}_{r_e} \int d^3r'_e \frac{\mu_{\upsilon\sigma',c\tau'}}{|r_e - r'_e|} F_{c\tau',\upsilon\sigma'}(r'_e, r'_e; R_b) \right], \quad (9.169)$$

where in the second term within the parentheses the integration can be performed over the whole space because the singularity of $|r_e - r'_e|^{-1}$ is integrable. By a partial integration, we have the expression of the long-range exchange term as

$$-\frac{4\pi}{3} (\mu_{c\tau,\upsilon\sigma} \cdot \mu_{\upsilon\sigma',c\tau'}) \int d^3r \, F^*_{c\tau,\upsilon\sigma}(r, r; R_a) F_{c\tau',\upsilon\sigma'}(r, r; R_b)$$

$$+ \int d^3r \, \text{div}_r \left(F^*_{c\tau,\upsilon\sigma}(r, r; R_a) \mu_{c\tau,\upsilon\sigma} \right) \text{div}_r \left[\int d^3r' \frac{\mu_{\upsilon\sigma',c\tau'}}{|r-r'|} F_{c\tau',\upsilon\sigma'}(r', r'; R_b) \right]. \quad (9.170)$$

The short-range exchange term is simply written as

$$V(c\tau r_0, \upsilon\sigma' r_0; c\tau' r_0, \upsilon\sigma r_0) \int d^3r \, F^*_{c\tau,\upsilon\sigma}(r, r; R_a) F_{c\tau',\upsilon\sigma'}(r, r; R_b) \quad (9.171)$$

and the Coulomb term is calculated by

$$-\int d^3r_e \int d^3r_h \, F^*_{c\tau,\upsilon\sigma}(r_e, r_h; R_a) \frac{e^2}{\varepsilon |r_e - r_h|} F_{c\tau,\upsilon\sigma}(r_e, r_h; R_b). \quad (9.172)$$

Combining three terms (9.170), (9.171) and (9.172), we can estimate the exciton transfer matrix element in (9.161).

It is to be noted that when the Coulomb term and/or the exchange term are of comparable magnitude to the energy difference between localized exciton states at R_a and R_b, the eigenstates should be mixed states of two localized excitons and a simple picture of exciton transfer between two sites cannot be applied. Thus we have to check the inequality

$$|J(R_a, R_b)| \ll |E_X(R_a) - E_X(R_b)|$$

before we apply the exciton transfer model. We can check numerically that the matrix element $|J(R_a, R_b)|$ is typically about several tens of μeV except for a very close pair of islands, whereas $|E_X(R_a) - E_X(R_b)|$ is about a few meV for typical sizes of islands. Thus the simple picture of exciton transfer can be applied safely.

In order to see the typical behavior of the exciton migration rate, we employ two quantum disk islands characterized by $L_z = 3$ nm and the lateral size parameters of $(a, b) = (20$ nm, 15 nm$)$ and $(30$ nm, 20 nm$)$ and consider the migration between two exciton ground states whose energy difference is 0.68 meV. The exciton migration rate depends on the distance between two islands, the geometrical configuration of two islands and on the direction of exciton polarizations. In Figs 9.31(a) and 9.31(b), the migration rate is plotted as a function of the center-to-center distance between two islands. In Fig. 9.31(a), two islands are aligned such that the longer axes of two ellipses are coincident with each other. In this configuration, the migration rate is larger for the exciton polarization along the longer axis than for the exciton polarization along the shorter axis because of the larger interaction through the surface charges. In Fig. 9.31(b),

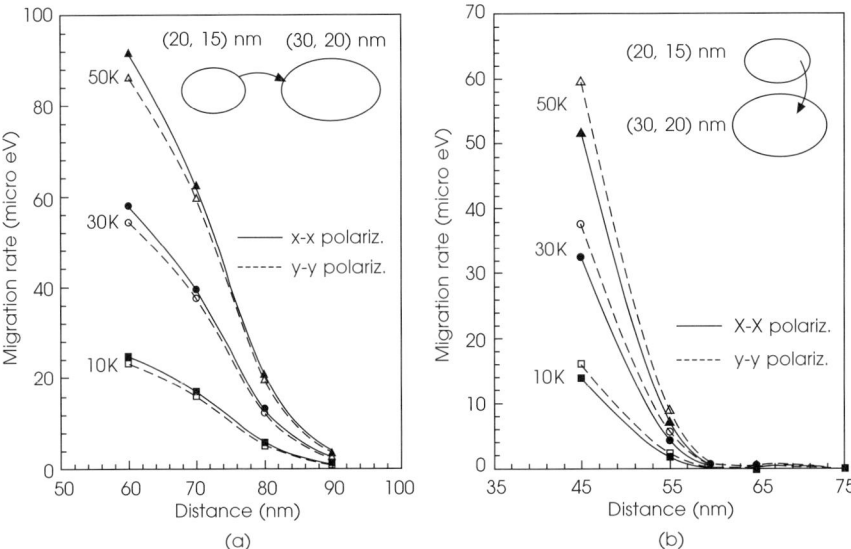

Fig. 9.31 Phonon-assisted migration rate from the exciton ground state in a quantum disk of $(a, b) = (20$ nm, 15 nm$)$ to the exciton ground state in a quantum disk of $(a, b) = (30$ nm, 20 nm$)$ is plotted as a function of the center-to-center distance between two quantum disks at temperatures of 10K, 30K and 50K. In (a) ((b)), two disks are aligned such that the longer (shorter) axes of two ellipses are coincident with each other and the distance is measured along the longer (shorter) axis. The exciton polarization in two disks is aligned along the x or y direction and this is indicated by $x-x$ or $y-y$ polariz., where $x(y)$ denotes the direction of the longer (shorter) axis of the ellipses (from Takagahara [54]).

458   T. Takagahara

two islands are aligned such that the shorter axes of two ellipses are coincident with each other. In this configuration, the migration rate is larger for the exciton polarization along the shorter axis than for the exciton polarization along the longer axis. The absolute magnitude of the exciton migration rate is about several tens of μeV.

We have also estimated the migration rate between the exciton ground state and the excited exciton states. The results are shown in Fig. 9.32 for the transition from the exciton ground state of an island with $(a, b) = (20$ nm, 15 nm$)$ to the second optically active exciton state in an island with $(a, b) = (30$ nm, 20 nm$)$. The configuration of two islands is the same as in Fig. 9.31(a). This migration process is associated with phonon absorption (~2.11 meV) in contrast to the case in Fig. 9.31. As a result, the migration rate is several tens times smaller than in Fig. 9.31(a). In addition, the distance dependence is not monotonic. This feature arises from the interference among three terms in (9.166) and may be sensitively dependent on the spatial profile of the exciton wavefunction.

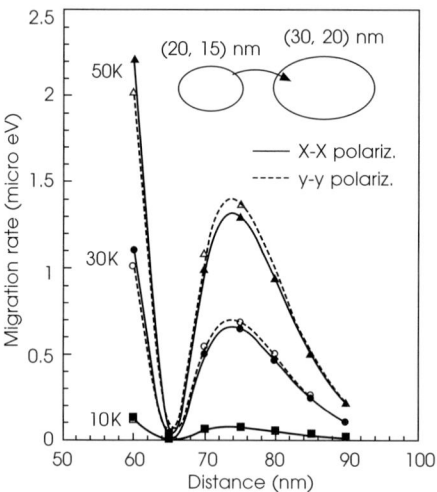

Fig. 9.32 Phonon-assisted migration rate from the exciton ground state in a quantum disk of $(a, b) = (20$ nm, 15 nm$)$ to the second lowest optically active exciton state in a quantum disk of $(a, b) = (30$ nm, 20 nm$)$ is plotted as a function of the center-to-center distance between two quantum disks at temperatures of 10K, 30K and 50K. The configuration of two disks is the same as in Fig. 9.31(a) (from Takagahara [54]).

From the above considerations we can identify the likely mechanism of the population decay of the exciton ground state as the combination of thermal activation to the excited exciton states within an island and phonon-assisted exciton migration to neighboring islands.

9.16 Recent progress in studies on exciton decoherence

Very recently, a very long exciton decoherence time (~ nanosecond) at low temperatures has been reported by Borri *et al.* [64] for InGaAs quantum dots and by Birkedal *et al.* [65] for InAlGaAs quantum dots. These measurements are based on the time-domain spectroscopy, e.g., four-wave mixing. In the previous reports, the homogeneous linewidth of excitons in quantum dots is about a few tens of µeV even at low temperatures and is one order of magnitude larger than the above values. The origin of this discrepancy is considered as follows. A quantum dot is very sensitive to its environments, especially to the charge distribution and the spectral position of excitons is temporally fluctuating. These were observed in phenomena of the blinking or intermittent emission from excitons [66] and the spectral wandering of the exciton emission [67]. Thus in the CW measurements the spectral fluctuation contributes significantly to the homogeneous linewidth. On the other hand, the measurements in the time-domain spectroscopy are carried out much faster than the time-scale of the spectral diffusion and the contribution from the environmental fluctuations can be much reduced. In fact, the measurements of the exciton linewidth in CdSe/ZnS core/shell nanocrystals based on the spectral hole burning were reported [68] and the dependence on the modulation frequency was demonstrated. The reported exciton decoherence time about 1 nanosecond can be considered as lifetime-limited. Thus the ultimate decoherence time would be about 1 ns, as far as optically active excitons in semiconductor quantum dots are used as qubits in the quantum state manipulation.

9.17 Theory of dephasing of nonradiative coherence

So far we have discussed a theory of exciton dephasing based on the exciton–phonon interactions. In general we are interested in the dephasing of quantum coherence between two levels, e.g., two different exciton levels. This coherence can be induced by the V-shaped or Λ-shaped Raman-type two-photon transitions [69]. If the two levels belong to the ground sublevels, a direct optical transition is forbidden between them and a long dephasing time of the coherence is expected. Here we extend the formulation in the last section 9.10 to the general case. Denoting the coherence between a pair of levels $|e_i\rangle$ and $|e_j\rangle$ by $|e_i\rangle\langle e_j|$, we consider the time-evolution with the total Hamiltonian:

$$\exp\left[-\frac{i}{\hbar}H^{\times}t\right]|e_i\rangle\langle e_j| = e^{-\frac{i}{\hbar}Ht}|e_i\rangle\langle e_j|e^{\frac{i}{\hbar}Ht}. \tag{9.173}$$

If this quantity shows a time-dependence such as

$$\sim e^{-\gamma_{ij}t - i\omega_{ij}t}|e_i\rangle\langle e_j| \tag{9.174}$$

with $\hbar\omega_{ij} = E_i - E_j$, where $E_i(E_j)$ is the energy of the level $|e_i\rangle$ ($|e_j\rangle$), we introduce the Fourier–Laplace transform by

$$\mathrm{Re}\left[\int_0^\infty dt\, e^{i\omega t} \langle e_i|\exp\left[-\frac{i}{\hbar}H^\times t\right](|e_i\rangle\langle e_j|)|e_j\rangle\right] = \frac{\gamma_{ij}}{(\omega - \omega_{ij})^2 + \gamma_{ij}^2} \tag{9.175}$$

and we can estimate the dephasing rate γ_{ij} of the coherence by the HWHM(Half Width at Half Maximum). In the case of the exciton dephasing, substituting $|e_i\rangle\langle e_j|$ by $|e\rangle\langle g|$ where $|e\rangle$ is an exciton state and $|g\rangle$ is the ground state, we reproduce the formulation in section 9.10. In this section also, we consider the dephasing due to the electron–phonon interactions and decompose the total Hamiltonian as

$$H = H_e + H_g + V, \tag{9.176}$$

where H_e denotes the electronic part, H_g the phonon Hamiltonian and V is the electron–phonon interactions. Then the Laplace transform is expanded as

$$\int_0^\infty dt\, e^{-st} \exp\left[-\frac{i}{\hbar}H^\times t\right]\rho_0 = \frac{1}{s + \frac{i}{\hbar}H^\times}\rho_0 = \frac{1}{s + \frac{i}{\hbar}H_e^\times + \frac{i}{\hbar}H_g^\times + \frac{i}{\hbar}V^\times}\rho_0$$

$$= \left\{\frac{1}{s + \frac{i}{\hbar}H_e^\times + \frac{i}{\hbar}H_g^\times} + \left(-\frac{i}{\hbar}\right)\frac{1}{s + \frac{i}{\hbar}H_e^\times + \frac{i}{\hbar}H_g^\times}V^\times \frac{1}{s + \frac{i}{\hbar}H_e^\times + \frac{i}{\hbar}H_g^\times}\right.$$

$$+ \left(-\frac{i}{\hbar}\right)^2 \frac{1}{s + \frac{i}{\hbar}H_e^\times + \frac{i}{\hbar}H_g^\times}V^\times \frac{1}{s + \frac{i}{\hbar}H_e^\times + \frac{i}{\hbar}H_g^\times}V^\times \frac{1}{s + \frac{i}{\hbar}H_e^\times + \frac{i}{\hbar}H_g^\times} + \ldots\bigg\}\rho_0, \tag{9.177}$$

where $\rho_0 = |e_i\rangle\langle e_j|\rho_{ph}$ and ρ_{ph} is the phonon density matrix at the thermal equilibrium. For example, the third term corresponds to the time-dependent expression:

$$\left(-\frac{i}{\hbar}\right)^2 \int_0^t dt_1 \int_0^{t_1} dt_2\, e^{-\frac{i}{\hbar}(H_e^\times + H_g^\times)(t-t_1)} V^\times e^{-\frac{i}{\hbar}(H_e^\times + H_g^\times)(t_1-t_2)} V^\times e^{-\frac{i}{\hbar}(H_e^\times + H_g^\times)t_2}\rho_0$$

$$= \left(-\frac{i}{\hbar}\right)^2 \int_0^t dt_1 \int_0^{t_1} dt_2\, e^{-\frac{i}{\hbar}(H_e^\times + H_g^\times)(t-t_1)}(V\, e^{-\frac{i}{\hbar}(H_e^\times + H_g^\times)(t_1-t_2)}(V(e^{-\frac{i}{\hbar}(H_e^\times + H_g^\times)t_2}\rho_0)))$$

$$- \left(-\frac{i}{\hbar}\right)^2 \int_0^t dt_1 \int_0^{t_1} dt_2\, e^{-\frac{i}{\hbar}(H_e^\times + H_g^\times)(t-t_1)}(V\, e^{-\frac{i}{\hbar}(H_e^\times + H_g^\times)(t_1-t_2)}((e^{-\frac{i}{\hbar}(H_e^\times + H_g^\times)t_2}\rho_0)V))$$

$$-\left(-\frac{i}{\hbar}\right)^2 \int_0^t dt_1 \int_0^{t_1} dt_2 e^{-\frac{i}{\hbar}(H_e^\times + H_g^\times)(t-t_1)} (e^{-\frac{i}{\hbar}(H_e^\times + H_g^\times)(t_1-t_2)} (V(e^{-\frac{i}{\hbar}(H_e^\times + H_g^\times)t_2} \rho_0))V)$$

$$+\left(-\frac{i}{\hbar}\right)^2 \int_0^t dt_1 \int_0^{t_1} dt_2 e^{-\frac{i}{\hbar}(H_e^\times + H_g^\times)(t-t_1)} (e^{-\frac{i}{\hbar}(H_e^\times + H_g^\times)(t_1-t_2)} ((e^{-\frac{i}{\hbar}(H_e^\times + H_g^\times)t_2} \rho_0)V)V).$$

(9.178)

Since H_e and H_g commute with each other, we have

$$e^{-\frac{i}{\hbar}(H_e^\times + H_g^\times)t} = e^{-\frac{i}{\hbar}H_e^\times t} e^{-\frac{i}{\hbar}H_g^\times t}. \tag{9.179}$$

In general, the electron–phonon interaction has a form given by

$$V = \sum_\alpha M_\alpha (b_\alpha + b_{-\alpha}^\dagger), \tag{9.180}$$

where b_α ($b_\alpha^\dagger$) is the second-quantized annihilation (creation) operator of phonons, the index α denotes the wavevector or mode index of phonons and M_α is the coupling operator among the electronic levels. For example, the first term in (9.178) can be written as

$$\sum_{\alpha,\beta} \left(-\frac{i}{\hbar}\right)^2 \int_0^t dt_1 \int_0^{t_1} dt_2\, e^{-\frac{i}{\hbar}H_e(t-t_1)} M_\beta e^{-\frac{i}{\hbar}H_e(t_1-t_2)} M_\alpha e^{-\frac{i}{\hbar}H_e t_2} |e_i\rangle\langle e_j| e^{\frac{i}{\hbar}H_e t_2}$$

$$\times e^{\frac{i}{\hbar}H_e(t_1-t_2)} e^{\frac{i}{\hbar}H_e(t-t_1)} e^{-\frac{i}{\hbar}H_g(t-t_1)} (b_\beta + b_{-\beta}^\dagger) e^{-\frac{i}{\hbar}H_g(t_1-t_2)} (b_\alpha + b_{-\alpha}^\dagger)$$

$$\times e^{-\frac{i}{\hbar}H_g t_2} \rho_{ph} e^{\frac{i}{\hbar}H_g t_2} e^{\frac{i}{\hbar}H_g(t_1-t_2)} e^{\frac{i}{\hbar}H_g(t-t_1)}. \tag{9.181}$$

Taking the trace of this expression concerning the phonons, we have

$$\mathrm{Tr}\, e^{\frac{i}{\hbar}H_g t_1}(b_\beta + b_{-\beta}^\dagger) e^{-\frac{i}{\hbar}H_g(t_1-t_2)} (b_\alpha + b_{-\alpha}^\dagger) e^{-\frac{i}{\hbar}H_g t_2} \rho_{ph}$$

$$= \mathrm{Tr}\, (b_\beta(t_1) + b_{-\beta}^\dagger(t_1))(b_\alpha(t_2) + b_{-\alpha}^\dagger(t_2)) \rho_{ph}$$

$$= \delta_{\alpha,-\beta} [n_\alpha e^{i\omega_\alpha(t_1-t_2)} + (1+n_\alpha) e^{-i\omega_\alpha(t_1-t_2)}], \tag{9.182}$$

where n_α is the phonon occupation number and $\hbar\omega_\alpha$ is the energy of phonons with the mode index α. Substituting these into (9.181) and again taking the Laplace transform, we have

$$\sum_\alpha \left(-\frac{i}{\hbar}\right)^2 \frac{1}{s + \frac{i}{\hbar}H_e^\times}$$

$$\times \left[M_{-\alpha} \left(\frac{n_\alpha}{s - i\omega_\alpha + \frac{i}{\hbar}H_e^\times} + \frac{1+n_\alpha}{s + i\omega_\alpha + \frac{i}{\hbar}H_e^\times} \right) \left(M_\alpha \frac{1}{s + \frac{i}{\hbar}H_e^\times} \rho_e \right) \right], \tag{9.183}$$

where $\rho_e = |e_i\rangle\langle e_j|$. In a similar way, the second, third and fourth terms in (9.178) can be written as

$$\sum_\alpha -\left(-\frac{i}{\hbar}\right)^2 \frac{1}{s+\frac{i}{\hbar}H_e^\times}$$

$$\times \left[M_{-\alpha}\left(\frac{n_\alpha}{s+i\omega_\alpha+\frac{i}{\hbar}H_e^\times}+\frac{1+n_\alpha}{s-i\omega_\alpha+\frac{i}{\hbar}H_e^\times}\right)\left(\frac{1}{s+\frac{i}{\hbar}H_e^\times}\rho_e\cdot M_\alpha\right)\right], \qquad (9.184)$$

$$\sum_\alpha -\left(-\frac{i}{\hbar}\right)^2 \frac{1}{s+\frac{i}{\hbar}H_e^\times}$$

$$\times \left[\left(\frac{n_\alpha}{s-i\omega_\alpha+\frac{i}{\hbar}H_e^\times}+\frac{1+n_\alpha}{s+i\omega_\alpha+\frac{i}{\hbar}H_e^\times}\right)\left(M_\alpha \frac{1}{s+\frac{i}{\hbar}H_e^\times}\rho_e\right)M_{-\alpha}\right], \qquad (9.185)$$

$$\sum_\alpha \left(-\frac{i}{\hbar}\right)^2 \frac{1}{s+\frac{i}{\hbar}H_e^\times}$$

$$\times \left[\left(\frac{n_\alpha}{s+i\omega_\alpha+\frac{i}{\hbar}H_e^\times}+\frac{1+n_\alpha}{s-i\omega_\alpha+\frac{i}{\hbar}H_e^\times}\right)\left(\frac{1}{s+\frac{i}{\hbar}H_e^\times}\rho_e\cdot M_\alpha\right)M_{-\alpha}\right]. \qquad (9.186)$$

These terms can be summarized as

$$\frac{1}{s+\frac{i}{\hbar}H_e^\times}\Sigma^{(2)}(s)\frac{1}{s+\frac{i}{\hbar}H_e^\times}\rho_e \qquad (9.187)$$

and $\Sigma^{(2)}(s)$ has a meaning of the self-energy in the second order as depicted in Fig. 9.33. The scheme of the double Feynman diagram is used to describe the time-evolution of the density matrix, where the upper (lower) line represents the time-development of the ket (bra) part of the density matrix. In a similar way, the higher-order terms in (9.177)

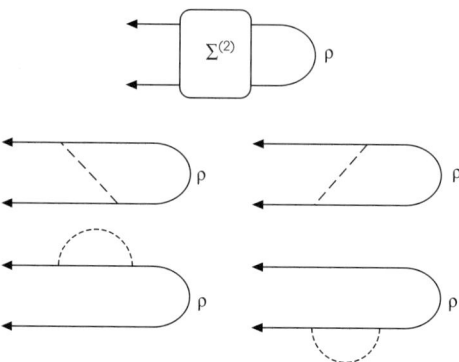

Fig. 9.33 The second order self-energy terms for the density matrix in the scheme of the double Feynman diagram. The dashed lines represent the exciton–phonon interaction.

Theory of exciton coherence and decoherence 463

can be calculated. The odd order terms with respect to $V^\times$ disappear because the trace over the phonon degrees of freedom vanishes. Thus the next higher-order terms are given by the fourth order terms whose typical diagrams are shown in Fig. 9.34. However, those diagrams in Figs 9.34(a)–(c) can be classified into the type of diagrams in Fig. 9.35 and can be included by renormalizing the propagator as

$$\frac{1}{s + \frac{i}{\hbar} H_e^\times - \Sigma^{(2)}(s)} = \frac{1}{s + \frac{i}{\hbar} H_e^\times} + \frac{1}{s + \frac{i}{\hbar} H_e^\times} \Sigma^{(2)}(s) \frac{1}{s + \frac{i}{\hbar} H_e^\times}$$
$$+ \frac{1}{s + \frac{i}{\hbar} H_e^\times} \Sigma^{(2)}(s) \frac{1}{s + \frac{i}{\hbar} H_e^\times} \Sigma^{(2)}(s) \frac{1}{s + \frac{i}{\hbar} H_e^\times} + \cdots \tag{9.188}$$

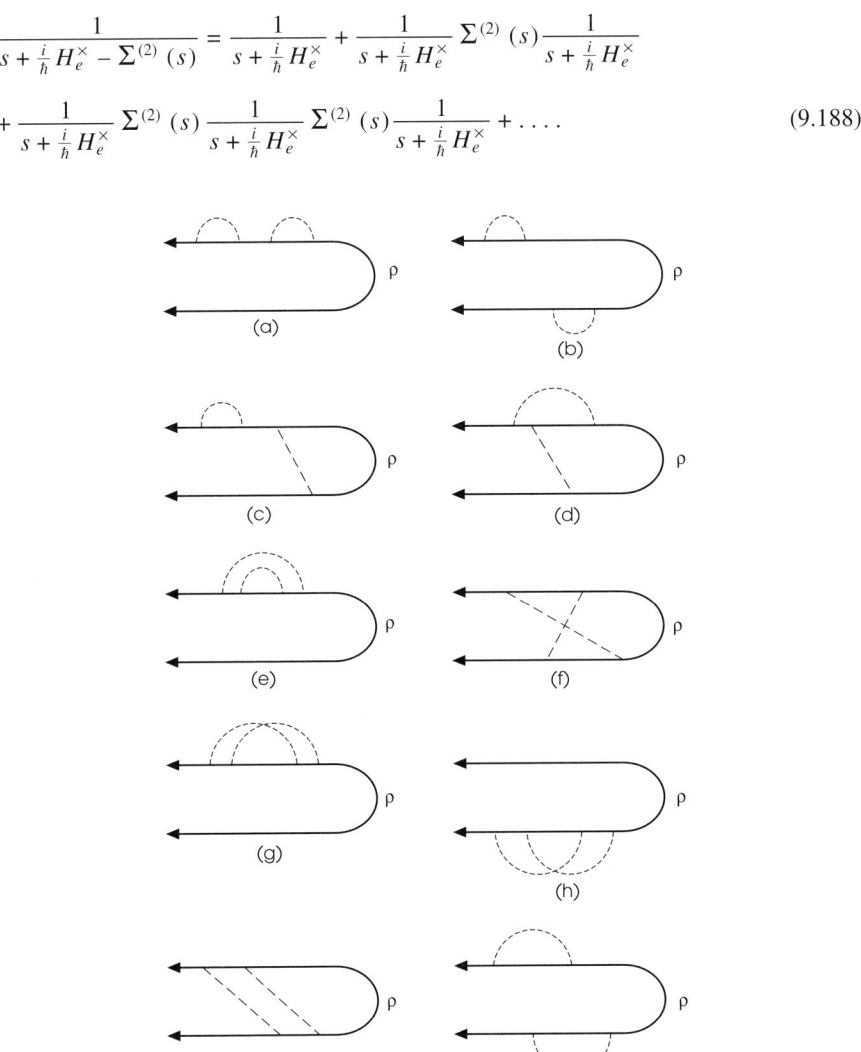

Fig. 9.34 Fourth order diagrams for the density matrix in the scheme of the double Feynman diagram.

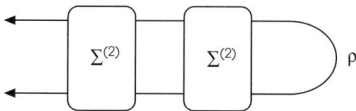

Fig. 9.35 A schematic representation of diagrams which are composed of series repetition of the second-order self-energy diagrams in Fig. 9.33.

Furthermore the diagrams in Figs 9.34(d)–(f) can be included by renormalizing the propagators in the self-energy diagrams $\Sigma^{(2)}$. Thus the irreducible fourth order diagrams are typically given in Figs 9.34(g)–(j). The systematic improvement of approximation can be carried out as follows. First of all, the second order self-energy part $\Sigma^{(2)}$ is calculated according to (9.187) and the renormalized propagator is calculated by (9.188). Then using this propagator the second order self-energy term $\Sigma^{(2)}$ and the fourth order self-energy term $\Sigma^{(4)}$ are calculated. Including these self-energy terms, we calculate the improved propagator as

$$\frac{1}{s + \frac{i}{\hbar} H_e^\times - \Sigma^{(2)}(s) - \Sigma^{(4)}(s)}. \tag{9.189}$$

In this way we can improve successively the propagator (Green function). The lowest reasonable approximation is given by

$$\int_0^\infty dt\, e^{-st} \exp\left[-\frac{i}{\hbar} H^\times t\right] \rho_0 = \frac{1}{s + \frac{i}{\hbar} H_e^\times - \Sigma^{(2)}(s)} \rho_e$$

$$+ \sum_\alpha \left(-\frac{i}{\hbar}\right)^2 \frac{1}{s + \frac{i}{\hbar} H_e^\times - \Sigma^{(2)}(s)} \left[M_{-\alpha} \left(\frac{n_\alpha}{s - i\omega_\alpha + \frac{i}{\hbar} H_e^\times - \Sigma^{(2)}(s - i\omega_\alpha)} \right.\right.$$

$$+ \frac{1 + n_\alpha}{s + i\omega_\alpha + \frac{i}{\hbar} H_e^\times - \Sigma^{(2)}(s + i\omega_\alpha)} \right) \left(M_\alpha \frac{1}{s + \frac{i}{\hbar} H_e^\times - \Sigma^{(2)}(s)} \rho_e \right) \Big]$$

$$- \sum_\alpha \left(-\frac{i}{\hbar}\right)^2 \frac{1}{s + \frac{i}{\hbar} H_e^\times - \Sigma^{(2)}(s)} \left[M_{-\alpha} \left(\frac{n_\alpha}{s + i\omega_\alpha + \frac{i}{\hbar} H_e^\times - \Sigma^{(2)}(s + i\omega_\alpha)} \right.\right.$$

$$+ \frac{1 + n_\alpha}{s - i\omega_\alpha + \frac{i}{\hbar} H_e^\times - \Sigma^{(2)}(s - i\omega_\alpha)} \right) \left(\frac{1}{s + \frac{i}{\hbar} H_e^\times - \Sigma^{(2)}(s)} \rho_e \cdot M_\alpha \right) \Big]$$

$$\begin{aligned}
&-\sum_{\alpha}\left(-\frac{i}{\hbar}\right)^2 \frac{1}{s+\frac{i}{\hbar}H_e^\times - \Sigma^{(2)}(s)}\left[\left(\frac{n_\alpha}{s-i\omega_\alpha + \frac{i}{\hbar}H_e^\times - \Sigma^{(2)}(s-i\omega_\alpha)}\right.\right.\\
&+\left.\frac{1+n_\alpha}{s+i\omega_\alpha + \frac{i}{\hbar}H_e^\times - \Sigma^{(2)}(s+i\omega_\alpha)}\right)\left(M_\alpha \frac{1}{s+\frac{i}{\hbar}H_e^\times - \Sigma^{(2)}(s)}\rho_e\right) M_{-\alpha}\bigg]\\
&+\sum_{\alpha}\left(-\frac{i}{\hbar}\right)^2 \frac{1}{s+\frac{i}{\hbar}H_e^\times - \Sigma^{(2)}(s)}\left[\left(\frac{n_\alpha}{s+i\omega_\alpha + \frac{i}{\hbar}H_e^\times - \Sigma^{(2)}(s+i\omega_\alpha)}\right.\right.\\
&+\left.\frac{1+n_\alpha}{s-i\omega_\alpha + \frac{i}{\hbar}H_e^\times - \Sigma^{(2)}(s-i\omega_\alpha)}\right)\left(\frac{1}{s+\frac{i}{\hbar}H_e^\times - \Sigma^{(2)}(s)}\rho_e \cdot M_\alpha\right) M_{-\alpha}\bigg]. \quad (9.190)
\end{aligned}$$

Using this expression, the dephasing rate of the quantum coherence between any two levels can be calculated, although only the electron–phonon interactions are taken into account. We are often interested in the dephasing processes under strong excitation, namely when a few electron-hole pairs are present in a quantum dot. Even under such a situation, the present formulation can be applied simply by extending the basis states, such as $|e_i\rangle$ and $|e_j\rangle$, to many particle states. These extensions are left for future studies.

9.18 Summary

We have discussed the Rabi splitting and Rabi oscillation of excitons in semiconductor quantum dots and clarified their new characteristic features. These coherent optical phenomena are not restricted to semiconductor quantum dots and the newly predicted features would be observed also in atoms and molecules. The universality of the phenomena ensures the possibility for application to the quantum state control and the quantum information processing by optical means. We have discussed also the mechanisms of exciton dephasing in semiconductor quantum dots and clarified that the pure dephasing is significant at high temperatures due to the exciton–phonon interactions. The exciton decoherence time at low temperatures is ultimately limited by the radiative lifetime about a nanosecond. If a single quantum-state operation can be carried out within 100 femtoseconds, we can achieve about 10,000 quantum operations within the decoherence time. This may be an advantage of the quantum state manipulation by using optical pulses. However, the quest for a material with longer decoherence time is still going on.

The spin degrees of freedom are considered as a promising candidate (see, for example, Chapter 6 of this volume by D. Gammon *et al.*).

Acknowledgments

The author would like to thank Professor H. Wang, Dr H. Kamada, Dr H. Htoon and Professor C.K. Shih for enlightening discussions during the course of this work.

References

1. Bennett, C.H., and DiVincenzo, D.P. (2000). Quantum Information and Computation, *Nature* **404**, 247–255.
2. Barenco, A., Deutsch, D., Ekert, A., and Jozsa, R. (1995). Conditional Quantum Dynamics and Logic Gates, *Phys. Rev. Lett.* **74**, 4083–4086.
3. Biolatti, E., Iotti, R.C., Zanardi, P., and Rossi, F. (2000). Quantum Information Processing with Semiconductor Macroatoms, *Phys. Rev. Lett.* **85**, 5647–5650.
4. Troiani, Filippo, Hohenester, Ulrich, and Molinari, Elisa, (2000). Exploiting Exciton–exciton Interactions in Semiconductor Quantum Dots for Quantum-information Processing, *Phys. Rev. B* **62**, R2263–R2266.
5. Chen, P., Piermarocchi, C., and Sham, L.J. (2001). Control of Exciton Dynamics in Nanodots for Quantum Operations, *Phys. Rev. Lett.* **87**, 067401.
6. Zare, R.N. (1998). Laser Control of Chemical Reactions, *Science*, **279**, 1875–1879.
7. Allen, L., and Eberly, J.H. (1975). *Optical Resonance and Two Level Atoms* (Wiley, New York).
8. Meekhof, D.M., Monroe, C., King, B.E., Itano, W.M., and Wineland, D.J. (1996). Generation of Nonclassical Motional States of a Trapped Atom, *Phys. Rev. Lett.* **76**, 1796–1799.
9. Yeazell, J.A., and Stroud Jr., C.R. (1988). Observation of spatially localized atomic electron wave packets, *Phys. Rev. Lett.* **60**, 1494–1497.
10. Heberle, A.P., Baumberg, J.J., and Köhler, K. (1995). Ultrafast Coherent Control and Destruction of Excitons in Quantum Wells, *Phys. Rev. Lett.* **75**, 2598–2601.
11. Marie, X., Le Jeune, P., Amand, T., Brousseau, M., Barrau, J., Paillard, M., and Planel, R. (1997). Coherent Control of the Optical Orientation of Excitons in Quantum Wells, *Phys. Rev. Lett.* **79**, 3222–3225.
12. Gammon, D., Snow, E.S., Shanabrook, B.V., Katzer, D.S., and Park, D. (1996). Homogeneous Linewidths in the Optical Spectrum of a Single Gallium Arsenide Quantum Dot, *Science* **273**, 87–90.
13. Brunner, K., Abstreiter, G., Böhm, G., Tränkle, G., and Weimann, G. (1994). Sharp-Line Photoluminescence and Two-Photon Absorption of Zero-Dimensional Biexcitons in a GaAs/AlGaAs Structure, *Phys. Rev. Lett.* **73**, 1138–1141.

14. Hawrylak, P., Narvaez, G.A., Bayer, M., and Forchel, A. (2000). Excitonic Absorption in a Quantum Dot, *Phys. Rev. Lett.* **85**, 389–392.
15. Bayer, M., Stern, O., Hawrylak, P., Fafard, S., and Forchel, A. (2000). Hidden symmetries in the energy levels of excitonic 'artificial atoms', *Nature* **405**, 923–926.
16. Htoon, H., Kulik, D., Baklenov, O., Holmes, Jr., A.L., Takagahara, T., and Shih, C.K. (2001). Carrier Relaxation and Quantum Decoherence of Excited States in Self-assembled Quantum Dots, *Phys. Rev. B* **63**, 241303(R).
17. Chen, Gang, Bonadeo, N.H., Steel, D.G., Gammon, D., Katzer, D.S., Park, D. and Sham, L.J. (2000). Optically Induced Entanglement of Excitons in a Single Quantum Dot, *Science* **289**, 1906–1909.
18. Bayer, M., Hawrylak, P., Hinzer, K., Fafard, S., Korkusinski, M., Wasilewski, Z.R., Stern, O., and Forchel, A. (2001). Coupling and Entangling of Quantum States in Quantum Dot Molecules, *Science* **291**, 451–453.
19. Becher, C., Kiraz, A., Michler, P., Imamoglu, A., Schoenfeld, W.V., Petroff, P.M., Zhang, Lidong, and Hu, E. (2001). Nonclassical Radiation from a Single Self-assembled InAs Quantum Dot, *Phys. Rev. B* **63**, 121312(R).
20. Michler, P., Imamoglu, A., Mason, M.D., Carson, P.J., Strouse, G.F., and Buratto, S.K. (2000). Quantum Correlation among Photons from a Single Quantum Dot at Room Temperature, *Nature* **406**, 968–970.
21. Solomon, G.S., Pelton, M., and Yamamoto, Y. (2001). Single-mode Spontaneous Emission from a Single Quantum Dot in a Three-Dimensional Microcavity, *Phys. Rev. Lett.* **86**, 3903–3906.
22. Bonadeo, N.H., Erland, J., Gammon, D., Park, D., Katzer, D.S., and Steel, D.G. (1998). Coherent Optical Control of the Quantum State of a Single Quantum Dot, *Science*, **282**, 1473–1476.
23. Rabi, I.I. (1937). Space Quantization in a Gyrating Magnetic Field, *Phys. Rev.* **51**, 652–654.
24. Rabi, I.I., Millman, S., Kusch, P., and Zacharias, J.R. (1939). The Molecular Beam Resonance Method for Measuring Nuclear Magnetic Moments: The Magnetic Moments of $_3Li^6$, $_3Li^7$ and $_9F^{19}$, *Phys. Rev.* **55**, 526–535.
25. Torrey, H.C. (1949). Transient Nutations in Nuclear Magnetic Resonance, *Phys. Rev.* **76**, 1059–1068.
26. Gibbs, H.M. (1973). Incoherent Resonance Fluorescence from a Rb Atomic Beam Excited by a Short Coherent Optical Pulse, *Phys. Rev. A* **8**, 446–455.
27. Bai, Y.S., Yodh, A.G., and Mossberg, T.W. (1985). Selective Excitation of Dressed Atomic States by Use of Phase-Controlled Optical Fields, *Phys. Rev. Lett.* **55**, 1277–1280.
28. Hocker, G.B., and Tang, C.L. (1968). Observation of the Optical Transient Nutation Effect, *Phys. Rev. Lett.* **21**, 591–594.
29. Hoff, P.W., Haus, H.A., and Bridges, T.J. (1970). Observation of Optical Nutation in an Active Medium, *Phys. Rev. Lett.* **25**, 82–84.
30. Cole, B.E., Williams, J.B., King, B.T., Sherwin, M.S., Stanley, C.R. (2001). Coherent Manipulation of Semiconductor Quantum Bits with Terahertz Radiation, *Nature* **410**, 60–63.
31. Schlzgen, A., Binder, R., Donovan, M.E., Lindberg, M., Wundke, K., Gibbs, H.M., Khitrova, G., and Peyghambarian, N. (1999). Direct Observation of Excitonic Rabi Oscillations in Semiconductors, *Phys. Rev. Lett.* **82**, 2346–2349.

32. Cundiff, S.T., Knorr, A., Feldmann, J., Koch, S.W., Göbel, E.O., and Nickel, H. (1994). Rabi Flopping in Semiconductors, *Phys. Rev. Lett.* **73**, 1178–1181.
33. Takagahara, T. (2001). Theory of Excitonic Rabi Flopping and Quantum Interference in Single Quantum Dots, in *Technical Digest of QELS 2001* (OSA, Washington, DC) Vol. 57, p. 19.
34. Kamada, H., Gotoh, H., Temmyo, J., Takagahara, T., and Ando, H. (2001). Exciton Rabi Oscillation in a Single Quantum Dot, *Phys. Rev. Lett.* **87**, 246401.
35. Stievater, T.H., Li, Xiaoqin, Steel, D.G., Gammon, D., Katzer, D.S., Park, D., Piermarocchi, C., and Sham, L.J. (2001). Rabi Oscillations of Excitons in Single Quantum Dots, *Phys. Rev. Lett.* (87), 133603.
36. Htoon, H., Takagahara, T., Kulik, D., Baklenov, O., Holmes, Jr., A.L., and Shih, C.K. (2002). Interplay of Rabi Oscillations and Quantum Interference in Semiconductor Quantum Dots, *Phys. Rev. Lett.* **88**, 087401.
37. Mollow, B.R. (1969). Power Spectrum of Light Scattered by Two-Level Systems, *Phys. Rev.* **188**, 1969–1975.
38. Cohen-Tannoudji, C., Dupont-Roc, J., and Grynberg, G. (1992). *Atom–Photon Interactions* (Wiley, New York).
39. Bonadeo, N.H., Chen, G., Gammon, D., Park, D., Katzer, D.S., and Steel, D.G. (1999). Single Quantum Dot States: Energy Relaxation and Coupling, in *Technical Digest of Quantum Electronics and Laser Science Conference* (QELS'99, Baltimore), QTuC5, p. 48.
40. Feynman, R.P., Vernon, Jr., F.L., and Hellworth, R.W. (1957). Geometrical Representation of the Schrödinger Equation for Solving Maser Problems, *J. Appl. Phys.* **28**, 49–52.
41. Gammon, D., Snow, E.S., Shanabrook, B.V., Katzer, D.S., and Park, D. (1996). Fine Structure Splitting in the Optical Spectra of Single GaAs Quantum Dots, *Phys. Rev. Lett.* **76**, 3005–3008.
42. Takagahara, T. (2000). Theory of Exciton Doublet Structures and Polarization Relaxation in Single Quantum Dots, *Phys. Rev. B* **62**, 16840–16855.
43. Selvakumar Nair, V., and Takagahara, T. (1997). Theory of Exciton Pair States and their Nonlinear Optical Properties in Semiconductor Quantum Dots, *Phys. Rev. B* **55**, 5153–5170.
44. See, for example, *Confined Electrons and Photons, New Physics and Devices*, edited by Burstein, E., and Weisbuch C. (Plenum, New York, 1995); *Microcrystalline and Nanocrystalline Semiconductors*, Materials Res. Soc. Symp. Proc. Vol. 358, eds., Collins, R.W., Tsai, C.C., Hirose, M., Koch, F., and Brus, L., (Materials Research Society, Pittsburgh, 1995); *Optical Properties of Semiconductor Quantum Dots*, by Woggon, U. (Springer, Berlin, 1997); *Quantum Dot Heterostructures*, by Bimberg, D., Grundmann, M., and Ledentsov, N.N. (John Wiley, Chichester, 1999).
45. Hegarty, J., and Sturge, M.D. (1985). Studies of Exciton Localization in Quantum-well Structures by Nonlinear-optical Techniques, *J. Opt. Soc. Am. B* **2**, 1143–1148.
46. Webb, M.D., Cundiff, S.T., and Steel, D.G. (1991). Stimulated-picosecond-photon-echo Studies of Localized Exciton Relaxation and Dephasing in GaAs/AlxGa1-xAs Multiple Quantum Wells, *Phys. Rev. B* **43**, 12658–12661.
47. Schultheis, L., Honold, A., Kuhl, J., Köhler, K., and Tu, C.W. (1986). Optical Dephasing of Homogeneously Broadened Two-dimensional Exciton Transitions in GaAs Quantum Wells, *Phys. Rev. B* **34**, 9027–9030.

48. Takagahara, T. (1985). Localization and Homogeneous Dephasing Relaxation of Quasi-two-dimensional Excitons in Quantum-well Heterostructures, *Phys. Rev. B* **32**, 7013–7015; Excitonic Relaxation Processes in Quantum Well Structures, *J. Lumin.* **44**, 347–366 (1989).
49. Hess, H.F., Betzig, E., Harris, T.D., Pfeiffer, L.N., and West, K.W. (1994). Near-field Spectroscopy of the Quantum Constituents of a Luminescent System, *Science* **264**, 1740–1745.
50. Fan, Xudong, Takagahara, T., Cunningham, J.E., and Wang, Hailin, (1998). Pure Dephasing Induced by Exciton–phonon Interactions in Narrow GaAs Quantum Wells, *Solid State Commun.* **108**, 857.
51. Huang, K., and Rhys, A. (1950). *Proc. Roy. Soc. (London)* A **204**, 406.
52. Duke, C.B., and Mahan, G.D. (1965). Phonon-Broadened Impurity Spectra. I. Density of States, *Phys. Rev.* **139**, A1965–A1982.
53. Kubo, R., and Toyozawa, Y. (1955). Application of the Method of Generating Function to Radiative and Non-radiative Transitions of a Trapped Electron in a Crystal, *Prog. Theor. Phys.* **13**, 160–182.
54. Takagahara, T. (1999). Theory of Exciton Dephasing in Semiconductor Quantum Dots. *Phys. Rev. B* **60**, 2638–2652.
55. Benisty, H., Sotomayor-Torres, C.M., and Weisbuch, C. (1991). Intrinsic Mechanism for the Poor Luminescence Properties of Quantum-box Systems, *Phys. Rev. B* **44**, 10945–10948.
56. Bockelmann, U. (1993). Exciton Relaxation and Radiative recombination in Semiconductor Quantum Dots, *Phys. Rev. B* **48**, 17637–17640.
57. Takagahara, T. (1993). Electron–phonon Interactions and Excitonic Dephasing in Semiconductor Nanocrystals, *Phys. Rev. Lett.* **71**, 3577–3580; Electron–phonon Interactions in Semiconductor Nanocrystals, *J. Lumin.* **70**, 129–143 (1996); and references therein.
58. Blacha, A., Presting, H., and Cardona, M. (1984). Deformation Potentials of k = 0 States of Tetrahedral Semiconductors, *Phys. Stat. Sol. (b)* **126**, 11–36.
59. *Physics of Group IV Elements and III-V Compounds*, Vol. 17a of *Landolt-Börnstein*, edited by Madelung, O., Schulz, M., and Weiss H. (1982) (Springer, Berlin).
60. Kubo, R., in *Fluctuation, Relaxation and Resonance in Magnetic Systems*, ed. D. ter Haar, (Oliver and Boyd, Edinburgh, 1962, p. 23).
61. Takagahara, T., Hanamura, E., and Kubo, R. (1977). Stochastic Models of Intermediate State Interaction in Second Order Optical Processes – Stationary Response I & II, *J. Phys. Soc. Japan* **43**, 802–816; Stochastic Models of Intermediate State Interaction in Second Order Optical Processes – Transient Response, *J. Phys. Soc. Japan* **43**, 1522–1528 (1977).
62. Takagahara, T. (1985). Localization and Energy Transfer of Quasi-two-dimensional Excitons in GaAs–AlAs Quantum-well Heterostructures, *Phys. Rev. B* **31**, 6552–6573.
63. Takagahara, T. (1993). Effects of Dielectric Confinement and Electron-hole Exchange Interaction on Excitonic States in Semiconductor quantum Dots. *Phys. Rev. B* **47**, 4569–4584.
64. Borri, P., Langbein, W., Schneider, S., Woggon, U., Sellin, R.L., Ouyang, D., and Bimberg, D. (2001). Ultralong Dephasing Time in InGaAs Quantum Dots, *Phys. Rev. Lett.* **87**, 157401.
65. Birkedal, D., Leosson, K., and Hvam, J.M. (2001). Long Lived Coherence in Self-Assembled Quantum Dots, *Phys. Rev. Lett.* **87**, 227401.
66. Empedocles, S.A., Norris, D.J., and Bawendi, M.G. (1996). Photoluminescence Spectroscopy of Single CdSe Nanocrystallite Quantum Dots, *Phys. Rev. Lett.* **77**, 3873–3876.

67. Empedocles, S.A., and Bawendi, M.G. (1997). Quantum-Confined Stark Effect in Single CdSe Nanocrystallite Quantum Dots, *Science* **278**, 2114–2117.
68. Palinginis, Phedon, and Wang, Hailin, (2001). High-resolution spectral hole burning in CdSe/ZnS core/shell Nanocrystals, *Appl. Phys. Lett.* **78**, 1541–1543.
68. Ferrio, K.B., and Steel, D.G. (1998). Raman Quantum Beats of Interacting Excitons, *Phys. Rev. Lett.* **80**, 786–789.

Index

II–VI semiconductors, 167, 282
III–V semiconductors, 167, 283

A-exciton resonance, bulk CdSe, 1–22
Absorption coefficients, 13–14, 28, 49
Absorption spectra:
  bulk CdSe, 6–7
  exciton dephasing rate, 436, 441–2
  half-width at half maximum, 436, 442, 446, 460
  lineshapes, 445–50
  quantum disks, 445, 447
  quantum dots, 290
  quantum wells, 50–3
  quantum wires, 55–63
  see also Optical density
Acoustic phonons:
  dephasing rate determination, 436
  exciton migration, 453–8
  exciton–phonon coupling, 385–8, 436, 442–4, 446, 448–9
  quantization, 366, 376, 389
  scattering, 89, 93, 103, 108–12
  wavevectors, 443–4, 449, 454–6, 461
Algorithms, quantum computing, 338, 339, 341–2, 350–4
Aluminium gallium arsenide, 31, 213–14, 216–17, 218, 372–3, 374
Anderson transition, 127, 133, 135
Anisotropic quantum disks, 444–5, 446
Anticommutation rules, 33
Antireflection coatings, 29, 34
Area theorem approximation, 351
Autler–Townes splitting, 59
Autocorrelation, 27–8, 29, 93, 126–31, 307, 309–10

Backscattering, resonant secondary emission, 94, 132–9
Band gaps:
  bleaching, 24, 31, 36, 388
  energies, 35
  GaAs, 24, 26, 29, 31–2
  renormalization, 35, 36, 81–2, 171, 439
  third harmonic, 23, 27, 29–30, 34–7
Band structure, 33, 378
Bandwidth filters, 10
Beer absorption length, 6
Bell state, 332
Bessel function, 65–6, 369
Bethe–Salpether equation, 136, 137
Biexcitons, 178–80
  II–VI semiconductors, 167, 178
  coherent nonlinear optical spectroscopy, 318–25
  continuous wave excitation, 319–25
  dephasing, 325–6
  dynamics-controlled truncation scheme, 182, 183
  GaAs, 187, 189–93, 195
  higher-order Coulomb correlations, 187–95, 198, 199
  lifetime, 326–9
  photoluminescence, 215
  quantum dots, 215, 281–2, 284–5, 312, 316, 318–30
  recombination, 319, 326–8
  resonant excitation, 320–3
  second-order Born approximation, 3
  transition dipole moment, 329–30
Bloch equations, 3, 5, 24–5, 33–4, 35, 40, 43–5
  Coulomb interactions, 172
  Hanle effect, 237, 268–9, 271

472   Index

Bloch equations (*Contd.*)
  see also Semiconductor Bloch equations
Bloch functions, 219, 222, 252, 256, 282
Bloch spheres, 26, 349–50
Bloch states, 140
Bloch vectors, 26, 27, 34, 406–9, 410–13, 415–20, 430
Blue shift, 195–6
Bohr radius, 92, 95–6, 213, 249, 270, 284, 449
Boltzmann equations, 63, 172–3
Born approximations, 3–4, 5, 12, 93, 98, 138, 147, 173, 178, 196
Bose fields, 103, 104, 105
Boson commutation rules, 101

Cadmium selenide:
  A-exciton resonance, 1–22
  bulk crystals, 6
  CdS/ZnS nanocrystal cavity QED, 366, 368, 375–90
  coherent nonlinear pulse propagation, 1–22
  Rabi flopping, 3, 4–6, 10, 11, 13, 14, 15–16, 18
Calcium fluoride prisms, 29, 30
Carrier–carrier interactions, 44, 54, 63–4, 383
Carrier–carrier scattering, 5, 13, 63, 68, 172–3, 177, 197
Carrier density, 28, 45
Carrier–light coupling, 44
Carrier–phonon interactions, 63–4, 171–2, 366, 376
Carrier-wave Rabi flopping (CWRF), 23–39, 40
  bulk CdSe, 3, 4–6, 10, 11, 13, 14, 15–16, 18
  carrier-wave reshaping, 77–9
  GaAs, 23–4, 26, 28–9, 40
  quantum wells, 67–70
  semiconductors, 79–83
  sub-optical carrier pulse propagation, 71–9

theory, 32–7, 40, 71–3
time-derivative effects, 75–9
see also Rabi flopping
Cavity QED:
  effective mode volume, 367, 370, 372, 376, 384–5
  Q-factor, 366, 367
  quantum dots with dielectric microspheres, 366–94
Center-of-mass momentum (COM) states
  acoustic phonon scattering, 108–10
  bulk semiconductors, 283–4
  Hamiltonians, 101–2, 116
  resonant secondary emission, 90, 94, 113, 139
  Schrödinger equations, 92, 95–9, 116, 147, 152
  wave functions, 141, 142, 212
Chirp parameters, 42, 70–1
Circular polarization, 145, 147
  co-circular polarization, 189–90, 191
  counter-circular polarization, 94, 193–6
  emission polarization, 238
  hyperfine effects, 252
  magnetic field effects, 235–6, 270–1
  many-body processes, 176–7
  quantum dots, 286
  trions, 250–1
Coherence effects:
  coherent limit, 182, 195
  femtosecond lasers, 24, 168
  four-wave mixing, 174
  quantum dots, 287–8, 395–470
  two-photon, 318–19, 320–5, 330, 331
Coherent exciton–light coupling, bulk CdSe, 1–22
Coherent multiple pulse breakup, 10–17, 18
Coherent nonlinear pulse propagation:
  bulk CdSe, 1–22
  suppression, 2, 3, 8–10
Coherent optical control, 306–10, 338, 395, 420–4, 425, 426
Coherent optical spectroscopy, 281–365

Coherent potential approximation (CPA), 98–9
Colloidal chemistry, core/shell nanocrystals, 366, 368, 375, 459
Conduction bands, 24, 26, 32–3, 140, 169, 181, 215, 443
Confinement:
  photonic, 368, 369–70
  potentials, 94–5, 109, 155–6, 213, 369
  see also Quantum confinement
Continuous wave lasers
  optical spectroscopy, 290–3, 296, 298–301, 336, 397, 459
  photoluminescence, 208, 319–20, 373
Core/shell nanocrystals, 366, 368, 375–90, 459
Correlation effects, 3, 5, 64–5, 70
  exponential, 96–9
  five-level systems, 427–9, 430–1
  four-level systems, 425, 426
  see also Autocorrelation; Coulomb correlations
Coulomb correlations
  coherent regime, 187–97
  experimental studies, 187–201
  four-particle correlations, 190, 191, 193–4, 197–9
  higher-order, 166–206
  incoherent densities, 197–9, 202
  many-body correlations, 167–70, 177, 178–87, 203
  ultrafast spectroscopy, 167–78, 202
Coulomb interactions
  biexcitons, 318
  charged carriers, 25, 32
  electron–hole pairs, 167, 168, 215, 283–5
  exchange, 225, 227
  exciton migration, 454, 456
  first-order, 195
  high-field effects, 40, 46, 47, 50, 53, 54–6, 58, 63, 66
  screening, 173, 176–8, 189
  two exciton-state entanglement, 332–3, 336
Coulomb renormalizations, 5, 82

Coulomb scattering, 90
Coupled oscillator model, 383–4
Cross-correlation signals, 7, 8, 10, 12, 17–18, 119, 307, 309–10
Cross-polarized emission, 140
Curl equations, 40, 42, 72–3

Decoherence, quantum dots, 288, 324, 339, 395–470, 459
Deformation-potential coupling, 108–9, 443
Delta function, 97, 113, 157–8, 191
Density matrix theory:
  coherent optical spectroscopy, 309, 351
  higher-order Coulomb correlations, 169, 180, 184, 186, 202
  phonons, 462
  quantum dots, 397–401, 403–4, 421–4
  resonant secondary emission, 89, 93, 101–8, 111, 115
  time-evolution, 462
  two-photon coherence, 427
Density of states (DOS), 53, 96–8
Dephasing:
  biexcitons, 325–6
  decoherence, 288
  density dependence, 177
  excitation-induced, 177, 325
  excitonic polarization, 3, 4, 5, 9–10, 17–18, 103, 168
  four-wave mixing, 174, 190–3
  Green function formalism, 436–42, 464–5
  nondiagonal, 64–5, 68, 106–7, 451–2
  nonradiative coherence, 459–65
  nuclear spin relaxation, 268
  phonon-induced, 17–18, 108
  pulse shaping, 351, 353–4
  pure dephasing, 450–2
  quantum dots, 431, 433, 435–6
  resonant secondary emission, 103, 108, 116, 121–3
  semiconductor nanocrystals, 366, 376, 385–90

474 *Index*

Dephasing (*Contd.*)
  temperature dependence, 445–50, 451
Dephasing times:
  biexcitons, 325–6
  coherence, 429, 459–65
  excitons, 55, 296–8, 300, 406, 436–42
  four-wave mixing, 174, 177
  high energy transitions, 36
  inhomogeneous broadening, 57, 81
  many-body interactions, 168, 172
  quantum dots, 403
Depolarization field, 225–7, 259–60, 305
Detuning, 48, 69, 388, 413, 415–20, 425
Deutsch–Jozsa problem, 338, 340–3, 350–4
Diamagnetic shift, 253, 255, 304–5
Dielectric constants, 14, 15, 34, 103, 150, 153, 178, 368
Dielectric microspheres:
  cavity QED of quantum dots, 366–94
  MBE-grown nanostructure composites, 372–5
  semiconductor nanocrystal composites, 366, 368, 375–90
  whispering gallery modes, 366, 368–72, 373–5, 382–5
Differential transmission (DT), 31, 290–3, 296–7, 301–4, 312–13
  biexcitons, 321–5, 327–9
  continuous wave, 296–7, 301–4
  detuning, 387–8
  strong-field, 313–16
  Zeeman coherence, 333–8
Dipole coupling, 367, 374–5, 384–5
Dipole–dipole interactions, 259–61
Dipole matrix elements, 23, 28, 33–4, 45, 80, 140, 386
Dipole moments, 15, 141, 344
  biexcitons, 329–30
  measurement, 295–8, 315–16
  operators, 396, 421
  transition, 225, 400
Disorder effects, 89, 90, 92, 101, 133, 135, 149

eigenstates, 93, 94–108, 109, 113–14, 139, 149–52
  level repulsion, 125–30
  potential variance, 155–6
Double Feynman diagrams, 462–4
Dressed exciton states, 401–2, 413
D'ykonov–Perel mechanism, 264
Dynamics-controlled truncation scheme (DCT), 3–4, 9, 10, 180–5, 187, 188–9, 195–202
Dyson equation, 151

Effective mass approximation (EMA), 92, 94, 140, 155, 283–4, 378
Effective mode volume, 367, 370, 372, 376, 384–5
Effective polarization model (EPM), 185–6, 187–95
Eigenfunctions, 54–5
Eigenstates:
  disorder, 93, 94–108, 109, 113–14, 139, 149–52
  dressed exciton states, 402
  electron–hole pairs, 284
  exciton migration, 456
  exciton–phonon coupling, 385–6
Elastic scattering, 90, 98, 101, 263, 282
Electro–magneto–excitons, 51–2
Electro–magneto–optical simulation, quantum wells, 40, 49–53
Electro–optical effects, 40
Electron–electron interactions, 44, 54, 63–4, 383
Electron–hole pairs:
  binding energies, 41, 178–9, 212, 241–2, 248–9, 323
  Bloch states, 140
  optically excited, 3, 43–4, 63–5, 168–70, 212
  ponderomotive energy, 63
  quantum disks, 445
  quantum dots, 284

recombination, 168, 244–5, 252, 268
transitions, 43, 44
Electron–hole plasmas, 171
Electron–hole wave packets:
  high-field effects, 45–63
  interferometry, 306–8, 403, 420–4, 425, 426
  nonperturbative, 40, 41, 46, 284
  real-space–time modeling, 40, 41, 45, 47–59
Electron–phonon interactions, 63–4, 171–2, 366, 376, 386, 460–5
Electronic spin:
  hyperfine fields, 256–7, 263, 265–8
  polarization, 210–11, 264, 271
  quantum dot optical spectra, 207
  relaxation, 256–7, 263, 264–5, 269
  rotation, 273–4
Electrostatic pictures, 226–7
Elliott–Yafet mechanism, 264
Emission spectrum, 396–401
Energy levels:
  crossing, 69
  dephasing of nonradiative coherence, 460
  emission spectrum, 396–401
  level repulsion signatures, 89, 93–4, 123–32
  nonradiative relaxation pathways, 420, 427
  quantum disks, 445, 446, 449
  quantum dots, 212–13, 216–17, 281, 286
  spacing distribution, 124–5
Enhanced backscattering (EBS), 89, 94, 132–9
Enhanced spontaneous emission, cavity QED, 366–8, 376–81, 384
Entangled states, 282, 330–8, 391, 425, 430
Epitaxial growth *see* Molecular beam epitaxy
Exchange coupling, 141–2
Exchange interactions, 220–1, 234, 253, 391, 455–6
Exchange (spin) splitting, 89, 94, 139–47, 207, 209, 219–22, 238–9

long-range, 222–7, 455–6
short-range, 456
Excitation induced dephasing (EID), 177, 325
Exciton–exciton interactions, 3, 8, 90–1, 108, 185, 189–95, 316–17, 318
Exciton ionization effects, 54
Exciton–phonon interactions, 2, 17, 18, 186, 442–4, 448–9
  dephasing, 385–8, 436–41, 446, 451
  resonant secondary emission, 89, 93, 103, 107–12
Exciton–polaritons, 3, 5, 149, 151
Excitonic polarization
  depolarization field, 225–7, 259–60
  effective polarization model, 185–6, 187–95
  nonthermal spin polarization state, 210
  phase-locked pulses, 430–3
  photoluminescence, 210, 223–5, 233–7, 238–9, 242–3, 249–55
  polariton theory, 149–51
  pseudo-spin, 233–40, 349–50
  time evolution, 15–17, 168, 173, 175
  wave packet interferometry, 420–4, 425, 426
  see also Dephasing
Excitonic structure
  beating structure, 424, 429, 430–1
  binding energies, 41, 178–9, 212, 241–2, 248–9, 323
  charged states, 240–52
  coherence length, 148
  decoherence, 288, 324, 339, 395–470, 459
  delocalization, 316–17
  density dependence, 177
  dressed states, 401–2, 413
  excited states, 451–2, 458
  fine structure, 207, 209, 219–40, 243–8, 271, 294–5
  heavy-hole states, 140, 149–50, 157, 198, 219, 221, 224–5, 231, 374, 445
  homogeneous broadening, 89, 90, 103, 122, 128–9, 209, 238

Excitonic structure (*Contd.*)
  inhomogeneous broadening, 89, 90, 122–4, 144, 152, 209, 217, 257, 263, 435
  interface disorder, 89, 90, 92, 94–108, 109, 113–14, 149–52, 285, 435
  intra-exciton ground state, 451–2, 458
  level repulsion, 89, 93–4, 123–32
  light-hole states, 140, 198, 224–5
  linewidth, 89–90, 388–90, 459
  Lorenz broadening, 128–9
  nanocrystals, 378
  negative charge, 241, 248–52
  optical density, 97–9, 110, 111–12, 114, 123, 133–4
  phonon-assisted exciton migration, 453–8
  population decay, 105–6, 109–10, 122, 152–4, 157–9, 397, 404, 452–8
  pseudo-spin model, 232–40, 349–50
  quantum disks, 444–51, 452–3
  quantum dots, 281–2, 286, 293–310, 395–470
  spin effects, 139–47, 149, 157, 158, 179
  see also Center-of-mass momentum states; Localized states
Excitonic trapping, 40, 65–71
Extreme electromagnetic field effects *see* High-field effects

Faraday configuration, 229, 234, 235, 236, 243–4, 246, 248, 250, 252–3, 270, 304
Fast Fourier transforms, 99
Femtosecond range, 409, 410–20, 430, 432
  mode-locked lasers, 23, 24–5, 27–8, 30–1, 35–6, 168
Fermi–Dirac distributions, 168, 172
Fermi golden rule, 436
Feynman diagrams, 462–4
Field-induced tunneling, 57
Field penetration depth, 375
Fine structure, 207, 209, 219–40, 243–8, 271, 294–5

Finite-difference time-domain method (FDRD), 49, 72–3
Five-level systems, 427, 431–3
Flip–flop transitions, 258–9
Foerster energy transfer, 142
Four-level systems:
  quantum dots, 286, 316, 336, 353–4
  wave packet interferometry, 420–4, 425, 426
Four-wave mixing (FWM), 91, 173–8, 185, 187–93, 197–8, 459
Fourier–Laplace transforms, 400, 436
Fourier transforms, 27, 49, 98, 99, 101, 109, 114, 137, 400, 436
Franz–Keldysh effect:
  bulk crystals, 46, 54
  dynamic, 59–63
  quantum wells, 50–1, 52
  quantum wires, 53–63
  static, 53–9
Free carriers, 108
Free-electron lasers, 40
Free-exciton resonance, bulk CdSe, 1–22
Free spectral range (FSR), whispering gallery modes, 371, 373
Frequency-modulated spectroscopy, 66–71, 459
Frustrated total internal reflection, 370–1, 374
Fused silica microspheres, 366, 368, 370–2, 382–5

g-factors, 227–32, 239, 246–8, 304
Gallium arsenide:
  band edge, 24, 26, 29, 31–2
  Bloch states, 140
  bound biexcitons, 187, 189–93, 195
  carrier-wave Rabi flopping, 23–4, 26, 40
  dephasing processes, 435
  enhanced resonant backscattering, 135
  four-wave mixing, 175, 187
  hyperfine interactions, 265–7

polariton effects, 148–9, 152, 158
polarization-dependent emission, 144–5
pump/probe spectroscopy, 195, 198–9
quantum dots, 281, 285–7, 293–4, 299, 304–5, 306, 327, 330
quantum wells, 212–14, 216–17, 220, 241–3, 372–3, 374
semiconductor–metal transitions, 23–4
spin relaxation, 270
trions, 241–3, 248
Germanium, four-wave mixing, 178
Green's functions, 97, 116, 150–2, 184
 exciton dephasing rate, 436–42, 464–5
 non-equilibrium, 178, 202
Gyration vector, 407

Hadamard transformation, 339–40
Half-width at half maximum (HWHM), 436, 442, 446, 460
Hamiltonians:
 center-of-mass momentum states, 101–2, 116
 dressed exciton states, 401–2
 electron–hole pairs, 43–4, 101–8
 emission spectrum, 396–401
 energy-level statistics, 124
 exchange interactions, 220–1, 234, 253
 exciton–phonon interactions, 92, 436–9, 442–4, 454, 460
 optical density, 98
 phonons, 186
 pulse shaping, 344–6, 348–9
 Rabi flopping, 25, 32
 spin, 253, 257, 259, 264
 time-evolution, 459–65
 trions, 247
 two-band semiconductors, 169
 Zeeman effect, 227–9, 230, 234
Hankel function, 369
Hanle effect, 207, 210, 237, 239–40, 268–71
Harmonic generation:
 high-field, 41
 second harmonic, 28
 third harmonic, 23, 27, 29–30, 34–7
Hartree–Fock approximations, 25, 45, 63, 177, 186, 195
 random phase approximation, 171–2, 173, 182–3
Heavy-hole states, 140, 149–50, 157, 198, 219, 221, 224–5, 231, 374, 445
Heisenberg equation of motion, 33, 44–5, 103, 105, 169–72
Heisenberg exciton operator, 101–5
High-field effects:
 semiconductor nanostructures, 40–88
 theory, 43–5
High-field harmonic generation (HHG), 41
High optical field effects:
 carrier-wave Rabi flopping, 71–83
 quantum wells, 63–5
High-Q regime, 366, 367, 376, 381–5
High-resolution spectral hole burning, 366, 386–90
Higher-order Coulomb correlation effects (HOCs), 166–206
Hole burning, 366, 386–90, 459
Homodyne detection, 290–2, 299, 312
Huang–Rhys model of F-centers, 436
Hyperfine interactions, 207, 209, 211, 252–63, 265–8, 273

Incoherence:
 higher-order Coulomb correlations, 197–9, 202
 optical spectroscopy, 299–300
 secondary emission, 90–2, 93, 111, 113–18, 121, 143–4
Indium gallium arsenide, 195–6, 201
Inelastic scattering, 90, 92, 101–8, 173, 263
Interface roughness, 89, 90, 92, 93, 96, 155–6, 213–15, 285, 435
Interferometry (IF), 27, 29, 30, 93, 119–20, 306–8, 403, 420–4
 see also Quantum interference
Ionization tunneling, 57

478 *Index*

Island structures *see* Molecular beam epitaxy; Quantum disks
Ivchenko–Pikus representation, 221, 228

k-space approach, 48, 64–6, 68, 116, 117, 120–3, 149
Kane dipole matrix element, 80
Kinetic equations, 93, 106, 107, 109, 111, 115, 121, 156, 259
Knight shifts, 262–3
Kramers doublet, 140
Kronecker symbol, 116

Ladder diagrams, 94, 136, 137, 196
Landau structures, 50–2
Laplace transforms, 400, 404, 405, 436–8, 460–2
Larmor frequency, 234, 266
Laser spectroscopy, 209, 215, 218, 241–2
Lasers:
    femtosecond mode-locked, 23, 24–5, 27–8, 30–1, 35–6, 168
    free-electron, 40
    MIR, 40
    self-phase modulation, 31
Lattice mismatch, CdSe–BaF$_2$, 7
Level repulsion signatures, 89, 93–4, 123–32
Light-hole states, 140, 198, 224–5
Light intensity:
    coherent nonlinear pulse propagation, 1–22, 23
    Rabi flopping, 23–4
Light scattering *see* Rayleigh scattering; Resonant light scattering
Lindblad operators, 351, 353
Linear absorption spectroscopy, 6–7, 290, 291–2, 295–8
Linear optical regime, real-space–time modeling, 40, 41, 45, 47–59
Linear polarization, 146, 234, 235–6, 238
Linhard dielectric function, 173

Local field effects, 25
Localized states:
    Anderson transition, 127, 133, 135
    exciton migration, 453–8
    fine structure, 207, 212–15, 256, 263–5
    Hanle effect, 268–71
    quantum disks, 444–5
    quantum dots, 293
    quantum wells, 435
    resonant secondary emission, 127–31, 133–5, 139–47, 148
    spin relaxation, 263–5
Logic gates, 340
Longitudinal acoustic modes, 443–4, 449, 455
Lorenz broadening, 128–9
Low-Q regime, 366, 376–81, 384
Luminescence *see* Photoluminescence
Luttinger parameters, 228, 264

Magnetic field effects, 40, 46, 189–90
    polar angles, 229–31, 246–8
    quantum dots, 304–5
    quantum wells, 49–53
    spin relaxation, 263, 264
    trions, 243–6, 250–2
    see also Hanle effect; Zeeman effect
Magnetic resonance, 209, 261–3
Magneto–excitons, 51–2, 304–5
Magnus expansion, 346
Many-body interactions, 5, 13, 14, 42, 46, 66, 70, 79, 311
    Coulomb correlations, 167–70, 177, 178–87, 203
Many-particle theory, 94, 135–7, 168–72, 182–4, 190, 191, 193–4, 197–201
Markov approximation, 104, 105, 107, 116, 156–7, 173, 196
Maximally crossed diagrams, 94, 136, 137
Maxwell–Bloch equations, 2, 5, 9, 12, 72–3
Maxwell equations, 32, 34
    curl equations, 40, 42, 72–3
    high-field effects, 43–5, 49

polariton effects in secondary emission, 94, 147, 149, 150
whispering gallery modes, 368–9
Maxwell–Schrödinger formalism, 147, 149–52
Mean-field effects, 3, 5–6, 12, 14, 25, 178
Michelson interferometer, 27, 29, 30, 306
Microcavities *see* Cavity QED
MIR lasers, 40
Molecular beam epitaxy (MBE):
  interface roughness, 90, 96, 213, 216, 285, 435
  quantum dots, 282–3
  quantum wells, 435
  semiconductor nanostructures, 366, 368, 372–5, 386
Multiple quantum wells (MQWs), 149–50, 154–5, 158
Multi-subband model, 58–63

Nanocrystals:
  CdS/ZnS cavity QED, 366, 368, 375–90
  dephasing, 366, 376, 385–90
  dielectric microsphere composite systems, 366, 368, 375–90
Nanostructures:
  high-field effects, 40–88
  MBE-grown, 366, 368, 372–5, 386
  see also Quantum dots; Quantum wells
Narrow bandpass filters, 7
Near-field scanning optical microscopy (NSOM), 128, 301, 302
Negatively charged excitons, 241, 248–52
Nondiagonal dephasing, 64–5, 68, 106–7, 451–2
Non-interacting approximation, 170–1, 174–5, 193
Nonlinear optical spectroscopy, 290–2, 298–304, 318–25
Nonlinear optics (NLO):
  excitonic trapping, 40, 65–71
  higher-order Coulomb correlations, 171–2, 176–7, 178–87

two-level systems, 4–6
Non-Markovian processes, 25, 196
Nuclear magnetic resonance (NMR), 262–3, 340–1
Nuclear spin:
  dipole–dipole interactions, 259–61
  hyperfine fields, 256–7, 265–8
  polarization, 207, 211, 252, 258–63, 274–5
  quantum dot optical spectra, 207
  relaxation, 256, 273–4
  self-polarization, 272
Numerical simulation, quantum computing, 338–43

Optical density (OD), 97–9, 110, 111–12, 114, 123, 133–4
Optical field effects:
  carrier-wave Rabi flopping, 71–83
  quantum wells, 63–5
Optical length, 131
Optical matrix element, 102, 116, 123, 125, 142–3, 145
Optical nuclear magnetic resonance, 261–3
Optical orientation, 210, 235–6, 237, 248–52
Optical polarization, 44, 47, 58, 61
  carrier-wave Rabi flopping, 26–7, 32–3
  counterpolarization, 94
  linear, 146, 234, 235–6, 238
  optical spectra, 210, 235–6
  quantum wires, 54
  resonant secondary emission, 94, 105–6, 139–47
  see also Circular polarization
Optical pumping, 207, 210, 249–50, 311–13
  see also Pump/probe spectroscopy
Optical selection rules (OSRs), 207, 210, 222, 252, 286–7, 330, 331
Optical spectroscopy:
  coherent, 281–365
  electronic and nuclear spin of quantum dots, 207–80
Optical Stark effect (OSE), 195–6

Optical transitions:
　carrier-wave Rabi flopping, 23, 26–36
　fine structure, 207, 209, 219–40, 243–8, 271, 294–5
　quantum dots, 286, 396–401
　radiative decay rates, 105–6, 109–10, 122, 152–4, 157–9, 397, 404, 452–8
　see also Dephasing
Oscillator strengths, 99–100, 152–3, 158, 208, 329–40
Overhauser effect, 207, 211, 253, 254–5, 257, 258–9, 261–3, 305

Pancharatnam screw, 27
Parallel processing, 341
Participation ratio, 130
Pauli blocking, 4, 5, 170, 172, 174–5, 177, 195, 198
Pauli lowering operator, 384
Pauli matrices, 221
Perturbation theory, 133, 135–9, 155–6, 284, 299, 322, 402–3, 405, 423
Phase-locked pulses:
　delay time, 403, 406–9
　lasers, 23, 24–5, 27–8, 30–1, 35–6, 168
　polarized exciton states, 420–4, 425, 426, 429, 430–3
Phase relaxation rate, 17–18
Phonon-assisted density matrices, 103–4
Phonon-induced dephasing, 17–18, 108
Phonon scattering see Acoustic phonons; Exciton–phonon interactions
Phonons:
　emission and absorption, 90, 104–5, 108, 111, 218
　exciton migration, 453–8
　population relaxation, 452–3
Photoluminescence:
　biexcitons, 318
　CdSe/ZnS nanocrystals, 377–9, 382–3
　diamagnetic shift, 253, 255, 304–5
　magnetic field dependence, 243–6, 270
　polarization, 210, 223–5, 233–7, 238–9, 242–3, 249–55, 268–9
　quantum dots, 207–19, 289–95, 319–20, 373–5, 377–83, 435
　quantum interference effect, 307–10
　spatial resolution, 216–17
　temperature dependence, 242–3, 248, 380–1, 435–6
　time-development, 271, 309–10, 377–9, 390, 409, 413, 417, 419, 423–4
　trions, 241–2
　versus Rayleigh scattering, 89–159
Photoluminescence excitation (PLE), 111–12, 216, 217–19, 289–90, 293–5, 296
Photon echo technique, 91, 386–7
Photon–exciton interactions, 102–3, 105–7, 147–8
Photonic confinement, dielectric microspheres, 368, 369–70
Photons:
　antibunching, 293–4
　dressed exciton states, 401–2
　exchange processes, 391
Piezoelectric coupling, 443, 448–9
Piezoelectric scattering, 108, 443, 448
Polar angles, 229–31, 246–8
Polaritons:
　beating, 3, 4, 8–10
　formation, 3, 5
　model, 147, 149
　resonant secondary emission, 89, 94, 131–2, 141–2, 147–59
Polarization-dependent secondary emission, 94, 105–6, 139–47
Polarization functions, 44, 64
Polaron shifts, 152–4, 385–6
Polystyrene microspheres, 366, 376–81
Ponderomotive energy, 63
Population decay, 105–6, 109–10, 122, 152–4, 157–9, 397, 404, 452–8
Population transfer, 455–7
　see also Quasi-adiabatic population transfer

Porter–Thomas law, 99–100
Potential correlation function, 96–9, 136, 146
Poynting theorem, 316
Prisms, 29, 30, 372–3, 382
Pseudo-spin model, 232–40, 349–50
Pulse propagation:
  bulk CdSe, 1–22
  coherent nonlinear, 1–22
  delay, 2, 7, 10, 13–14, 301–3, 406
  linear to nonlinear transition, 8
  pulse intensity, 2, 8–9, 10, 14, 18
  sub-optical carrier pulse propagation, 71–9
  subpicosecond, 1–22
  temperature effects, 17–18
  velocity, 2, 13–14
Pulse shaping, 3, 15, 18, 343–50
Pump/probe spectroscopy, 173–7
  biexcitons, 179
  coherence, 402–3
  dynamics-controlled truncation scheme, 201
  effective polarization model, 185–6
  exciton–exciton correlations, 193–6
  incoherent populations, 198–9, 202
  spectral hole burning, 387–9
Purcell factor, 376, 377, 379

Q-factors, 366, 367, 371–2, 374
Q-spoiling, 374–5
Q-switching, 24
Quantum computers, 281, 287–8, 318, 332, 338–54
Quantum confinement:
  quantum disks, 444–5
  quantum dots, 211, 212–14, 249, 281–2, 283, 284, 285, 318, 435
  quantum wells, 212–14, 219–22
Quantum disks, 444–51, 452–3, 457–8
Quantum dots:
  arrays, 281
  biexcitons, 281–2, 284–5, 312, 316, 318–30
  cavity QED, 366–94
  coherent optical spectroscopy, 281–365
  decoherence, 288, 324, 339, 395–470, 459
  energy levels, 212–13, 216–17, 281, 286
  energy relaxation, 287–8, 300, 303–4, 433, 435
  exciton coherence, 395–470
  modeling, 285–7
  molecules, 281
  natural, 211–15, 240–3
  optical properties, 283, 288
  passivation, 282–3
  photoluminescence, 207–19, 289–95, 319–20, 373–5, 377–83, 435
  quantum entanglement, 282, 330–8, 391, 425, 430
  self-assembled, 208, 269
  single, 215–19
  single exciton states, 293–310, 313–17
  size estimation, 295, 446
  spin in optical spectra, 207–80
  synthesis, 282–3
  two-qubit, 338–54
Quantum information *see* Qubits
Quantum interference, 395–6, 402–6, 424–31
  see also Interferometry
Quantum state diffusion (QSD) equation, 253–4
Quantum wells:
  carrier-wave Rabi flopping, 79–81
  coherent optical control, 420
  electro-magneto-optical simulations, 49–53
  excitonic trapping, 65–71
  four-wave mixing, 188–9
  high-field effects, 40, 45–63
  high optical field effects, 63–5
  multiple, 149–50, 154–5, 158
  polariton effects, 147–59
  quasi-adiabatic population transfer, 40, 65–71
  resonant secondary emission, 94, 101–2, 108–17, 128, 133–4, 147–59

Quantum wires, high-field effects, 40, 45–63
Quasi-adiabatic population transfer (QAPT), quantum wells, 40, 65–71
Qubits, 272, 287, 318, 332, 338

Rabi energy, 67, 82, 344, 348–9, 351
Rabi flopping, 24–6
  see also Carrier-wave Rabi flopping
Rabi frequency, 4–5, 10, 13, 14, 18
  Coulomb correlations, 171
  electron–hole wave packet modeling, 45, 47–8
  envelope pulse area, 28, 30
  light frequency, 23–4
  sub-optical-carrier pulse propagation, 73
Rabi oscillations:
  decay, 316–17
  quantum dots, 310–17, 395–6, 402–6, 429
Rabi rotations, 340, 341–2, 345
Rabi splitting, single quantum dot, 396–401, 424
Radiative decay rates, 105–6, 109–10, 122, 152–4, 157–9, 397, 404, 452–8
Random matrix theory (RMT), 93, 124, 127
Random phase approximation, 171–2, 173, 182–3
Rashba field, 264
Rayleigh scattering:
  non-ergodic, 114
  time-resolved, 94, 102, 106, 118, 122–3, 130–1, 152, 154
  versus luminescence, 89–159
Real-space–time modeling, 40, 41, 45, 47–59
Red shift, 195–6, 241–2, 446
Refractive index, 14, 16, 31, 34, 49, 103, 374
Resonant excitation, 320–3, 339
Resonant light scattering, 371–2, 382–3
  see also Rayleigh scattering
Resonant secondary emission, Rayleigh scattering versus luminescence, 89–159
Rotating wave approximation (RWA), 24, 25, 32, 44, 64, 105, 169, 344, 397

Runge–Kutta method, 72–3
Rydberg atoms, 41, 62, 170, 178

Scanning tunneling microscopy (STM), 213, 214
Schrödinger equations:
  center-of-mass momentum states, 92, 95–9, 116, 147, 152
  disorder, 92, 94–100
  Hamilton operators, 116
  high-field effects, 47
  Maxwell–Schrödinger formalism, 147, 149–52
  nonlinear, 185
  population distribution, 66
  whispering gallery modes, 369
Secondary emission, 89–159
  coherent, 90–2, 93, 111, 113–20, 121–2, 131
  cross-polarized, 140
  damping, 137, 139
  enhanced resonant backscattering, 94, 132–9
  frequency-resolved, 120–3
  incoherent, 90–2, 93, 111, 113–18, 121, 143–4
  polariton effects, 89, 94, 131–2, 141–2, 147–59
  polarization-dependent, 94, 139–47
  radiative rates, 152–4, 157–9
  spin dependence, 89, 94, 139–47
  time dependence, 89, 103, 113–19, 120–1, 130–1, 143–4, 152, 154
Second-harmonic generation, 28
Second-order Born approximation (SOBA), 3–4, 5, 12, 173, 178, 196
Self-consistent second Born approximation (SCSB), 93, 98, 138, 147
Self-energy, 135–9, 148, 439–42, 462–4
Self-induced transmission, 6, 8, 10–17, 60
Self-phase modulation, 31
Semiconductor Bloch equations (SBE), 45, 47, 56

coherent, 64, 79
higher-order Coulomb correlations, 171–2, 176, 178, 185, 196–7
many-body systems, 79
ultrashort pulses, 83
see also Bloch equations
Semiconductor core/shell nanocrystals, 366, 368, 375–90, 459
Semiconductor–metal transitions, GaAs, 23–4
Semiconductor–microsphere systems, 366–94
Shallow impurities, 212, 269–70
Signal-to-noise ratio, 2, 7
Silica microspheres, 366, 368, 370–2, 382–5
Single exciton states:
    coherent optical control, 306–10, 338, 420–4, 425, 426
    optical spectroscopy, 293–305
    Rabi oscillations, 313–17
Six-particle correlation, 184, 199–201
Six-wave mixing, 199–201
Slowly varying envelope approximation (SVEA), 12, 25, 32, 40, 71–2, 74–5, 83
Sommerfeld factor, 53, 170
Space–time modeling, 40, 41, 45
Speckle analysis, 89, 92, 93, 114
    resonant secondary emission, 122–3, 130–1, 132, 140, 142, 154
    speckle measurement, 117–20
Spectral hole burning, 366, 386–90, 459
Spectroscopy:
    coherent optical spectroscopy, 281–365
    frequency-modulated, 66–71, 459
    techniques, 288–93
    ultrafast, 167–78, 173–8, 202
see also Absorption spectra; Photoluminescence; Pump/probe spectroscopy; Transmitted spectra
Spin:
    degeneracy, 209, 219, 221, 295
    hyperfine interactions, 207, 209, 211, 252–63, 265–8, 273
    lifetime, 238–40, 252, 263

optical pumping, 207, 210
polarization, 215, 233–7, 252, 256
precession, 227–8, 234, 237, 238, 239, 257, 259, 263–71
projections, 248–9
pseudo-spin model, 232–40, 349–50
quantum dot optical spectra, 207–80
qubits, 272, 338
see also Electronic spin; Nuclear spin
Spin beating, 94, 140, 143, 269, 271
Spin-dependent secondary emission, 89, 94, 139–47
Spin flip processes, 179, 237, 252, 264, 269, 379
Spin–orbit interactions, 209, 211, 228, 252, 263–5
Spin relaxation, 140, 158, 207, 210–11, 228, 237–9, 256, 263–71, 273–4
Spin–spin interactions, 256
Spontaneous emission, cavity QED, 366–8, 376–81, 384
Stark effect, 46, 195–6
Static fields, 40
Stokes shift, 90, 111–12
Sub-optical carrier pulse propagation, 71–9
Subpicosecond pulse propagation, bulk CdSe, 1–22
Surface polaritons, 148

Takagahara's model, 303
Temperature dependence
    exciton dephasing rate, 442, 445–50, 451
    homogeneous linewidth, 388–90, 459
    photoluminescence, 242–3, 248, 380–1, 435–6
    population decay, 452–3
Tetrahertz fields, 40, 46–63, 196–7
    extreme high-frequency, 52–3
Third-harmonic generation, 23, 27, 29–30, 34–7
Three-level systems, 335–6, 399, 403–6
Time domains, 49, 72–3, 113–17, 459

Time-resolved techniques, 271
  photoluminescence, 271, 377–9, 390
  Rayleigh scattering, 94, 102, 106, 118, 122–3, 130–1, 152, 154
Transfer-matrix technique, 149, 455
Transient differential transmission, 290–3, 298, 301–4
Transmitted fundamental wave, 23, 31
Transmitted spectra:
  bulk CdSe, 15–16, 17–18
  GaAs, 23, 29
  see also Differential transmission spectra
Transparency, tunneling-induced, 57
Transverse acoustic modes, 443–4, 449, 455
Transverse electric modes, 368–9, 370
Transverse magnetic field modes, 369, 370
Trions, 207, 209, 215, 240–52, 268, 272–3
Tunneling:
  field-induced, 57
  transparency, 57
Two-exciton-state entanglement, 330–8, 425, 430
Two-level systems:
  Bloch vector model, 406–9
  carrier-wave Rabi flopping, 23–4, 25–6, 34–6, 42, 73–9
  coherent optical control, 306–10, 338
  exciton Rabi splitting, 396–401
  nonlinear optics, 4–6
  quasi-adiabatic population transfer, 42, 65–6
  Rabi oscillations of quantum dots, 311–13
Two-particle diagrams, 135–7
Two-photon coherence, 318–19, 320–5, 330, 331, 424–31, 459
Two-subband model, 55–8, 64–5

Ultrafast spectroscopy:
  Coulomb correlations, 167–78, 202
  measurement techniques, 173–8

Valence bands, 24, 26, 32–3, 140, 169, 181, 215, 443
Voigt configuration, 229, 245–6, 248, 269, 270

Wannier equations, 172, 284, 454–5
Wannier–Stark ladder states, 196
Ward identity, 136–7
Wave equations, 3, 4–5
Wave functions, 95–7, 282, 386, 444–5, 451, 458
  center-of-mass momentum states, 129–30, 141, 142–3, 158
  Fourier-transformed, 102
  localization length, 124–5
  speckle analysis, 118
Weak-memory approximation, 156–7
Whispering gallery modes, 366, 368–72, 373–5, 382–5
White-noise disorder, 137–8

X-ray transients, 41

Zeeman coherence, 333–8
Zeeman interactions, 227–32
  fine structure, 207, 209, 222
  quantum dots, 304–5
  splitting, 210, 245–6, 251, 252–4, 258–9, 261, 267
Zener tunneling, 24
Zero detuning, 48, 69, 388
Zinc blende:
  CdS/ZnS nanocrystal cavity QED, 366, 368, 375–90
  higher-order Coulomb correlations, 176–7
  hyperfine interactions, 256, 264
Zinc selenide, higher-order Coulomb correlations, 188–9, 193–5, 197–8, 199, 200